Springer Biographies

The books published in the Springer Biographies tell of the life and work of scholars, innovators, and pioneers in all fields of learning and throughout the ages. Prominent scientists and philosophers will feature, but so too will lesser known personalities whose significant contributions deserve greater recognition and whose remarkable life stories will stir and motivate readers. Authored by historians and other academic writers, the volumes describe and analyse the main achievements of their subjects in manner accessible to nonspecialists, interweaving these with salient aspects of the protagonists' personal lives. Autobiographies and memoirs also fall into the scope of the series.

William Sheehan

Parallel Lives of Astronomers

Percival Lowell and Edward Emerson Barnard

 Springer

William Sheehan
Flagstaff, AZ, USA

ISSN 2365-0613   ISSN 2365-0621   (electronic)
Springer Biographies
ISBN 978-3-031-68799-0   ISBN 978-3-031-68800-3   (eBook)
https://doi.org/10.1007/978-3-031-68800-3

This Springer imprint is published by the registered company Springer Nature Switzerland AG
The registered company address is: Gewerbestrasse 11, 6330 Cham, Switzerland

If disposing of this product, please recycle the paper.

To Michael Armstrong and David Baron—
Amicitiae nostrae memoriam spero sempiternam fore

Preface

The wealthy, Harvard-educated, businessman-turned-Orientalist-turned-astronomer Percival Lowell is one of the best known astronomers of the late nineteenth and early twentieth centuries. Founder of his own observatory in Flagstaff, Arizona, he became a legend in his lifetime because of his brilliant prose and captivating arguments for the existence of intelligent life on Mars. These arguments were based on fleeting visions, in the thin clear air of Northern Arizona, of "fine lines and little gossamer filaments" on the surface of the planet—the famous canals of Mars. Though Lowell claimed they were unmistakable to any observer with training and skill in good seeing conditions, he failed to convince many of his colleagues—notably the celebrated astronomer Edward Emerson Barnard, who in contrast to Lowell's advantages was born into grinding poverty, had two weeks of formal education as a child but received a practical education in photography as apprentice to a photograph gallery, and who, through sheer determination and integrity, rose to become one of the most accomplished observers of all time. He never managed to make out the canals as Lowell did.

Though the question of whether there are canals on Mars in the artifactual sense has been resolved by spacecraft images, the controversies about their existence remain one of the most instructive episodes to historians and philosophers of science. We now know that the canals were artifacts of the eye-brain-hand system—an illusion, indeed, but a complex one, in which different observers' "personal equations" entered in. Long fascinated with the question of how some saw canals and others did not, I have decided that perhaps the best approach is to compare and contrast the perspectives of individuals who stood prominently on opposite sides of the question, and that is what I propose to do here. Rather than writing a biography of either Lowell or Barnard by himself (which in any case has already been done), or an in-depth analysis of the controversies about the canals and their place in the history of science (which has also already been done), I have decided to take up the "Parallel Lives" approach used by the Greek biographer Plutarch and consider Lowell vs. Barnard. The result is, I hope, a work that will break the usual molds for biography and history and prove useful to anyone interested in the history of Mars observations in particular and in nineteenth- and twentieth-century American astronomy in general.

Above all, I have been drawn to these two individuals because they represented an era when astronomers were focused on understanding worlds that stood at the boundary of perception and comprehension. Work at the edge always illuminates the human mind—both its greatness and its limitations. I believe that nowhere more than in the present case do the history and philosophy of astronomy so insistently call for the resources of the biographer.

Flagstaff, AZ, USA   William Sheehan
May 1, 2024

Contents

About the Author

William Sheehan retired after 30 years in practice as a psychiatrist, earned his MD degree (1987) and completed his psychiatric residency (1992) from the University of Minnesota Medical School. He has been a lifelong amateur astronomer and a historian and author of books on astronomy and history of astronomy including: *Planets & Perception* (Tucson, 1988), *Worlds in the Sky* (Tucson, 1992), *The Immortal Fire Within: The Life and Work of Edward Emerson Barnard* (Cambridge, UK, 1995), *The Planet Mars* (Tucson, 1996), *In Search of Planet Vulcan* (with Richard Baum) (New York, 1997), *Epic Moon* (with Thomas A. Dobbins) (Richmond, VA, 2001), *Mars: The Lure of the Red Planet* (with Stephen James O'Meara (Amherst, NY, 2001), *Transits of Venus* (with John Westfall) (New York, 2004), *Galactic Encounters* (with Christopher Conselice) (Springer, 2015), *Celestial Shadows: Eclipses, Transits, and Ooccultations* (with John Westfall, 2015), *Camille Flammarion's Planet Mars* (with Sir Patrick Moore) (Springer 2015), *Discovering Pluto: Exploration at the Edge of the Solar System* (with Dale P. Cruikshank) (Tucson, 2018), *Neptune: From Grand Discovery to a World Revealed* (Springer, 2021), and *Discovering Mars* (with Jim Bell) (Tucson, 2021). He has also written *Mercury*, *Venus* (with Sanjay Limaye), *Jupiter*

(with Thomas A. Hockey), and *Saturn* for Reaktion Books' Kosmos Series. He is a longtime contributing editor for *Sky & Telescope*, a frequent contributor to *Astronomy*, the *Journal for Astronomical History and Heritage*, *Scientific American*, and other journals, and a prolific contributor of essays to the Biographical Encyclopedia of Astronomers. He is also a fellow of the John Simon Guggenheim Memorial Foundation (2001) and a recipient of the Gold Medal of the Oriental Astronomical Association (2004). He is a past member of the IAU's Working Group on Planetary System Nomenclature, and asteroid Sheehan (16037) has been named in his honor.

1

The Flower and the Bee

Contents

> *[I]t seems to me that we should rather be the flower than the Bee…. Now it is*
> *more noble to sit like Jove [than] to fly like Mercury—let us not therefore go*
> *hurrying about and collecting honey-bee like, buzzing here and there impatiently*
> *from a knowledge of what is to be arrived at: but let us open our leaves like a*
> *flower, and be passive and receptive—budding patiently under the eye of Apollo*
> *and taking hints from every noble insect that favors us with a visit.*
> *—John Keats to John Hamilton Reynolds, February 19, 1818.*

A Tale of Two Observers

In 1894, Mars came to a favorable opposition, a little farther from the Earth than it had been in 1892, but much higher above the horizon for observers in the Northern Hemisphere where most of the major observatories were located.[1] For months increasing in brightness as it drew steadily closer to the Earth, at the time of its mid-October opposition, when it approached within 40-million miles (64-million kilometers), it loomed as "a great red star that

© The Author(s), under exclusive license to Springer Nature Switzerland AG 2024
W. Sheehan, *Parallel Lives of Astronomers*, Springer Biographies,
https://doi.org/10.1007/978-3-031-68800-3_1

rises at sunset through the haze about the eastern horizon, and then, mounting higher with the deepening night, [blazed] forth against the dark background of space with a splendour that ... rivals the giant Jupiter himself. Startling for its size, the stranger looks the more fateful for being a fiery red. Small wonder that by many folk it is taken for a portent" [1].

The planet's rather manic behaviour—becoming now fiery red, the color which above all others seizes the attention, then sinking into dim obscurity every 2 years, 2 months—had made a deep impression even in ancient times. It was known to the Babylonians as Nergal, the "Star of Death"; later and inevitably, appearing like a drop of blood in the sky, it became associated with war and bloodshed. It was identified with the war-god Ares by the ancient Greeks, with Mars by the Romans. Its motions were strangely erratic: most of the time it moved in a forward direction, but around the time of opposition it moved backward like a crab (as the astronomer Kepler had put it), after which it resumed its forward motion. Was it alive, possessed of voluntary will? Plato thought so; perhaps it was pushed along by gods and angels. At last it was shown that its backward motion was an illusion, due to the fact that the Earth in its orbit around the Sun caught up with and passed the slower-moving outer planet. As it did, its position among the stars shifted according to the perspective lines of the observer. Caught in one sensational trick, it would nevertheless continue, long afterward, to be a planet of illusion.

The most compelling illusion that followed the disentangling of its motion appeared with the increasing perfection of telescopes in the seventeenth and eighteenth centuries. The features of its globe—whitish patches on the poles, dark grey or blue-greenish patches resembling seas set against a background of ochre continents that came into view and passed away with an Earth-like, 24 h 40 min rotation period—strongly suggested the planet might be a distant Earth.

In the pre-photographic era of planetary observation, an observer faced significant challenges that are not so easy to appreciate today. The historian Omar W. Nasim has written of attempts at this time to visualize the nebulae:

Thanks to their utterly strange and enigmatic character ... these objects ... continually demanded special attention. The challenge in particular was to visualize them, since other means—like description or numbers—simply failed or were clumsy in the face of the indescribable. Exactly what these astronomers were visualizing was for the most part unknown. On top of that, the visual products or the work that went into them rarely were governed by any generally accepted standards specific to nebulae as scientific objects.... [This required] procedures of observation [to assist] in making out what an observer saw (over many nights and days of looking and inscribing) and in gradually stabilizing the phenomena into something visualized in a particular way that could be used by theoreticians, natural philosophers, fellow observers, and others… [2].

Planets were perhaps somewhat less strange and enigmatic than nebulae. There was always the Earth itself to compare them to—and the comparison was often implicitly or explicitly made. However, some of the other points Nasim makes apply to planets no less than to nebulae, and in many ways were even more severe, given that minute planetary details were only accessible for fleeting moments, and this stretched to the limit what we might now call the digital-image capture rate of the visual system, the capacity of short-term memory to hold information without distortion and decay, and the skill of the hand to enter into an observing log book a sketch of what was just seen. The latter process of course depended greatly on the particular virtuosity and skill of the observer as the wielder of stylus, paint, and paper. In the end, any one sketch had limited weight; but cumulatively the sketches contributed, as for the nebulae, to stabilization of the phenomena into something visualized in a particular way—a stabilization embodied as a map or as images, to quote Nasim once again, "used as proxies for an object, or a means of 'virtually witnessing' what otherwise could be seen only through large telescopes and by but a few."[2]

A few of the stages in the stabilization of images of Mars are represented in the gallery shown here. These examples culminate in what was to be, for a time, the definitive image of the planet, and the most influential: that produced in 1877–78 by Giovanni Virginio Schiaparelli, a highly respected professional astronomer at the Brera Observatory in Milan. Schiaparelli not only produced a map based on careful measurement of specific points of the surface to insure the exactness of positions and forms of the main features; he also devised an ingenious and beautiful system of nomenclature based on names from classical mythology and the Bible which after a short time became definitive (and still serves as the basis of that used today). In addition, his map introduced an utterly astounding finding: the planet seemed to be crisscrossed by a system of *canali*—channels, or canals—which in some cases, as Schiaparelli found subsequently, appeared to double (*geminate*) according to some inscrutable law. In due course other astronomers, having "virtually witnessed" the planet through Schiaparelli's drawings and maps, confirmed some or most of his findings and did not hesitate (as he himself did) to embrace the stunning prospects they seemed to open up: that this neighbor planet was the most probable abode of life, including intelligent life, in the Solar System. By Mars's opposition of 1894, popular interest soared to unprecedented heights in what has been referred to as the "Mars furor."

A Miscellany of Views of Mars, pre-1894

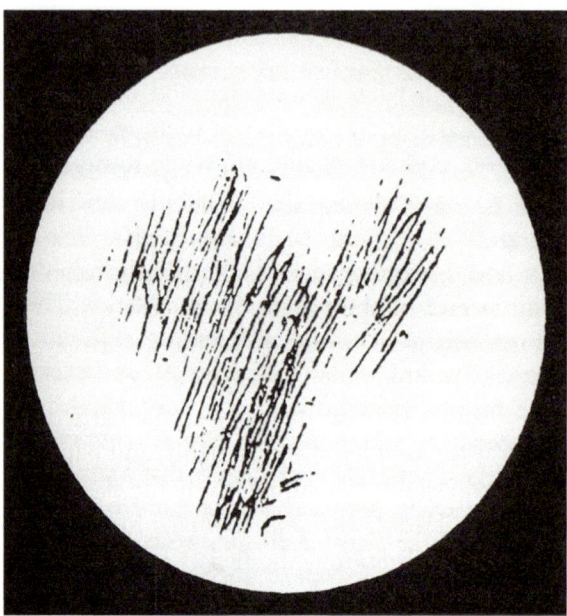

Christiaan Huygens's celebrated sketch of the "Hourglass Sea," or Syrtis Major, made on November 28, 1659, with an optically excellent 2-inch refractor. Note that in this and all representations of Mars in this book, unless otherwise noted, South is at the top, corresponding to the inverted image view in astronomical telescopes. (From: Camille Flammarion, La Planète Mars, *vol. 1 (1892))*

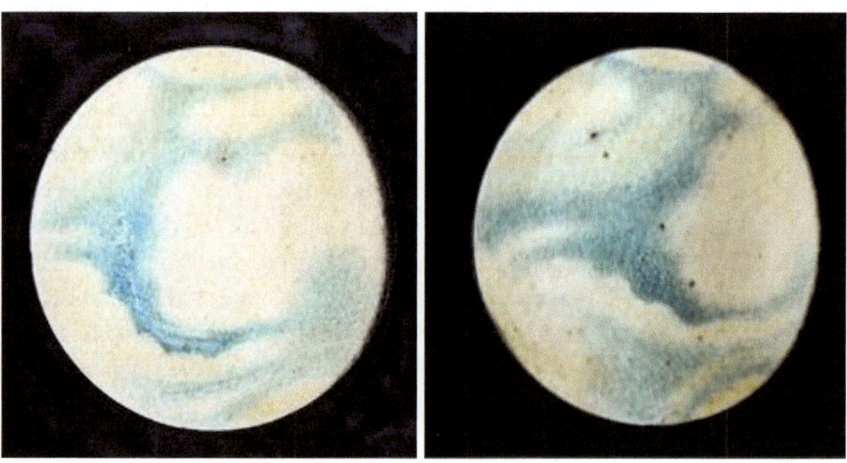

Chalk pastel renderings of Mars by Father Angelo Secchi, made on June 11 and 13, 1858, with the 9-inch Merz refractor of the observatory of the Collegio Romano, Rome. (Credit: William Sheehan Collection)

Sketch of Mars, September 1, 1877, 14 h 20 m UT, by professional artist and amateur astronomer Nathaniel Green, made with a 13-inch Newtonian reflector at Madeira. (From: "Observations of Mars, at Madeira, in August and September 1877," Memoirs of the Royal Astronomical Society, vol. 44, 1877–79)

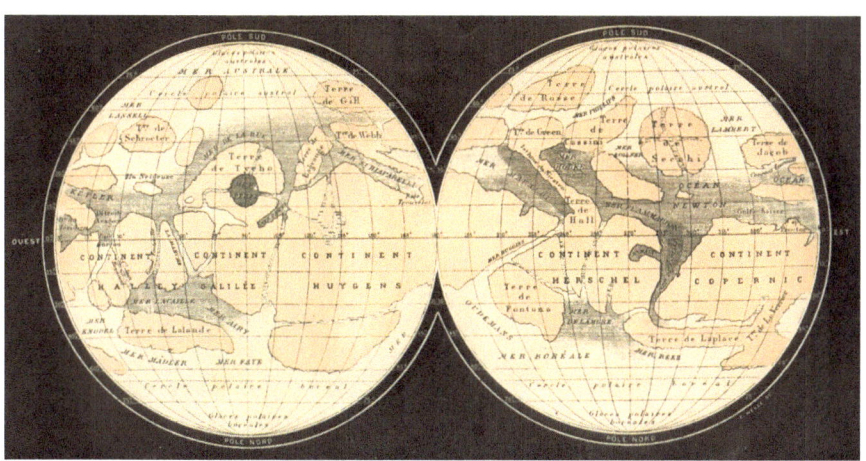

A map of Mars by Camille Flammarion, based on drawings 1862–64 by the English astronomer William Rutter Dawes. (From: Camille Flammarion, Les Terres du Ciel *(1884))*

During the opposition of 1894, hundreds of telescopes were turned toward the sky and millions of readers pondered the findings in newspapers, magazines, and scientific journals. Of all the many Mars-watchers to scrutinize the planet in that memorable year, two in particular concern us here. One, Percival Lowell, was a Boston Brahmin, possessed of extraordinary wealth, a Harvard education, extensive world travel and the means to borrow a good telescope (an 18-inch refractor) and set up his own private observing station on a mesa he called "Mars Hill" in Flagstaff, Arizona. The site had been chosen specifically for its allegedly calm air and undisturbed images. The other, Edward Emerson Barnard, grew up "as the poorest of poor boys in a 3rd-rate Southern town (Nashville, Tennessee) just after the war which ruined the South."[3] He went to work at an early age in a photographic studio, to support his widowed and sickly mother. There, he served a long and fruitful apprenticeship in photographic techniques. He also, by practicing the strictest economy, was able to save up enough money to buy a small telescope, which he wielded so effectively, especially in discovering new comets, that he was eventually invited to join the Lick Observatory, on Mt. Hamilton, near San Jose, California, as a member of its original staff.

A pen sketch by Peter Ross Calvert, Barnard's colleague and future brother-in-law, depicting him alongside the 5-inch Byrne refractor he used to observe Jupiter in 1879–80 and to discover his first comets. It was with this small telescope that Barnard learned his way around the heavens and developed the extraordinary observational skills he later applied at the eyepiece of the 36-inch Lick refractor, the largest refractor in the world at the time. (Credit: Special Collections and University Archives, Jean and Alexander Heard Library, Vanderbilt University)

Percival Lowell, sketching a planet (probably Venus) at the eyepiece of the 24-inch Clark refractor at Lowell Observatory, in 1896. (Credit: Lowell Observatory Archives)

During the summer and autumn of 1894, these two men scrutinized the planet Mars. One was Lowell who was, to use Keats's analogy cited in the epigraph, "bee-like, buzzing here and there impatiently from a knowledge of what is to be arrived at." The other, Barnard, "opened his leaves like a flower, was receptive, and budded patiently under the eye of Apollo."

Up until then, these men had met only once, and then fleetingly. Despite both being eager for knowledge of the Martian world, they worked independently of one another, with no knowledge of the other's purposes and activities, at observatories some seven hundred miles apart. Lowell was a highly charismatic figure on the lecture stage, possessed of great literary skill and known to be a sharp mathematician, but without any significant experience as an observer. He announced in a talk even before departing for Flagstaff what he expected to find when he began his astronomical studies (hence he was indeed the "honeybee"). "The amazing blue network on Mars hints that one planet besides our own is actually inhabited now," he said.[4] Spending the month of June in Flagstaff getting his observational bearings, he then left for Boston, leaving the observatory in the care of his active assistants William H. Pickering and Andrew E. Douglass. He did not return until the last part of August and the first part of September. After 2 weeks, finally, from early October to late November, he returned again, around the time of Mars's opposition.

Barnard remained on Mount Hamilton, and studied Mars whenever he could with the Observatory's mighty 36-inch refractor. He was assigned two nights a week, Fridays and Sundays. He was a highly experienced and judicious observer, with a sensitive eye, and possessed of a fairly high level of visual imagination and artistic skill. In his irregular education, he had largely missed out on mathematics, which may have been an advantage as it allowed his visuo-spatial powers to develop without interference. In contrast to Lowell, Barnard had no developed views about what he would find on Mars that summer.

These two men present contrasts of personality, training and approach as great as any between the noble Greeks and Romans famously compared by Plutarch in his second century book of *Parallel Lives*. Developing the contrasts will occupy us as we proceed through the pages of a long book. For now, let us consider the two men at a single moment in time, the morning of September 3, 1894. That morning, Lowell made a final go at Mars before catching the train back to Boston, and he would not again return to Flagstaff and observing until October. He had by now a fairly well-developed and elaborate theory about the planet whose colors in the telescope reminded him of "lambent saffron." The colors also seemed to his eye to resemble the pine forests and Painted Desert as seen from the nearby San Francisco Peaks. They presented both a scene of "robin's-egg blue and roseate ochre," brilliant "in the flood of sunshine from out of a cloudless burnished sky" and rivalling the tints of a "fire-opal" [4]. In the 4 weeks in June, and again in the 2 weeks just concluding, he had grasped canals, fragmentarily, several at a time; they were fugitive, shy visitants, that appeared and then disappeared as quickly as they came. The Martian reality revealed itself to him only in furtive "glimpses." The details were not so numerous as to entice him away from the practice of sketching them on disks only 2 inches to the diameter of the planet—a practice from which he never departed, up to his final Mars observations in March 1916.

On September 3, 1894, Lowell was seated at the eyepiece between 7:30 and 8:30 a.m., according to what we would now call Mountain Standard Time, and Mars was rising into the morning sky. He began to scrutinize the hemisphere of Mars then in view, centered on the large dark roundish area known as Solis Lacus and the broad canal Agathadaemon. The brightish area around Solis Lacus (known as Thaumasia—the Land of Wonders) appeared threaded with numerous canals—Lowell marvelled at the "effect"—and he also "had a glimpse showing Ganges [canal] double, but do not credit it." It was a provocative if tentative impression, but Lowell did not linger long; he would soon stand on the platform in town and catch the train back to Boston. After Lowell departed, Douglass took over the telescope, and continued the observations for another hour or so.

Despite being rushed, Lowell's impressions regarding the planet, made on the basis of 6 weeks of observing, had already crystallized into unshakeable convictions. He had no doubt that Mars was a world on which with the melting of the polar cap (the southern was then tilted Earthward), the dark areas faded, just as vegetation would be expected to do in passing from spring verdure to late-summer sere. Meanwhile, the canals—and they were exactly what the name implied—conveyed water from the poles toward the equator, and made the strips of the adjoining desert bloom. For Lowell, it was impossible to see all this and not conclude that Mars was an inhabited world. The Martians represented an advanced civilization whose great engineering works had been constructed, on the grandest scale imaginable, in order to stave off impending doom as their planet gradually but inexorably dried up. As he trundled down Mars Hill to the train station the next day, Lowell must have done so with confident, assured steps. He was now sure of what he had come to Flagstaff to find out, and believed himself in possession of a great discovery—a discovery that he would cling to tenaciously and without a doubt until his death 22 years later.

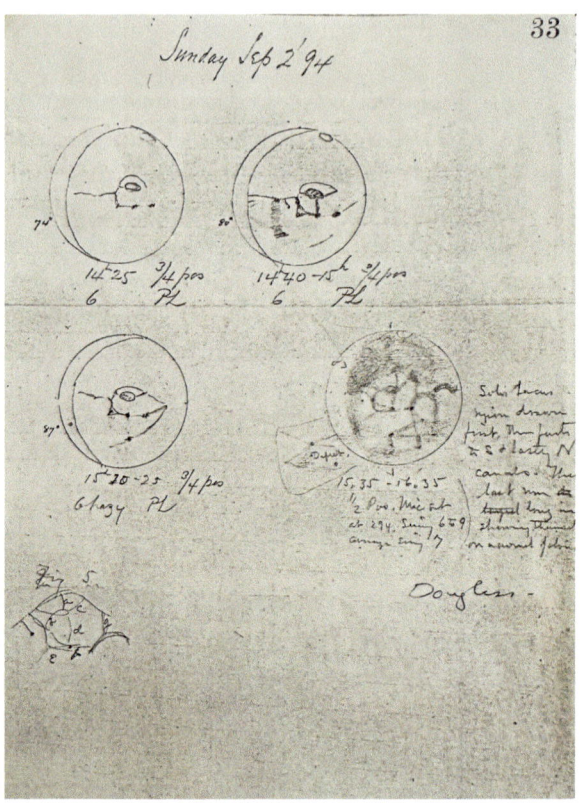

A page of the logbook recording the Mars observations of Percival Lowell and Andrew E. Douglass after sunrise, Sunday morning, September 2, 1894. Lowell's "fragmentary" views capture details revealed in glimpses around the Solis Lacus. The broad smudgy "canal" extending northward is the Ganges, which Lowell suspected of being double. Douglass's more finished drawing follows in the bottom right, along with his small and minutely detailed diagram of the canals around Solis Lacus. (Credit: Lowell Observatory Archives)

At the same exact time, Barnard was at the 36-inch refractor of the Lick Observatory on Mt. Hamilton, some 700 miles to the west. All that summer he had been watching as Mars grew closer to the Earth, and as it approached he began seeing so much detail that he increased the size of the disks in his drawings from the 2 inches Lowell used, to 3 inches, and to finally 5 inches as he did tonight. The strange features, light and dark, had mystified him, but after sunrise on the mornings of September 2 and 3, the drama of discovery reached a new pitch of excitement. Though Flagstaff was vaunted for its calm air, at Mt. Hamilton, in late summer, the seeing conditions are often extraordinary—then, if ever, come perfect images—as laminar airflow from the ocean sweeps like a warm breeze over the isolated mountaintop and sharpens the usually jumpy and blurry image to the stillness of a steel-engraving. Whereas Lowell usually used 370 X, Barnard often boosted the magnifying power to 1000 X or more. All alone in the great dome, throughout whole nights, Barnard would study the planet, but the best views (as for Lowell) were usually after sunrise, as Mars rose into the blue sky. At these times, the air often became steady and the amount of detail beggared belief. Barnard made sketch after sketch, and he recorded in his observing log book:

> The region of the lake of the Sun [Lacus Solis or Solis Lacus as it is now generally known] has been under review. There is a vast amount of detail… I however have failed to see anything of Schiaparelli's canals as straight narrow lines. In the regions of some of the canals near the Lacus Solis there are details—some of a streaky nature but they are broad diffused and irregular and under the best conditions could never be taken for the so called canals.[5]

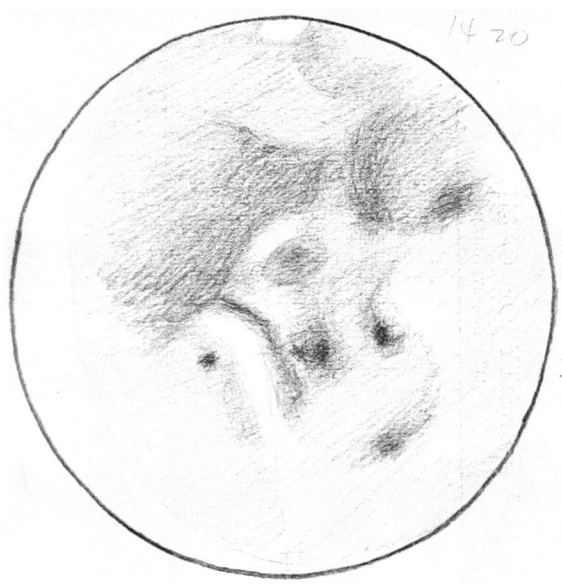

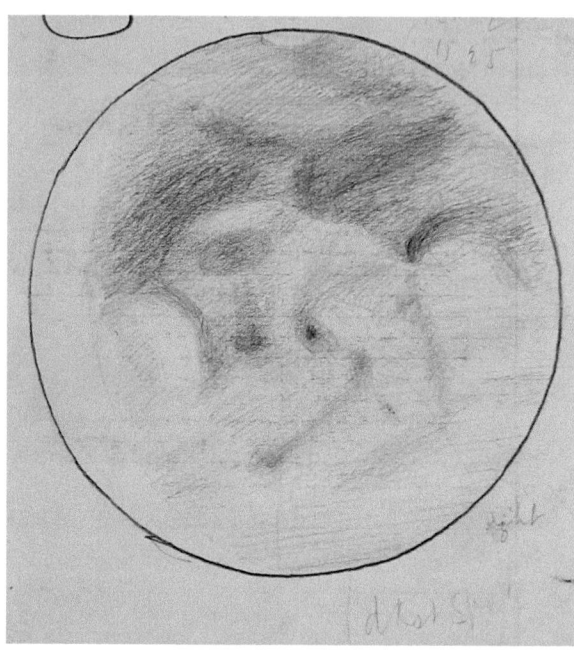

Two sketches by Barnard, finished about an hour apart with the 36-inch refractor on Sunday morning, September 2–3, 1894. The upper one was made between 14 h 5 m and 14 h 20 m UT; the lower between 15 h 12 m and 15 h 25 m UT. A number of interesting details are visible: The Solis Lacus is the ellipsoidal feature to the left of and above center; the broad bright band between dark markings on either side is the canal Ganges (suspected to be a double canal by Lowell, but Barnard shows it correctly); the bent marking above the Ganges is the classical canal Agathodaemon, now known to be part of the dark-floored Valles Marineris canyon system; the dusky diagonal sash running below center is the "canal" Ulysses; the dusky patches at either end are in the locations of the shield volcanoes, Ascraeus Mons and Arsia Mons, which appear dark when, as at this season, they are not hooded by clouds. (Credit: Historical Collections Project, Lick Observatory)

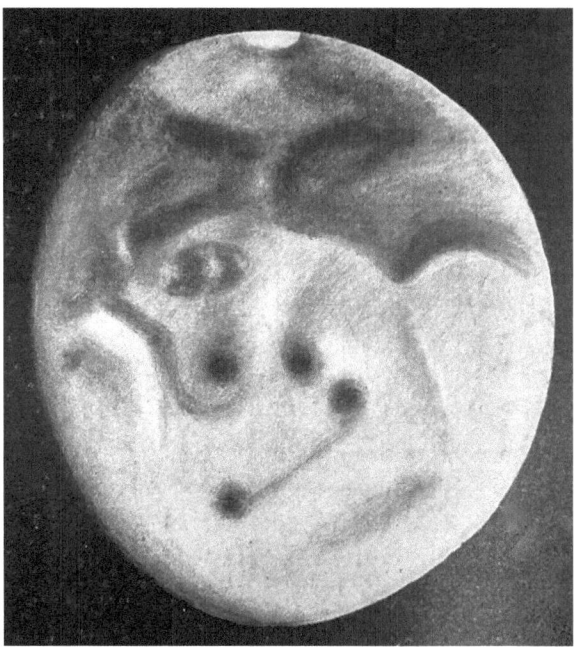

Barnard's sketch of September 3, 1894 done up for publication and published as a frontispiece to his article "The South Polar Cap of Mars" (1903) Astrophysical Journal, *vol. 17, no. 4. Clearly, the boldness of the features has been greatly exaggerated
Apart from four drawings (including this one) published on a very small scale in the 1895 edition of Chambers's* Story of the Solar System, *of which the originals have now been lost, and his drawings showing the shrinking of the South Polar Cap, no other Barnard drawings from 1894 were ever published*

Two men, two mountains, two views of Mars. How could they possibly be so different? Perhaps it was simply, as the astronomy historian Richard Baum has written, "In the romantic description of Lowell, color was to his purpose not detail; Barnard, who had survived a grey and austere childhood, referred more matter-of-factly to the appearance alone."[6] An astute observation, and bound to contain an element of truth. But it is not the whole truth.

As Plato grasped in the Analogy of the Cave, what we see are only shadows; the real world—for Plato, the world of ideas—lies ungraspable beyond our reach. The same idea is perhaps dimly adumbrated by modern neuroscientists who suggest that truth and fiction (however defined) overlap, that both are products of our own brain processes, and that we operate—not in direct contact with what is really and truly there, which is unknowable—but only through models which we construct with our brains, a kind of "controlled hallucinations," [5] which are tested and modified through experience. That way—such is the hope of the scientific method—the shadows on the wall of the cave may hopefully more nearly resemble and correspond with the figures casting them, figures over there beyond the glare which we can never turn our faces to look upon or comprehend directly.

Astronomers have long discussed the fact that different observers often differ sharply in their results. These differences in part can be ascribed to factors such as the conditions at their observing sites, their respective telescopes, and their procedures of observation. In addition, however, the differences include those in the characters of men themselves. As has been well said, "The aperture of the telescope is not the only thing to be taken into account. There is also the man at the small end."[7] Idiosyncrasies of personality, particular training and experience, distinctive aptitudes and deficiencies, and expectations all enter in. In addition, memory is important. As noted by the famous amateur painter Sir Winston Churchill:

> It would be interesting if some real authority investigated carefully the part which memory plays in painting. We look at the object with an intent regard, then at the palette, and thirdly at the canvas. The canvas receives a message dispatched usually a few seconds before from the natural object. But it has come through a post office *en route*. It has been transmitted in code. It has been turned from light into paint. It reaches the canvas a cryptogram. Not until it has been placed in its correct relation to everything else that is on the canvas can it be deciphered, is its meaning apparent, is it translated once again from mere pigment into light. And the light this time is not of Nature but of Art [6].

Different observers do not, of course, rely upon the same means of "transmission in code." Some, perhaps more verbally oriented, depend primarily on symbols as a form of shorthand. Others are more visual, retentive of the way things are massed into shapes. In any case, from the moment observers set eye to eyepiece, they are "compelled to dwell on the individual lines and points and to grasp their… connections."[8] Then transference from the realm of the mental to the realm of the physical must be effected to a permanent record through the medium of stylus and graphite, ink or paint onto paper. Thus artistic style and skill enter in. The light is now not of Nature but of Art. Of course, the scholarly literature on how all this works is enormous.[9] We cannot follow it in detail but note here that it is only through the final stage—when Nature becomes Art—that "the educated eye become[s] the watchful adviser of the investigating mind."[10] A particularly useful concept in this regard involves the artist's "mark making"—the different lines, dots, marks, patterns and textures an artist uses to create an artwork. One artist may be loose and gestural—this describes Barnard, whose style was quite free and expressive—while another may be controlled and neat and restrained—Lowell. Clearly, the one would tend to produce an image of a planet more naturalistic in style; the other a tighter more precise diagram-like impression.[11]

So, what is it that we are doing, ultimately, when we look at Mars? We are establishing perceptions in a framework of underlying belief. For Barnard, this was a naturalistic surface, reminiscent of the scenery around Mount Hamilton, "broken by canyon, slope and ridge." [8, p. 166] For Lowell, a system of lines and dots resembling, it would not be too far-fetched to suggest, a map of the canals and locks and mills of the textile mills of Lowell, Massachusetts; at any rate, something marked by the "appearance of artificiality."[12]

The "Personal Equation": A Brief History

It used to be common knowledge that there could be no higher standard of evidence (and none more damning that could be used in a court of law) than "eyewitness testimony." A friend of mine from high school, who became an evangelical minister, told me that his faith in the truth of Christianity was founded on the *fact*—his word—that there were eyewitness accounts by people who were there, and that they wrote these accounts down soon afterward. This, of course, is almost certainly not the case, but even were it true, just how reliable is eyewitness testimony anyway?

Damnably poor, it turns out. A great deal of rather sophisticated research has borne this out. Psychologist Elizabeth Loftus has written that despite its sway with juries, "… eyewitness testimony is not always reliable. It can be flawed simply because of the normal and natural memory processes that occur whenever human beings acquire, retain, and attempt to retain information" [9].

It is more than just flaws in the memory processes, however. Humankind's interminable contentions over questions of religion, philosophy, politics and everything else undermine the naïve notion that we all see things more or less the same way, though even preschool children (most, anyway) begin to grasp that other people are separate psychological beings, and no two think alike [10]. In general, it takes a long while and special circumstances to realize this fully. Many adults never do get the hang of it.

Rather, the persistence of the early infantile notion is attested by the notion that everyone should embrace a set of common dogmas, and that departures are not to be tolerated. Hence those who think differently may be proscribed as "heretics." The idea that we think alike—or ought to—also undergirds the belief in "absolutes," prevalent, of course, in dogmatic religion though they have also played a role in science; e.g., Isaac Newton's ideas of "absolute space" and "absolute time." Similarly, Newton's friend and close ally Edmond Halley seems to have believed—when formulating his famous method of determining the solar parallax by observing the contacts between the disk of Venus and the limb of the Sun—that these contacts could, at least in principle, be determined with the absolute precision of constructions in Euclidean geometry. Experience soon gave the lie to expectation. Observers who attempted to implement Halley's method at the transits of 1761 and 1769 (by which time Halley himself was dead) found it remarkably difficult to agree among themselves, even about such seemingly straightforward matters as finding the exact moments when the Sun's limb and the limbs of Venus came into contact [11]. Thus at the transit of June 3, 1769, Captain James Cook and his astronomer Charles Green, observing at Tahiti, differed from botanist Daniel Solander, stationed at Morea, by over a minute in their estimates of the interior contact at ingress, and by almost 2 min at the exterior contact at egress.

The experience of these observers was hardly atypical: all observations humans make, it turns out, are surrounded with clouds of uncertainty, irreducible "error bars." In astronomy, because of its requirement for extreme accuracy in its observations, this realization came earlier than in other disciplines. Though the mathematical theory of errors is relatively recent, the basic realization that the senses are not entirely trustworthy goes back at least as far as Plato and his Analogy of the Cave. The point was stated with particular clarity by Francis Bacon in his *Novum Organum* of 1620:

> Man's sense is falsely asserted to be the standard of things: on the contrary, all the perceptions, both of the senses and the mind, bear reference to man and not to the universe; and the human mind resembles those uneven mirrors which impart their own properties to different objects … and distort and disfigure them [12].

Suffice it to say, astronomers' quest for the absolute at the transits of Venus proved unsuccessful: the distance from the Earth to the Sun could not be determined, using Halley's method, with the accuracy expected.[13] Nor were the observations of Venus's contact points uniquely difficult, it turned out. Astronomers hoped to observe the positions of stars to as high a degree of accuracy as possible in order to, in fulfilment of King Charles II's warrant in founding the Royal Observatory at Greenwich in 1675, to find "the longitude of places for perfecting navigation and astronomy." The star positions would establish a reference frame against which the motions of the Moon and planets could be tracked, and the vexing problem of longitude at sea finally solved—the latter depended on the fact that, since the Moon moves through some 13 degrees of celestial longitude every night, it can serve as a clock: if the movements are sufficiently well-known in advance and published in almanacs for Greenwich time or Paris time, comparison of the local time with the Greenwich or Paris time will give the wanted result. This was the basis of the "method of lunars," which depended ultimately on both the accuracy of predictions of the Moon's motions and the existence of accurate star positions "in the Moon's way." The latter were to be discovered by specialized observers using meridian circles or transit instruments (where the term transit here refers to that of a star across a series of vertical wires in an eyepiece rather than of a planet across the Sun). In the famous "eye and ear" method introduced by the the Reverend James Bradley, the third Astronomer Royal after John Flamsteed and Halley, the observer notes as accurately as possible the distance of a star from one of these vertical wires in terms of the ratio of two beats of a clock heard immediately before and after the star's crossing [13]. The assumption was that seeing the distance from the wire and hearing the beat of the clock were simultaneous events, with the mental operations needed to make these judgments happening instantaneously (or, as the expression went, with the "speed of thought"). As with the determination of the contacts of Venus with the solar limb during transits, all this seemed completely straightforward and uncomplicated.

The Gautier Meridian Circle, acquired in 1896, at the National Observatory of Athens. This type of instrument is fixed along the meridian and can only move around a horizontal axis perpendicular to the meridian plane. It is used to determine the coordinates of celestial bodies as the rotation of the Earth causes them to drift across a slit that opens to the sky. A standard method of measurement involved the eye-and-ear method, as described in the text. (Photograph by William Sheehan)

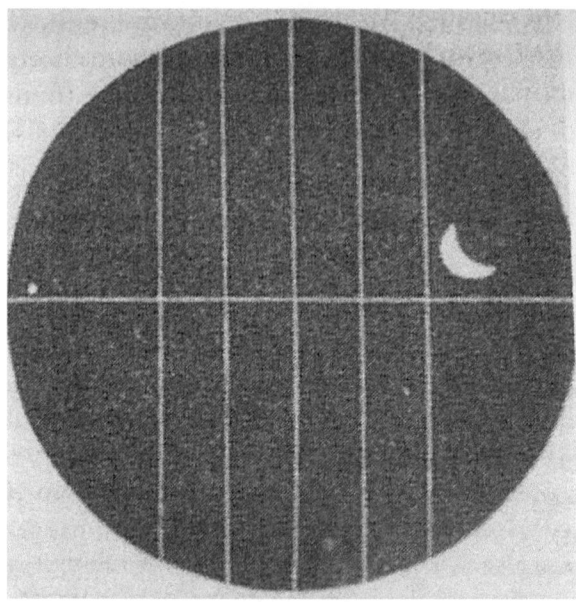

A star and Venus crossing the wires in the field of view of a meridian or transit tele-scope. (From: George F. Chambers, Astronomy *(New York: D. Van Nostrand Company, 1913), p. 297)*

Again, however, there were unexpected complications, and even well-trained observers did not always agree. The most famous case involved David Kinnebrook, assistant to the fifth Astronomer Royal at Greenwich, Nevil Maskelyne. Kinnebrook was dismissed from his post for having "commenced a vitious way of observing the times of the Transits too late" [14]. In August 1795, Kinnebrook's timings were half a second late compared with Maskelyne's. By January of the following year, the discrepancy had increased to 8/10 s. In dismissing Kinnebrook, Maskelyne's view was aligned with received notions of expertise in which the observations of those with more experience and prestige were favored over those of comparative novices. He still held stead-fastly to the Enlightenment view that one could achieve expertise through training and diligent attention to make observations connected to truth, and so reveal the true state of the world.[14] His diagnosis of the problem was that Kinnebrook had not been quick enough in fixing the places of stars at the beat of the clock, leading him to assign stars "too backward a place in the tele-scope." Yet there was an alternative view, which was suggested by Kinnebrook himself in a letter to his father:

Dr. Maskelyne observed several stars last night, and compared them with my observations, and the rate differed about a second from the rate deduced by me. He spoke to me this morning and said that he was sorry to lose a diligent Assistant, but he was under necessity, owing to the great difference of our observations, of getting some other person in my room. I told him that I had been particularly careful in making my observations and I had no reason to suppose that the difference arose from any defect in my eyes. Dr. Maskelyne said that many observations had erred in some cases above a second from his observations. I told him that he had as much reason to suppose that his observations had erred in quantity from mine.[15]

Who shall decide when doctors disagree? Was the superior experience of an aging (and possibly hard of hearing) Maskelyne to be trusted over the testimony of the younger, more acute, but overworked and weary assistant? Who was to be trusted in the perception of colors? John Dalton and his brother, or nearly everyone else? (Dalton was the first person to recognize color blindness a year before Kinnebrook was dismissed, and in the process revealed the surprising fact that color names tended to be quite inexact and subjective [16].) People experience the feeling of certainty, as, for instance, regarding the co-occurrence of two events such as the beat of the clock and the arrival of a star on a wire, differently. Also vision—and presumably much of the rest of the data provided by the senses—is experienced differently by distinct individuals. Why not other things? Thus, in Rowan University philosopher Matthew Lund's words, these developments "called into question the presupposed alignment of the phenomenology of perception with the world."[16]

Later, this difference was given a name: the personal equation. The concept came out of the same branch of astronomy in which Maskelyne and Kinnebrook had been active—positional astronomy. After his dismissal, Kinnebrook returned (with a character reference provided by Maskelyne) to his native Norfolk, and assumed a position as usher at Gresham School. By 1801, his health was failing; he could no longer stand the rigors of boisterous schoolboys; and, difficult as it must have been, wrote to Maskelyne for other work. Maskelyne employed him as a piecework computer for the *Nautical Almanac*. He died in May 1802, at the age of 30. Maskelyne himself died in 1811. The whole Maskelyne-Kinnebrook incident might have been forgotten but for its being brought to the attention of the German astronomer Friedrich Wilhelm Bessel, Director of the Prussian King Friedrich Wilhelm III's observatory at Königsberg, in 1816. Bessel was then engaged in the substantial task of creating a fundamental star chart based on Bradley's observations at the Royal Observatory, Greenwich. Though most of this work was finished by

1813, publication was delayed until 1818 largely due to the Napoleonic wars raging at the time. The result was worth waiting for, however, for it represented something wholly new—a _Fundamenta Astronomiae_, as Bessel entitled his book. Sir John Herschel later wrote that it "affords the first example of the complete and thorough reduction of a great series of observations, grounded, in the first instance, on a rigorous investigation, from the observations themselves, of all the instrumental errors, and carried out on a uniform plan, neglecting no minutiae which a refined analysis and perfect system of computation could afford" [17]. The old notion of direct Euclidean access to celestial phenomena was rendered obsolete forever.[17] Bessel next employed his analysis to the Maskelyne-Kinnebrook affair; he strongly suspected that the difference in the measurements was not merely a case of a single "vitious observer" but rather denoted something more profound since, as he noted, even when informed of his "error," and thus had motivation to attempt to correct his technique, Kinnebrook had failed to do so; paradoxically, his error had actually increased after he had been informed of it. This suggested that, whatever its basis, it was probably involuntary.

In following up his hunch, Bessel not only requested a copy of Maskelyne's complete observations from Greenwich, he also investigated whether the difference between the observations of Maskelyne and Kinebrook had counterparts in the work of other experienced observers. He and an assistant, Henrik Johan Walbeck, both employed Bradley's eye and ear method to make transit observations of two sets of five stars on alternate nights for five nights, and found that their results differed even more than the 0.8 s separating Kinnebrook's results from Maskelyne's, as Bessel announced in 1823. "We ended the observations with the conviction that it would be impossible for either [Bessel or Walbeck] to observe differently," Bessel concluded, "even by only a tenth of a second" [18].

"It was fortunate," writes the experimental psychologist and historian Edwin Boring, "that the difference [between Bessel and Walbeck] was so large, for it stimulated Bessel to further work...." [19] Over the next several years Bessel compared his observations with those of a number of other observers, and found that systematic differences like those between Maskelyne and Kinnebrook were ubiquitous; they were involuntary—Bessel himself thought that they must be psycho-physiological, based on the observers' inability to compare impressions on the eye and ear with each other "in an instant," with different observers using different times "for carrying over the one impression upon the other."[18] They could not be eliminated, but they could be quantified and corrected for. Observatories henceforth began to publish the averaged differences for their observers, so that the errors, thus identified, could be taken account and eliminated just like any other source of error from the calculations. Soon the term "personal equation," first introduced by the Reverend John Pond, Maskelyne's successor as Astronomer Royal, in 1833, began to be used to refer in general to all such sources of error. Astronomers themselves were generally not very interested in understanding how the personal equation was produced. Though it seemed it might have something to do with the finite, and measurable, velocity with which nerve impulses travelled (as was first demonstrated by Herman Helmholtz in 1850),[19] astronomers were generally only interested in refining their methods of measurement to eliminate the observer as far as possible. Already, in 1843, François Arago, director of the Paris Observatory, assigned two observers to make the measurements instead of only one. One observer, at the telescope, shouted out the moment when the star in question was bisected by the wire; the other observer, standing at the clock, estimated and recorded the time between adjacent beats. Soon afterwards, about 1850, the entire method was completely automated; instead of shouting out, the observer at the telescope simply pressed a key to produce a mark on a rotating-drum chronograph, which effectively reduced the personal equation to a simple (and eminently measurable) reaction time.

This was an important development, quite new in defining the nature of science itself. It represented an effort that would continue apace during the remainder of the nineteenth century to eliminate the human element—the subjective—as far as possible from scientific observations. In doing so, astronomers took a step toward the valorization of "objectivity" as their ideal. According to Lorraine Daston and Peter Galison,

> Scientific objectivity has a history. Objectivity has not always defined science. Nor is objectivity the same as truth or certainty, and it is younger than both…. To be objective is to aspire to knowledge that bears no trace of the knower—knowledge unmarked by prejudice or skill, fantasy or judgment, wishing or striving. Objectivity is blind sight, seeing without inference, interpretation, or intelligence… [20].

As part of this process, experimental psychology emerged as an attempt to observe—objectively—the observer. Studies of the personal equation thereby became—impersonal. In short order, an "absolute personal equation" was defined that depended only upon the reaction time—a measure of the speed of transmission of nerve impulses from the brain to the hand. The historian of experimental psychology Edwin Boring writes that

The most active period of experimentation upon the personal equation belongs to the [18]60s and [18]70s, the period of the birth of physiological psychology. It was already plain that at bottom the problem is psychological, that expectation, preparation, and attention are factors in the explanation.[20]

Planetary Observation: "Personal Equation" vs. Artistic Interpretation

To follow the details of these developments further would take us far afield, and in any case, a full treatment is available elsewhere.[21] Much of the discussion centered on how an observer going over from hearing to seeing (and thus attended primarily to the clock) differed from one who went over from seeing to hearing (and attended primarily to the wire). Maskelyne was of the first type, Kinnebrook the second. In addition, it was found that even for a highly trained and disciplined observer, the personal equation was highly susceptible to the conditions in which the observations were made, particularly whether or not the observer was well rested. Constant, and fatiguing, attention of observers affected their results, as was demonstrated in studies by astronomer Adolph Hirsch at the Neuchâtel Observatory in Switzerland (devoted entirely to providing the Swiss clock industry with precise determinations of time). Hirsch found that a fatigued, albeit disciplined observer could only produce faulty observations.[22] Kinnebrook (as well as Maskelyne's other assistants) performed many more observations than Maskelyne did, and later at night, and under worse conditions. Being clearly overworked and chronically fatigued, they were likely to produce faulty observations.

Automated systems such as those introduced in the case of transit observations had a clear advantage as they were not susceptible to fatigue. Similarly, the problem Nasim has considered at length—of recording, by visual means, the forms of the nebulae—was largely solved by the introduction of photographic techniques. From the 1880s, gelatine bromide or dry plates were produced that were sensitive enough to be used for astronomical purposes, insofar as they were able to build up a faint image over time. Thus, they largely (though not entirely) replaced the eye as the detector for faint objects such as nebulae (and also comets).[23] George Phillips Bond labored at the Harvard 15-inch Merz refractor, often in bitterly cold conditions, between 1859 and 1863 to produce a definitive drawing of the Orion Nebula—"he made scores of drawings, in white on black, and the reverse, in colors, etc.," while preparing it, wrote Edward S. Holden, "each of which was revised and re-revised many times" to produce "the most accurate drawing that has been made, even as a map, and as a picture … decidedly the best representation of a single celestial object which we have by the old methods."[24] Yet all of this effort was to be superseded by a single exposure of 137 min on a gelatine bromide dry plate, taken by Henry Draper at his private observatory at Hastings-on-Hudson New York on March 14, 1882, which showed the Orion Nebula "for nearly purpose incomparably better" than Bond's meticulously crafted engraving.[25]

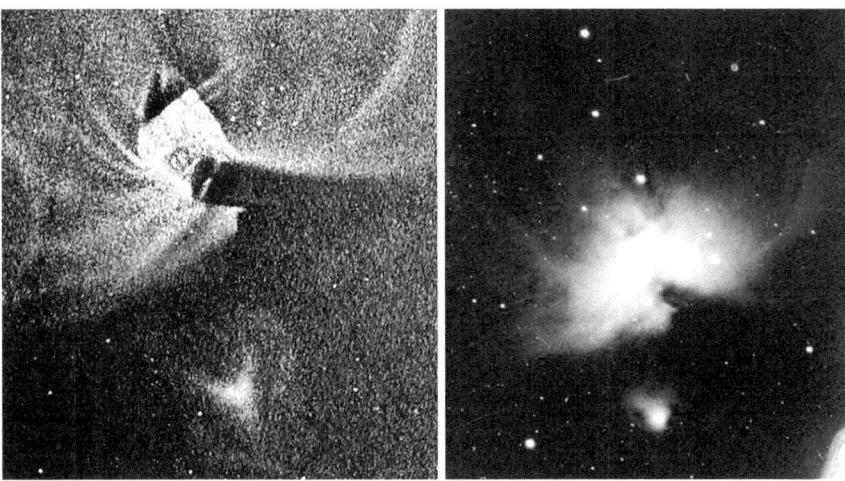

Left, G.P. Bond's carefully prepared rendering of the Orion Nebula using the Harvard College Observatory's 15-inch refractor took 4 years, between 1859 and 1863. Right, Dr. Henry Draper's photograph of the same object on 14 March 1882 took only 137 min. (Credits: Left, George Phillips Bond, "Observations on the Great Nebula of Orion," ed. Truman Henry Safford, Annals of the Harvard College Observatory (Cambridge, MA, 1867), vol. 5, frontispiece. Right, Public Domain via Wikimedia Commons)

The Great Comet of 1882, recorded on a gelatine bromide dry plate by David Gill at the South African Astronomical Observatory, Cape Town, on November 7, 1882. Public Domain via Wikipedia Commons

In planetary observations, the plate fell short, and—for another century—the eye retained its supremacy. Here it was agility rather than sensitivity that counted, and the eye, if not instantaneous, was at least able to take an impression in a fraction of a second. Even the most sensitive photographic emulsions used in the late nineteenth century required several seconds to capture an image of a planet, Mars for instance. Also, the light-gathering power of the telescope (a function of aperture) did not appear to be the determining factor as to what was seen as had been the case in the study of nebulae. It was even argued by some that the smaller instruments used by amateurs were sometimes or even often better for planetary work than the giant instruments wielded by Grand Amateurs and professional astronomers at great observatories. Thus, the English amateur William Frederick Denning argued in 1891:

> … We must judge of large glasses by their revelations; their capacity must be estimated by results. We often meet with glowing descriptions of colossal telescopes: their advantages are specified and their performances extolled to such a degree that expectation is raised to the highest pitch. But it is not always that

such praise is justified by the facts. The fruit of their employment is rarely prolific to the extent anticipated, because the observers have been defeated in their efforts by impediments which inseparably attend the use of such huge constructions.[26]

The impediments involved the restless state of the atmosphere, and Denning argued—not uncontroversially—that, except on the rare occasions when the atmosphere assumed an unusual state of quiescence perhaps one night in fifty, the advantages of the large instrument were offset by its greater susceptibility to the agitation and blurring of the air.[27]

Because of the incessant agitation of the air, images of a planet are not, as one sometimes imagines, comparable to still-lifes, where the details can be filled in at leisure. Instead they resemble a scene in violent motion. The planet is kind of a galloping horse, its movements too fast for the eye to follow. In the case of the horse, these movements could be articulated in the end only with the stop-action photography of Eadward Muybridge. By taking special measures Muybridge was able to make the photographic plate faster than the eye [26]. However, there were obviously no trip wires for planets.

Under the usual conditions of seeing, planetary details come out only in a flash, like an image illuminated for a moment by an electric spark or displayed by a tachistoscope, an ingenious device introduced in 1859 for the study of perceptions in brief durations by the German physiologist A.W. Volkmann. In Volkmann's original "falling door" tachistoscope, a falling screen briefly opens one or both of two apertures in an oblong box, allowing the observer a peek at an image in the rear of the box. The observer trying to make out details on the surface of a planet—say, Mars—faces a similar "tachistoscope effect" [27]. Owing to the rapid contortions produced by the atmosphere, the details are only intermittently visible. At first sight the planet resembled a "buzzing, blooming confusion," as psychologist Williams James phrased it regarding the way the world appears to a newborn [28]. Early observers of Mars, such as Johann Hieronymus Schroeter, could not always convince themselves they were looking at permanent features rather than the ephemeral forms of a shell of cloud. Wilhelm Beer and Johann H. von Mädler, who produced the first map of the planet, found that "ordinarily some time elapsed before the indefinite mass of light resolved into an image with recognizable features."[28] Even the great Mars observer Giovanni Schiaparelli noted that, when he first began his long study of the planet, "at first I didn't know how to orient myself at all."[29] With sustained attention and practice, the various dark and light areas gradually took shape—this was a case similar to what Nasim has described in the case of the nebulae as "stabilizing the phenomenon into something visualized in a particular way"[30]—and various features came into view, of which the finer ones were often fleeting visitants, revealed only in glimpses. Naïve observers—even if intelligent and practiced in other kinds of observation such as of cross-sections in the microscope—usually found the experience of studying Mars disorienting. Thus, Percival Lowell's friend Edward Sylvester Morse, a specialist in brachiopods who spent 6 weeks in May-June 1905 observing with the 24-inch Lowell refractor on almost every evening, wrote:

> Imagine my surprise and chagrin when I first saw the beautiful disk of Mars through this superb telescope. Not a line! Not a marking! The object I saw could only be compared in appearance to the open mouth of a crucible filled with molten gold. Slight discolorations here and there and evanescent areas outline for the tenth of a second, but not a determinate line or spot to be seen [29].

With practice, Morse came to see canals, a few at a time. The whole network was never visible at once but had to be pieced together from bits and pieces revealed in flashes. This was a result of the fact that the fovea—the small depression within the retina where the density of cones is greatest and visual acuity highest—is so small. We see the world through the tiny roving pinhole of foveal consciousness and so, as another Lowell friend and sometime visitor, George Russell Agassiz, put it after spending the month of May 1907 with Lowell in Flagstaff:

> It must not be imagined that any drawing represents what the observer sees the moment he looks through the telescope. Instants of exceptional seeing flash out, here and there, at different spots on the planet. It is not till the same phenomena repeat themselves in the same way, in the same place, a great number of times, that the observer learns to trust these impressions. One has to keep one's mind constantly at the highest pitch to catch and retain what the eye sees.
>
> It is like looking at a Swiss landscape from a high Alp, with the summer clouds sweeping about one. Now the mist roles away, revealing a bit of the valley, and shuts in again in a moment; while in some other spot the clouds break away, and disclose a jagged summit, or a portion of a shining glacier [30].

In an obituary for Lowell written in 1916, Princeton University astronomer Henry Norris Russell captures perfectly the difficulties of observing Mars:

> Under ordinary circumstances, Mars is a heartbreaking object for the observer. The larger and less interesting details upon his ruddy disc are indeed visible telescopically on any good night, but the finer markings are very delicate, and

flash into view only by glimpses in the too rare moments when the ceaseless turmoil of the atmosphere through which we must look dies down and permits us to see the planet with relative little blurring of its finer lineaments.... There is, however, a remarkable diversity in the descriptions of their appearance given by those who had studied them at first hand....

How can these extraordinary differences between the testimony of eye-witnesses be reconciled?...

To the writer, after a somewhat careful investigation, only one explanation appears reasonable, namely, the influence of what is call the "personal equation." [31]

Russell cautiously puts the term personal equation in quotation marks, in recognition of the fact that, properly speaking, the term ought to be used only for the classic eye-and-ear method of determining transits of stars across the wires of a transit instrument. Nonetheless, Russell notes that in some respects they share marked similarities. Thus:

In the far more complex case of observations of planetary details, concerning which all observers agree that they can be seen only in the best moments for a few seconds at a time, similar principles are doubtless to be applied [as in the classic case of the telegraph key determination].[31]

Pursuing a similar line of thought, Edward Walter Maunder, a spectroscopic assistant at the Royal Observatory, Greenwich, wrote in 1900 that Maskelyne and Kinnebrook's disagreements might not consist of simple mechanical or physiological factors as usually assumed. Instead they might reveal profound differences in personality and temperament, predisposing one observer in one way, another in another:

A very slight knowledge of character will show that [the determination of transits] will require different periods of time for different people. It will be but a fraction of a second in any case, but there will be distinct difference, a constant difference, between the eager, quick, impulsive man who habitually anticipates, as it were, the instant when he sees star and wire together, and the phlegmatic, slow-and-sure man who carefully waits till he is quite sure that the contact has taken place, and then deliberately and firmly records it. These differences are so truly personal to the observer that it is quite possible to correct for them, and after a given observer's habit has become known, to reduce his transit times to those of some standard observer [32].

Maunder has intuited something fundamental in his distinction between the "eager, quick, impulsive man" and the "phlegmatic, slow-and-sure man." Manifested at once in transit observations, where it is only necessary to determine *that* the star transited the wire at such and such a time (*when* it did so), it enters also into planetary observations, though in a somewhat more complicated way. The first thing the eye discovers is *that* a detail of some kind is present, and the brain inquires *where*. This involves one type of cortical processing that calculates the position, direction, and orientation of the object. The further discovery of the shape and form of the object—in other words, the *what*—involves a different stream of cortical processing, which carries out a different set of calculations.[32]

So far as it goes, this analysis supports Russell's analogy to the classical personal equation in planetary observations. Some observers—the eager, quick, impulsive ones—are more interested in *where* something revealed in a glimpse may be located; others—the more phlegmatic, slow-and-sure ones—discount what is revealed in the glimpse, and stand by patiently until they can make out not just *where* but *what* has been revealed.

What then? The next thing belies the analogy to the personal equation, for now instead of merely pressing a telegraph key, the observer needs to undertake a far more complicated method of recording. It is at this point that the process of planetary observation diverges decisively from the eye-and-ear method and requires artistic interpretation. At an early stage of perception, the eye uses saccadic movements to test hypotheses—beliefs—about how the sensory data it encounters are caused. It makes predictions, and if the predictions are not verified—if a saccadic movement overshoots or undershoots the target—the error is corrected and a new saccadic movement, i.e., round of hypothesis testing, beings. The explorations and hypothesis testing of the eye at this early stage of building up the visual scene have their counterpart in the creation of a drawing. As art historian David Rosand has remarked, "The drawn mark is the record of a gesture, an action in time past now fixed permanently in the present; recalling its origins in the movement of the of the draughtsman's hand, the mark invites us to participate in that recollection of its creation." Thus, the act of drawing involves "probing," "groping," "grasping," "exploration" [33]. In the hand (literally) of the artist, the drawing reflects a particular—a this and no other—mode of seeing and knowing. And so we have, coming back to the case of Mars, differences as marked as those of a Schiaparelli or a Nathaniel Green who, though summarizing their impressions during the same season of observations

(1877–78), have produced drawings so unlike that they can hardly be believed to represent the same planet. Here the difference cannot be attributed to that inherent in the personalities of the observers—a quick impulsive man on one side and a phlegmatic slow man on the other. Both of these observers were painstaking, careful, and deliberate in their methods. The difference between them lies in something more pervasive, in the different goals of the observers and their underlying beliefs as to in what essentials the reality of the features in the Martian scene were to be best defined. The Rev. T.W. Webb wrote in 1880:

> At first sight there is more apparent difference in their results than might have been expected. It is not surprising that in the case of minute details each should have caught something peculiarly his own; but there is a general want of resemblance that is not easily explained, till, on careful comparison, we find that much may be due to the different mode of viewing the same objects, to the different training of the observers, and to the different principles on which the delineation was undertaken. Green, an accomplished master of form and color, has given a portraiture, the resemblance of which as a whole, commends itself to every eye familiar with the original. The Italian professor, on the other hand, inconvenienced by color blindness, but of micrometric vision, commenced by actual measurement of sixty-two fundamental points, and carrying on his work with most commendable pertinacity, has plotted a sharply-outlined chart, which, whatever may be its fidelity, no one would at first imagine to be intended as a representation of Mars. His style is as unpleasantly conventional as that of Green indicates the pencil of the artist; the one has produced a picture, the other a plan… [34].

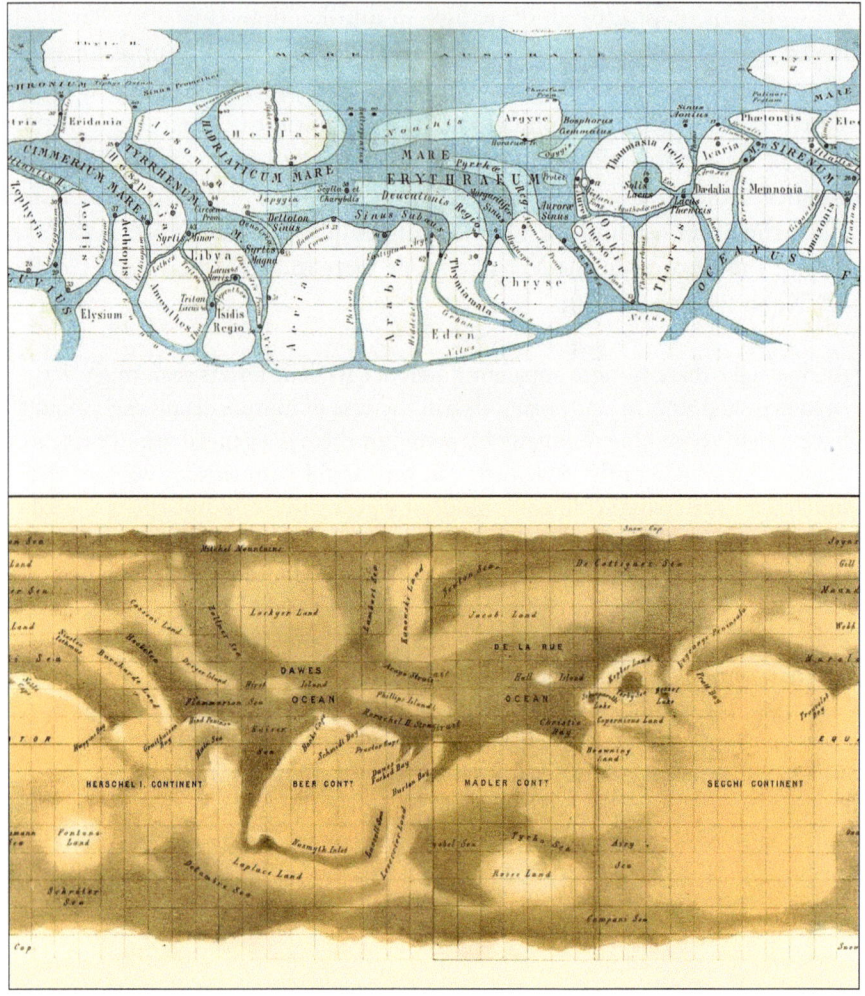

Giovanni Schiaparelli's map of Mars compared to Nathaniel Green's, both 1877–78.
(Credit: Images in William Sheehan Collection processed by Joel Hagen)

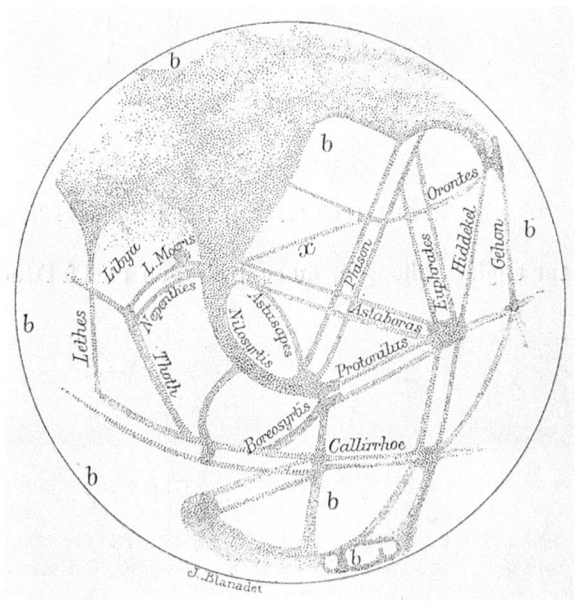

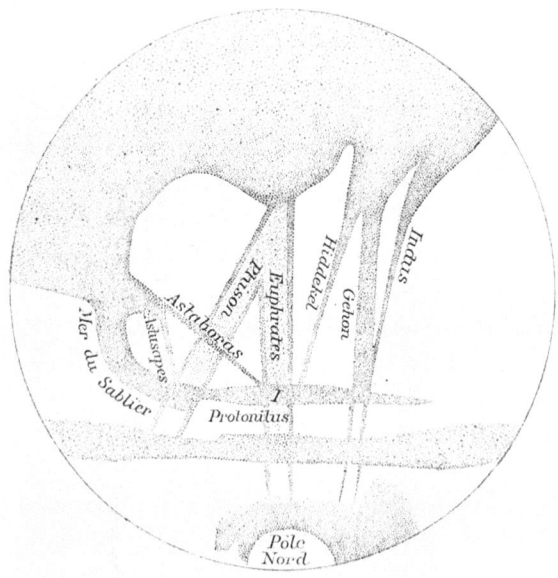

Nearly simultaneous drawings of Mars, made an hour apart on June 3, 1888, by Giovanni Virginio Schiaparelli with the 19-inch refractor at the Brera Observatory in Milan and Henri Perrotin with the 15-inch refractor at the Nice Observatory on the Côte d'Azure. (From: Camille Flammarion, La Planète Mars, pp. 550–551)

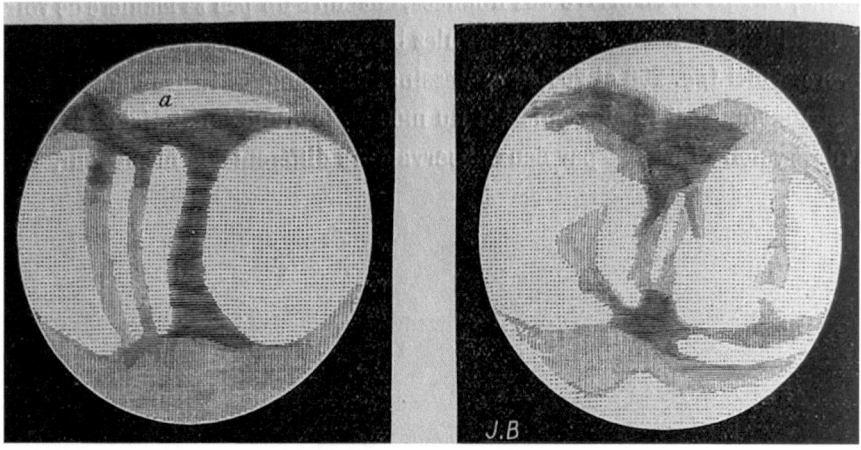

Drawings of Mars by (left) Edward S. Holden and (right) James E. Keeler, with the same instrument, the Lick 36-in. refractor, made 15 min apart on July 27, 1888. In comparing these, Camille Flammarion exclaimed, "In each of these two sketches, the Hourglass Sea crosses the middle of the disk, from above to below. But what differences in aspect!" From: Flammarion, La Planète Mars, *p. 552*

Predictive (Bayesian) Brains

When details in an image are only fleetingly and partially revealed, it is necessary—as visual physiologists such as Volkmann with his tachistoscope had begun to work out in the mid-nineteenth century—for the brain to fill in whatever is missing (as in the well-known case of the blind spot) with the imagination. We obviously don't (though we often believe we do) experience *objects* directly, whether the chair on the other side of the room, or the planet on the other side of the telescope. They do not *impose* their properties on us. To the contrary, sensory information is so incomplete in general that it can hardly serve as an adequate guide to what is really there. As the British psychologist Richard L. Gregory writes:

> Does it convey all that we need to know about an object in order to behave toward it appropriately? At once we see the difficulty—the continuous problem the brain has to solve. Given the slenderest clues to the nature of surrounding objects we identify them and act not so much according to what is directly sensed, but to what is believed. We do not lay a book on a "dark brown patch"— we lay it on a table. To belief, the table is far more than the dark brown patch sensed with the eyes; or the knock with the knuckle, on its edge. The brown patch goes when we turn away; but we accept that the table and the book remain [35].

Small wonder it is so easy to be misled, and to find ourselves in realms of illusion and erroneous surmise. No wonder there can be such great individual differences in what various observers see. Indeed, the remarkable thing is perhaps that we agree as much as we do. Perceptual psychologists have created numerous examples of ambiguous figures where something looks now like one thing and now like another. According to Gregory, these pictures "tend to remain stable until the eyes move to a different region."[33] Looking at certain regions tends to favor one perception over others; this underscores the importance of directed attention—the ability to voluntarily manage the focus of our thoughts. In the case of Mars, one observer may focus on the lighter-hued "deserts" and see "canals," another on the darker "seas" and see "canyon, slope and ridge." Moreover, selected attention being a scarce and finite mental resource, when placed under continual demand, the ability to focus wanes, leading to "directed attention fatigue." This fatigue may explain why a spontaneous perceptual change often occurs precisely when one's more strenuous attempts relax, and results in the sudden revelation of unexpected alternatives.

It is clear enough, even without going deeply into the details, that the human eye's near-instantaneity, which had always been taken for granted, cannot be justified. A great deal goes on unnoticed in the visual world's *mise en scene*. The eye had been fast enough for the slow medieval world of traditional human purposes, but it was unable to follow a world traveling at dizzying speed out the window of a moving train or a telescopic image constantly disrupted by fluctuating air. Both artists and planetary astronomers became "impressionists," attempting, to the best of their ability, to transcribe the less than stable visual reality presented to them. They were discovering the eye's difficulty in following phenomena that occur at more than cinematic speed. This same difficulty lay at the bottom of magician's tricks where, "Now you see it, now you don't!" and "the hand is quicker than the eye," are stocks in trade [36]. The planets—in the pre-spacecraft and pre-charge couple device (CCD) era—were indeed magicians, master illusionists. As the pioneering planetary astronomer Gerard P. Kuiper pronounced in the 1950s:

> Visual observation is superior to photography in the resolution of fine detail… The reasons for this superiority have often been debated; my impression is that it is due in part to the fact that the eye "works in series" with the brain and in part to the remarkable qualities of the eye as a recording instrument. The eye sees the equivalent of thousands of "snapshots" in succession, and the brain sorts out this information, tries to reject all that appears useless (including seeing effects), and integrates the bits that appear interesting. Thus a mental picture is built up that contains more than any one photograph could possibly show.

But will this mental picture tell the truth? There is obviously no assurance that it will. An observer who passionately believes that something is true will, under marginal observing conditions, be tempted to observe and integrate until he obtains the impression he hopes for. Or he may, under such conditions, arrive at a new conclusion; his mind may play with his conclusion and decide that it is interesting. Thereafter renewed observation, still under marginal conditions, may confirm the result over and over again. Yet it may be quite wrong. I am personally convinced that this has happened on certain surface features on Mars... [37].

In the CCD and video imaging era, the eye's capture rate has been found to be only 5 to 15 images per second. This is why video looks like a seamless continuum—25 to 30 frames per second is above the threshold that is required for a phenomenon that psychologists call *flicker fusion* of a moving image to occur. Flicker fusion refers to the eye-brain combination that fills in portions of the picture that are absent for very short intervals. During this process noise is averaged out and frames of limited quality are smoothed to create the perception of a sharp, vivid picture [38].

In the visual era of planetary observation, a great deal of imagination had to be used to fill in the missing parts. It used to be assumed that the way the brain builds its models of the visual world was from the bottom up, through the amassing of low-level cues supplied by line-detectors, edge-maps, and so on in the retina itself and in specialized areas of the visual cortex. More recently, the concept of a the "predictive" brain—or Bayesian brain—has become influential. Broadly based on the Bayes's Theorem, first recognized by the eighteenth-century English clergyman Thomas Bayes, it was applied to perception by the great nineteenth-century German physiologist Hermann Helmholtz. According to Helmholtz, perception involves a bi-directional cascade of cortical processing (top-down vs. bottom-up). The brain does not grasp the world directly. Instead it is a "phantastic organ," where what we perceive is a kind of 'controlled hallucination.'"[34] The "controlled hallucination" (which might be a theory or something we simply imagine might be true) makes a claim about reality. To each claim we assign some degree of belief (prior credence). If we are rather sure of ourselves (I'm sitting on a chair just now), then the prior credence is close to unity. If we have no reason whatever to believe a claim (unicorns exist), the prior credence will be close to zero.

Bayesian reasoning is closely related to what is referred to as *abductive* reasoning, and distinguished from the more familiar *deductive* reasoning and *inductive* reasoning. In *deductive* reasoning (as in math), conclusions are reached by logic alone: so long as the premises are true, the conclusions will be correct. *Inductive* reasoning reaches its conclusions from a series of observations so that—to give the familiar example—the Sun has risen in the east for all recorded history, therefore it is highly likely that it will rise in the east tomorrow (though consider a case where inductive reasoning is wrong: the chair that was there in the corner for the last 10 days is missing today). Both deductive and inductive reasoning involve forward reasoning: they proceed from causes to effects. The third kind of reasoning, *abduction*, involves working backwards from observed effects to their most likely causes, and can be described as "inference to the best explanation." A typical example is the classical case of Ptolemy reasoning backward from observed patterns in planetary phenomena to their causes in epicycles. Whether the reasoning is correct or not does not at this point enter in: Ptolemy's epicycles do not exist, but they did provide, for a time, an optimal inference about what was actually going on in the heavens that would "save" (or explain) the phenomena (that is, sensory inputs). In the same way, Lowell's planet-wide irrigation system, or Barnard's naturalistic surface of canyon, slope and ridge provided optimal inferences for those observers about Mars; i.e., based on their background knowledge about Mars, they asked what possible explanation was best accounted for this data. Obviously, they made very different inferences, and this gets right to the heart of the complexity of the matter.

In Bayesian reasoning, the prior credence—belief, or expectation—is important. It might be, for instance, that the existence of intelligent life on Mars is almost certain. Then the prior credence would be around unity. On the other hand, it might be almost certain that there is no intelligent life on Mars. Then the prior credence would be around zero. That is the departure point for Bayesian reasoning, but it does not all end there. Instead, it seeks out evidence about these claims, and updates these credences on the basis of what turns up. (The revised credences are referred to as "posterior credences.")

To give some straightforward examples: Consider the expectations regarding a coin toss. For a fair coin, the prior credence should be 0.50—that is, we know there is a 50 percent chance the coin will turn up heads, and 50 percent it will turn up tails.[35] Or consider a game of poker. Statistically, it can be shown that a five-card hand will be "nothing" about 50 per cent of the time, one pair about 42 percent, and a flush (five cards of the same suit) less than 0.2 percent of the time. Something close to these should represent the prior credences of a player at the start of the game. However, as soon as the opposing player discards a certain number of cards and draws replacements, new information becomes available. Thus, if a player draws just one card, she is unlikely to have one pair; however, she is quite likely to have two pair or four of a kind. The updated beliefs (posterior credences) are calculated based on this new information (likelihoods) of what the other player holds. Bayes's Theorem states that the case of proposition X (say, the player has four of a kind) in the light of observation D (the player draws one card) is

(Credence in proposition X given observation D) $\propto$
    (Likelihood of observation D given proposition X) times (prior credence in proposition X),

where the symbol $\propto$ means proportional to.

It turns out, according to the latest and best neurological thinking, that the brain also makes predictions based on *prior beliefs* (which are neurologically top-down) regarding expected sensory inputs (neurologically bottom-up). Often the predictions ("expectations," "best guesses") agree with the sensory data; nonetheless, sometimes they disagree, in which case the percept or inference is said to be illusory or false. Neuroscientist Anil Seth writes:

> … the brain is continually generating predictions about sensory signals and comparing these sensory predictions with the sensory signals that arrive at the eyes and the ears—and the nose, and the skin, and so on. The differences between predicted and actual sensory signals give rise to prediction errors. While perceptual predictions flow predominantly in a top-down (inside-to-outside) direction, prediction errors flow in a bottom-up (outside-to-inside) direction. These prediction error signals are used by the brain to update its predictions, ready for the next round of sensory inputs. What we perceive is given by the content of all the top-down predictions together, once sensory prediction errors have been minimized—or "explained away"—as far as possible.[36]

Some observers hold tenaciously to their prior beliefs (sometimes because the prior belief reflects something they desperately want to be true). In that case they will become very good at explaining away the prediction errors. In other words, they become adept at squeezing data to fit with their prior beliefs (a tendency sometimes referred to as confirmation bias). Such observers are extremely unlikely to change their prior beliefs (Lowell on Mars was like this). Other observers are more "bottom-up"; they start with less strongly held prior beliefs, and are more receptive to sensory data and more apt to entertain shifts in credences (Barnard was like this).

Lowell, even before he went to Flagstaff to observe, said about the Martian canals in a lecture before the Boston Scientific Society, "The most self-evident explanation from the markings themselves is probably the true one; namely, that in them we are looking upon the result of the work of some sort of intelligent beings."[37] Thus, his prior credence regarding intelligent life on Mars was, if not quite unity, nonetheless very high—perhaps .90, .95, or even .99. This implied that it would prove difficult to alter his views, and so it did: throughout his career, he showed a very strong tendency to squeeze observations to conform to his prior credence, and to explain away prediction errors.

Barnard, on the other hand, from his first view of Mars through a telescope, cautioned against "Rashly Drawn Conclusions." He wrote, "It is well to fetter the wings of our fancy and restrain its flights. It is quite possible we may have formed entirely erroneous ideas of what we actually see…. Man is too quick at forming conclusions. Let him but indistinctly see a thing, or even be undecided as to whether he does actually see it and he will then and there set himself to theorizing, and build immense castles of conjecture on a foundation, of whose existence he is by no means certain."[38] This describes the position of someone with a relatively low prior credence regarding intelligent beings on Mars. Though it's hard to give a number, perhaps 0.20, 0.10 or even less would not be very far from the truth. Nothing he observed made him revise this upward. The point is that the priors of Lowell and Barnard—and their relative weighing of new information—set them on a definite course throughout the remainder of their careers.

It may be useful to generalize here a bit with an example from outside of astronomy, but not without relevance to planetary observations. In a 1940s-era psychological experiment, subjects were asked to identify a series of playing cards on short and controlled exposures with a tachistoscope [40]. This experiment strongly impressed the philosopher of science Thomas Kuhn and contributed to his well-known ideas about paradigm-shifts and the way science sometimes advances. He wrote:

> Many of the cards were normal, but some were made anomalous, e.g., a red six of spades and a black four of hearts… After each exposure the subject was asked what he had seen, and the run was terminated by two successive correct identifications…. Even on the shortest exposures, many subjects identified most of the cards, and after a small increase, all the subjects identified them all. For the normal cards these identifications were usually correct, but the anomalous cards were almost always identified, without apparent hesitation or puzzlement, as normal…. With a further increase of exposure to the anomalous cards, subjects did begin to hesitate and to display awareness of anomaly [41].

The moment of transition—from seeing the card expected to seeing the card revealed—tends to give rise to (as other tachistoscope experiments showed) "decidedly unpleasant feelings of tension and unrest which later subside when a final, stable configuration is revealed" [42]. This metastable state (the precariousness of the equilibrium and the ease with which it is apt to be undone triggering the feelings of tension and unrest) is referred to by predictive-brain researchers as "surprisal."[39] It often occurs at the moment when what is rather similar to a phase-transition in a physical system—from ice to liquid water, or liquid water to vapor—occurs, and may (or may not) represent the moment in which an observer undergoes a "change of mind."

An example is given in the following report of Schiaparelli, who after years of recording single and double canals suddenly experienced surprisal in his views of the planet with a 19-inch refractor on the nights of June 2, 4, and 6, 1888:

> What strange confusion! What can all this mean? Evidently the planet has some fixed geographical details, similar to those of the Earth.... Comes a certain moment, all this disappears to be replaced by grotesque polygonations and geminations which, evidently, attach themselves to represent apparently the previous state, but it is a gross mask, and I say almost ridiculous.[40]

In the case of the playing card experiment, observers who tended to be more top-down oriented—and good at explaining away prediction errors—were presumably likely to experience the anomalous cards as normal; others, more bottom-up oriented, were presumably more likely to notice them and to experience a surprisal. Note that glimpses are sufficient to confirm a prior belief, expectation, or bias, but a good close look is needed to effect a change (in Kuhn's terms, to shift a *paradigm*). This becomes obvious in the case of the canals of Mars, but it is also attested in cases like the Loch Ness monster or UFOs; those who are already convinced of a proposition are much more apt to see confirmation in marginal perceptual data (e.g., blurry photographs) than someone who is sceptical. Lowell, corresponding to Maunder's quick, impulsive man, was satisfied to have the where of a thing—the vision of a line with such and such an orientation in such and such a place, as revealed in a glimpse, a momentary flash of detail. Barnard, the more careful, sure man, wanted to get beyond the where to the what; detail revealed only in glimpses was not to be relied on. Thus—though for precisely opposite reasons—neither Lowell nor Barnard shifted his prior credences in the light of what he saw at the telescope eyepiece.

Thus do we lay our scene. One observer is not interchangeable with another. Lowell and Barnard could hardly present greater differences of approach, and nowhere do they show up more glaringly than in the study of Mars.

Notes

1. The title of this chapter is based on that of a talk by the author given at the History Session of the Astronomical Society of the Pacific at Lowell Observatory in 1994.
2. Nasim, *Observing by Hand*, p. 9.
3. Edward Singleton Holden to E.B. Knobel, June 16, 1893; Yerkes Observatory Archives. Quoted in William Sheehan [3].
4. Percival Lowell, talk to the Boston Scientific Society, given on May 22, 1894; Boston *Commonwealth*, May 26, 1894.
5. E.E. Barnard, observing log book; Lick Observatory.
6. Richard Baum to William Sheehan, personal communication; August 12, 1994.
7. W.H. Steavenson, on dressing down H. Percy Wilkins over his credulity about the existence of an arch bridge on the Moon. Quoted in Thomas A. Dobbins and Richard M. Baum (January 1998), "O'Neill's Bridge Remembered," *Sky & Telescope*, pp. 105–108:p. 108.
8. These comments by Julius von Sachs in his *History of Biology* (1875) describe observations with a microscope, but the principle is, of course, the same for telescopic observations. Quoted in Nasim, *Sketching the Nebulae*, p. 15.
9. None is a better starting point than E.H. Gombrich [7].
10. Von Sachs, quoted in *Sketching the Nebulae*, p. 15.
11. I am grateful to Tony Misch, Director of the Historical Collections Project of Lick Observatory, for sharing this highly relevant insight.
12. Lowell, *Mars and Its Canals*, p. 365.
13. See Sheehan and Westfall, *Transits of Venus*.
14. For a more complete history of eighteenth and nineteenth century theories of observation, see: Christoph Hoffmann [15].
15. David Kinnebrook, Jr., to David Kinnebrook, Sr., January 20, 1796.
16. Matthew Lund to William Sheehan, personal communication, April 22, 2022.
17. Of course, the notions of "absolute space" and "absolute time" would survive for a while yet—until the Einsteinian revolution of the early twentieth century.
18. Bessel, *Astronomische Beobachtungen*, p. vii.
19. Boring, *Experimental Psychology*, p. 143.
20. Boring, *Experimental Psychology*, p. 146.
21. Significant findings emerged from the famous laboratory for experimental psychology founded by Wilhelm Wundt at the University of Leipzig, which showed that both the reaction time and the visual-auditory "complication" introduced in Bradley's eye and ear method depended crucially on the attentional predisposition of the observer. This attentional predisposition is now described as "prior entry," a phenomenon of selective attention in which an

event that arrives in a channel to which we are attending—say the visual channel—is perceived as earlier than a concurrent event arriving in a channel to which we are not attending—say the auditory channel. A switch of attention from one channel to another typically takes around 300 milliseconds. See: Adam Reeves and George Sperling [21].

22. According to Adolph Hirsch, as noted in Jimena Canales [22, pp. 190–191].
23. Though photographic plates did build up an image over time, it should be mentioned that "reciprocity failure" was real, and linearity of response only existed to a point, after which increasing the exposure further didn't produce an additional gain.
24. Edward Singleton Holden [23], Appendix I, Addendum, p. 227.
25. Holden, "Monograph," p. 227. The qualification—"for nearly every purpose"—was just, as it must be said in defense of Bond that his drawing shows features seen in the Hubble Space Telescope's "sharpest view of the Orion Nebula" not seen in Draper's photo, since the eye can detect subtle contrast variations that were burned out of the photographic emulsion (i.e., streamers and wing tails). As an aside, it may be useful to point out that the advantage of the plate was not in its greater sensitivity than the eye to photons. The rods of the eye have 50 percent quantum efficiency, the photographic emulsions used in the early days of astronomical photography less than 1 percent of 1 percent. Photography's advantage was rather its ability to accumulate photons over long periods of time. The eye is rapidly fatigued as the photopigments of the retina are used up. The plate is, by comparison, relatively indefatigable. The eye is the hare that jumps out ahead of the slower tortoise only to be overtaken and completely vanquished at last.
26. William F. Denning [24 , p. 33]. For details about the small versus large telescope controversy, see: John Lankford [25].
27. Denning, *Telescopic Work*, p. 33.
28. Sheehan, *Planets and Perception*, p. 45.
29. G.V. Schiaparelli (1996) *Astronomical and Physical Observations of the Axis of Rotation and the Topography of the Planet Mars, First Memoir, 1877–1878.* Translation by William Sheehan (San Francisco, the Association of Lunar and planetary Observers), A.L.P.O. Monograph Number 5, p. 1.
30. Nasim, *Sketching the Nebulae*, p. 5.
31. Though Russell puts quotation marks around "personal equation," he clearly relates what he is describing in the case of planetary observations to the classical one of transit observations, as he adds (ibid., p. 782): "In the far more complex case of observations of planetary details, concerning which all observers agree that they can be seen only in the best moments for a few seconds at a time, similar principles are doubtless to be applied."
32. The detection of position, direction and orientation of an object is owing to what is called the M-stream of cortical processing. The M-stream begins in the retina with receptors known as M-type retinal ganglion cells; they will be

stimulated by—to take specific examples—seeing leaves rustling, perhaps meaning there is a tiger on the prowl but more likely just the effect of a gust of wind, or by a detail that appears suddenly, in a flash, on Mars. These cells send information via the M stream to the visual cortex at the back of the head and then on to other cortical areas dedicated to motion detection. This motion-detection system is designed to respond rapidly and decisively. However, this makes it susceptible to what are referred to as Type I errors—false positives. In evolutionary terms, these errors are a small price to pay when misconstruing the movement of a bush as produced by a gust of wind rather than a tiger on the prowl is likely to lead to much graver consequences than the other way around. The other system of visual processing begins with P-type retinal ganglion cells, which send information via the P stream to the visual cortex and on to cortical areas that are not motion-sensitive but instead perform calculations regarding the location and shape/form of the object. That is, they address the question of *what* the object is. They make a slower more measured response. They are susceptible, nevertheless, to Type II errors, or false negatives; in other words, to missing the tiger and being devoured, end of story.

Neuroanatomically, the M-stream is located dorsally in the cortex and involves the parietal association cortex and superior and middle temporal visual association cortex. The P-stream is located ventrally, in the inferior temporal visual association cortex. As might be expected, damage to the dorsal visual association cortex results in deficits in spatial orientation, motion detection and in guidance of visual eye tracking movement, whereas damage to the ventral visual cortex produces deficits in complex visual perception tasks, attention and learning and memory. See Valentin Dragoi, Visual Processing: Cortical Pathways, *Neuroscience Online*, Department of Neurobiology and Anatomy, University of Texas Medical Center at Houston. https://nba.uth.tmc.edu/neuroscience/m/s2/chapter15.html

33. Gregory, *Intelligent Eye*, p. 39.
34. Seth, *Being You*, p. 79.
35. For a highly accessible introduction to Bayesian reasoning, with examples, see: Sean Carroll [39].
36. Seth, *Being You*, pp. 110–111.
37. Lowell, "Lowell Observatory: Percival Lowell before the Boston Scientific Society, May 22, 1894." Boston *Commonwealth*, May 26, 1894.
38. E.E. Barnard (1880). "Mars; its moons and heavens." Unpublished manuscript, Barnard. Edward Emerson. Papers, Special Collections and University Archives, Jean and Alexander Heard Library, Vanderbilt University.
39. For a general discussion, see: Andy Clark (2013), "Whatever next? Predictive Brains, Situated Agents, and the Future of Cognitive Science," *Behavioral and Brain Sciences*, pp. 1–73. On "surprisal," in the technical sense, see p. 6: "Surprisal names the sub-personally computed implausibility of some sensory

state, given a model of the world. Entropy, in this information theoretic rendition, is the long-term average of surprisal, and reducing information-theoretic free energy amounts to improving the world model so as to reduce prediction errors, hence reducing surprisal (since better models make better predictions)."

40. G.V. Schiaparelli to F. Terby, June 7, 1888; in: G.V. Schiaparelli (1963) *Corrispondenza su Marte* (Pisa: Domus Galilaeana), vol. 1 (1877–1889), pp. 203–204:p. 204. Translation by W. Sheehan.

References

1. Lowell P (1895) Mars. Boston, Houghton, Mifflin and Company, p 1
2. Nasim OW (2013) Observing by hand: sketching the nebulae in the nineteenth century. University of Chicago Press, Chicago/London, pp 5–6
3. Sheehan W (1995) The immortal fire within: the life and work of Edward Emerson Barnard. Cambridge, Cambridge University Press, p 1
4. Lowell P (1906) Mars and its canals. New York, Macmillan, p 149
5. Seth A (2021) Being you; a new science of consciousness. New York, Dutton, p 79
6. Churchill WS (1950) Painting as a pastime. McGraw-Hill, New York, pp 28–29
7. Gombrich EH (1961) Art and illusion: a study in the psychology of pictorial representation. Princeton University Press, second edition revised, Princeton
8. Barnard EE (1896) Micrometric measures of the Ball and Ring system of Saturn, and measures of the diameter of his Satellite Titan.... with some remarks on large and small telescopes. Mon Not R Astron Soc 56:163–172
9. Loftus EF (1996) Eyewitness testimony. Harvard University Press, Cambridge, MA, pp 6–7
10. Gopnik A, Meltzoff AN, Kuhl PK (1999) The scientist in the crib: what early learning tells us about the mind. New York, William Morrow, p 50
11. Sheehan W, Westfall J (2004) The transits of Venus. Prometheus, Amherst, p 194
12. Bacon F (1620) Novum Organum. In: Dick HG (ed) Francis Bacon: selected writings. Modern Library, 1955, New York, p 470
13. Sheehan W (2013) From the transits of Venus to the birth of experimental psychology. Phys Perspect 15:130–159
14. Maskelyne N (1799) Observed transits of the fixed stars and planets over the Meridian. In the year 1788. In: Astronomical observations, made at the Royal Observatory at Greenwich, vol 3, p 319
15. Hoffmann C (2007) Constant differences: Friedrich Wilhelm Bessel, the concept of the observer in early nineteenth-century practical astronomy and the history of the personal equation. Br J Hist Sci 40(3):333–365
16. Dalton J (1798) "Extraordinary facts relating to the vision of colours, with observations" (read October 31, 1794). Manchester Lit Philos Soc Mem 5(Pt. 1):28–45

17. Herschel SJ (1847) A brief notice of the life, researches and discoveries of Friedrich Wilhelm Bessel. London, George Barclay, p 5
18. Bessel FW (1823) Astronomische Beobachtungen auf der Universitäts-Sternwarte zu Königsberg, vol 8, pp iii–viii
19. Boring EG (1957) A history of experimental psychology, 2nd edn. Appleton-Century-Crofts, New York, p 135
20. Daston L, Galison P (2010) Objectivity. New York, Zone Books, p 17
21. Reeves A, Sperling G (1980) Attention gating in short-term visual memory. Psychol Rev 93:180–206
22. Canales J (2001) Exit the frog, enter the human: physiology and experimental psychology in nineteenth-century astronomy. Br J Hist Sci 34:173–197
23. Holden ES (1882) Monograph of the central parts of the nebula of Orion. In: Washington astronomical observations for 1878. US Government Printing Office, Washington, DC
24. Denning WF (1891) Telescopic work for starlight evenings. Taylor and Francis), London, pp 29–30
25. Lankford J (1981) Amateurs vs. professionals: the controversy over telescope size in late Victorian science. Isis 72:11–28
26. Solnit R (2003) River of shadows: Eadweard Muybridge and the technological wild west. Penguin, New York
27. Sheehan W (1988) Planets and perception: telescopic views and interpretations, 1609–1909. University of Arizona Press, Tucson
28. James W (1890) The principles of psychology, vol 1. Dover reprint of original edition, New York, p 488
29. Morse ES (1906) Mars and its mystery. Little, Brown & Company, Boston, p 80
30. Agassiz GR (1907) Mars as seen in the Lowell refractor. Popular Sci Monthly 71:275–282
31. Russell HN (1916) Percival Lowell and his work. The Outlook (December 6):781–783
32. Maunder EW (1900) The royal observatory, Greenwich: a glance at its history and work … with many portraits and illustrations, etc. Religious Tract Society, London, pp 176–177
33. Rosand D (2002) Drawing acts: studies in graphic expression and representation. Cambridge, Cambridge University Press, p 2
34. Webb TW (1880) Planets of the season: Mars. Nature 34:213
35. Gregory RL (1970) The intelligent eye. London, Weidenfeld and Nicholson, p 11
36. Macknick SSL, Martinez-Conde S (2010) Sleights of mind: what the neuroscience of magic reveals about our everyday deceptions. Picador, New York
37. Kuiper GP (1955) On the Martian surface features. Publ Astron Soc Pac 67(398):271–283
38. Massey S, Dobbins TA, Douglass EJ (2000) Video Astronomy. Sky Publishing, Cambridge, MA, p 10

39. Carroll S (2016) The big picture: on the origins of life, meaning, and the universe itself. Dutton, New York, pp 69–83
40. Bruner JS, Postman L (1949) On the perception of incongruity: a paradigm. J Pers 13:206–223
41. Kuhn TS (1962) The structure of scientific revolutions. University of Chicago Press, Chicago, pp 62–63
42. Flavell JH, Draguns J (1957) A microgenetic approach to perception and thought. Psychol Bull 54(3):197–217

2

Astronomy from the Top Down: Percival Lowell

Contents

> *[Lowell and I] are not one. This is because his standpoint of pure science is too high to allow of that intimacy which means soul sympathy. I have tried to study from the bottom up what he has observed from the top....*
>
> —*Lafcadio Hearn to Basil Hall Chamberlain, January 14, 1893*

© The Author(s), under exclusive license to Springer Nature Switzerland AG 2024   **45**
W. Sheehan, *Parallel Lives of Astronomers*, Springer Biographies,
https://doi.org/10.1007/978-3-031-68800-3_2

Master of all he surveyed, Percival Lowell looks down on the Grand Canyon from Yavapai Point on the South Rim (where the geology museum is now). To the left, with the picture boundary dissecting it, is Bright Angel Canyon, while the butte right above his head is O'Neill Butte, where the Kaibab Trail runs today. Photograph taken by Edward S. Morse, May/June 1905. (Credit: Lowell Observatory Archives)

Ancestors

Percival Lowell was the quintessential "top-down" astronomer. He never looked at things from the ground level, always from on high. The American literary critic Christopher Benfey writes, in introducing Lowell's 1891 ascent of the sacred Japanese mountain Ontaké, in order to study Shinto trances, "Percival Lowell loved heights. Standing on a jagged peak above the clouds, with the humdrum human world forgotten and the stars for aristocratic company, Lowell felt right at home." [1] Lowell's book *Occult Japan* (1894), in which he described this ascent, was finished just as he prepared to set out for the new heights of "Mars Hill" in Flagstaff, as he called the mesa where he would found his observatory and from which he would attempt to climb a much higher mountain yet—the planet Mars. Here, too, he was to be "a sort of high priest and hardy pilgrim" (says Benfey), "who must travel alone to distant and elevated shrines."[1] As Lowell himself was to put it in *Mars and Its Canals* (1906), the astronomer

> … must abandon cities and forego plains. Only in places raised above and aloof from men can he profitably pursue his search, places where nature never meant him to dwell and admonishes him of the fact by sundry hints of a more or less distressing character. To stand a mile and a half nearer the stars is not to stand immune.
>
> Thus it comes about that today besides its temples erected in cities, monasteries in the wilds are being dedicated to astronomy as in the past to faith; monasteries made to commune with its spirit, as temples are to communicate the letter of its law… Primitive, too, they must be as befits the still austere sincerity of a cult, in which the simplest structures are found to be the best [2].

So, says Benfey, "the solitary pilgrim sipping tea on Ontaké and the astronomer high on Mars Hill are one."[2]

Lowell was born on March 13, 1855, at 131 Tremont Street in Boston, in the heart of historic downtown Boston, with a good view of the golden-domed State House across from Boston Common, half a block from the Park Street Congregational Church and just around the corner from the residences of his two grandfathers, John Amory Lowell and Abbott Lawrence, at no. 7 Park Street and no. 8 Park Street respectively. The frontage of the particular building in which Lowell was born was taken over by Shephard Norwell & Company in 1865, and in 1913 replaced by the 8-story building in which the Shephard Stores were housed until 1937. (The building still exists, and more recently has been converted into expensive flats and condos—expensive because the location remains eminently desirable.) The birthplace would have been just about the point from which painter Childe Hassam's "Boston Common at Twilight," looking southwest along Tremont Street toward Boylston, was later painted. The splendor of his natal surroundings suggest the splendor of his situation, which was rather like that of the heiress Milly Theale in Henry James's *Wings of the Dove*—when thinking of the Lowells and the social circles in which they moved, one cannot help thinking of Henry James. To slightly paraphrase, Lowell's "range was … immense; he had to ask nobody for anything, to refer nothing to anyone; his freedom, his fortune, and his fancy were the law; an obsequious world surrounded him, he could sniff up at every step its fumes." [3]

Percival grew up, in short, in staggering wealth, and acquired a sense of entitlement—of superiority to the rest of the world—and tended to dismiss loftily all notions except his own.

Childe Hassan, "Boston Common at Twilight," 1885–86. Suzanne and Terrence Murray Gallery, Museum of Fine Arts, Boston. (Credit: Public Domain courtesy of Wikimedia commons)

Tremont Street, 1895, looking toward Park Street Church. Boston Common on the left. Lowell's birthplace would have been roughly across the street from where the farther streetcar is parked. (Credit: Public Domain, courtesy of Boston Public Library)

Park Street, ca. 1860, looking towards Tremont Street. On the left is the Amory-Ticknor House, and the two houses on the right (Nos. 8 and 7 Park Street) were the residences of Percival's two grandfathers, Abbott Lawrence and John Amory Lowell. Both of the latter were eventually taken over to form the Union Club, to which Lowell's brother Lawrence later belonged. It was odd, he said, to be served dinner in what had once been his grandmothers' bedrooms. (Credit: Photograph by Southward and Hawes; public domain, courtesy of Library of Congress/Harvard University Graduate School of Design)

The Lowell family fortune, based on textiles, depended on the cotton produced by slaves on plantations in the Southern states. The family name had long become synonymous with the high-caste and well-bred Boston aristocracy known, in humorous literary allusion to the highest caste of Hindu society, as "Brahmins." At a time when such things mattered, Percival could boast that on his father's side, the lineage could be traced back to a cousin of William the Conqueror; on his mother's side, even farther back, to an associate of Richard Coeur de Lion. However, despite some moderately distinguished antecedents, the real founder of the family fortune was Francis Cabot Lowell, who had begun his career on Long Wharf, Boston, as a trader of silks from China and hand-spun and hand-woven textiles from India, and who in 1810—in what was to become a Lowell family tradition—attempted to recover from a bout of serious ill-health with a long (2 year) excursion overseas. England was his destination, and there he visited the cotton mills ("Dark Satanic mills" William Blake called them) of Manchester, Birmingham, and Leeds that were leading England into the Industrial Revolution. Though the exportation of either machines or plans was strictly forbidden under British law and punished by heavy penalties, Lowell simply memorized the plans, and returned home with them. With a brother-in-law he secured a charter for "The Boston Manufacturing Company," purchased the property of an unsuccessful paper and spinning mill at Waltham on the Charles River, where the current ran a little faster to provide water power, and with the help of an inventor named Paul Moody, set about the re-invention of machinery until they had developed a power loom and related machinery that would permit the combination of all the steps in the production from cotton of cloth—picking, carding, spinning, warping, and weaving—under a single roof. Within a few years the Waltham mill was producing thirty miles of cotton cloth everyday and paying dividends to the stockholders of ten percent or better.[3]

By the time of Francis's death in 1817, he was dreaming of a creating a self-contained manufacturing center that would include factories, canals to provide waterpower to drive the power looms (needless to say, any essay on Percival Lowell needs to emphasize the canals), as well as provide acceptable housing and opportunities for education and culture for workers, primarily young women ("mill girls") recruited from rural areas to textile mills where they could earn more money and live a cultured life in the city. They lived a highly regimented life, living in boardinghouses and held to strict hours and a moral code presided over by an older woman called a "matron," while typically working 80 h a week in the often insalubrious conditions of the mills. Usually they lasted only a few years before returning home to marry, or to migrate West.

Within a few years Francis's associates abandoned the Waltham operation and established their textile mills at East Chelmsford at the confluence of the Concord and Merrimack rivers. The city that resulted—renamed Lowell in Francis's honor—proceeded to spring up, as the poet John Greenleaf Whittier noted, "like the enchanted palaces of Arabian Tales, as it were in a single night." He added:

"A stranger, in view of all this wonderful change, feels himself, as it were, thrust forward into a new century; he seems treading on the outer circle of the millennium of steam engines and cotton mills. Work is here the patron saint. Everything bears his image and superscription. Here is no place for that respectable class of citizens called gentlemen, and their much vilified brethren, familiarly known as loafers. Over the gateways of this new world Manchester glares the inscription, 'Work or die.'" [5]

From the grime-filled existence of workers in the textile mills and the back-breaking toils of slaves in the cotton fields, Percival Lowell's silver spoon was fashioned. The so-called Boston Associates—who included in the first generation Lowells, Jacksons, Cabots, Appletons, and Lawrences, as well as, in the second, more Lowells, more Cabots, more Jacksons and also Higginsons, Russells, Gardners, Gorhams, Putnams, Duttons—controlled a fifth of America's cotton production by 1850.[4] By 1876—when Percival Lowell graduated from Harvard—the population of Lowell, Massachusetts, had just reached fifty thousand, and the city was the nation's leading producer of cotton textiles. A number of the Boston Associate families became tightly knit and interbred through the marriages of so many siblings and of the numerous second marriages, so that everyone was related to a dozen other families. They consolidated their fortunes by such means, and lived on a steady diet of privilege that, for the male scions at least, included a Harvard education. They were the American aristocracy and were thoroughly convinced, on the British model at the time, that, as Barbara W. Tuchman said of the latter, "prolonged retention by one family of education, comfort and social responsibility was natural nourishment of 'superior fitness.'" [6]

Lowell textile mills in 1850. The mill buildings line the Northern Canal, part of the great canal system devised to use the dammed Merrimack River to supply power to run the mill machinery. The problem of labor was solved by employing young women (usually single), known as "Lowell girls," who could be paid less than men and were housed in company-owned dormitories run by older women known as "matrons." Though the laborers typically worked 80 h a week in often harsh conditions, the Waltham-Lowell system was regarded as enlightened for its time (at least early on), and encouraged these women to educate themselves and pursue intellectual activities that included attending lectures by the likes of Ralph Waldo Emerson and John Quincy Adams. However, in the years before the Civil War, the system collapsed, as women were drawn away from the mills to other economic opportunities (including nursing), and the system became increasingly reliant on foreign and child labor largely supplied by Irish immigrants escaping from the Great Famine. By the 1850s the Lowell system was considered a failed experiment. (Credit: Public Domain courtesy of Wikipedia Commons)

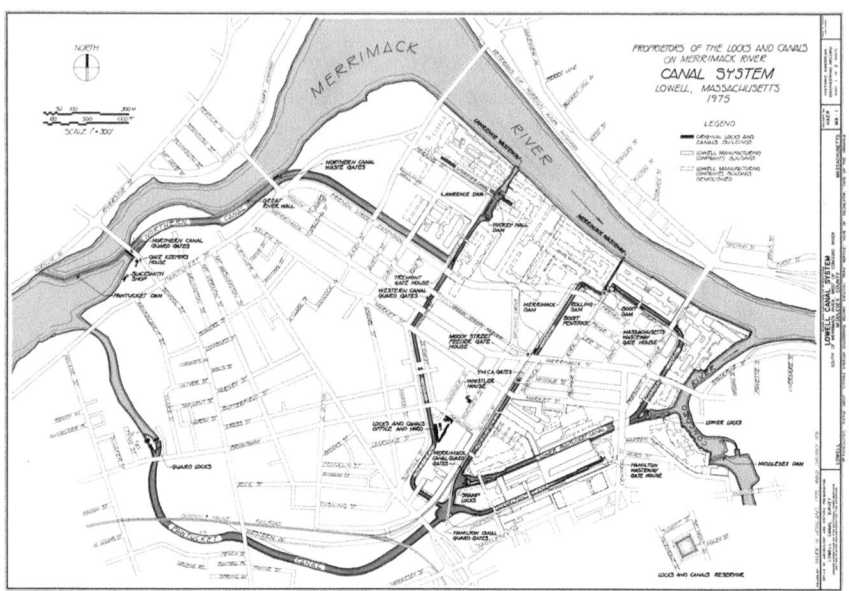

The Locks and Canals Historic District of the Lowell National Historical Park includes all of Lowell's 5.6-mile canal system, associated dams, locks, gatehouses, millyards, and corporate boarding houses. The system was begun in the 1790s with a transportation canal, the Pawtucket Canal, used to float logs from the sawmills of New Hampshire down to the Merrimack River. In the early 1820s, the Associates of Francis Cabot Lowell bought up the old canal, widened and deepened it, and used it as a direct power source for their textile mills. By the late 1840s, the canal system reached its fullest development and produced as much power as possible. The Pawtucket dam diverted the entire Merrimack (during periods of lower flow) into this canal system, whose components included the Northern Canal along the Merrimack River at upper left, and the Pawtucket Canal, along the bottom of the map. Perhaps the canals of Lowell helped to inspire—possibly at a subconscious level—Percival Lowell's theories of canals of Mars. (Credit: Public Domain, Lowell National Historical Park)

Nowhere was this concentration of wealth more successful in its results than in Percival Lowell's own family. Percival's father, Augustus Lowell, was the only child of John Amory Lowell's second marriage (both to cousins). Percival's mother, Katharine Bigelow Lawrence (known in the family as "Kitty"), was the daughter of John Amory Lowell's business partner, Abbott Lawrence, minister to the Court of St. James's from 1849 to 1852.[5] In the union of the Lowell and Lawrence families two of the greatest textile fortunes of New England were combined (with the Lawrences also having a textile city in Massachusetts named for them).

Augustus Lowell, remembered as a "slightly built, tight-lipped, impeccably dressed businessman [who] epitomized bourgeois productivity and expected it from his children." (From the Proceedings *of the* American Academy of Arts and Sciences, *vol. 37 (1901))*

Percy and his mother, 1855. (Credit: Lowell Observatory Archives)

Early in Augustus Lowell's married life, he worked under Colonel Henry Lee, his cousin, at the counting room of Bullard and Lee, and being fast on his legs, 5 min before the closing hour of the Exchange, some hundreds of yards away, he would arrive just in time to do business. The turning point in his career occurred during the panic of '57, when he left the counting firm to assist his father in the latter's State Street office, and both through the panic and the war years that followed he served as president or treasurer of ten cotton companies, and director of as many more and of several banks. It was under him that the family pocket book "reached its most distended proportions." [9]. In the years after the war, "the difficulty of labor management reached its limit, and the paternalistic attitude of the owners was worn away…. Although a philanthropist in motive and practice, Augustus was a martinet in mill management."[6] The original Lowell-Waltham system broke down; with the coming of the Civil War, girls moved back to their farms or took positions that men had left when they joined the army, and after the war no longer needed employment in the mills; increasingly the textile manufacturing operations came to depend on foreign and child labor—especially Irish immigrants who had escaped the Potato Famine of the 1840s and who had flocked to Massachusetts. The Lowell (and Lawrence) mills exploited these individuals, who now were increasingly male, from the lower classes and made them dependent, permanently, on the low-paying jobs. The mills became just as exploitative as the ones in the English Midlands that the Lowell and Lawrence families had originally hoped to distinguish themselves from. The Waltham-Lowell system was no longer profitable, and collapsed, while the new generation of mill managers, including Augustus, were seen as regarding "profits rather than people [as] their primary, even sole, concern." [10]

Augustus was a rather prim, driven, stoical individual, who preferred work to pleasure, and in the margins of time not devoted to business served in a number of philanthropic roles (as a good Brahmin was expected to do). He was for many years trustee of the Lowell Institute, which sponsored public lectures in Boston as called for in the will of Francis Cabot Lowell's son John Lowell, Jr. The latter had given up business for travel as a way of assuaging the grief of losing his wife and two children within a few months of one another and, at age 39, himself became gravely ill during a camel trip across the Egyptian desert; he died just after arriving in Bombay, leaving his fortune to support this philanthropic endeavor.[7] Augustus's trusteeship of the Institute was inspired, and probably the best thing he did.[8] He was also treasurer and Vice President of MIT, treasurer of the American Academy of Arts and Sciences, sometime member of the Harvard Corporation and an officer of the Massachusetts Historical Society.

He prided himself on his severe self-discipline and what was at the time referred to as "will power." Even as he got on in years, his hardiness was attested by his "habit to rise at four o'clock in the summer and at four-thirty in the winter. The before-breakfast hours in the summer he employed in a swim at the family's country home in Lynn, Massachusetts. He "went in swimming every summer day until he was 80; and he never wore an overcoat!" recalled his daughter (and Percival's oldest sister) Elizabeth." [12] He also shared the general interest in horticulture that ran in the family for generations, and devoted a good part of the leisure hours of his workaholic life in his garden, where he showed a softer, less formidable side. As his daughter Amy (the poet) recalled, he "covered it with beautiful and exotic flowering shrubs brought from all parts of the world, and many rare and lovely flowers."[9] It was this garden which served as the seedbed for many of her most inspired lyrics (e.g.,

"In a Garden," "The Garden by Moonlight"). On the other hand, Augustus was a rather bigoted conservative, and refused to allow a volume of Shelley "much less of Darwin" in the family's 7000-volume library.

Though Percival Lowell would one day clash mightily with his strong-willed father (and eventually concede to him respect, even grudging admiration), he was from first to last almost excessively devoted to his mother. His younger brother Lawrence thought that it was from Katharine—Kitty—that Percival acquired "sociability, ease of companionship, and charm."[10] Though as a woman she was denied the opportunity to achieve a higher formal education, her father—himself unable to attend Harvard because thrown into textile manufacturing at a young age, something he would always regret—encouraged the development of her intellect. Thus, she learned to speak seven languages, play five instruments, and sing. Percival himself seems to have shown no particular inclination toward music—nor, for that matter, a taste for and ability in art, something which was to prove a serious handicap indeed during his career as an astronomer (and suggests, perhaps, that he already had only a casual interest in the evidence of his eyes). But clearly, Kitty was the emotional center of his life. His earliest letter to be preserved, written at age ten, was to her, and for long periods he wrote her nearly every day, even when traveling, right up until her death in 1895. The dependency, however, went two ways. Though by all accounts Kitty was intelligent, kind, and matronly, according to the mores of the time her will was always absolutely subservient to Augustus's, and as an invalid from her early thirties, she must often have tried to live vicariously through the adventures of the eldest son to whom she was so attached and whom she usually addressed "My darling Percy." Of their relationship, one of Percival's sisters, also a Katharine, recalled that Percival was "peculiarly fond of his mother" and "his tender solicitude for her was unfailing…. We used to say that he could do anything with her, and she spoiled him."[11]

Halfway Up a Turning Staircase

Though Percival was their first-born, others followed rapidly, beginning with Abbott Lawrence a year and a half later.[12] His arrival prompted Augustus and Kitty's move into large quarters, first to Park Square on the other side of Boylston Street at Charles Street, next-door to Augustus's half-sister Sue Sohier. Here, Percival had his first sight of a comet. "Consciously," he later wrote, "I came into this world with a comet, Donati's Comet of 1858 being my earliest recollection and I can see yet a small boy half way up a turning staircase gazing with all his soul into the evening sky where the stranger stood."[13] The winding staircase that furnished Lowell's introduction to wonders astronomical contrasts sharply with the "wagon bed" from which E. E. Barnard witnessed a great comet a few years later, in war-ravaged Nashville.

Park Square, 1850 Bird's-eye view by John Bachmann. The tidal Charles River basin is in the foreground, the railroad depot to the right of center, the Park Street Church on Tremont Street and Park Avenue and Boston Common to the left. (Credit: Public Domain, courtesy of Boston Public Library, Norman B. Leventhal Map Center)

Comet Donati, as it appeared on October 5, 1858. The Big Dipper is to the right of t he comet, and the bright star near the head is Arcturus. (From: E. Weiß, Bilder-Atlas der Sternenwelt—eine Astronomie für Jedermann, 1888)

Percival Lowell, right, and his younger brother Abbott Lawrence Lowell, the future president of Harvard and Percival's first biographer; in a formal photo probably taken in the Park Square residence where the family lived during the 1850s. (Photo credit: Lowell Observatory Archives)

The Lowells did not remain at Park Square for long; during the early years of the Civil War, they were at 81 Mount Vernon Street (the same street where the future astronomers Edward C. and William H. Pickering lived). At this time Percival was sent to a dame school kept by Miss Fette. Percival spent summers on the "Gold Coast" of Beverly, Massachusetts, where Augustus acquired a summer house and where Brahmins who lived in the neighbourhood between Beacon Street and Commonwealth Avenue in Back Bay enjoyed one another's exclusive company.

After the death of young Roger (twin of Elizabeth), the family fled Beverly and, in the spring of 1864, "there came a sudden change," according to Percival's brother Lawrence. "[Mother] was far from well, and losing ground so fast that his father was advised to take her abroad for a complete change as her only chance, a heroic remedy which proved in time to be successful. So the family sailed in the *Africa*, a paddle-wheel steamer of 2500 tons with the sails of a full-rigged ship,—the father with an invalid wife, four children aged from nine to two, a nurse sea-sick all the time; and in addition the care of three more children of a friend in Europe, with a nurse who was well, but bereft of sense. However, they arrived safely, spent the summer in England, and, as all Americans did in those days, went to Paris for the winter." [15]

THE CUNARD UNITED STATES MAIL STEAMER "AFRICA."

The Cunard steamer Africa, *on which the Augustus Lowell family sailed for Europe in the spring of 1864. In 1855, the Harvard Observatory's Bond Chronometric expedition, which used chronometers made by William Bond & Son, the Bond family's private business, had sailed across the Atlantic on the* Africa. *At the time longitude differences between various stations were determined by observing celestial objects at each and transferring chronometers from one to another—in this case, the transfer occurred between Cambridge, Massachusetts, and Liverpool, England. The expedition was led by Philip Sidney Coolidge, who was known as a keen-eyed observer of Saturn at Harvard during this era. (Credit: Public Domain, courtesy of Rich Schmidt)*

With Lawrence and a cousin, George P. Gardner, Percival was enrolled for two winters in a French Boarding School kept by a Mr. Kornemann, spending all week at the school except Sundays where (despite the presence of a few English boys) he enjoyed full immersion in a French atmosphere and learned the language like a native speaker. By 1866, Kitty's health had improved enough for Percival's parents to propose to go to Italy; but Percival at the time was so ill at ease in travel that he remained behind at the famous boarding school run by the Silligs at Vevey, Switzerland. From there he wrote to his mother his first preserved letter, in which he showed off the French in which he was, at age ten, already fluent: "For Papa, Lawrence and Katie, 1000 *baisers chacun, et gardez 10,000 pour vous même.*"[14] Within another year, he seems to have been fluent in Latin as well.

Only in the summer of 1866 did the family return to Boston. Augustus found that the estate at Bromley Vale, Roxbury, where his father John Amory Lowell had lived and where he had spent much of his childhood, was about to be "mutilated" by the B&P Railroad at the corner of the property; its days being clearly numbered, Augustus sold it for development (except for a small house in which John Amory Lowell's two maiden sisters, Anna and Amory, remained until their deaths), and a year later acquired an estate at the corner of Heath and Warren Streets in Brookline (70 Heath Street). An old colonial house that had stood there had recently been torn down, and replaced with a brownstone edifice. There the Lowells could enjoy the pleasures of a more private and moderately rusticated existence, while often maintaining town houses in Boston for the winter, "successively abandoned as soon as the growth of the city made them unattractive."[15] Because of the seven Lowells who eventually inhabited it (Augustus and Kitty, Percival, Lawrence, Katharine, Elizabeth, and Amy), the brownstone edifice came to be known as "Sevenels," and painted, resembled the dome Percival would later build for his telescope on Mars Hill, looking "in … magisterial whiteness … like a wedding cake." [16] It was set far back on a wooded park, with stables on the left and a sunken garden on the right, and lay only 10 min by trolley from Boston Common, 30 min from Harvard Yard. The interior appointments reflected the rather somber bourgeois tastes of the time, which comported well with Augustus's rather stiff and suffocating personality: "The … rooms were heavy with massive furniture and Oriental rugs, dark with thick draperies drawn against the daylight, details which reflected Augustus Lowell's serious side. Dinners were served elegantly and punctually on fine silver and fresh napery, with flowers cut routinely from the garden, and everything was polished and waxed to the point of perfection. The chores were performed by a large staff of house servants .... Dinner-table conversation, even when the family dined alone, was intense and often brilliant."[16]

Sevenels, the Augustus Lowell House at 70 Heath Street and Warren Street, Brookline, Massachusetts, which after Augustus's death in 1900 was acquired by the youngest of the Lowell sibship, Amy Lowell, and served as her residence until her death in May 1925. The upper is a front entry view, and the lower a garden view. Note the flat roof from which Percival made his first telescopic observations. (Credit: Public Domain, courtesy of Digital Commonwealth; Massachusetts Collections online)

Here, beginning at age 12, Percival spent his summers and winters. Of these times Percival's brother Lawrence recalled, "Father drove us into town and out again each day, he going to his office and the children to school. On the road he talked on all subjects and we learned much in this way."[17] "School" was for 1 year a private day school kept by a Mr. Fette, brother of his childhood teacher, then 5 years at Noble's Classical School, a boys' day school founded in 1866 as a preparatory school for Harvard by George Washington Copp Noble. (Informally known as "Nobles," it still exists as Noble and Greenough, no longer in Boston but in Dedham, and now a co-ed day and boarding school.) The school's motto, from the *Aeneid*, is *Spes Sibi Quisque*, which emphasized individualism. Lowell did well in classics, and exhibited his deftness in turning out Latin hexameters by composing a poem of several hundred lines about the loss of a toy boat in a shallow pond made by the melt-ing snow on the lawn at Heath Street. It is his first recorded literary triumph. However, rather curiously, Noble did not regard him to be so strong in English Composition and Mathematics, subjects in which he received tutoring before he went to Harvard. Apparently the effort was worthwhile, since these were subjects in which Percival would excel. In addition to his academic pursuits, Lowell learned from the coachman employed at Sevenels to ride bareback with a halter for a bridle, and though he was never particularly interested in riding at the time, his skill would later resurface in an all-consuming passion for polo that emerged later.

These years also saw the emergence, apparently without any external stimu-lus and quite on his own, of Lowell's passion for astronomy. As no. 8 Park Street will, at least for the astronomically oriented, forever be remembered for Percival's first astronomical memory of Donati's Comet, the house at Heath Street will be remembered as the place where, on the house's flat roof, Percival had his first view of the planet that would come to dominate his life, through a 2 1/4-inch refractor received in gift from his mother. The date of this obser-vation is somewhat uncertain; he himself once said "I may say that my interest in astronomy did not date, as you suppose, from 1894 but 1870 when I used to look at Mars with as keen interest as now."[18] But Mars was nowhere near opposition (the most favorable occasion for its observation) at any time in 1870. His brother distinctly says that Percival "recalled that with it he had seen the white snow cap on the pole of Mars crowning a globe spread with blue-green patches on an orange ground."[19] This would be a difficult observa-tion in any case with such a small instrument, and though perhaps augmented by retrospective embellishment, it could only have been made near opposi-tion; this means probably 1871, when opposition occurred on March 20. Since this was a week after his sixteenth birthday, it seems that perhaps the telescope was given as a birthday gift.

Lowell's first telescope, a 2 1/4-inch telescope, on a pillar and claw stand, received as a gift from his mother. It cannot have provided a very satisfactory view of Mars when Lowell first turned it on the planet from the flat roof of Sevenels, probably at the opposition of March 1871. Here it is displayed on a table on the porch of the "Baronial Mansion." (Courtesy: Lowell Observatory Archives)

Harvard

A year after he first looked at Mars through a telescope, he entered Harvard, as well prepared as any young man of the time could be, and possessed of a notion deeply embedded in him by his father that, as Lawrence would put it, "every self-respecting man must work at something that is worth while, and do it very hard. In our case it need not be remunerative, for [Augustus] had enough to provide for that; but it must be of real significance."[20] At Harvard Percival was assigned Room 21, Holworthy Hall. He soon discovered that, despite his lack of distinction in the subject at the Nobles school, he was—as many men in the Lowell line had been—gifted in mathematics.

His mathematics teacher and the greatest influence on him at the time was Benjamin Peirce, who had already been teaching mathematics at Harvard for 40 years. In contrast to many leading American mathematicians and scientists of his generation who received some of their education in Europe, Peirce was entirely American trained. During his own undergraduate years at Harvard, he had been mentored by Nathaniel Bowditch, and regularly read the proof-sheets of Bowditch's translation of Laplace's *Traité de mécanique celeste*. Since 1831 he had been tutor of mathematics, then professor, then professor of mathematics and astronomy, and finally, in 1842, Perkins Professor of Mathematics and Astronomy, an endowed chair which he held until his death. He wrote a number of text-books,[21] which were considered "original and mathematically elegant but were rather too demanding for the American students of the time [who] found them too concise."[22] Many students found his style of lecturing difficult, with one student complaining that "he lectured without stopping for questions and filled the blackboard with a mass of scribblings." Nevertheless, the former pupil admitted that "the best students … were able to appreciate his quirks and were inspired by his enthusiasm for mathematics." [17] Peirce himself recognized this, and rather than attempt to make his lectures more accessible, decided that only the more talented and dedicated students should be allowed to continue with mathematics beyond the first year [18]. He also condoned slavery, embracing the old "Mudsill" theory that it served a useful purpose if gave an aristocratic elite consisting of Peirces and Lowells the cultured idleness needed to pursue scientific enquiry.[23]

Benjamin Peirce (1890–1880), showing him as he looked near the end of his life and at about the time Percival had him as a mathematics teacher. Public Domain courtesy of Wikipedia Commons

Not only was Percival inspired by him, so was younger brother Lawrence who also attained the highest honors in mathematics and regarded Peirce "as having the most massive intellect with which I have ever come into close contact, and as being the most profoundly inspiring teacher that I ever had." He later recounted his experience under the great man in the classroom:

> … He expected and received close and rapid attention in class, and hard, though not extensive, work outside. We read his Analytic Mechanics, Briot and Bouquet on Elliptic Functions, Tait and Hamilton on Quaternions; while his direct instruction consisted mainly, but not wholly, in solving problems by writing on the blackboard that covered the end of the room with a series of equations which we copied into our notebooks.
>
> As soon as he had finished the problem or filled the backboard he would rub everything out and begin again. He was impatient of detail, and sometimes the result would not come out right; but instead of going over his work to find the error, he would rub it out, saying that he had made a mistake in a sign somewhere, and that we should find it when we went over our notes. Described in this way it may seem strange that such a method of teaching should be inspiring; yet to us it was so to the highest degree. We were carried along by the rush of his thought, by the ease and grasp of his intellectual movement.[24]

Though astronomy classes were certainly offered at Harvard, it doesn't appear that Percival took any, but he certainly kept up his interest, and read at least one book on the subject, Richard A. Proctor's *Saturn and Its System* (1865), which he had received from his mother. Doubtless he was also aware of the great controversy that had thrust Peirce conspicuously into the public eye—his 1847 claim that the French and British astronomers U. J. J. Le Verrier and J. C. Adams, who were credited with the mathematical discovery of Neptune before the planet was seen by J. G. Galle in a telescope, had not, in fact, actually predicted the position of the planet at all [19]. Peirce contended "that the planet Neptune is not the planet to which geometrical analysis had directed the telescope; that its orbit is not contained within the limits of space which have been explored by geometers searching for the course of the disturbances of Uranus; and that its discovery by Galle must be regarded as a happy accident." [20] This particular incident is worth mentioning in showing that Peirce was, as later researchers have agreed, "characteristically carried away and overstated his case,"[25] as Percival Lowell was often to be. In addition, Lowell's later interest in using methods based on those of Le Verrier and Adams to attempt to calculate the position of "Planet X," a planet yet farther out than Neptune, was likely seeded. This late-in-life obsession was in part—perhaps in large part—an attempt to emulate his old master.

Percival had managed to become Peirce's pet, and the aging professor—with only a few years left to live—not only declared him one of the best mathematicians of all the students to have come under observation but hinted that, if Percival would devote himself thereto, he could succeed Peirce in his chair. Percival, however, declined. He refused, he said, "because I preferred not to tie myself down—not because mathematics had not always appealed to me as the thing most worthy of thought in the world." [22] One notes that already he was warmly attached to the aristocratic privilege of having complete freedom to do as he pleased.

Lowell the undergraduate was busy playing the field, academically, and rather than specialize as Peirce had done, adhered to the letter of then-president Charles W. Eliot's ideal of a harmonious and balanced education by taking an equal number of courses in sciences and humanities, including in the classics, physics, and history. His success was especially marked in English Composition. Since this, with mathematics, had been a subject in which Noble had thought him not as strong as he could be, Lowell now allowed himself to gloat that the headmaster had in fact misjudged him. His range of interests and accomplishments, impressive as they were, did not yet lead him to experience the "Lowell problem," in which his "amalgam of disparate parts"[26] would eventually, as they were bound to do in an era of increasing specialization, come into conflict, require painful choices, and produce a crisis.

All that lay ahead. At the moment success at anything seemed to come as easily as blowing on his hands. As his excellence in mathematics had led to second-year honors, so his achievement in English Composition led to the award of the Bowdoin Prize for his essay "The Rank of England as a European Power between the Death of Elizabeth and the Death of Anne." He was elected to Phi Beta Kappa and granted a part in the commencement, in which he spoke on "The Nebular Hypothesis,"

a topic seen as highly original and adventurous at the time (if a little dry, especially for a hot summer day, as would be noted by at least one member of his audience).

In addition to setting the hounds of his talents after the fox of academic glory, he engaged in other pursuits typical of the Brahmin college set. Like his father and younger brother, he was apparently fleet of foot, and competed, rather casually one gathers, in athletic events. Also, "he was constantly that year in dancing parties in Boston; and, being naturally sociable, and strongly attached to his friends, he made many in college."[27] All in all he left a brilliant record, and one taken note of on the other side of the Charles River, where his second cousin, the eminent poet and professor of languages at Harvard James Russell Lowell, declared him "the most brilliant man in Boston." [23].

An Interlude of Vocational Diffusion

By the time he left Harvard, Percival was a confident mathematician, a skillful writer of English prose, and a committed Darwinian—or perhaps more precisely, a Spencerian, an enthusiastic follower of the then highly fashionable and influential British philosopher Herbert Spencer. Basing himself only loosely on Darwin, Spencer laboriously promulgated through ten volumes what he referred to as his "System of Synthetic Philosophy," in which he attempted to establish the universality of what he called the "principle of evolution" to all the other sciences (biology, psychology, sociology, even, according to his scheme, morality). The basic idea was that all the structures of the universe develop from a "simple, undifferentiated homogeneous state" (e.g., the Laplacian nebula) to a complex, differentiated heterogeneity"—the terms were Spencer's own—while being accompanied by a process of greater integration of the differentiated parts. Despite (or rather because of) its diffuseness, or at least elastic adaptability, Spencer's vision enthralled Lowell, who later would employ it in his own theories of the development of human societies and later, to the very end of his life, to the planets of the Solar System. Though when first enunciated the Spencerian framework seemed—as William James put it—to "enlarge the imagination and set free the speculative mind," [24] well before the end of Lowell's career its assumption, "that current conditions were the predictable outcome of a single principle working gradually and inevitably from past to present," had come to seem overbroad and less compelling, not least because it was too deterministic and left no room for "the cataclysm, chance events, or choice, which a new generation of intellectuals would regard as essential components of the universe."[28]

Percival Lowell's Harvard graduation photo. In this photo, and others of the period, he assumes an unemotional haughtiness and assurance reminiscent of figures by the sixteenth-century painter Bronzino (Agnolo di Cosmo). (Credit: Harvard University Archives)

Percival's advanced views can hardly have pleased Augustus, who probably would have agreed with the apocryphal Victorian woman who is said to have said about Darwin's theory that man was descended from apes, "I hope it isn't true, but if it is, that no one finds out about it."[29] But disagreements between the father and his eldest scion about such ideas were the least of their problems. Augustus planned—or rather, simply assumed—that Percival would now pursue a Boston-based and investment-centered life and marry into a Brahmin family (ideally a cousin, as he and his grandfather had done) to further consolidate the family fortunes. However, Percival instead embarked on a series of long postponements of commitments and final decisions. In other words, to use one of Henry James's favorite phrases, he "hung fire."

His principal Harvard chums Barrett Wendell, Frederic J. Stimson, and Ralph Curtis were themselves, severally, engaged in such moratoria. They were all Anglophiles, and as Barbara W. Tuchman has said of the upper-crust social set in England at the time, "As a group they were particularly literate, self-consciously clever and endlessly self-admiring. They enjoyed each other's company in the same way that an unusually handsome man or woman enjoys preening before a mirror."[30] They were often found haunting the Boston Athenaeum and St. Botolph's—the latter an exclusive gentleman's club at 2 Newbury Street, on the west side of the Public Garden that, in contrast with the older Somerset and Union clubs with their preference for politics and public affairs, introduced the latest fashions in the arts, and hosted a long-running series of exhibits that introduced the works of avant-garde figures such as the American impressionists Dennis Miller Bunker and Dodge Macknight. At other times a life of sensation and physical exertion provided an escape from intellectual effort, and according to Lawrence, Percival came to regard the Dedham polo ground for several years from the summer of 1887, when he acquired a polo pony, "as his chief resource for recreation and diversion in this country until he built his Observatory in Arizona."[31]

The Dedham Polo Club, August 1887. Few in New England had seen or played polo at the time, but several upper-class Bostonians including William F. Weld, Frederic J. Stimson, Samuel D. Warren, Herbert Maynard and Percival Lowell (second from right), on gathering at George Nickerson's home in Dedham, decided it was the perfect place to play polo. According to his brother Lawrence, Percival, "with his great quickness and furious energy, soon forged ahead, leading the list of home handicaps in the club with a rating of ten, and becoming the first captain of the team." (Credit: Public Domain, courtesy of Dedham Polo and Country Club)

In addition, Lowell and his brilliant, affluent, restless circle of friends enjoyed the luxury of experimenting with careers in science, literature, and the arts, searching for a life that provided more personal fulfilment than simply adding to the already bloated family purse or serving public institutions. They also tried to overcome the puritanical repressions and inhibitions for which Boston was known by means of alcohol and other drugs, including opium. In the end, the chief stimulant they discovered was travel.

Europe—The Grand Tour

A year after he graduated, Lowell and yet another cousin, Harcourt Amory, made an 8-month peregrination through Europe and the Middle East. Lowell's letters to friends back home, especially to Wendell, indicate vagaries of mood and prevarications about his ultimate direction. Though one suspects that he may have been merely adopting a Byronic pose—he was "Childe Percival," affecting an attitude of ennui, misanthropy, and self-absorbed melancholy—it was not entirely a pose. One senses a real oppression—of being imprisoned in accounting rooms of his grandfather's office on State Street, or surrounded by the dark heavy furniture and draperies of a soul-smothering Brahmin domesticity. He wrote to Wendell on January 16, 1877, the year that was to see Schiaparelli begin the mapping Mars:

> I hope that Chastity Hall [i.e., Holworthy Hall] is, as in old times, the scene of many a good time and that you like my old qualities…. [As for myself,] I have … become decidedly misanthropic and, with the exception of a few friends, should not feel many pangs at migrating to another planet – or ceasing to exist – were either plan practicable.[32]

Whether Lowell paid any particular heed to the bright reddish object that stood forth in the heavens that year, brighter than it would be for a decade and a half, is nowhere recorded, and it would be another 17 years before he succeeded in fashioning a practicable plan for migrating, not in body but in mind, to that other planet.

Drifting along with ennui and a sense of purposelessness bordering at times on despair—but unwilling to give up his complete freedom to do as he pleased—Lowell was often "blue" (as, for that matter, were his similarly affluent and well-connected chums). "The blue I think can cheer the blue,"[33] he told Wendell. "We would spend the evenings together laughing at what is gay, smiling at what is sad." He advised Wendell to marry or, if he could not do that, to "live … the life of an opium-eater."[34]

Wendell opted for the first alternative. He married, and embarked on a literary career (he later taught literature and composition at Harvard, where one of his students was the future poet Wallace Stevens). Stimson also married and went into law, while Curtis—with whom Lowell would later spend much time bumming around Europe—moved in with his parents, relatives of the painter John Singer Sargent, in the Palazzo Barbaro in Venice and tried his own hand at painting; lacking Sargent's genius, he eventually gave up painting and—despite probably being homosexual—married an heiress.[35]

For Lowell, there was no such easy resolution of his difficulties. During what has been called "the age of the bachelor," Lowell delayed marriage indefinitely and embraced the complete freedom of single status. As he wrote to Wendell later that year from Sevenels, "I am fast settling down into an old bachelor. It is not the lot I should have picked out for myself but we cannot always control the decrees of Fate."[36] Though he did not say so, one suspects that a good part of his reluctance had to do with the fact that emotionally the most important person in his life remained his mother.

Though discontented with the conventional Brahmin prospects, he did not totally neglect his duties. Manfully, he took his turn working in his grandfather's State Street office managing trust funds and serving as treasurer of a cotton mill and bleachery. He may have hoped to realize the tonic benefits once asserted by James Russell Lowell who once quipped, "There is no better ballast for keeping the mind steady on its keel, and saving it from all risk of *crankiness*, than business," [26] but the tonic that had done wonders for so many Lowell predecessors failed to have the desired effect on Percival's spirits. At some level he recoiled from doing as his father's and earlier generations of Lowells had done, piling fortune upon fortune like Ossa on Pelion. At the same time, he was being pressured into a marriage scheme involving a Brahmin woman. Long unidentified, she is now known to have been Rose Lee, a daughter of George Cabot Lee, a leading investment banker and member of the State Street firm of Lee, Higginson & Co. She was also the sister of Alice Hathaway Lee, who had just married a dynamic up-and-coming New York Assemblyman named Theodore Roosevelt. We know very few details, unfortunately; but certainly from the standpoint of the Lowell family, she would have seemed a perfect match for the eldest son of Augustus Lowell. At first Percival seems to have agreed to go ahead with it. Suddenly, he panicked. The man who had been unwilling to tie himself down to marriage with the chaste muse of mathematics did not agree to enter into marriage to an ordinary flesh-and-blood woman, however advantageous it may have looked from a social or financial point of view. Perhaps he worried that the result would be one resembling that of his parents, whose relationship was traditional and, probably, rather loveless. Given the fact that at the time he only loved himself, this seems likely. Whatever the exact circumstances, Percival abruptly broke off the engagement.[37] The scandal was almost unmentionable; all trace of it seems to have been expunged from the family records apart from an odd reference in letters to his sister Bessie,[38] then in Europe with Augustus and Kitty, to whom he sent an anguished note: "I am so blue myself that I would I could speak with some of you. It grieves me to have brought so much sadness to all of you. But I have had it too…. Everyone seems to have been happy while I alone was wretched."[39]

He had committed, from the standpoint of polite Brahmin society, social suicide. The Brahmin had suddenly become an Untouchable. Indeed, from this distance in time, it is hard to appreciate just how unthinkable—and utterly selfish—Percival's abandonment of a fiancée under the circumstances must have seemed, but perhaps a hint of it appears in a letter Harvard President Charles W. Eliot wrote a decade later to Harvard College Observatory director Edward C. Pickering, at the time Percival was setting up his observatory: "Mr. Percival Lowell is undoubtedly an intensely egoistic and unreasonable person…. Fortunately he is generally regarded in Boston among his contemporaries as a man without good judgment. So strong was this feeling a few years ago that it was really impossible for him to live in Boston with any comfort."[40]

Go (Far) East, Young Man!

The Rubicon having been crossed, Percival took the next step toward the achievement of freedom. He resigned from the family business. Though it was hardly the irrevocable step it may have seemed at the time, and he would later return to the family business in various roles, he had for the time being cut himself loose, and bought time. Once again he was hanging fire. His younger brother Lawrence, who was equally fast on his legs and shared his older brother's highest honors in mathematics, took advantage of the situation by leaping headlong into the breach, and deftly taking over the roles Percival had abdicated.

Though, as Tolstoy famously said, "Happy families are all alike; every unhappy family is unhappy in its own way,"[41] many of the Lowells, ancestral, contemporary, and future, belonged to unhappy families burdened with the weight of family blood, and all the privilege and expectation that implied. In addition, among many of the Lowells, traits recurred in similar forms, including abilities in literature, math, and business. There was also very likely a genetically determined tendency to manic-depressive illness [27]. The poets—and, in a minor way, this included Percival—seem to have been particularly touched.

The first known example of what were to be several mental breakdowns in Percival's life, the "crisis" of 1882 provided at least temporary relief—secondary gain—from the impossible demands being placed on him, and access to the at least marginally socially acceptable role of invalidism that his mother had occupied for many years. Perhaps his inability to press on was explained in terms of the fashionable diagnosis of that era, neurasthenia, by which was meant a rather vague constellation of symptoms including fatigue, anxiety, headache, heart palpitations and depressed mood. It was supposed to be almost exclusively a condition experienced by members of the upper class, and especially among American businessmen, so much so that it was sometimes referred to as "Americanitis" (though there were many sufferers in Victorian Britain too).[42] The underlying cause was posited to have been brought on by overworking the brain leading to exhaustion of the central nervous system's reserves. Percival certainly checked the right boxes. Though his mother doubtless provided what support she could to her "darling Percy," Augustus—not the most empathic or psychologically aware individual under any circumstances—was livid. He reacted to his eldest son's rebellion by asserting an emasculatingly close surveillance over the prodigal's financial affairs until well past his 30th birthday. But the punishment was less draconian than it might have been; family was still family, and far from being disinherited, Percival's inheritance was delayed, not cut off. He could still expect, as all the Lowell children did, to inherit a sum of $100,000, worth the equivalent of over $3,000,000 today. (Put in some perspective, at the same time the fledgling astronomer E. E. Barnard would pay $380 for a 5-inch refractor, an amount that came to two-thirds of his annual salary as an assistant in the photograph gallery where he worked in Nashville.) Living on his investments alone, Percival could expect to live a very comfortable existence indeed, and by 1900, the year of his father's death, had managed to grow his fortune to $500,000 (some $18,000,000 in today's value).

The usual cure for neurasthenia was the "rest cure," which required resources and so was the exclusive purview of the wealthy. (The lower class was supposed to be too insensitive to experience such things, and if they did suffer from nervous system collapse, they would be relegated to publicly funded asylums rather than to exclusive and expensive sanitoriums or spas.) The nostrums recommended for the latter (not former) often included change of scene requiring, somewhat inconsistently, strenuous travel and exercise.

Usually the travel and exercise took wealthy Americans to Europe. Taking a page from Shakespeare's haughty Roman war-lord Coriolanus who said,

> "… I turn my back;
>   There is a world elsewhere"
>   *Coriolanus* III.iii

Percival, disillusioned with puritanical Boston, resumed the well-worn theme of travel and set out for a "world elsewhere." However, instead of Europe, something different, and for the era far more original, suggested itself.

The Far East

A free lecture series given during the winter (January and February) of 1882, the year of crisis, at the Lowell Institute by zoologist Edward Sylvester Morse, suggested the contours of treatment that would lead to Lowell's recovery. The Institute was a kind of college-without-walls, endowed by one of Percival's late cousins. At the time of Morse's lectures the sole trustee was Augustus Lowell. The talks were held in a stout neo-Renaissance building on Boylston Street, designed to match the architectural style of the adjacent Museum of Natural History. It was the first structure built for the new Massachusetts Institute of Technology, which eventually—in the early 1900s—outgrew its original quarters and expanded into ten buildings in the Copley Square area, and still later (in 1916) moved across the Charles River to its present campus.

According to writer David Baron, "Morse was an engaging lecturer, ambidextrous and artistic, and able to draw sketches at the blackboard with both hands simultaneously." (In fact, he first came to the attention of Louis Agassiz at the Museum of Comparative Zoology at Harvard because of his skill as a draughtsman).[43] He had been invited to speak about the world of feudal Japan, whose culture and traditional ways of life were fast vanishing with the modernization associated with the Mejii Restoration. "Going to Japan is like visiting a new world—another planet," he said. His own interest in that part of the world dated to 1877, when he went in search of brachiopods, but soon his interest in the country's history and culture exceeded even that in trochozoan animals. His Lowell Institute lectures on Japan were a palpable hit. ("No other of the several winter courses has been so thronged," reported the Boston *Evening Transcript*.[44] The audience included not only Percival but at least one other emotionally vulnerable young Brahmin, William Sturgis Bigelow. The latter was so stirred by the romance of the East that he did not hesitate to waste his Harvard medical degree and 5 years of medical studies in France with Louis Pasteur to set sail with Morse and the historian of Japanese art Ernest Fenollosa for Tokyo. The following May of 1883, Lowell followed, though he arrived too late to meet up with Morse; the latter had returned to the United States in February and would never again visit the "Land of the Rising Sun." In Tokyo, Lowell stayed in the residence building of the U.S. legation (in the Ropponki area). Though he remained for only 3 months, he almost immediately hit upon the great discovery (or glib generalization) that he would later attempt to develop fully and systematically. This discovery was at first deduced from his necessarily superficial exposure to the structure of their language. "Again, perhaps," he wrote to his mother on June 8, "a key to the Japanese is impersonalism":

> Forced upon one's notice first in their speech, it may be but the expression of character. In the Japanese language there is no distinction of persons, no sex, no plural even. I speak of course of their inflected speech. They have pronouns, but these are used solely to prevent ambiguity. The same is true of their genders and plurals. To suppose them, however, destitute of feeling, as some have done, I am convinced would be an error. The impersonalism I speak of is a thing of the mind rather than the heart. I suggest rather than posit.[45]

After spending a few weeks in Tokyo, sufficient to inspire a lifelong appreciation of Japanese art, architecture, and gardens, he embarked, with a Western companion, across the mountains to the other side of the country. The rural diet this son of privilege experienced was meager; during a decade of travel in the Far East, he would never cease to complain of a diet rich in rice, vegetables, and—but only among the better-off Japanese—eggs and fish, which scandalously included "no milk, no butter, no cheese, no bread, and almost no meat."[46] Nevertheless, he found aesthetic nourishment in the dramatic scenery and romanticized the people who, despite their rudimentary ways of life, seemed to have been almost unaffected by contact with foreigners such as himself.

On arriving at the famous (but then badly deteriorated) Himeji Castle near Kobe, he struck the high romantic note—largely based in nostalgia, and not without a note of sadness that the romance was soon to pass away, making it, meanwhile, all the more enticing:

> We mounted through some seven barnlike rooms, up Japanese ladders to the top story. Sitting by the window and looking at the old feudal remains below, the moat with its stagnant slime and the red dragon flies skimming its surface,

the old walls, the overgrown ramparts where now the keeper tries to grow a crop
of beans, all tended to carry my thoughts back to the middle ages, or was it only
to my own boyhood when the name *middle ages* almost stood for fairy land?
And yet all this had been a fact, even while I had been dreaming of it. My
dreams of Western feudalism had been co-existent with Eastern feudalism. So it
was only eleven years ago that the last Daimio of the place left the castle of his
ancestors forever.[47]

No sooner had he returned to Tokyo (on August 13, 1883) than he was invited
to accompany a Special Mission from Korea to the United States as its Foreign
Secretary and Counsellor. For the "Hermit Kingdom," which had zealously
pursued a policy of avoiding contact with the outside world and long refused
to allow its citizens to travel outside the country except on diplomatic mis-
sions to China or Japan, this was something of a sea-change. Lowell did not
immediately agree. William Sturgis Bigelow informed Augustus Lowell that
"after two days of unconditional refusal and one of doubt,"[48] he finally gave
his consent. Clearly, Percival was anxious as to what his father might say.
Seeing this, Bigelow continued, "He distrusts himself too much, he has great
ability, he has learned Japanese faster than ever saw any man learn a language—and he only needs to be assured that he is doing the right thing to
make a success of anything he undertakes, whether science or diplomacy."[49]

In the end, Percival succeeded brilliantly. Accompanying the Koreans from
Tokyo to San Francisco, he and his delegation crossed to New York City,
where that September they turned up at the Fifth Avenue Hotel, led by Prince
Min Yong Ik, nephew to the Queen of Corea (as it was then spelled), and met
by the then President of the United States, Chester A. Arthur. Their ground-
breaking diplomatic mission accomplished, the entourage (including Lowell)
toured the United States including the White House, West Point and the
textile mills of Lowell and Lawrence, then returned the way they had come.
In gratitude for his service to them, Percival was invited to go to Korea with
them, as a guest of King Gojong and the Crown Prince. He agreed, and after
a stopover in Nagasaki, he established himself in Seoul within a group of
buildings making a part of the Foreign Office, of which he was, of course,
now formally a member. He described the setting:

From the street you enter a courtyard, then a garden, and so on, wall after wall,
until you have left the outside world far behind and are in a labyrinth of your
own. Before you, lies a garden; behind another surrounded by porticoes.
Courtyards, gardens, porticoes, rooms, corridors in endless succession until you
lose yourself in the delightful maze.[50]

Lowell, second from the right, in Tokyo in 1883. He is standing next to Okakura Kakuzo, later a noted Japanese scholar and art critic. He had learned English at a missionary school and at 15 entered the newly renamed Tokyo Imperial University, where he studied under the Harvard-trained art historian Ernest Fenollosa. In his later career he decried the Western caricaturing of the Japanese, and promoted a critical appreciation of traditional forms, customs and beliefs. The others shown are, left to right, Fenollosa, (probably) Basil Hall Chamberlain, and William Sturgis Bigelow. (Credit: Lowell Observatory Archives)

He received many gifts, but the most valuable returns he brought back with him were of course his impressions, which involved a kind of "storing up for yourself riches above the reach of fickle fate,—what the moths and rust … cannot touch,"[51] as he had told his sister Bessie on an earlier occasion when she was traveling in Europe—and his photographs. Photography was, he had found, "a very catching epidemic," as the hand-held camera and gelatin bromide dry plates (then bringing about a revolution in astronomy) had replaced the cumbersome old boxes on tripods—allowing photographers to develop their pictures at their leisure rather than having to do so in a darkroom. Lowell still found developing photographs laborious and continued to use a tripod, but overcame the difficulties with aplomb in Korea, where most of his subjects had never seen a camera, and produced an album of fifty-three photographs for the King of that country, of which a selection of which would be published in his first book. David Strauss has written:

The artistic quality of Lowell's photographs is evident on first inspection. The images are sharp, an effect that is enhanced by the contrast between light and dark areas. Lowell was a master at framing his pictures so that the focal images were nicely framed by interesting backdrops of mountains or city walls. He strove to capture a broad range of subjects, including royalty and common folk, landscapes and cities, interiors of homes and courtyards, street scenes and shops. Even though most of his subjects had never before seen a camera, they appear remarkably unposed in the photographs....[52]

Inevitably, Lowell's photographs have something of a voyeuristic quality—he stands apart and looks, without (usually) entering into the scene. A number of them show women, including one, probably a geisha, captured in a state of dishabille. East Asian women were for him somewhat academic subjects, admired for their "naturalness" and delicacy, but his interest only went so far—"I don't fancy them sufficiently to take to myself a fair Jap partner," he told Stimson.[53] Nonetheless, he intimated to his Boston friends that he was enjoying sexual conquests at will. Thus, in a long excerpt from a letter to Stimson, quoted by Greenslet, he laid out his fascination for a Japanese woman he encountered in Nyeno park (Ueno Park in central Tokyo), calling her "the most beautiful Japanese woman I had ever seen taking care of course to particularize the Japanese part of it" (the latter clarification perhaps necessary to make make the comment passable to an American female companion with whom he was in company at the time). He later encountered his "Sweet Smile" again, though as he was then "with bold bad men." "... I only murmured to myself," he continued,:Go to but I will know that fairy. Always be it remarked in a distant, quiet gentlemanly way."[54] He eventually claimed "Miss Sweet Smile" as a conquest: of his camera. She was, he found, "a virgin—to photography. It is confidently reported that the other kind existeth not in the island."[55] Despite a good deal of *double entendre* using photographic terms to suggest otherwise—"we all know how difficult it is to prove a negative"; "I left her washing in her tub"—on this occasion the relationship seems to have remained platonic. On the other hand, another, with a Korean singing girl called "The Fragrant Iris," did not, apparently, end with a photograph; at least, as David Baron discovered, word at the U.S. Legation in Seoul held that she bore his child after he left.[56]

Back in Boston, he was now determined to follow friends like Stimson and Wendell on a literary career, which the set-pieces he had been writing for them was nerving him up to and would serve to rescue him at least temporarily from the moratorium that had followed on the collapse of his Brahmin roles. He had begun to outline a first and second book, in which he hoped to analyze in terms of the "impersonalism" concept the characteristic differences between the civilizations of Eastern Asia and Western Europe. In a letter penned from Bombay to his Harvard chum and fellow writer Frederic J. Stimson, he expressed his Narcissus-like literary ideal, one that would have greatly amused his later critics of his writings on Mars had they only known of it:

> Somebody wrote me the other day apropos of what I may or may not write, that facts not reflections are the thing. Facts not reflections indeed. Why that is what most pleases mankind from the philosopher to the fair; one's own reflections on or from things. Are we to forego the splendor of the French salon which returns us beauty from a score of different points of view from its mirrors more brilliant than their golden settings. The fact gives us but a flat image. It is our reflections upon it that make it a solid truth. For every truth is many-sided. It has many aspects. We know now what was long unknown, that true seeing is done with the mind from the comparatively material supplied by the eye.[57]

Further reflection, and the attempt to actually wrestle with the demands of composition of a longish work, led to his admission of self-doubts. In a letter to his mother from Paris on October 7, 1884, he wrote, "As for me, I wish I could believe a little more in myself. It is at all times the one thing needful. As it is I often get discouraged. You will—said Bigelow the other day to me in Japan. There will be times when you will feel like tearing the whole thing up and lighting your pipe with the wreck. Don't you do it. Put it away and take it out again at a less destructive moment."[58]

Lowell's First Book: Chosön

That autumn he returned to Boston, where he was to make his headquarters for the next 4 years. With the warm support and constructive criticisms of his family, and especially of course his mother, writing was at least somewhat easier, and Lowell would later, at least partially, get over his tendency to labor over it obsessively; in any event, the self doubts vanished, rarely to return again. He returned to the usual rounds of the social scene—among other things, venturing up to New York City in April 1885 to stand up for Edward Wharton (of 127 Beacon Street) at his wedding to Edith Newbold Jones of Newport, Rhode Island. One of the things that appealed to Edith—who later had some choice things to say about Boston—was that Teddy had no desire to go on living in his hometown, about which she said, "It was not until I went to Boston on my marriage that I found myself in a community of wealthy and sedentary people seemingly too lacking in intellectual curiosity to have any desire to see the world." At the time of her marriage, Teddy was "a jolly chap, a big, handsome, fit, moustachioed sportsman, 'full of fun,' who liked fast travel (cars would be a passion), dogs and horses, fishing, shooting, wine." Besides which he had been "a not-very-successful Harvard student" and "utterly unliterary," was jobless and not especially well-off—dependent on his mother for an allowance, and not coming into his inheritance until he was sixty.[59] Nevertheless, they tried to make a go of it, but the marriage was ultimately unsuccessful, and would end in divorce in 1913. Percival's friendship with the Whartons was a testimony to the rather eclectic circle of his acquaintance—Teddy undoubtedly came from the passionate-for-polo-playing set. Though he had by then a portly pocketbook of his own (having succeeded regaining access to the inheritance his father kept from him after the Rose Lee debacle), he was aware of being a financial inferior to many members of his set, and when, in 1887, he visited Teddy and Edith Wharton in their Newport house, Pencraig Cottage, he remarked, "I dined with seventy millions [of dollars] one evening, yet were less in number than the Muses.... I sat between Cornelius Vanderbilt and Mrs. Somebody Burden, somebody's burden indeed." As for Newport generally, he added, "of all the inanities of which man is capable I believe that town can show the most utter [sic.] Plutus whom one might well mistake for Pluto reigns supreme." Despite any sense of financial inferiority he might have felt to such people as an ordinary garden-variety millionaire, he could always assert his charm and learning as superior to theirs.[60]

By November 1885 he had finished the preface of his first book, *Chosön— the Land of the Morning Calm—A Sketch of Korea*. It is today the rarest of Lowell's Far Eastern books but still of interest as among the very first books about a country that had been referred to as "the hermit nation" (by William Eliot Griffis) and which had as recently as 1876 signed the first treaty (with Japan) by which it had deigned to acknowledge the existence of the outer world. For a long time the book remained the most thorough and detailed account of Korea based on a direct encounter by a Westerner. It established Lowell's mature style—ornate, polished, sometimes highly emotional, syntactically complicated. Even then such a style was beginning to seem a little old-fashioned, and it risked (and in Lowell's case, often succumbed) to becoming "precious." Though any style of course depends on vocabulary, on the minute particles of color and singularity of single words and phrases, Lowell strove to build up to large, cumulative, massed effects. This is particularly evident in his creation of highly elaborate set-pieces to which his flowery style was suited, such as the following passage from *Chosön*, which describes the succession of tree-flowers that blossom during the long Korean spring, beginning with the plum-tree and culminating with the cherry:

… The year begins for man when it begins for Nature; and the earth awakens from her winter's slumber with a blush, for it is in tree-flowers that she shows her return to feeling.

The plum-tree is the first to bloom,—not the edible plum, but that species which is known in Japan as *ume*. By the end of January it begins to blossom,—a pretty pinkish-white flower. It is quite beautiful in itself; and then from being the first, it is specially prized. It is not easy to convey to the Western mind an idea of the mingled love and admiration the far-Oriental lavishes upon it…

Poetry and painting vie with each other in their attempts fittingly to praise the flower. Sonnets innumerable are written in its honor, and have been from dim antiquity. … Its name is one of the commonest of the flower-names of girls. The glory of the tree vanishes with its flower, for it bears no fruit.…

Early in April the cherry-tree comes into bloom; and of all the superb succession of flowering trees and shrubs it is the finest. It is all flower,—one mass of blossoms, —and flower is all that it is, for its fruit is not worthy the name…

In Korea the sight is fine, but in Japan it is even finer…. [E]ach kind of tree, as its turn brings it round, is made the occasion of a festival… The blossoming of the cherry-tree is one of the great events of the year. To see it is a sensation… You feel as if the earth had decked herself for her briadal, and you had somehow been bidden to the wedding… [32].

In general, the book was well received, with Lafcadio Hearn saying of it, "How luminous and psychically electric is Lowell's book." [33] Not least of its felicities were its maps and 25 full-page photogravures of Lowell's photographs.

Lowell with camera and attendants, in Korea, 1883. (Courtesy: Lowell Observatory Archives)

Lowell with his household servants, in Korea, 1883. (Courtesy: Lowell Observatory Archives)

One of the cameras belonging to Lowell was the Premo, manufactured in Rochester, New York in the late 1890s. It exhibits features including the folding bellows and 4 x 5-inch plates similar to that of the one Lowell used in Japan and Korea. (Credit: Lowell Observatory Archives)

After seeing *Chosön* through the press, Lowell followed up with an article, "A Korean Coup d'Etat," for the November 1886 *Atlantic Monthly*, while in addition to his writing serving part of his time as acting treasurer of the Lowell Bleachery and haunting the Dedham Polo Club. Among more intellectual pursuits, he belonged, at his mother's behest, to a drama group called the Place aux Dames, who held a soiree on January 27, 1887, at the Walter Cabot home in Brookline. In addition to Lowell, Harvard law student John Jay Chapman, and Minna Timmons, Chapman's romantic interest (and later wife), were in attendance. Chapman convinced himself that Lowell, who had apparently encouraged a reputation among his friends as having made sexual conquests at will in Japan, was using the Place aux Dames as a pretense to seduce Minna, and in a jealous rage, savagely beat Lowell with his cane. Thereafter, full of remorse and realizing Lowell's innocence, Chapman retired to his lodgings and deliberately burned his hand—so badly that it had to be amputated. Strauss says, "This incident developed from the fevered imagination of John Jay Chapman rather than from any evidence of Lowell's capacity for sexual conquests, it does suggest that his reputation as a lady's man was well established in Boston."[61] Interestingly, the incident does not seem to have dampened Mrs. Lowell's enthusiasm for the group, and she later recruited Percival's friends Wendell and Stimson to the effort.

Impersonalism: The Soul of the Far East

Less sordidly, Lowell also pressed on with his next book, *The Soul of the Far East*, which he finally got off his hands at the end of 1888. Though smaller in size and type, and lacking the illustrations of the comparatively sumptuous *Chosön*, was to be the most celebrated of Lowell's writings on the Orient. The book had a decisive effect on Hearn. "Gooley!" he wrote to his friend George Gould, "I have found a marvellous book—a book of books!—a colossal, splendid, godlike book. You must read every line of it. For heaven's sake don't skip a word of it."[62] Inspired by Lowell, Hearn himself would set sail for Japan in 1891, marry a Japanese woman, and spend the rest of his life there. Though perhaps no one was more impacted by Lowell's writings than Hearn, he was hardly alone, and there can be no doubt those writings—especially *The Soul of the Far East*—had, as historian David Strauss has documented, a "decisive impact" on the late-nineteenth century American mind [34]. That impact has, for various reasons, softened over time, and seems rather perplexing to the late twentieth-century and early twenty-first century reader. Though Strauss for one has attempted to argue for its continued relevance, it probably remains the case that most scholars tend to give Lowell's work short shrift as arrogant, shallow, and dated. It is of interest to the biographer of Lowell; less so to the student of Japan.

What seem the book's greatest faults today but its greatest virtues when it first appeared are its "philosophizing tendencies," manifested in Lowell's unwillingness to just present the facts but to allow his penchant for his own "reflections on or from things"—call it an "egotistical sublime." To a large degree in everything he wrote, but especially here, he loved to clothe the facts in the irradiations of his own grand personality, and for many readers, this tendency was what gave his writing its greatest charm. He never struggled, as philosophers like Sir William Rowan Hamilton and others going back to Plato's Allegory of the Cave have struggled, after an ideal of objectivity, something which involves the difficult, though perhaps not entirely futile, effort of distinguishing "that which belongs to or proceeds from the object known, and not from the subject knowing." [35] The facts about Japan without Lowell's attributions hardly held any interest for him. Perhaps not surprisingly, then, he found the key to the Japanese character within only two weeks of setting foot in Japan (it was little more than a tachistoscope flash's worth of seeing). The key was the "impersonalism he had described in his letter to his mother. This rather vague but elastic concept grew on him as his acquaintance with Far Eastern peoples increased, until what may have started out a tentative suggestion needing further research had become a dogmatic assertion. His brother Lawrence would later define it as "the comparative absence, both in aspiration and in conduct, of diversified individual self-expression"[63] he found in Orientals relative to Westerners of strong individuality (such as one Percival Lowell). Around this concept Lowell, as typical of his "top-down" approach—his tendency to put the theoretical cart before the observational horse—proceeded not only to assert the impersonalism as a feature but to deduce it as a necessary consequence of the same ready-made Spencerian (i.e., quasi-Darwinian evolutionary) cloak which he had embraced at Harvard. Evolution, according to Spencer's famous definition, was "integration of matter and concomitant dissipation of motion; during which the matter passes from an indefinite incoherent homogeneity to a definite coherent heterogeneity." Understood in those terms, the people of the East seemed to Lowell to lack individual personality; they were feminine (i.e., lacking in ego), unconscious; their societies, though quaint and in many ways artistic, were homogeneous, backward. Westerners were individualistic, possessed of masculine virtues, assertiveness, disciplined intellect, conscious awareness; their societies were heterogeneous, advanced. Despite their artistic attainments, the Japanese were, in Lowell's estimation, without imagination. "Imagination," he wrote, "the Japs lack it. Unless they acquire it, they will vanish off the face of the earth and leave our planet the eventual possession of the dwellers where the day declines." A passage which leads his biographer Ferris Greenslet to remark: "Both perception and prediction are at the level of genius."[64]

It is not hard to see in these formulations Lowell's attempt to solve the problems of his own divided nature. As cultural historian T. J. Jackson Lears sees it, in this, his most popular book, Lowell

> inflated his prejudices into principles… Instead of idealizing the "childish" elements in Oriental culture, Lowell ridiculed them—for important psychic reasons. Elevating consciousness over unconsciousness, separate selfhood over oceanic unity, Lowell reaffirmed his allegiance to conventional definitions of male adulthood…. Femininity, childhood, and the unconscious merged in Lowell's vision of. Premodern character; all represented tendencies he needed to repudiate…. Identifying dependence and loss of conscious will with the "infe-

rior" Japanese, Lowell revitalized his commitment to autonomous Western manhood. A racist condescension toward Oriental character helped create a negative identity against which Lowell could define a stronger sense of self [36].

This, of course, marked his full recovery from the mental collapse that had occurred after the breakup of his relationship with Rose Lee and his rejection of the roles demanded of the first-born male heir of a Brahmin family. The West—and the masculine roles advocated for him by his father—had now routed completely the Eastern and feminine side of his nature shared with his mother.

Lowell saw the East as capable of progress, but it was to be achieved, if at all (just then in the early stages of the Meiji restoration Japan was beginning to modernize) it would only on the basis of its people's genius for imitation, which was possessed even by (especially by) children and savages. But if it did not adapt—if it did not assimilate the scientific, technological, philosophical, political, legal and aesthetic ideals of the West—the East would cease to be. As he concluded in *The Soul of the Far East*:

> That impersonality is not man's earthly goal they unwittingly bear witness; for they are not of those who will survive. Artistic, attractive people that they are, their civilization is like their own tree flowers, beautiful blossoms destined never to bear fruit; for whatever we may conceive the far future of another life to be, the immediate effect of impersonality cannot but be annihilating. If these people continue in their old course, their earthly career is closed. Just as surely as morning passes into the afternoon, so surely are these races of the Far East, if unchanged, destined to disappear before the advancing nations of the West. Vanish they will off the face of the earth and leave our planet the eventual possession of the dwellers where the day declines [37].

It is notable, by the way, that Hearn, after living for a short while in Japan and coming into contact directly with the *facts* of the place, became disillusioned with Lowell's ethnocentric and condescending point of view, and thought that in *The Soul of the Far East* Lowell had missed "the most essential and astonishing quality of the race: its genius for eclecticism." [38] Later he would prefer Japanese life and thought to those of the West:

> [E]verything considered, there is a charm about Japanese life and thought, about their way of taking life and enjoying it, so deliciously natural, that only to be in its atmosphere a while is like a revelation of something we Westerners never suspected… To escape out of Western civilization into Japanese life is like escaping from a pressure of ten atmospheres into a perfectly normal medium. I

must also confess that the very absence of the Individuality essentially character-istic of the Occident is one of the charms of Japanese social life for me: here the individual does not strive to expand his own individuality at the expense of every one else. According to a French thinker, that is the great law of modern life abroad. Here each can live as quietly in the circle of himself as upon a lotus-blossom in the Gohuraku: the orbs of existence do not clash and squeeze each other out of shape. Now would not this be also the condition of life in a per-fected humanity?[65]

"The Fancy Took Me to Go to Noto"

Perhaps rather more appealing to the modern reader is the book Lowell wrote on his next visit to Japan, *Noto: An unexplored corner of Japan* (1891), an adventure story whose impressions of places Lowell encountered seem fresh and well-observed even today.[66]

Lowell had set out from San Francisco Bay in December 1888, as soon as he had seen off *The Soul of the Far East*, and arrived in Yokohama Bay on January 8, 1889. Rather than stay in the residence building of the US legation in Tokyo as he had on previous occasions, he hired a house belonging to Masujima Rokuichiro, a British-Japanese international lawyer, diplomat, and legal advisor of the Japanese Ambassador to London, and a figure strongly identified with the modernizing program of the Meiji government.[67] He had been introduced to Rokuichiro by Basil Hall Chamberlain, one of the leading Japanologists of the time whose own sojourns in the Far East had begun in the pursuit of recovery from a nervous breakdown occasioned by the stress of working in a British bank to which he found he was unsuited. (Lowell and Chamberlain must have found a great deal to commiserate about.) Chamberlain helped organize a May 1889 trip to the Noto Peninsula in Ishikawa Prefecture, which had begun, Lowell tells us, while he was idly scan-ning a map:

> The fancy took me to go to Noto. Scanning, one evening, in Tokyo, the map of Japan, in a vague, itinerary way, with the look one first gives to the crowd of faces in a ballroom, my eye was caught by the pose of a province that stood out in graphic mystery from the western coast. It made a striking figure there, with its deep-bosomed bays and its bold headlands.[68]

The studied nonchalance with which he described its expedition contrasts with the furious pace Lowell set once he had made up his mind to realize his fancy. Indeed, there were similarities in is approach to that he later demonstrated in his headlong expedition to Flagstaff to study Mars in 1894. His guide, known only as "*Yeijiro*," played the role that W.H. Pickering and A.E. Douglass would in his astronomical adventure; he used a guide-book (the second edition of Murray's) which served as Camille Flammarion's book would do for Mars; and in place of telescopes, makeshift domes, and workmen to hoist apparatus up hill and assemble it, he engaged a large contingent of Japanese to serve as porters and servants, acquired all necessary equipment and means of transport (rail, jinrikisha, boat) and last but not least—having learned from his earlier experience in traveling through the Japanese countryside that Westerners "are not inwardly contrived to thrive solely on rice and pickles"—vast quantities of Western food such as canned milk.[69]

Basil Chamberlain's jinrikisha, with servants. Though Lowell did travel at times by jinrikisha during his Noto expedition, he more commonly traveled by rail. Lowell Observatory Archives

He set a hectic pace to Noto in 1889, as he would to Mars in 1894. Though gone from Tokyo for only two weeks (he had to return precisely on schedule because he had already arranged his sailing from San Francisco from Yokohama Bay), he kept extensive notes that in due course he wrote up as a series of articles and eventually a book—as he would later do with Mars.

The first leg was by rail from Nagano to Naoetsu, where he hired a jinriki-sha (literally, "man-powered car") to cross the narrow seashore and flat delta on the way to Noto. At Arayama Pass, he stopped at one of two tea houses and looked out for the first time on the actual place that had looked so alluring on the map. The map had promised the prospect of "deep-bosomed bays"; the reality was woefully flat. Indeed, his description of the scene from this point bears comparison with his later telescopic views of Mars:

> Panoramic views are painfully plain. They must needs be mappy at best, for your own elevation flattens all below it to one topographic level. Field or woodland, town or lake, show by their colors only as if they stood in print; and you might as well lay any good atlas on the floor and survey it from the lofty height of a footstool…. No pains, evidently, had been spared by the inhabitants to make their map realistic. There the geometric lines stood in ludicrous insistence; any child could have drawn the thing as mechanically.[70]

After this disillusionment, Lowell set out for what had actually been his main goal, to cross the Tateyama range by means of the Harinoki Toge (pass). Becoming as avid an alpinist as he had been a polo player and seeking another trophy of virility, he hoped to emulate the legendary feat of a sixteenth-century daimyo, Sasa Narimasa.

Unfortunately, in May the pass is still snowbound, and though Lowell had been warned at Toyama not to make the attempt at this time of year, "with the fatal faith of a man in his guidebook," he wrote, "we ignored the native forebodings." He made it only part way. His descriptions of this harrowing crossing mark the dramatic climax of his book:

> We now began to enter the snow in good earnest, incipient glacier snow, treacherously honeycombed. It made, however, more agreeable walking than the boulders. The path had again become precipitous, and kept on mounting, till of a sudden it landed us upon an amphitheatral arena, dominated by high, jagged peaks. One unbroken stretch of snow covered the plateau, and at the center of the wintry winding-sheet a cluster of weather beaten huts appealed pitiably to the eye. They were the buildings of the Riuzanijita hot-springs; in summer a sort of secular monastery for pilgrims to the Dragon peak. They were tenanted now, we had been told, by a couple of watchmen. We struck out with freer strides,

while the moon, which had by this time risen high enough to overtop the wall of peaks, watched us with an ashen face, as in single file we moved across the waste of level white.[71]

"It was very disheartening to turn back," he admitted at last, "but it had to be done." And yet he had, in his own mind, achieved at least a partial success. "Probably none I know will ever tread where I was treading then, nor I ever be again in that strange wild cleft, so far out of this world," he wrote; "and yet, if years hence I should chance to wander there alone once more, I know the ghost of that romance will rise to meet me as I pass."[72] In the event, the first European to achieve this crossing would be the British alpinist Walter Westin, who made the attempt in August.

Though Lowell presented his adventures as exploration, he was actually, in fact, a tourist. He claimed a kind of lofty solitude, but his trail—blazed by means of rail, jinrikisha, boat, basha, and even by foot—was always in company of *Yeijiro* and his porters. Also, as always, his impressions were based on fleeting vistas—seeing in glimpses, like a scene illuminated by a lightning flash; the tachistoscope effect in action once again. He never stayed anywhere long enough to make a deep study; he breezed past each scene, trusted to his guidebook, scanned the highlights, jotted the impressions, captured his photographs. Landscapes were vividly described but—as in traditional Japanese landscape paintings—people are hardly present. Lowell seems to have had virtually no curiosity about other human beings, at least other human beings who were not members of his family or of his well-heeled social set. Japanese people are cardboard cutouts. How could they do otherwise, given their "impersonalism" and "homogeneity"? It would not have occurred to him to pay any more attention to the typical Japanese as individuals than it would have to the "bobbin' girls" toiling in the textile mills of Lowell and Lawrence or to any of the large staff of domestic servants who carried out their routines at Sevenels in going about their smooth, unvarying, invisible—and faceless—routines.

In one revealing passage in *Noto*, he describes the Japanese passengers on the train from Nagano to Naoetsu, who wore Western clothes in awkward ways, as "specimens" into which he might stick pins as one does in butterflies. In general, his portraits of the Japanese are utterly lacking in interiority. Lowell regards them from his usual superior vantage point. He is devoid of, and seemingly largely incapable of, empathy—the ability to enter into another person's world on his own terms. Like his panoramic view of the Noto peninsula as seen from the Arayama Pass, that of the Japanese people was "mappy at best." It is hardly surprising that his literary persona partakes of the same aloofness and superiority. He is the grandée who looks down from on high rather than the person who participates in that experience at the everyday human level. Among the most frequently occurring words in Lowell's vocabulary are "alone" and "lonely." In Japan, Lowell always dines alone, never in company of the Japanese. Japan was for him, as Mars Hill was later to be, a great solitude. This was recognized by Hearn, who came to strongly admire and identify with the Japanese people and wrote to Chamberlain: "What you say about Mr. Lowell's being probably less intimate with the common people than I now am, is, I think, true. Certainly so large a personality as his would find it extremely difficult – probably painful – to adopt Japanese life without reserve, its costumes, its diet, its life upon the floor, its interminable small etiquette, its everlasting interviews with people who have nothing to say but a few happy words…. He is what the French would call *une envergure trop vaste pour ça*; and for so penetrating and finely trained an intellect, the necessary sacrifice of one's original self would be mere waste."[73] Chamberlain, too, became increasingly disenchanted with Lowell's whole approach—his tendency, he complained, to "joke one off" whenever anyone questioned his method.[74] He also characterized Lowell's method as being "to argue down deductively from general notion, such as the supreme virtue of 'modernity,' the 'impersonality of Orientals,' etc.—he has only 3 or 4—and then the facts to cite the preconceived idea, seasoning the whole with verbal fireworks."[75]

Back in Boston, Lowell returned to his rounds of business and polo, and at the end of January 1890, sailed with Ralph Curtis for Europe, where they visited Spain. He was next in Tokyo in April 1891, with George Agassiz, grandson of the Swiss-born Harvard naturalist Louis Agassiz. Together, they contrived at a July–August expedition to Mount Ontaké, with the principal objective being to study Shinto religious rites. During their ascent, Lowell saw a pilgrim fall into a trance. He became so fascinated that he threw himself into the esoteric studies of fox possessions and trances generally, taking some persons as subjects to his house in Tokyo and observing and testing the reality of the trance states (Hearn would refer to his approach as "Mephistophelian") by sticking pins into his subjects. In order to account for a phenomenon that would perplex more profound investigators such as William James, Pierre Janet, and Freud, who were defining such phenomena in terms of the concept of hysteria and suggestion, Lowell trotted out impersonalism once again. "Where the individuality was so weak" as in the Japanese, he wrote, "it was easy to accommodate the possessing spirit. The soul of the possessed could be easily snuffed out."[76] The same weakness of personality, and aptitude for imitation, also seemed to account for the apparent inconsistency of how the

Japanese at the very moment he wrote were moving headlong into an industrialized nation state and great power that a few years later would stun the Russians by defeating them in the Russo-Japanese War. Their susceptibility to foreign ways was little different from the trance-like behavior of the Shinto pilgrims, a childlike yielding to the suggestions of a stronger power—as he concluded in his book on Shinto trances, *Occult Japan: or, the Way of the Gods* (1894):

> It is hardly an exaggeration to say that Japan at this moment is affording the rest of the world the spectacle of the most stupendous act ever seen, nothing less than the hypnotization of a whole nation, with its eyes open. Forty million of folk there are now innocent freaks of foreign suggestion. It is not simply imitating of foreign customs, but the instant unassimilated character of the imitation that stamps the national state of mind as kin to hypnosis, and gives to both their cousinly touch of caricature.[77]

So did the strongly individualistic Percival Lowell, like a honeybee "hurrying about and collecting … buzzing here and there impatiently from a knowledge of what is to be arrived at," succeed in finding in Japan that which he expected to find (in Bayesian brain terms, he showed great skill in turning prior beliefs into posterior beliefs, with a minimum of prediction errors or, to use the technical term, "surprisals"). He found what he expected, and expected what he found.

From Occult Japan to Mars

While Lowell was traipsing the Far East, a steady stream of letters and packages arrived at Sevenels, "full of lengthy descriptions of exotic and wondrous Oriental landscapes," containing "pictures and prints, paper fans and fishes, miniature toys and bright bric-a-brac," which had a tremendous effect on his youngest sister Amy—19 years his junior, so much younger that he called himself her "stepbrother" and referred to her as "Postscript." What Percival sent her was far more interesting than anything available to her at school, where the curriculum did not include such things as classical languages as (it was argued, and her family agreed) such studies were not necessary for girls. He was "a folk hero, comparable to the heroes of the storybooks she loved to read: a vision out of the 'Little Rollo' series; or perhaps an embodiment of Sir Launfal, the knight in shining armor poised smartly upon the broad back of a white steed.'"[78] When later Amy became an accomplished and famous poet, many of her poems took their inspiration from, first, her father's garden in the back of Sevenels, and the Japanese art, architecture, and gardens Percival had described to her from Japan. He himself tried his hand at writing poetry, which was but a short step from his usual florid and highly wrought prose style. In 1891, he wrote a poem for the Harvard branch of Phi Beta Kappa, "Sakura no Saka" ("Cherry Tree Blossom"), which develops the central theme of *The Soul of the Far East*, and in 1893 wrote a poem "Ontake," for the short-lived "Authors' Club" which he co-founded with friends, which compresses the story of *Occult Japan* into a narrative poem that begins:

> Ten thousand feet above the world where men
> Live out the common level of their lives,
> The closer thus to stand to heaven's self…
> Sacred Ontaké solitary towers….
> In front the slope in one full sweep falls off
> Into a sea of cloud that veils its foot
> From contact base, and isolates the peak

To undisturbed communion with the sky.

Once more his fingers would sweep the lyre, in the poem "Mars" (which however was never published), his poetic aspirations seemingly being killed by the ridicule of what Lafcadio Hearn called Lowell's

> quite too dreadful poem (so-called) on Azuma-yama—a mountain which sates its fury
> 'On some Japanese students sent
> From Tokio to investigate
> That mountain's interesting state'!!!
> etc., etc., etc.[79]

In this case, even Lowell's friends failed to rally around him; his and Chamberlain's friend Walter B. Mason remarked that he "should not have thought it possible to commit literary suicide with such thoroughness and dispatch," while after much soul-searching Chamberlain himself advised Lowell to cut poetry and devote his efforts to writing prose.[80]

This incident was not alone, and convinced Percival to abandon his effort to devote himself full-time to a literary career. In addition—with the rise of Lafcadio Hearn as the leading interpreter of Japanese culture to American

audiences—he realized it was time for a change. After 10 years of living intermittently in the Far East, and thinking and writing about it, the blossom was
off the cherry-tree. What had first drawn him to the East when he was first
fired by Morse's lectures in 1882 had been its mystery. In *Chosön*, he had written about the allure of Korea:

> To most minds there lurks a certain charm in the mysterious. The very fact that
> secrecy wraps a subject as with a mantle renders us all the more eager to tear
> away the veil. The possession of this feeling is at once an exciting cause and a
> sanction to knowledge. We realize its power as regards persons, things, events;
> less commonly is it a motive force to the study of a whole nation, and yet it is in
> this connection that I would call upon it now. I ask you to go with me to a land
> whose life for ages has been a mystery,—a land which from time unknown has
> kept aloof, apart, so that the very possibility of such seclusion is itself a mystery,
> and which only yesterday opened its gates....[81]

His 1892 expedition to Japan was cut short by the tragic death on the Fourth
of July of his brother-in-law, Alfred Roosevelt, a millionaire Wall Street banker
who had married his sister Katharine Lowell. From the beach house on Long
Island where his family enjoyed cool breezes off the Long Island Sound, Alfred
commuted by train to his Manhattan office. One morning, on attempting to
board a moving train, he slipped, and was dragged and thrown against an iron
abutment. His right ankle was crushed beneath the wheels, and he was unable
to survive his injuries and the amputation of his leg. Word of his tragic loss
reached Percival in Tokyo, with relatives urging him to return home; and so,
on October 22, he returned to Boston, where after a suitable period of mourn-
ing, he resumed polo, attended cultural exhibitions, and graced the "event of
the season," Sarah Bernhardt performing at the Tremont Theater. In addition—and this was premonitory of the change in direction he was about to
take—he spent a significant part of his time at meetings of the Boston Society
of Natural History (next to the MIT building where the Lowell Institute lectures were held), including reconnecting with his old friend Morse before
returning to Japan for the last time.

At this time he must have been silently asking himself: Where was another
land which had kept itself aloof and apart? Another mystery?

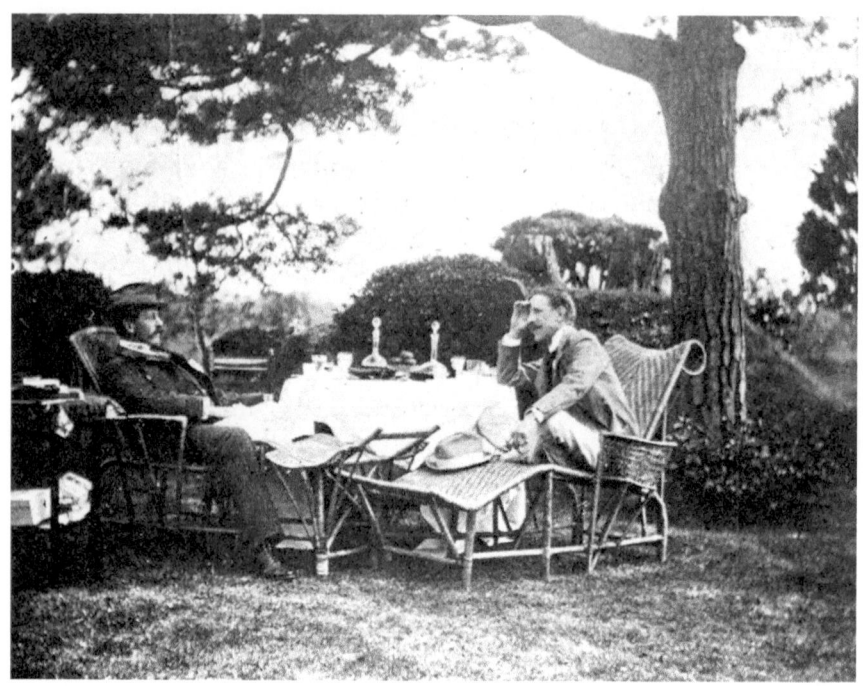

Percival Lowell and Ralph Curtis; de jeuner in the garden of the 18-room house Lowell rented in the Akasaka area of Tokyo, 1893. (Courtesy: Lowell Observatory Archives)

Lowell in his garden in Tokyo, probably during his last trip to the Far East, 1893. (Credit: Lowell Observatory Archives)

He would, of course, find that world in the mysterious planet Mars, which had seemingly lay hopelessly outside the grasp of human science until 1877, when the Italian astronomer Giovanni Virginio Schiaparelli had partially torn away the veil and revealed the possibility of an inhabited world.

Lowell's interest in astronomy went far back, and astronomical references and imagery are scattered throughout the Far Eastern books and if gathered together would make a long list. In discussing the climate of Korea in *Chosön*, he had gone so far as to embark upon a rather frightfully complicated and thorough analysis of the astronomical factors involved, that included the differential heating of the Earth in middle latitudes versus the poles at various seasons; in a footnote he even included a pair of integral signs! While in Boston in 1890, he had discussed Mars with Harvard astronomer (and fellow member of the Appalachian Mountain Club) William H. Pickering, who was about to set out on a two-year sabbatical in Arequipa, Peru, for the purpose of setting up the Boyden station of the Harvard College observatory and observing Mars. A year later he told his brother-in-law William Lowell Putnam, II, who had married his sister Bessie and had been among the family members who had urged his return after Alfred Roosevelt's tragic death, that he wanted to write a kind of treatise on cosmogony, "a philosophy of the cosmos, with illustrations from celestial mechanics."[82] Then, in 1892, just before setting out on the last trip to Japan, he had asked the Harvard College Observatory Director Edward C. Pickering (William's brother), "Could you kindly tell me [what are] the most modern charts or drawings of Mars?... Are Schiaparelli's, Terby's, procurable?"[83] At the same time he acquired a six-inch Clark refractor which he lugged with him to Japan and used to observe not Mars, now well past its summer of 1892 opposition that had produced a significant "boom" of interest in the planet, but Saturn.

Lowell in 1892 was a formidable if rather enigmatic figure: aloof and aristocratic, a rebel against the notion endorsed by his father and later stated clearly by Calvin Coolidge that "the business of America is business," an eminently eligible bachelor who had fled from possible emotional attachments, someone fascinated with using aliens (the Koreans, the Japanese) as raw matter for poetic reverie, was about to set his sights on another alien world, and allow himself to indulge a far more compelling reverie from which neither he—nor most of the rest of the world—would wake in his lifetime.

Notes

1. Benfey, *The Great Wave*, p. 187.
2. Benfey, *The Great Wave*, p. 188.
3. For a detailed study of Lowell, Massachusetts by an architectural historian, see: John P. Coolidge [4].

4. The year, of course, of the Compromise concerning slavery that attempted, unsuccessfully, to heal the festering wound that the U.S. Constitution had inflicted on the country with the "peculiar institution."

5. Known as a vigorous and successful advocate of New England business interests, Abbott Lawrence became active in politics. He was elected to the U.S. House of Representatives, serving two terms (1835–37 and 1839–40). He became a close friend of Daniel Webster, unsuccessful Whig candidate for the presidency in 1836, and like Webster was a "Cotton Whig." In the words of S. Foster Damon, these men were "aware that the problem of slavery confronting America was not a simple one: statecraft and economics, the legal points of the Constitution, and justice to property owners, were also involved." See: S. Foster Damon [7]. They opposed the abolition of slavery, and thus were opposed by the "conscience Whigs," who favored its abolition; the split would eventually doom the Party. In the run-up to the 1849 presidential election, Lawrence was a leading candidate for vice president on the Zachary Taylor ticket, but lost out on the second round to Millard Fillmore (curiously, his brother Amos refused to vote for him); had he been chosen instead of Fillmore, he would, since Taylor served only a year and a half before his death, have become the 13th president of the United States. Though Lawrence was offered any position in the Taylor cabinet he wanted, he refused them all, and instead opted to become Minister to Great Britain, serving at the Court of St. James's from 1849 to 1852. He may have served as the inspiration for the "gentleman with gold-sleeve buttons" in Hermann Melville's 1876 novel *The Confidence-Man: His Masquerade*, as argued in: William Norris [8].

6. Greenslet, *Lowells*, p. 320.

7. On the Lowell Institute, see: Edward Weeks [11].

8. According to Greenslet, *Lowells*, pp. 233–235: "The demi-decade from 1837 to 1842 must have been anxious and overburdened for John Amory Lowell, yet it was precisely within these years that he launched the great undertaking of the Lowell Institute which had been entrusted to his sole charge by John, Jr., completing his last will and testament on the banks of the Nile…. As executor of John, Jr.'s, will and trustee of the fund it created, he had managed matters so well that when the Institute opened the half of the estate devoted to it amounted to a round quarter of a million dollars, which, invested in mill stocks and short-term loans to merchants, would yield an annual income of not less than twenty-five thousands dollars. This permitted very generous remuneration of the lecturers… In their selection the trustee had advice from specialists… During the forties in addition to the recurrent presentation of the Evidences of Christianity, usually by Doctor Palfrey, later supplemented by Mark Hopkins and others, avid audiences heard Jeffries Wyman on Comparative Anatomy, Charles Lyell on Geology, George Glidden on Ancient Egypt, Jared Sparks on American History, Asa Gray on Botany, O.M. Mitchell on Astronomy, George Hilliard on Milton, and, as a grand climax, Louis

Agassiz, M.D., specially imported for that purpose, in three annual courses of twelve lectures each, always repeated, on the Plan of Creation in the Animal Kingdom, Ichtyology, and Comparative Embryology. The Agassiz courses created nothing less than a *furore*...." It seems that it was the influence of these lectures that led Abbott Lawrence to decide to endow the Lawrence Scientific School at Harvard in 1847, while with income from the gift almost immediately the President and Fellows elected Agassiz Professor of Zoology and Geology at Harvard.

9. Damon, *Amy* Lowell, pp. 29–30.
10. Greenslet, *Lowells*, p. 347.
11. As noted in David Strauss [13].

Percival's sister Katharine reported that Percival took all of his mother's belongings when the estate was divided among the children. Nor was Percival's mother the only woman in the family to fuss over little Percy. His maiden aunt Anna Cabot Lowell (1808–1884), John Amory Lowell's sister, showed great interest in him. And a trace of her affection is found in an inscription on his first birthday in a children's book, *A Treasury of Pleasure Books*: "Percival Lowell: from his loving Aunt Annie, March 13, 1859." The intermarriage of so many cousins create pitfalls for the biographer because there are so many almost identical names; it is important to note that this Anna Cabot Lowell is not to be confused with Anna Cabot Jackson Lowell (1811–1874), who belonged to a different branch of the Lowell family. This Anna was a remarkable woman who founded a school for the purpose of educating her own children then began taking in the children of friends and relatives. She was adored by an entire generation of young women including Ralph Waldo Emerson's daughters, and wrote numerous books including *Theory of Teaching* (1841), *Edward's First Lessons in Grammar* (1843), *Gleanings from the Poets, for Home and School* (1843), *Edward's First Lessons in Geometry* (1844), *Outlines of Astronomy, or the World as It Appears* (1850), *Thoughts on the Education of Girls* (1853), etc. Among the students she taught was Percival's youngest sister Amy, who was "totally indifferent to classroom decorum. Noisy, opinionated, and spoiled, she terrorized t he other students and spoke back to her teachers." See Heyman, *American Aristocracy*, p. 164. Anna Cabot Lowell was also the mother of the famous Charles Russell Lowell, Jr., scholar, mechanic, traveller, railroad treasurer, ironmaster, and commander of a fighting brigade of cavalry in the Civil War. He was known in the family as *Beau Sabreur*, and died tragically of a sniper's bullet at the Battle of Cedar Creek in October 1864, at the age of only 29.

Yet another remarkable woman in the family was her sister-in-law, Mary Lowell Putnam, who married Samuel Putnam whose sister was Elizabeth Putnam who married John Amory Lowell and was the mother of Augustus. She spoke many languages and lived in Europe and befriended the liberal intellectuals of the 1848 reform movements in France and Hungary. She was the mother of William Lowell Putnam killed at Ball's Bluff in October 1861 during the Civil War. For details, see: Carol Bundy [14].

12. The details regarding Percival's siblings were as follows:

Abbott Lawrence, the future president of Harvard and Percival's future biographer, born December 13, 1856;

Katharine, born 1858, who afterward married banker Alfred Roosevelt (first cousin of Theodore, the future president of the United States); and after his death in 1891, married Thomas James Bowlker in 1902;

Elizabeth (known in the family as Bessie), born in 1862, and her twin, Roger. Roger died before his second birthday. Elizabeth married the son of another Boston scion, William Lowell Putnam, a name partner in the law firm of Putnam, Putnam & Bell, and was an amateur poet and painter as well as an activist for prenatal care;

May, born in 1870; lived 1 day;

Amy, born in 1874, junior to Percival by almost 20 years and known in the family as "Postscript," who would become an avant-garde poet, dressed in men's clothes, wore a pince-nez, smoked cigars, and perhaps most notoriously, openly cohabit with Ada Dwyer Russell, a former actress and divorcee.

13. P. Lowell, "Comets," text dated March 10, 1910; Lowell Observatory Archives.

14. Greenslet, *Lowells*, p. 347. The Institution Sillig at Vevey, in the shadow of the Alps, had been founded in 1836, and educated the Princes Victor and Louis Napoleon and the American financier J. Pierpont Morgan. As described by Max Sillig, a future headmaster of the school (and grandson of the original headmaster), the methods used contrasted with those of the Prussian system at the time. "I believe that with a certain class of boys you can get better results by means of gentle persuasion and good example than by means of the harsh methods of German discipline. Harsh methods kill the individuality and friendly spirit. Although I am a First Lieutenant in the Swiss Army I do not believe in bringing up young boys as soldiers. Willing effort is better than forced effort." See: "American Schools Do Not Educate, Says Swiss Master." The New York *Times*, January 11, 1914.

15. Damon, *Amy Lowell*, p. 30. According to Damon, p. 30n, the Boston Directories show that Augustus lived at 7 Pemberton Square (1852–53), 131 Tremont Street (1854–57), 10 Park Square (1858–61), 61 Mt. Vernon Street (1862–63), 70 Heath Street Brookline (1864–77), 97 Beacon Street and Brookline (1878–79), and 171 Commonwealth Avenue and Brookline (1880–1900).

16. Heyman, *American Aristocracy*, p. 161.

17. A. Lawrence Lowell, *Biography*, pp. 4–5.

18. P. Lowell to Hector MacPherson, Jr. November 16, 1903; Lowell Observatory Archives.

19. A. Lawrence Lowell, *Biography*, p. 5.

20. A. Lawrence Lowell, *Biography*, p. 5.

21. *Elementary Treatise on Plane Trigonometry* (1835), *First Part of an Elementary Treatise on Spherical Trigonometry* (1836), *An Elementary Treatise on Algebra: to which are added Exponential Equations and Logarithms* (1837), *An*

Elementary Treatise on Plane and Solid Geometry (1837), *An Elementary Treatise on Plane and Spherical Trigonometry*, and two volumes on *Curves, Functions, and Forces* (vol. 1 appearing in 1841 and Vol. 2 in 1846).

22. "Benjamin Peirce," MacTutor History of Mathematics. https://mathshistory. st-andrews.ac.uk/Biographies/Peirce_Benjamin/

23. A mudsill was the lowest threshold supporting the foundation for a building. The "mudsill theory" had first been articulated in a speech on March 4, 1858, by James H. Hammond, a Democratic Senator from South Carolina and a Southern plantation owner, and held that there must be, and always has been a lower class or underclass (menial workers whether called slaves or not) for the upper class and the rest of society to rest upon. Abraham Lincoln argued forcefully against the mudsill theory in a speech in Milwaukee, Wisconsin, in 1859, on the grounds that mudsill advocated concluded that "all laborers are necessarily either hired laborers, or slaves," since to them, "Nobody labors unless somebody else, owning capital … induces him to do it." Obviously, Lincoln failed to convince the likes of Peirce.

24. R.C. Archibald (1925) *Benjamin Peirce, 1809–1880, In: Biographical Sketch and Bibliography, … with Reminscences by President Emeritus Charles W. Elliot, President A. Lawrence Lowell, Professor Emeritus W.E. Byerly and Chancellor Arnold B. Chace* (Oberlin, Ohio: The Mathematical Association of America), p. 4.

25. Nathan Reingold (ed.) [21]. Peirce's claim that Neptune was not the planet predicted by Le Verrier and Adams was long remembered, and viewed with skepticism even by some of the Harvard undergraduates. Thus, in 1859, when William Cranch Bond, the first Director of the Harvard College Observatory, died, Peirce threw his hat into the ring to succeed him in the role; whereupon Horace P. Tuttle, an assistant at the Observatory who was then achieving brilliant record as a discoverer of comets, entered in his log book: "The man who said, 'Neptune is not the planet to which geometrical analysis directed the Telescope' is endeavoring to obtain the Directorship of the H[arvard] C[ollege] Observatory," followed by a large question mark. H.P. Tuttle, February 25, 1859; Harvard College Observatory Archives.

26. Heymann, *American Aristocracy*, p. 134.

27. A. Lawrence Lowell, *Biography*, p. 6.

28. As noted by Strauss, *Percival Lowell*, p. 267.

29. I first heard of this particular quotation from David Lewis-Williams at the University of Witwatersrand in Johannesburg, South Africa, in September 2001. On attempting to track it down, I found that it has been quoted (in somewhat different forms) by a number of authorities including Richard Leakey, Nicholas Humphreys, and various others, and attributed to the Bishop of Worcester's wife in the 1860s and the Bishop of Birmingham's wife in the 1880s. The earliest citation I have found is by Robert Formon Horton, who in an 1893 text, "attributes it not to a Bishop's wife but to a "timid and

decorous spinster": "The Church swarms with people who have no spiritual sinew, and whose lungs cannot breathe the invigorating air of Truth: they take up the cry of that timid and decorous spinster who, on heaving an exposition of the Darwinian theory that men are descended from apes, said, 'Let us hope it is not true, or if it is, let us hush it up.'" See: Robert Formon Horton [25].

30. Tuchman, *The Proud Tower*, p. 586.

31. A. Lawrence Lowell, *Biography*, p .50.

32. Percival Lowell to Barrett Wendell, January 16, 1877. Houghton Library, Harvard University.

33. Ibid.

34. Ibid.

35. Curtis's parents first rented the Palazzo in 1881, and 4 years later bought the top floors for only $13,500 (about $400,000 in today's dollars). It became the hub of American life in Venice, its regular visitors including not only Sargent but Henry James (who first visited the Palazzo in 1887 and finished writing his *Aspern Papers* there and later described its ballroom in the *Wings of the Dove*), the poet Robert Browning, the art collector Isabella Steward Gardner (who had married a cousin of Percival Lowell), and painters James McNeill Whistler and Claude Monet. The heiress Curtis married was Lisa de Wolfe Colt, of the Colt revolver fortune. Though in fact Curtis gave up painting before he married Colt, Edith Wharton later wrote a short story, "The Verdict," in which a successful young artist marries a wealthy widow and forsakes his career—a story that enraged Mrs. Curtis.

36. P. Lowell to Wendell Barrett, November 8, 1877. Houghton Library, Harvard University.

37. It may be that eventually more will be known, as unprocessed papers on the subject exist at the Massachusetts Historical Society. David Baron, personal communication to William Sheehan, April 27, 2024.

38. P. Lowell to Katharine Lowell, July 19, 1882. Houghton Library, Harvard University. Katharine herself was then engaged to marry Alfred Roosevelt (Theodore's cousin), and did so in December of that year.

39. P. Lowell to Augustus and Katharine Bigelow Lowell, August 2, 1882. Houghton Library, Harvard University. Teddy Roosevelt was highly incensed by his future cousin-in-law's behavior, and wrote in a letter (now in private hands), "Your conduct has been such as only a mean-spirited and cowardly blackguard could be guilty of."

40. Charles W. Eliot to Edward C. Pickering, November 22, 1894. Director's correspondence, Harvard University Archives.

41. In *Anna Karenina*.

42. See Janet Oppenheim [28].

43. David Baron (forthcoming), *The Martians* (New York: Liveright), chapter 1. [Page unavailable. Manuscript was reviewed in draft form. Book title may change upon publication.]

44. Baron, *The Martians*, chapter 1.
45. A. Lawrence Lowell, *Biography*, pp. 9–10.
46. A. Lawrence Lowell, *Biography*, pp. 10–11.
47. A. Lawrence Lowell, *Biography*, p. 11.
48. A. Lawrence Lowell, *Biography*, p. 12.
49. A. Lawrence Lowell, *Biography*, p. 12.
50. A. Lawrence Lowell, *Biography*, p. 14.
51. A. Lawrence Lowell, *Biography*, p. 9.
52. Strauss, *Percival Lowell*, p. 80.
53. Strauss, *Percival Lowell*, p.35.
54. Greenslet, *Lowells*, p. 352.
55. Greenslet, *Lowells*, pp. 352–355.
56. Baron, *The Martians*, chapter 11.
57. A. Lawrence Lowell, *Biography*, p. 30. As chauvinistic as it now seems, and
    began even to seem to European students of Japanese culture such as
    Chamberlain and Hearn even at the time, Lowell's "individualism" vs.
    "impersonalism" dichotomy is not entirely plucked from thin air. The differ-
    ences in these cultures have now been thoroughly studied; individualistic
    societies (such as the United States) emphasize autonomy, personal achieve-
    ment, uniqueness, and the needs and rights of the individual; collectivist cul-
    tures, like the Japanese, espouse harmony, interdependence, and conformity,
    where the needs of the community guide behavior. People from the U.S. (on
    average) are more likely to use first-person-singular pronouns, and to organize
    memory around events rather than in terms of social relations. The Japanese
    (on *average*) do the opposite. These differences even reflect different workings
    of the brain and body. It has been suggested that the reason for the differences
    is that whereas American history has been defined by an expanding western
    border settled by similarly tough, individualistic pastoralist pioneers, Far
    Eastern collectivist societies have developed in relation to rice farming, which
    demands massive amounts of collective labor to turn mountain into terraced
    rice paddies through the use of massive and ancient irrigation systems. It
    seems the Martian canal-builders, did they exist, would more likely have
    resembled the impersonal Japanese than the individualistic Americans. See:
    Robert M. Sapolsky [29]. Also: Richard E. Nisbett [30].
58. A. Lawrence Lowell, *Biography*, p. 31.
59. Hermione Lee [31]. It has been suggested also that a contributing factor to
    the failure of their marriage seems also have been Teddy's ambiguous sexual-
    ity. Says Lee (p. 77), "The marriage was apparently unconsummated for 3
    weeks, and always thereafter the physical relationship between the pair … was
    agonized."
60. Strauss, *Percival Lowell*, p. 51. The Vanderbilts, in addition to their New York
    City mansion, were soon to set about building "the Breakers," the grandest of
    Newport's summer "cottages" which symbolized the family's social and finan-

cial preeminence in the late Golden Age. Edith Wharton would of course have agreed with Percival as to the moral vacuity and rot of such people, which she satirized in books such as *The House of Mirth.*

61. Strauss, *Percival Lowell,* p. 36.
62. Bisland, *Life and Letters,* vol. 1, 116–117.
63. A. Lawrence Lowell, *Biography,* p. 32.
64. Greenslet, p. 357.
65. Lafcadio Hearn to Basil Hall Chamberlain, May 22, 1891. In: Bisland, *Life and Letters,* vol. 1, p. 9. Hearn would go even further later. In a letter to Chamberlain, January 14 1893, he wrote: "Now, to me, the most beautiful, the most significant, the most attractive point of Japanese character, is revealed by the very absence of that personality to which Mr. Lowell's book [*The Soul of the Far East*] points to as an Oriental phenomenon.... [E]ven now my knowledge is trifling. Still, it teaches me this:

    1. That the lack of personality is to a great extent voluntary, and that this fact is confirmed by the appearance of personality, strongly and disagreeably marked, where the social and educational conditions are new, and encourage selfishness.
    2. That every action of Japanese life ... [has been] regulated by the spirit of self-repression for the sake of the family, the community, the nation,—and that she so-called impersonality signifies the ancient moral tendency to self-sacrifice for duty's sake...

[This] was the highest possible morality from any high religious standpoint,—Christian or pagan,—the sacrifice of self for others...."

66. As I found on repeating his itinerary with Japanese friends, Masatsugu Minami and Tadashi Asada, 105 years later. Two years earlier, Minami had retraced and documented part of Lowell's route, as detailed in "Lowell Road Report 1." https://www.kwasan.kyoto-u.ac.jp/~cmo/cmomn3/LProads1. htm. See also W. Sheehan (2005) "To Mars by Way of Noto," *Sky & Telescope* (December), pp.108–111. See also William Lowell Putnam, III (2006) Letter, *Sky & Telescope* (April), p. 12.
67. Lowell's arrival shortly preceded the adoption of the new Meiji Constitution on February 12, 1889. On the same day, the education reformer and former Japanese ambassador to the United States, Mori Arinori, was assassinated by an ultra-nationalist enraged by Mori's alleged failure during a visit to the Ise Shrine 2 years before to follow religious protocol of remove his shoes before entering and pushing a sacred veil with a walking stick. Mori's assassination deeply affected Lowell, who wrote a moving account of the assassination: P. Lowell [39].
68. P. Lowell, *Noto: an unexplored corner of Japan.* Boston, Houghton-Mifflin, 1891, p. 1.

69. Lowell, *Noto*, p. 9.
70. Lowell, *Noto*, pp. 103–104.
71. Lowell, *Noto*, p. 175.
72. Lowell, *Noto*, p. 202.
73. Lafcadio Hearn to Basil Hall Chamberlain, May 22, 1891. In: Bisland, *Life and Letters,* vol. 1, p. 9.
74. Basil Hall Chamberlain to Lafcadio Hearn, January 10, 1893. In: Kazuo Koizumi (ed.) (1936) *Letters from Basil Hall Chamberlain* (Tokyo: Hokuseido Press), p. 3.
75. Basil Hall Chamberlain to Lafcadio Hearn, January 10, 1893, Koizumi, *Letters*, p. 2.
76. Lowell, *Noto*, p. 237.
77. Percival Lowell [40].
78. Heyman, *American Aristocracy*, p. 164. Katharine Lawrence Lowell also tried to push Amy into more traditional directions for a young girl, without success. According to hers authorized biographer Foster Damon, *Amy Lowell*, p. 60: "One of her early friends testifies: 'Mrs. Lowell thought Amy should be more feminine and tried to interest her in dolls and sewing, things which Amy abhorred. It was an act of self-sacrifice that after I had met with a slight accident, Amy came every day to play papers doll with me because she knew I liked him." The influence of Percival's Oriental letters and gifts is attested in some of Amy's later poetry, especially marked in Amy Lowell (1919) *Pictures of the Floating World* (New York: Macmillan) of which her poem "Paper Fishes" is characteristic:

> "The paper carp,
> At the end of its long bamboo pole,
> Takes the wind into it mouth
> And emits it at its tail.
> So is man,
> Forever swallowing the wind."

79. Quoted in Strauss, *Percival Lowell*, p. 78.
80. Strauss, *Percival Lowell*, p. 78.
81. Lowell, *Chosön*, p. 11.
82. A. Lawrence Lowell, *Biography*, p. 60.
83. P. Lowell to E.C. Pickering, September 10 and November 7, 1892. Pickering. Edward Charles. Papers, Harvard University Archives.

References

1. Benfey C (2003) The great wave: gilded age misfits, Japanese eccentrics, and the opening of old Japan. Random House, New York, p 177
2. Lowell P (1906) Mars and its canals. Macmillan, New York, pp 7–8
3. James H (2006) The wings of the dove. In: Novels 1901–1902. The Library of America, New York, p 330
4. Coolidge JP (1942) Mill and Mansion: architecture and society in Lowell, Massachusetts, 1820–1865. Columbia University Press, New York
5. Whittier JG (1845) The City of a day. In: The stranger in Lowell. Waite, Pierce & Co, Boston
6. Tuchman BW (2012) The proud tower. In: The guns of august and the proud tower. Library of America, New York, p 579
7. Foster Damon S (1935) Amy Lowell: a chronicle. Houghton Mifflin, Boston/New York, pp 29–30
8. Norris W (1976) Abbott Lawrence in the confidence-man: American success or American failure? Am Stud 17(1 (spring)):25–38
9. Greenslet F The Lowells and their seven worlds. Houghton Mifflin, Boston/New York, p 320
10. Shi DE (1985) The simple life: plain living and high thinking in American culture. University of Georgia Press, Athens, pp 93–98
11. Weeks E (1966) The Lowells and their Institute. Little, Brown, Boston
12. Crawford MC (1930) Famous families of Massachusetts, vol 1. Little, Brown, & Co, Boston, p 146
13. Strauss D (2001) Percival Lowell: the culture and science of a Boston Brahmin. Harvard University Press, Cambridge, MA, p 14
14. Bundy C (2005) The nature of sacrifice: a biography of Charles Russell Lowell, Jr., 1835–64. Farrar, Straus and Giroux, New York
15. Lawrence Lowell A (1935) Biography of Percival Lowell. New York, Macmillan, p 2
16. David Heyman C (1980) American aristocracy: the lives and times of James Russell Lowell, Amy Lowell, and Robert Lowell. Dodd, Mead, & Co., New York, p 161
17. Ackerberg A (1999) Benjamin Peirce. In: American national biography, vol 17. Oxford University Press, Oxford, pp 250–251
18. Auspitz JL (1994) The wasp leaves the bottle: Charles Sanders Peirce. Am Scholar 63(4, (Autumn)):602–618
19. Hubbell SJG, Smith RW (1992) Neptune in America: negotiating a discovery. J Hist Astron 23(4):263–286
20. Peirce B (1847) Notice of the computations of Mr. Sears C. Walker, who found that a star was missing.... In: *Proceedings of the American Academy of Arts and Sciences*, vol. 1 (1846–48), pp 57–65

21. Reingold N (ed) (1985) Science in nineteenth century America: a documentary history. University of Chicago Press, Chicago, p 136
22. Lowell P (1907) The canals of Mars, optically and psychologically considered: a reply to professor Newcomb. Astrophys J 26(3):131–140
23. Leonard L (1921) Percival Lowell: an afterglow. Richard G. Badger, Boston, p 25
24. James W (1904) Herbert Spencer. Atl Mon 94:104
25. Horton RF (1893) Verbum Dei: The Yale lectures on preaching, Lecture IV: The Bible and the Word of God. New York/London, Macmillan, p 132
26. Lowell JR (1890) New England two centuries ago. In: Literary essays, the writings of James Russell Lowell in ten volumes, vol 2. Houghton, Mifflin and Co., Boston/New York, p 9
27. Jamison KR (2017) Robert Lowell: setting the river on fire, a study of genius, mania, and character. Alfred A. Knopf, New York
28. Oppenheim J (1991) Shattered Nerves: doctors, patients, and depression in Victorian England. Oxford University Press, New York/Oxford, p 1991
29. Sapolsky RM (2023) Determined. Penguin Press, New York, pp 74–76
30. Nisbett RE (2003) The geography of thought: how Asians and Westerners think differently… and why. Simon and Schuster, New York
31. Lee H (2007) Edith Wharton. New York, Alfred Knopf, p 74
32. Lowell P (1886) Chosön: the land of the morning calm. Ticknor and Co., Boston, pp 26–27
33. Bisland E (ed) (1906) The life and letters of Lafcadio Hearn, vol 1. Houghton-Mifflin Co., Boston, p 459
34. Strauss D (1993) The 'Far East' in the American mind, 1883-1894: Percival Lowell's decisive impact. J Am-East Asian Relat 2(3, (Fall)):217–241
35. Ratliff F (1965) Mach bands: quantitative studies on neural networks in the retina. San Francisco, Holden-Day, p 209
36. Jackson Lears TJ (1981) No place of grace: Antimodernism and the transformation of American culture, 1880–1920. University of Chicago Press, Chicago/London, p 235
37. Lowell P (1888) The soul of the Far East. Houghton, Mifflin and Co., Boston, pp 225–226
38. Hoyt WG (1976) Lowell and mars. University of Arizona Press, Tucson, p 19
39. Lowell P (1890) The fate of a reformer. Atlantic Monthly 56:680–693
40. Lowell P (1894) Occult Japan: or, the way of the gods. Houghton, Mifflin and Co.), Boston, p 288, 367

3

Astronomy from the Bottom Up: Edward Emerson Barnard

Contents

It is astonishing how excellent an observer he is. He is like Sir W. Herschel for seeing and noting what is new. But his reasoning on what he sees is very apt to be quite wrong. This is not only from a defective education, it is in his nature. He is like Sir W. Herschel cut into two parts, and only the observing faculty left… His education is unfortunately very limited; incredibly so one might say. He does not seem able to perfect it—or even desirous to do so. There is not a single one of his comet observations (all of which are most carefully made at the telescope) which is entirely correct in the reductions. Something is always wrong—parallax, refraction, reduction to apparent date—something.

© The Author(s), under exclusive license to Springer Nature Switzerland AG 2024   **109**
W. Sheehan, *Parallel Lives of Astronomers*, Springer Biographies,
https://doi.org/10.1007/978-3-031-68800-3_3

—E.S. Holden to E.B. Knobel, June 16, 1893

Barnard and an unidentified man, possibly his employer John H. Van Stavoren, posing as Oliver Twist and Fagin, in an image taken at Van Stavoren's photograph gallery in about 1866. The sore visible on his right jaw would lead to permanent disfigurement. (Credit: University of Chicago Photographic Archive [apf6-04139], Hanna Holborn Gray Special Collections Research Center, University of Chicago Library)

Hard Times

Edward Emerson Barnard was born on December 16, 1857, in Nashville, Tennessee, a city better known today for its music, not astronomy.[1] His parents, Reuben and Elizabeth, had just moved to Nashville from Cincinnati, but Edward never met his father, who died 3 months before his birth. His mother, Elizabeth Jane née Heywood, already 42 years old, was originally from Kentucky. In contrast to patrician Percival Lowell, who could trace his ancestry on his maternal side back to the Norman conquest, Barnard was more like the charity boy Noah Claypole in Dickens's *Oliver Twist*, who could trace his genealogy all the way back to his parents.

His parents' move to Nashville may have been motivated by hope of finding work, but Reuben's death left Elizabeth in a precarious position as her only known employable skill was modelling flowers in wax. In addition to having an infant to care for, she also had to care for Edward's feeble-minded older brother, Charles.

Despite her difficult circumstances, Elizabeth was quite well-educated for someone of her social class. Probably she was unaware of the fact that she shared the name of the granddaughter and last direct descendent of William Shakespeare; but her literary bent is attested in the fact that she chose Edward's middle name, Emerson, in honor of the famous New England writer Ralph Waldo Emerson. When Edward was very young she taught him to read from the Bible. Other books which ranked among his youthful favorites included (unusually) a volume of the writings of the Jewish historian Josephus, which, he later recalled, "had a fascination equaled only by the story of Robinson Crusoe, the *Arabian Nights*, and *Nicholas Nickleby*." In addition, "an old volume of Scientific Discovery and Invention which I read among my first books perhaps had some influence in turning my mind to the wonders of science."[2]

Nashville in the 1850s was the largest and most important American city south of the Ohio River, with the exception of New Orleans. It had grown between the turn of the century and the eve of the Civil War from a frontier town consisting of some sixty or eighty families to a city of nearly 17,000, of whom 4000 were black slaves. The slave market stretched from the intersection of Fourth Avenue, North and Charlotte Avenue to the Public Square, just two streets over from the photograph gallery where Barnard later worked. Young Barnard would certainly have been familiar with what went on there, though he never mentioned any of this in his writings. The slave market did a brisk business through traders with names such as Glover and Boyd, Dobbs and Porter, James and Harrison, and Lyles and Hitchings. Of course, the shame of slavery did not belong exclusively to the South; the pro-slavery party had strong supporters in the North as well—including "cotton Whigs" such as Augustus Lowell and Abbott Lawrence, whose textile mills long depended on cotton produced by slave labor in Southern plantations.

Barnard never seems to have concerned himself much with slavery or politics in general. For that matter, he doesn't ever say much about religion, either. Presumably his views about man or God reflected those held in the place where he grew up, though his single-minded focus on astronomy seems to have kept him from taking any more interest in worldly affairs than necessary.[3]

During the years before Barnard was born, wide turnpikes and busy railroad lines radiated from Nashville, and the coiling Cumberland River, draining Cumberland Gap on its way to the Mississippi, was crowded with river traffic carrying goods from the city's foundries, ironworks, and small manufactures to Memphis and New Orleans. In addition, it was a center of education, with a female academy and Nashville University, and its free public school system was the first in the South. There were also a theater and an opera, and one observer commented that there was as much fashion in Nashville as in New York.

The 1850s had begun with the Convention of Southern States, which convened in the Nashville State Capitol in June 1850 with the express purpose of forming a southern sectional party to protest the attempt then being made in Washington by President Zachary Taylor's administration to exclude southern men with their slaves from the territories recently held by Mexico. The Convention adjourned without conclusion, though plans were made to reconvene in November if Congress did not respond to Southerners' demands. It never did; when Taylor died suddenly, his successor, Millard Fillmore, rushed to push through Congress the Compromise of 1850, which proposed popular sovereignty as the way of deciding the issue of slavery in the territories. This policy unraveled with the Kansas/Nebraska Act, the Dred Scott decision, and the election of Abraham Lincoln in 1860. Seven Southern states seceded after Lincoln's election, and four more, including Tennessee, after Fort Sumter in April 1861. The confederate government tried to claim Missouri and Kentucky too, to make up the magic number of thirteen; however, they failed, and the Confederate States remained eleven.

Tennessee immediately became a major theater of the war. After a short siege in February 1862, Union General Ulysses S. Grant captured Fort Donelson, which had been built to defend vital Confederate supply lines where they crossed the Cumberland River, and Nashville was laid open to the invading Union troops. It was immediately captured and would remain under Union occupation for the duration of the war.

Against this background, the Barnard family attempted to eke out a miserable existence. Edward recalled long afterward that his early youth was "so sad and bitter that even now I cannot look back to it without a shudder."[4] Elizabeth, Charles and Ed lived briefly at a place near the old toll-gate on the Lebanon Pike south of the city, and later moved to the bank of the Cumberland a mile or so above the old waterworks. During these terrible years, the family was often on the verge of starvation. Once, a steamer loaded with provisions was sunk as it was trying to get into the city Ed Barnard and others, gaunt with hunger, watched sorrowfully as the provisions from the steamer floated down river. Some people put out in small boats to salvage what they could.

Young Barnard, already a strong swimmer, dove into the river and managed to rescue part of a box of crackers, which his mother whipped into batter and made into cakes.

A Comet and a Wagon Bed Under the Stars

Barnard's earliest memories of the stars were associated with the war and tinged with the intense loneliness and sadness of that period of his life. In addition—like Percival Lowell on the winding staircase in the mansion at Park Street—they involved comets. "When I was very small I saw a comet," he wrote, "and I have a vague remembrance that the neighbors spoke of this comet as having something to do with the terrible war that was then desolating the South" [1, p. 275]. Just which one it may have been is uncertain as there were several brilliant comets which swept across the sky during the late 1850s and early 1860s. Lowell had seen scimitar-tailed comet Donati in October 1858. Barnard's comet was probably the Great Comet of 1861, which had been discovered as a Southern Hemisphere object by farmer and amateur astronomer John Tebbutt at Windsor, New South Wales. It did not emerge into Northern skies until June 30, when it appeared spectacularly against the evening twilight, its head brighter than Jupiter and its tail stretching across some 105 degrees of sky. Though Barnard, being four, had only a dim memory of it, something of the spectacle it presented is recorded in the accounts of soldiers watching it from their camps in the lead-up to the First Battle of Bull Run (July 21). On the Fourth of July, a soldier from the Second Wisconsin wrote, "We have been visited for a week past by a very large comet which at full day appears very bright and transparent; late at night the tail stretched nearly to the Zenith while the star [head of the comet] was near the horizon." Two weeks later another soldier, Charles Johnson, serving with the 9th New York Volunteer Infantry near Fortress Monroe, Virginia, wrote: "I watched the comet, wondering if that mysterious little visitor was not perhaps at the same time watched by eyes that would beam gladly into mine; and I composed quite a number of beginnings of addresses to the curious thing, whatever it may be. But the comet is now tired of his visit to these regions of space, or disgusted it may be with the appearance of things on this side of our planet, for he is now leaving in seemingly greater haste than he came, with his tail between his legs, for the unknown regions out yonder."[5]

The Great Comet of 1861. (Credit: E. Weiß, Bilderatlas der Sternenwelt, 1888)

During these terrible years of the war, young Barnard would lie out in the open air on summer nights, flat on his back in an old wagon bed, and watch them as they passed. This, he recalled long afterward, "helped to soften the sadness of my childhood."[6] Among the stars he noticed was "a very bright one, which during the summer months shone directly overhead in the early hours of the evening."[7] It would be many years before he learned its name—Vega.

On Barnard's seventh birthday, December 16, 1864, fighting between the Union and Confederate armies came within only a few miles of Nashville, and he could hear the frightful sound of the cannons booming in the distance. General John Bell Hood, known for his impetuosity, had been trying to invade the north—or at least to get to the Ohio River and perhaps annex Kentucky—in order to distract and disrupt the campaigns of the Union generals Grant and Meade in Virginia and Sherman in Georgia. Tennessee was left to the defense of General George Thomas (the "Rock of Chickamauga"). Hood wanted to fight the Union forces in Tennessee on realizing they had become geographically divided between the army of Major General John M. Schofield, tasked by Thomas to delay Hood as much as possible, and Thomas's at Nashville. In the battle of Franklin, Hood made a frontal assault on the Federal lines; multiple assaults were repulsed, with over 6000 Confederate casualties including a large number of key Confederate generals. Schofield withdrew from Franklin during the night and marched into the defensive works around Nashville, which included a 7-mile-long semi-circular Union defensive line on the south and west sides of the city protecting Nashville against attacks from those directions, and several forts, of which the largest was Fort Negley. Hood encamped in front of the south line of Union fortifications, and—in what would prove to be a serious strategic error— diverted three brigades to attack the Nashville and Chattanooga Railroad between Nashville and Murfreesboro in the hope of drawing Thomas out. Thomas stayed put until he was ready to attack on his own terms, and on December 15 and 16—Barnard's seventh birthday—smashed Hood's army almost completely. For all practical purposes, the war was over, with Lee's surrender to Grant at Appomattox looming only a few months away.

The end of the war did not bring much immediate improvement to the Barnard family's situation, as Nashville remained under Union occupation, serving as a major supply base for Union forces in Tennessee during the Reconstruction period (the last Union troops were not removed until 1877). The city was, of course, economically depressed, and many of the citizens, including the Barnards, often lived hand to mouth. For a brief period of 2 months Barnard attended public school—it was the only formal education he would receive until much later he went to Vanderbilt as a Special Fellow—but he was withdrawn either because of his mother's failing health or because of a cholera epidemic which broke out during the summer of 1866.

The city had some experience of the dread disease. An epidemic in 1849 had claimed the life of the former President of the United States, James K. Polk. There were others in 1850 and 1854, but that of 1866, which peaked about August 1, was the worst so far. In part it owed its fury to the large influx of population into Nashville during the years of the Union Occupation, which included an influx of fugitive slaves from surrounding areas in Tennessee seeking protection of the U.S. Government behind the Union lines. According to a contemporary account:

> This scourge, as usual, in paying its respects to the cities of the earth, has held high carnival in Nashville….
>
> We were in a high state of preparation for cholera. During three years of the late civil contentions, Nashville presented many attractions for every species of vagabondism; as Tacitus says, was the case of Rome upon all occasions. They lived upon the offal of a large army, and when the army was gradually withdrawn, like subsiding water after an inundation, they were left behind, a heterogeneous deposit of humanity, dabbling in the mud, and living and dying as best they could.
>
> Does any one, acquainted with the habits of cholera, wonder that it did not overlook so tempting a banquet? [2].

By the end of September, the cholera outbreak was gratefully in retreat, but in the meantime some eight hundred citizens had succumbed to the dread disease including, almost, Ed Barnard.

Nashville, looking idyllic from a distance, during the Union occupation. The State Capitol is prominent in the distance, the Cumberland River on the right. (From: Harper's Pictorial History of the Great Rebellion, 1866)

A Job for a Street Urchin

Toward the end of 1866, Elizabeth came across the photograph of a man she thought she recognized from her better days in Ohio. Deciding to call on him, she found that he was indeed the same man, John H. van Stavoren. He operated a photograph gallery in the second and third stories of a brick building at the southeast corner of Union and Cherry Streets (now Union and Fourth Avenue). By sheer luck, van Stavoren was just then rather desperately looking for a boy to run errands for him and to take charge of guiding an immense solar camera, called *Jupiter*, which he had mounted on the studio roof. It was said to be one of the largest such cameras in existence at the time and served mainly to make life-sized enlargements on silvered paper from negatives by means of a large condensing lens, but it had other uses. An artist could also use it to paint from the image. Barnard's employer often used it in this way for portrait-painting.

In form the camera was a large rectangular box. Attached to the front and projecting just beyond the edge was a piece of sheet metal through which a small hole had been bored. As the Sun's rays passed through this hole, a bright spot was formed on a screen several feet away. By holding this bright spot in a fixed position on the screen, the image formed by the condensing lens remained steady for the painter, but because of the Sun's steady movement across the sky the camera had to be panned slowly in order to remain in position. This was accomplished means of a pair of hand wheels, one turning the camera to the west, the other tilting it up or down. It was not a job in which any adult would have lasted for long. Child labor laws, of course, being non-existent, van Stavoren had tried several small boys to guide the giant camera. They were able to reach the hand wheels only by standing atop a step ladder. However, the effects of tedium and the warm sun caused them to keep falling asleep, and led to their dismissal. His inability to find anyone suitable had become a source of "much apprehension and annoyance" to van Stavoren. In a third person account, Barnard recounted how the string of failures was finally broken:

> It was necessary to keep it moving precisely with the motion of the sun. If it deviated much from the solar motion the intense heat collected by the great condensing lens and brought to a focus would touch the wooden part of the instrument and set it and the house on fire. The boys that had previously attended the great camera had nearly burned the house up several times by going to sleep in the warm sunshine. "But will your boy keep awake?" asked the photographer after explaining the difficulties. "My son will not go to sleep" replied the mother with confidence. And he never did go to sleep while on duty. Through summer's heat and winter's cold he stood upon the roof of that house and kept the great instrument directed to the sun. It was sleepy work and required great patience and endurance for one so young, and at this distant day he realizes that this training doubtless developed those qualities—patience, care and endurance so necessary to an astronomer's success.[8]

As for the way that van Stavoren had decided on *Jupiter* as the name for his giant camera, Barnard continued:

> … The photographer was certainly not learned in astronomy. He knew there were planets and he knew the names of some of them. He liked to name everything he had. Since this great instrument's main duty was to do obeisance to the sun, what name more appropriate than that of one of the planets? It was a mere selection at random, and the photographer had incidently [sic.] named it "Jupiter"! Little thought the poor child when he was put in charge of "Jupiter" that his own name would forever be linked with that of the mighty planet which it represented.[9]

John van Stavoren, left, and James W. Braid, a man of extraordinary technical abilities and young Ed Barnard's mentor, with the Jupiter camera used for making enlargements of photographic prints, about 1867. (Credit: Special Collections and University Archives, Alexander Heard Library, Vanderbilt University)

So Barnard, the future discoverer of the fifth satellite of Jupiter, came to be hired as an attendant and satellite to the great *Jupiter* camera. He was just nine. He would remain at the photograph gallery, in one capacity or other, until he was twenty-five. The stepladder on which he stood in order to reach the hand wheels guiding the solar camera provided the first rungs of a brilliant career. Climbing from the bottom up, he attained to vertiginous heights.

Most assistants assigned to such a mindless task as cranking the hand wheels of the giant camera would have become numb and somnolent. Barnard, however, remained alert, and found diversion for his mind. For some reason he decided to track the Sun by marking its positions during the year in relation to the ringing of the noon bell of St. Mary's Catholic Church nearby. In this way he independently discovered the "equation of time."[10]

Ed Barnard, photographer's assistant. These photos were taken probably in 1866 or 1867, soon after he was hired to work at van Stavoren's Photographic Studio guiding the Jupiter Camera. (Credit: Special Collections and University Archives, Alexander Heard Library, Vanderbilt University)

Barnard in 1868, age 11. *Credit: University of Chicago Photographic Archive [apf6-04135], Hanna Holborn Gray Special Collections Research Center, University of Chicago Library*

Barnard was now the principal bread winner for the family—for himself, his ailing mother and feeble-minded brother. When he was about eleven, they moved to a poor area of South Nashville along the river known as "Varmint Town," because of the hardness and toughness of the people who lived there. It seems to have been located near the limestone bluffs along the Cumberland River, and adjacent to, or overlapping with, "Black Bottom," another area with a bad reputation which was physically at the lowest point in elevation and so highly prone to flooding (hence the name). Both areas had considerable African-American populations and were regarded as deplorable slums, with numerous saloons, brothels, and overcrowded dilapidated buildings. They were certainly a far cry from the white wedding-cake mansion in Brookline into which Percival Lowell was moving at the time. Despite the modesty (if that's the word) of the place in which he and his family were dwelling at the time, Barnard had to put in very long hours at the photograph gallery in order to pay the rent. He was presumably performing other tasks than just guiding the *Jupiter* camera as it was often dark when he made the long walk (about 1 ½ miles) home. He must often have been depressed and lonely during this sad time, but there was one ray of light, for on these walks he regularly encountered a man whose name was, as he later learned, Joseph S. Carels, and from whom he always counted on a word of encouragement or at least a smile:

> His smile lighted up my heart, for years he never ceased to greet me. Soon I learned to my awe that he was assistant Postmaster! Had he been President his position would not have appeared higher and more exalted to me—and that he should notice me and should stop to speak to me—I could not understand it and I cannot understand it to this day—unless it was indeed an inborn desire in him to sympathize with the friendless and wretched for friendless and wretched I was in those days if any one was ever friendless and wretched.[11]

Some of the early portraits of Barnard taken at the photograph gallery show him to have been very pale and thin, with dark rings under his eyes. He was often sickly, and developed an infection—probably a parotid gland abscess—on the right side of his face. For a long time, it refused to heal, and eventually left a permanent scar. Out of concern, Carels would always ask Barnard how that sore was coming along. To Barnard's niece Mary Calvert, this episode always "seemed eloquent of the loneliness of his life in those days."[12]

In addition to Carels's kindness, Barnard also derived some cheer during those lonely walks from the stars, his faithful, though still anonymous, friends. He later wrote:

> I often noticed in my long walks homeward in the early night an ordinary yellowish star which, to my surprise, seemed to be slowly moving eastward among the other stars. This attracted my attention because in all the time I had noticed the stars, though they came and went with the seasons, they seemed all to keep to their same relative positions. This one must be quite different from the others though it resembled them…. I watched it night after night and saw that though it moved eastward with reference to the other stars, it also partook of their general drift westward and was finally lost with them in the rays of the sun. In later years when I had become more familiar with astronomy it occurred to me to look up this moving star, and I found then that what had attracted my boyish attention was the wonderful ringed world of Saturn.[13]

A sickly looking Barnard, aged 12, probably about the time he encountered Joseph Carels on his long walks from the photograph gallery to the home in "Varmint Town." It was on these walks that he independently "discovered" the planet Saturn. (Credit: University of Chicago Photographic Archive [apf6-04132], Hanna Holborn Gray Special Collections Research Center, University of Chicago Library)

Another astronomically significant event occurred the following summer, on August 7, 1869, when he experienced his first solar eclipse. The path of totality swept across North America from Alaska to North Carolina. Though not total at Nashville, Barnard later wrote that "the sun was so nearly hidden that the spectacle presented some of the awe and sublimity of the total phase," and increased his "wonder at the phenomena of nature."[14]

The view to the east from Poole's photograph gallery at Union and Cherry, about 1871. This is a street young Barnard must often have walked along, including during the lonely nights when he walked home through Black Bottom to Varmint Town and was cheered by the friendly postmaster Joseph Carels. There were at least two other photograph galleries nearby (including Larcombe's shown here), so that competition for business was intense. Tennessee Virtual Archive

John van Stavoren was not a successful businessman. His business had failed in Cincinnati, and in Nashville he had continued to struggle until, in 1871, he gave up. His studio and equipment were purchased at a chancery court sale by a man named Rodney Poole, who renamed it "Poole's Photograph Gallery." The *Jupiter* camera was dismantled, but Barnard and the rest of Van Stavoren's employees were retained. Not only did Poole continue Van Stavoren's line of life size portraits in ink or crayon, and pastel and water color work, but he introduced a new department specializing in outdoor photography.

Meanwhile, the Barnards moved again, to a place on the Franklin Pike near Forts Moulton and Negley. The forts, along with others around Nashville, had been built largely by black laborers who had fled to them for refuge, and after the end of the War, they had remained as African-American districts. One of these, called New Bethel, was just beyond where the Barnards were living, and when in 1873 Nashville suffered another cholera epidemic, he noted that "the cholera made dreadful havoc among the colored people."[15]

Photographer's assistant Barnard in 1873, when he and his mother lived on the Franklin Pike near Forts Moulton and Negley. (Credit: University of Chicago Photographic Archive [apf6-04131r], Hanna Holborn Gray Special Collections Research Center, University of Chicago Library)

At this time Barnard's chief ambition was to become a sign painter, though a little later he formed a higher ambition—to become an artist. He was completely self-taught but his mother, the modeler of flowers in wax, had been artistic, and she must have encouraged his efforts. His dreams of becoming an artist were short-lived, however, as in 1875, Poole hired a Yorkshire immigrant, Peter Ross Calvert, as a retoucher and colorist, and on seeing his skill, Barnard immediately realized that he would never be more than a fair amateur.

Fort Negley was two miles from the photograph gallery, and in order to be closer to work, he took a room by himself on the top story of the St. Charles Hotel (N.B. Hamilton proprietor), at 35 North Market Street (now 2nd Avenue North). It was presumably not a hotel in the proper sense, but more of a boarding house, and seems to have been chosen both for its location and its affordability. Here one of the most dramatic events of the now-19-year-old Barnard's life took place, and one which, finally, was to give him a clear sense of direction. One late summer night, an acquaintance of his—perhaps one of the rough sort that he had known from Varmint Town, and whom Barnard knew as a "born thief"—dropped by his room asking for a loan. The thief, aware that Barnard was fond of books, left as a security for the return of the money a book (doubtless stolen). Reluctantly Barnard took it up, certain that he would never see his money again.

On examining it, Barnard found the book to be the second volume of an edition of the Rev. Thomas Dick's *Works*. Dick, the son of a small linen manufacturer in Scotland, had been trained in the same trade, but on seeing a brilliant meteor at age nine, he became passionate about astronomy, and henceforth read every book on the subject he could find even while working at his loom. Eventually he set out on a career in which he served briefly as a minister (from which he was dismissed for having an affair with a servant) and then as a teacher mainly in Dundee, Scotland, where he settled and began to devote himself largely to writing about the possibility of reconciling religion and science. On first peering into the volume, Barnard found that the first part consisted of sermons on covetousness, with horrible accounts given of the lives of misers. One of these was the Englishman Edward Nokes, who at his death was found to be possessed of the very large sum of between five and six thousand pounds. "In order to save the expense of sharing he would encourage the dirt to gather on his face to hide in some measure this defect. He never suffered his shirt to be washed in water; but, after wearing it till it was intolerably black, he used to wash it in wine to save the expense of soap. I must have groaned aloud when I saw this,"[16] Barnard recalled. Fortunately, the second part of the book contained more appealing fare than sermons on covetousness, including a number of the astronomical writings for which Dick was most famous, notably his *Celestial scenery, or, The Wonders of the planetary system displayed* (1838). Dick was a strong proponent at the time of the plurality of worlds, the belief that every planet of the Solar System is inhabited. (Rather optimistically, he computed that the Solar System contained 21,894,974,404,480 inhabitants!) Barnard probably read all of this carefully later on, but of greatest immediate interest were a few rudimentary star charts that Dick had included. As soon as he discovered them, Barnard, in great excitement, rushed to compare with what he could make out in the small patch of sky visible from the open window of his apartment. What happened next was that

> In less than an hour I had learned the names of a number of my old friends; for there was Vega and the stars in the Cross of Cygnus and Altair and others that I had known from childhood. This was my first intelligent glimpse into astronomy. It is to be hoped that my sins may be forgiven me for never having sought out the rightful owner of that book in all these years.[17]

This book, which he happened upon by chance, "awakened a thirst for astronomical knowledge which …never ceased to be controlling," wrote his closest friend from his later Lick Observatory days, Sherburne Wesley Burnham. "He had never seen either a telescope or an observatory. All he knew of the literature of the subject was found in this old book" [3, p. 193]. Barnard acknowledged this in a note he affixed to his copy of the book, which was later bequeathed to the library of the Yerkes Observatory: "This was the first book in the astronomical library of Edward E. Barnard into whose hands it accidentally came as an unsought pledge of a borrower in 1876. It gave him his first chart of the stars and was an important factor in determining his future scientific career."

Given all that he went on to achieve, Barnard had been a remarkably slow starter. At the same age, nineteen, at which Barnard was first learning the names of a few bright stars, Lowell was already reading about Laplace's Nebular Hypothesis and studying mathematics under Peirce. It was hardly, however, for lack of a sharp inquisitive mind; it had been only lack of opportunity that had held him back, and now that the opportunity had been given to him, he went forward like a shot.

Barnard in 1876. This shows him as he looked when he acquired (as surety of a loan to a thief) a volume of the Works of Thomas Dick, which included star charts, and began observing the phases of Venus with a small telescope. (Credit: University of Chicago Photographic Archive [apf6-04133r], Hanna Holborn Gray Special Collections Research Center, University of Chicago Library)

Braid Gives Barnard a Telescope

Barnard had no Peirce to mentor him, and always remained rather shaky in math, but he did find a remarkable mentor in science in a Scottish immigrant, James W. Braid. He had been hired in 1870 as van Stavoren's chief photographer and remained in the position after the gallery changed hands and Poole took over. Braid was keenly interested in science, especially electricity and electrical apparatus, and with the easy-going Poole's acquiescence he set up a small shop for his experiments in a back room of the photograph gallery. He was one of the first to fully recognize Barnard's potential, and he did everything he could develop it. For a while the two collaborated in running a "View Wagon," a traveling dark room for Outdoor Photography which roamed about Nashville looking for business. Braid was the operator who took the photographs while Barnard printed from the negatives.

Barnard, at center, poses with mentors J.W. Braid (with top hat at left) and Peter Ross Calvert (on right), in a photograph taken at Poole's Photograph Gallery about 1880. Barnard married Peter's sister Rhoda in 1881. (Credit: Public Domain, Special Collections and University Archives, Alexander Heard Library, Vanderbilt University)

As soon as he learned of Barnard's newfound enthusiasm for astronomy, Braid for a time turned his attention from induction coils, etc., to astronomical instruments, and instead got up for his young associate a rudimentary telescope fashioned from a simple tube and one-inch spyglass lens he had found lying in the street on a trip to the Union army barracks in north Nashville [4]. Though an extremely rudimentary affair, which Barnard later remembered as "a paper tube and lenses that looked as if they had been chipped out of a tumbler by an Indian in the days of the Mound Builders," it played its part, and according to Barnard, "it filled my soul with enthusiasm when I detected the larger lunar mountains and craters, and caught a glimpse of one of the moons of Jupiter."[18]

It must have been from about this time that a customer asked Poole how the business was going and received the reply, "With an operator crazy about electricity and a printer crazy about astronomy, I'm afraid my business is going to the devil!" [5, pp. 34–35]

A little later Braid, browsing in the shop of Charles Schott, Sr., a German-born instrument-maker, noticed the tube of an old ship's spyglass hanging on the hall. It had no lenses, and Schott had bought it for old brass and intended to cut it up. "I made an offer to buy it," Braid recalled, "and when I told him I wanted it for Ed. Barnard, he let me have it for two dollars."[19] Fitted up with a lens with an object glass of 2 1/2-inch aperture acquired from Queen & Co of Philadelphia, using an eyepiece made out of "the wreck of an old microscope" which gave a power of about 38x, and mounted on a good tripod which had formerly served as a surveyor's instruments stand, Barnard used it to make his first astronomical observations worthy of the name. He found Venus, approaching its greatest elongation east of the Sun and holding forth brilliantly in the evening sky, irresistible, and from the roof where the *Jupiter* camera had once stood, he set up his telescope and managed to succeeded in making out the planet's phase, which give it the appearance of a small half-moon. He later wrote that this "made a more profound and pleasing impression" than the celebrated discovery of the fifth satellite of Jupiter.[20] Burnham, too, would attest: "Such excitement comes but once in a lifetime, although the enthusiasm and interest in the subject may never be abated."[21]

Ed Barnard, with the View Wagon for Outdoor photography that he and James W. Braid carted around Nashville looking for customers. (Credit: Special Collections and University Archives, Jean and Alexander Heard Library, Vanderbilt University)

The Splendors of the Milky Way

After the discovery of the star charts in Dick's book and the acquisition of a small telescope, Barnard would never look back. Astronomy absorbed him completely. Even the 2 ¼-inch telescope—identical in aperture with the one Percival Lowell had received in gift from his mother—failed to satisfy for long, and he began saving up for a better one, practicing the strictest economy (though presumably not quite to the extreme of Edward Nokes) and receiving a generous advance on his salary. Thus, in the winter of 1876, he was able to fulfil his heart's desire and obtain an excellent 5-inch equatorially mounted refractor made by John Byrne of New York City. Byrne had listed it for $550, but as a favor to Braid—or perhaps because he was in desperate need of the money; despite his excellent skill as an optician he was an alcoholic and always teetering on the edge of bankruptcy—Byrne agreed to part with it for only

$380 (equivalent to about $3000 today). This was a full two-thirds of Barnard's annual salary at the photograph gallery at the time, and Barnard swallowed hard and "with the last cent I had in the world besides going heavily into debt to make up the requisite price,"[22] bought it. (It is certainly worth noting, by the way, that he thus equipped himself with a better telescope than Percival Lowell had at the time.)

Barnard began observing with it immediately, and recorded his first impressions—or ecstatic transports—on the first night in an undated typewritten sheet preserved in the Vanderbilt University Archives:

> The first clear night after receiving my large telescope, I sat out on the roof of a three-story house [presumably the St. Charles Hotel where he was living at the time] all night long, surrounded by ice and snow, the night being bitterly cold. After exploring the wonders of the moon until it sank from view beneath the western horizon my telescope sought the Milky Way. Here amid the splendors of that mighty zone of stars, I spent hour after hour sweeping among its marvellous fields of glittering suns, never wearying of the wonders constantly presented with each movement of the telescope, but gaining additional enthusiasm as the night drew apace. Nor did I forget the many double stars and clusters I had learned with my smaller instrument for they were each examined and I wondered at the beautiful contrasts of color in some of the binary systems and the myriads of stars revealed in the clusters that I had but dimly seen before with that small telescope. But from these lesser lights my telescope constantly swung back to the Milky Way, again to gaze on the "broad and ample road where dust is stars." So enraptured was I with these glimpses of the Creator's works that I heeded not the cold nor the loneliness of the night. And when the approaching dawn began to whiten the eastern skies, I sought out the great planet Jupiter, then only just emerging form the solar rays, and beheld with rapture his four bright moons and vast belt system. But when the dawn had paled each stellar fire the coldness of the night forcibly impressed itself upon me and I retired from the field of glory…

The skies over Nashville were pristine, and not only was the Milky Way brilliant, but the faint hazy light of the gegenschein—now known to be a reflection of sunlight from interplanetary dust—was visible. Though it had been seen before, Barnard, without any knowledge of earlier observations, independently discovered it, thinking it might be a large comet before learning of its true nature. The fact that it could be made out from downtown Nashville attests both to the darkness of the skies and to Barnard's exceptional eyesight.

Barnard's reference to Jupiter just emerging from the solar rays allows dating of the observation with some accuracy; the planet had just passed superior conjunction with the Sun on December 5–6, 1876, around which date it would have been lost in the solar glare, but thereafter as its separation from the Sun increased it would have rapidly become visible. Probably he would have found it fairly easy to see in the morning twilight by the end of December, and this must have been when this magnificent first observation was recorded.

"I Had a Good Cry"

Barnard was justly proud of his telescope, which was superior to any in Nashville with the sole exception of the 6-inch refractor of the recently founded Vanderbilt University. He sometimes set it up on Capitol Hill, where he enjoyed an unobstructed view, or on the rooftop of Poole's (Fourth and Union) where at times the roof of the building was so crowded with guests that he worried that his precious instrument might be accidentally knocked off. Once the guests departed, he began his own solitary vigils, and his colleagues would often come to work seeing him still working diligently at the telescope.

Barnard in 1877, at the time he met Simon Newcomb at the AAAS meeting at the state capitol in Nashville. (*Credit: University of Chicago Photographic Archive [apf6-04137], Hanna Holborn Gray Special Collections Research Center, University of Chicago Library*)

He and his telescope soon became celebrated around Nashville. In August 1877, when the American Association for the Advancement of Science (AAAS) held its meeting in the Capitol building—it was the first such meeting held in a southern city since the Civil War. He was encouraged by friends, including the City Treasurer Anson Nelson whose wife was an active member of the Baptist Church and had taught Barnard a Sunday School class, and Albers Roberts, editor of the Nashville *American*, to attend the meeting. Presiding over the great affair was Simon Newcomb, who like Barnard himself had risen from a humble background. A native of Nova Scotia, he studied arithmetic largely on his own, and had learned how to take the cube roots of numbers by the time he was six. As soon as he could, he left Nova Scotia in search of better opportunities elsewhere, moving to Salem, Massachusetts, where his father, an itinerant school teacher, was based, and taught country school in Maryland for 2 years while continuing intensive studies of mathematics, astronomy, political economy and religion. Eventually, in 1857, he obtained a position as a computer (a functionary in charge of calculations) at the Nautical Almanac Office, then in Cambridge, Massachusetts. The following year he enrolled at the Lawrence Academy of Science and received a Bachelor of Science degree. During this period, Newcomb studied mathematics with Peirce, and was often invited to the latter's home. In 1875, he was offered the Directorship of the Harvard College Observatory (a position that Peirce himself had once craved), but declined because it required him to make observations rather than concentrate on the theoretical work which was his forte. Shortly before the meeting in Nashville, he had been named Director of the Nautical Almanac Office, now in Washington, D.C., where he pushed forward a massive project of re-investigating the motions of the Moon and planets. As a result, the precision of the tabulated data in the *American Ephemeris and Nautical Almanac* was greatly improved, and that same year, 1877, he received for his efforts the gold medal of the Royal Astronomical Society.

Newcomb was thus perhaps the most celebrated astronomer in America, and also had a reputation as an arrogant, intimidating figure, "more feared than liked" [6]. It was with this grandee of astronomy that Barnard, securing an introduction from Nelson and Roberts, who had attested to him the young man's "ardent and faithful work" with the telescope, that Barnard spent about 20 min in a side room of the Capitol. Barnard asked how a young man with a small telescope might make himself useful to astronomy. Though Newcomb recalled many years later that he "did not for a moment suppose that there was a reasonable probability of the young man doing anything better than amuse himself," but feeling it his duty to encourage him in some way, he suggested that

> there was only one thing open to an astronomical observer situated as he was, and that was the discovery of comets. I had never even looked for a comet myself, and knew little about the methods of exploring the heavens for one, except what had been told me by H.P. Tuttle [a noted observer of comets at Harvard]. But I gave him the best directions I could, and we parted [7].

Barnard, however, remembered an additional detail. The great man had also asked Barnard whether he knew any mathematics. Barnard answered, honestly, that he did not, whereupon Newcomb fulminated that if he ever hoped to succeed in astronomy he ought to "lay aside that telescope and master mathematics."[23] Barnard, of course, was crushed, and as Peter Calvert reported, "Barnard was much depressed at the unsympathetic words of the great man and after the interview, as he confessed to Mr. Braid, he got behind one of the columns and had a good cry" [8]. Another account of the meeting added that Barnard's tears were "because he had been commanded by one whom he reverenced to part with the thing which meant everything to him and also in pursuit of his profession, to pursue a course which he might find it impossible to successfully follow."[24]

Barnard was nothing if not resilient, and instead of "nurturing his wound" he immediately rebounded into action. Possibly the very next day he hired out of his meager purse a mathematics tutor, and for good measure also began to take lessons in French, burning the candle late into the night in these studies rather than the (for him) far more congenial occupation of viewing the heavens with his telescope. He continued in this course for a while, but soon decided that Newcomb's injunction had been too severe and decided on a compromise—he would devote all clear and moonless nights to his "ardent and useful work with the telescope," the rest to the study of mathematics.

Aside from Newcomb's visit, a main highlight of the AAAS meeting had been a demonstration by the famous Scottish inventor Alexander Graham Bell of the recently invented telephone. Bell set up a telephone line between Polk Place and the A.G. Adams at the adjoining corner of Seventh Avenue and Union Street (three blocks from the photograph gallery); Mrs. James K. Polk, widow of the late president of the United States, successfully received the message. Soon afterward, Braid, whose father had known Bell personally in Scotland, set up his own telephone line from the photograph gallery and the Hotel St. Cloud. This was the prelude to a far more ambitious attempt by Braid and a young friend, James Ross, which involved talking by telephone by utilizing the telephone lines between Nashville and Bowling Green, a distance of seventy miles. Barnard was a witness to the experiment, and noted that "for an hour conversations were carried on, every word being clearly heard… Ross [at Bowling Green] played on the guitar and sang. It was a triumph in every way. Up to that time long distance conversation (if I remember correctly) had been possible over only one or two miles, so that Mr. Braid's success was far ahead of the others" [9]. The newspapers the following morning gave an account of the experiment, and even Bell was impressed, and offered Braid a contract to establish the telephone service in Nashville. Braid, for whatever reason, declined, and the opportunity passed to Ross. His days at the photograph gallery were numbered, however, and within a year of the telephone experiment, he handed in his resignation to Poole and set up his own business which within a few years had become the largest business dealing in electrical equipment of its kind in the Southern United States. Among the firm's early accomplishments was to bring electric lighting to Nashville in 1879, thus laying the groundwork for the plague of light pollution that has now hidden from view all but a few of the brightest stars.[25] Braid and Barnard remained in contact in later years, and whenever Barnard was in Nashville to lecture on astronomy, Braid operated the projector with which Barnard showed lantern slides of his astronomical photographs.

The Calvert Brothers also did well in later years. In 1896 they bought Poole's business with another man named Sam Taylor; Taylor left the studio a year later, and the firm became "Calvert Bros." Ebenezer was remembered as a quiet, reflective individual, Peter as a genial extrovert who easily made friends. Devout Baptists, they never opened their studio on Sunday, and Ebenezer wouldn't even ride on a streetcar on the sabbath day. When Vanderbilt University's Chancellor James H. Kirkland requested that photographs of graduating classes be made on Sunday, the Calverts refused though it meant a considerable financial loss. The studio remained in the Calvert family until 1964, and 10,000 of their glass plate negatives were donated to the Tennessee State Library and Archives.[26]

The Tennessee State Capitol building in the distance, where Barnard met Simon Newcomb. The Baptist church is shown in the foreground. (Credit: Tennessee Virtual Archive)

Simon Newcomb glares from a photograph taken about 1905, some 30 years after he met Barnard in Nashville. The years did not soften him, though he came to appreciate the magnificent achievements of the green and untested enthusiast who had approached him at the State Capitol. (Credit: Public Domain, Library of Congress Prints and Photographs division)

From a photograph taken in 1877, showing Barnard—with a new, steelier expression in his face, and recovered from the discouraging advice of Simon Newcomb—began his intensive studies of the night sky. (Credit: University of Chicago Photographic Archive [apf6-04140], Hanna Holborn Gray Special Collections Research Center, University of Chicago Library)

Observations of the Planets: Jupiter and Mars

Barnard was eager to establish his reputation as an astronomer, and began publishing some of his observations. His first publication was of the transit of Mercury of May 6, 1878, which he observed front of the Capitol Building. Until then he had, as Newcomb had expected, largely been about amusing himself, but now he made a concerted effort to do something that would in addition be regarded as "useful."

He stepped up the pace of his astronomical work the following year, taking a more serious and systematic approach to his work with the telescope which may, in part, have been owing to his attainment of a more stable and emotionally supportive personal situation. After several years of living alone in the St. Charles Hotel, he now began renting for $9.00 a week (worth about $220 today) a large brick "house with a mansard roof" at 1919 Patterson Avenue in west Nashville, near the recently founded Vanderbilt University. His invalid mother moved in with him. At about the same time his closest friend at Poole's, Peter Calvert, introduced him to Rhoda Calvert, Peter's older sister. The Calverts had come from Morley, some 5 miles southwest of Leeds, and famous for its textile industry—notably the cloth, shoddy, worn by both sides in the American Civil War—and as the birthplace of the future prime minister, H.H. Asquith. Their father was a bookkeeper and their mother a dressmaker. Peter was the first to emigrate to the United States, in 1873, followed soon after by Ebenezer, also hired as an artist at Poole's, and Rhoda. Rhoda was 13 years older than Barnard, but the difference in age did not seem important; they were soon spending much of their time together, and she proved invaluable to him by lending him assistance with the necessary but time-consuming tasks of keeping the household running. In addition, she relieved him of the responsibility of caring for his mother, sparing him precious time which was not spent as a photographer for his astronomy. It must have been clear that Elizabeth's health, which had never been good, was failing fast, and that she would not have much longer to live. Perhaps realizing this, Barnard was, if only unconsciously, looking as much for a mother-substitute as a wife. In Rhoda, he found that person.

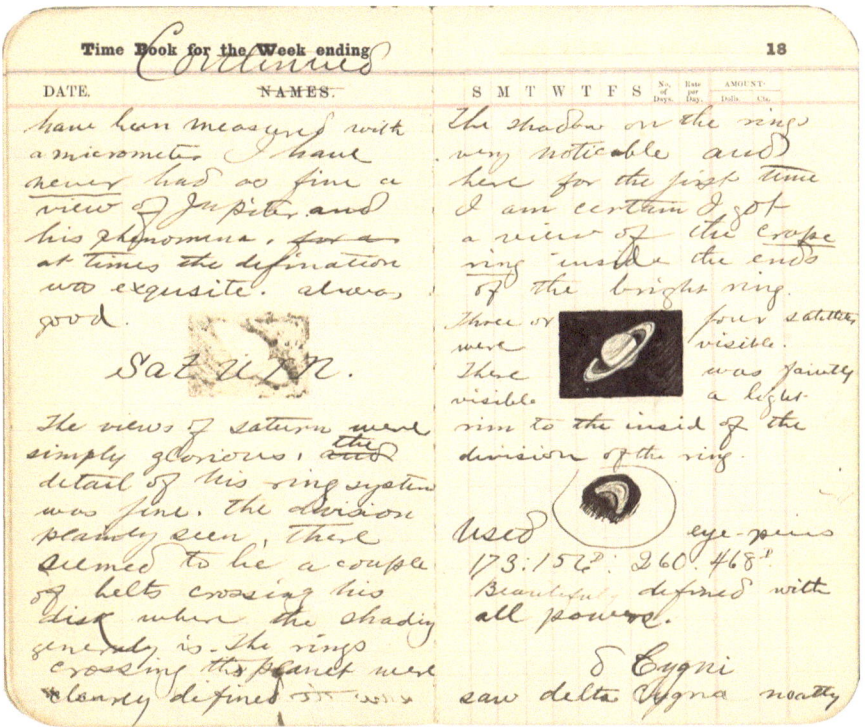

A page from Barnard's observing log book from 1881, showing a sketch of Saturn and rings as they appeared in the 5-inch Byrne refractor. He noted: "The views of Saturn were simply glorious." (Credit: Special Collections and University Archives, Jean and Alexander Heard Library, Vanderbilt University)

A rather dandyish looking Barnard, left, and his future brother-in-law Peter Ross Calvert, in a posed photograph taken at what was now Rodney Poole's photograph gallery, about 1880–81. This is from the time when Barnard was beginning to spend nights sweeping the skies for comets. (Credit: University of Chicago Photographic Archive [apf6-0414], Hanna Holborn Gray Special Collections Research Center, University of Chicago Library)

Group photo. In the row above, Peter Calvert, his wife, and an unidentified woman, and below, Rhoda Calvert and Barnard. They were married in January 1881. (Photo credit: Pubic Domain, William Sheehan collection; processing by Joel Hagen)

Their plans took time to mature, but meanwhile Barnard had become enraptured with the ever-changing cloud patterns on Jupiter. The famous Great Red Spot had just come into prominence, and in 1879 he began to observe and sketch the intricate details visible on the magnificent planet night after night, usually continuing until 2 a.m. With only one neighbor near, the loneliness of the place, he noted, "oftentimes impressed me with a kind of dread, for I was out at all hours of the night." Once, on pausing to rest his eyes for a moment from the telescope, he was, he continued,

… horrified to see two glaring, greenish-red balls of fire a few feet away in the obscurity of the bushes. Cold chills played up and down my backbone, and I was too frightened to move. These balls of fire came slowly toward me. A supernatural horror seemed to bind me to the spot as a nightmare holds one in its pitiless grasp. I could not have moved if I had known the foul fiend himself was behind those lights. Just at the point when I felt that I must collapse, as the hateful lights came close to me, I felt the warm touch of the tongue of some animal licking my cold hand; and in the obscurity I saw a great dog wagging his tail in a friendly manner. The relief was tremendous, and a warm flow of blood seemed to infuse my veins. It proved to be a great and fierce bulldog belonging to my neighbor. This dog had always looked so savage, with his cruel teeth, that I had not attempted to cultivate his friendship. Why he had not attacked me in the darkness and what made him seek this friendship, I do not know; but from that time on he was my good friend and made it a habit to lie down near me every night I was out observing. His friendship was a blessing, for I no longer felt the nameless dread of the night with his powerful form at my feet.[27]

It may have been remembrance of Barnard at this time that inspired Alfred E. Howell, a nephew of his Sunday school teacher, to recall, "Many the time I have noted his hollow eyes and faded cheeks and wist not that he was to be world famous from his vigils of the night, after an all-day's work at Poole's photograph gallery, his hands still stained with the chemicals" [10].

The large brick "house with the mansard roof," at 1919 Patterson Avenue in west Nashville, where Barnard moved with his mother. After his marriage to Rhoda Calvert, sister of his friend Peter Calvert at the photograph gallery in January 1881, she moved in with them. In this view, Elizabeth and Rhoda are seen seated on the front porch. The twin towers of Old Main (now Kirkland Hall) of Vanderbilt University looms in the distance, with Old Central to the left. In the front yard of this house, Barnard observed Jupiter in 1879–80 and discovered his first comet in May 1881. (Credit: William Sheehan Collection)

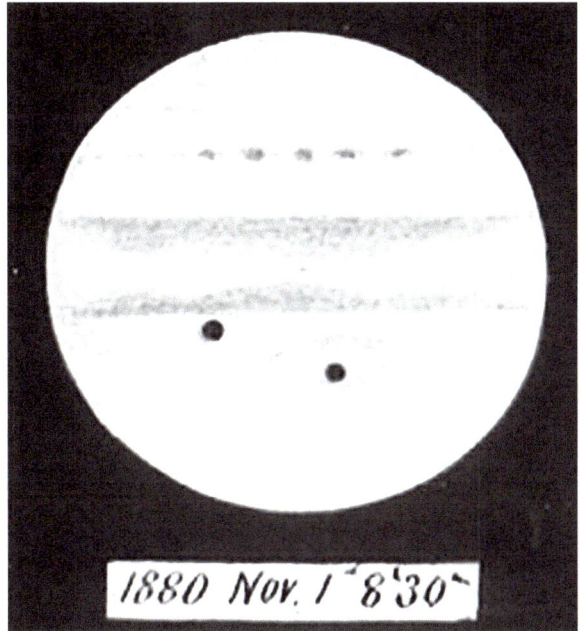

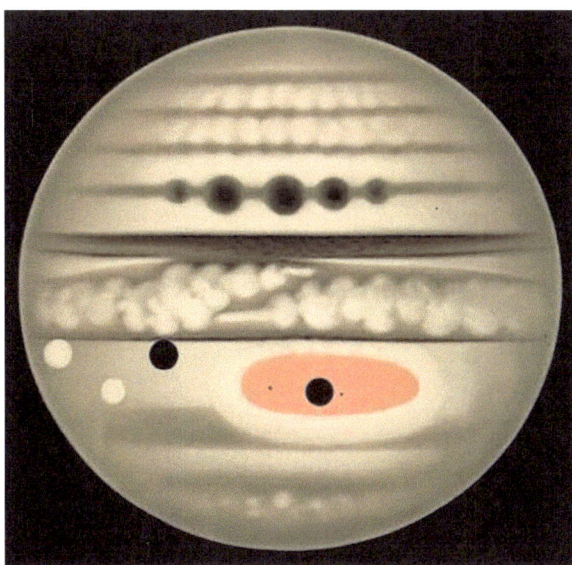

Near simultaneous observations. Above, E.E. Barnard, with the 5-inch Byrne refractor, 173x, on November 1, 1880. Below, a chromolithograph by the famous astronomical artist E.L. Trouvelot, same date, and time using the 15-inch refractor of the Harvard College Observatory. The cumulus cloud appearance especially in the Equatorial Zone shown by Trouvelot is a stylization, and Barnard has more accurately rendered the cigarform shape of the Great Red Spot, which was very prominent at the time. Trouvelot chromolithograph courtesy New York Public Library

After Jupiter, Mars became the favorite object of Barnard's telescopic attentions, and the November 1879 opposition was a favorable one. With the 5-inch refractor, he noted, "lucid spots at, or near, the poles," and followed their changes, which marked them as "in all probability snow and ice." The dark markings generally had "definite outlines, and seemed to have a slight greenish tinge," suggesting they might be oceans and seas, while the rest of the surface seemed of a "brick-dust color."

These are quotations from a monograph on Mars and its recently discovered satellites that Barnard set out to write in 1880, as the planet was receding from the Earth. *Mars; His Moons and His Heavens* was an ambitious project he intended to publish by subscription from his friends and leading citizens of Nashville. Though Barnard produced a handwritten manuscript of 144 pages, he failed to publish it, but it contains much of interest. The Martian "canal" furor was still several years from its height. Nevertheless, there was already much discussion of the possibility of some kind of life on the planet, and Barnard suggested (humorously) that perhaps some of the "Martialites" might fly through the air, while others, like mermen and mermaids, might swim about in the ocean. However, the 22-year-old writer cautioned against "Rashly Drawn Conclusions":

> … it is well to fetter the wings of our fancy and restrain its flights. It is quite possible we may have formed entirely erroneous ideas of what we actually see. The greenish grey patches may not be seas at all, nor the ruddy continents, solid land. Neither may the obscuring patches be clouds of vapor. Man is too quick at forming conclusions. Let him but indistinctly see a thing, or even be undecided as to whether he does actually see it and he will then and there set himself to theorizing, and build immense castles of conjecture on a foundation, of whose existence he is by no means certain…[28]

The House that Comets Built

It seems likely that Barnard wrote this in the hope of earning money. If this had been the goal, it failed. At this time, he and Rhoda had already made plans for their marriage, which took place on January 27, 1881. Though she had before this doubtless been a regular visitor to the household, they were to be a three-some until Elizabeth's death in 1884.

Meanwhile, in addition to whatever he was earning at the photograph gallery, a new scheme to improve his financial situation surfaced. A patent-medicine vendor named H.H. Warner of Rochester, New York, who had made a fortune selling such nostrums as the "safe" kidney and liver cure pill, began offering a $200 prize (equivalent to about $6000 today) for the discovery of each new comet by an American. At long last Barnard had an incentive to follow Newcomb's advice to search for comets, and he was soon searching the skies systematically—sweeping sequentially across horizontal zones of the sky and then with the completion of each sweep raising the telescope slightly and sweeping another zone partially overlapping with the first—in search of the elusive but potentially lucrative quarry.

Almost at once he found something, which attests to the efficiency of his method and the depth of his familiarity with the night sky. On May 12, 1881, he came across a comet; however, he was unable to recover it the following night and a run of cloudy weather kept it out of reach. Warner had stipulated that any claims to the prize had to be submitted exclusively to Lewis Swift, another active comet-seeker at the time and director of an observatory Warner had established in Rochester. When Swift failed to confirm the comet, he communicated data concerning its position to half a dozen other astronomers, including Seth Carlo Chandler, Jr., a Boston actuary who worked part-time doing calculations at the Harvard College Observatory and William Robert Brooks of Phelps, New York. Alas, the comet eluded detection, and the prize was never claimed. Far from being discouraged, however, this near miss only caused Barnard to redouble his efforts, and in September he found another comet; this time it was confirmed, and Barnard claimed his first Warner prize. He would win it eight times in all before the prize was discontinued after Warner suffered huge financial losses in speculative mining investments and his patent-medicine empire collapsed.

Though overshadowed by his later achievements, especially in the photography of the Milky Way, Barnard's efforts as a discoverer of comets still inspire awe. Before he turned increasingly to other lines of work in the early 1890s, he established a record of discovering 16 new comets as well as recovering three periodic comets, good for third place among visual discoverers of all time and surpassed only by Jean Louis Pons of the Marseilles Observatory and Brooks.[29] Using these modern techniques, one of Barnard's comets—a periodic comet which he found on a plate taken in October 1892; it was the first ever discovered by photography but was long believed to have been irretrievably lost—was recovered in late 2008 and named 206P/Barnard-Boattini, bringing the number of his comet discoveries to 17.)

As soon as the money from the first Warner prize was in hand, the question arose of what to do with it. Barnard would recall:

After due deliberation it was decided that we would try to get a home of our own with it. I had always longed for such a home where one could plant trees and watch them grow up and call them our own. So we bought a lot with part of the money, which was on what was afterwards called Belmont Avenue, but which was not then even a road. It was hard to find the lot after it was bought, for it was out in the open common. The place was in the midst of a scattered settlement of negro shanties, where the negroes had 'squatted' after the war, though on a beautiful rising ground which I had selected in part because it gave me a clear horizon with my telescope.

After some saving and some borrowing, and mainly a mortgage on the lot, we built a little frame cottage where my mother, my wife, and I went to live. Those were happy days, though the struggle for a livelihood was a hard one, with working from early to late for a bare sustenance (and the hope of paying off the mortgage), and sitting up all the rest of the twenty-four hours, hunting for comets.[30]

"Comet House" (807 Belmont Avenue), which Barnard built in 1882 with money received for his comet discoveries. Rhoda is on the porch (at left) and his mother Elizabeth is seated in the chair. The twin towers of "Old Main" on the Vanderbilt campus are visible in the distance. After Barnard moved to Vanderbilt, Rhoda's younger brother Ebenezer Calvert acquired this house, and built an addition in order to accommodate his wife and four daughters (of whom the eldest, Mary Calvert, later became Barnard's assistant at Yerkes Observatory). The house survived until 1964, when it fell victim to the development of "Music Row" and was torn down to make way for a parking lot. (Credit: Special Collections and University Archives, Jean and Alexander Heard Library, Vanderbilt University)

Though he was far from exclusively concerned with comets and published many other observations in journals such as the *Sidereal Messenger*, published by W.W. Payne at the Goodsell Observatory in Northfield, Minnesota, it was his comet discoveries, inevitably, that made him famous. In order to encourage his work several prominent citizens in Nashville, led by Judge John M. Lea, resolved led to build an observatory, "where he might carry on his nightly researches with more convenience, and more safety to his health" [11]. The foundation stones were in process of being hauled into the front yard when Barnard himself appeared, and insisted that they be removed; though thanking the would-be-donors for their generosity, he feared that a sense of gratitude for the gift would tie him down and prevent him from taking whatever better and more advantageous position might develop, even if it should call him away from his native city, as in fact happened.

Soon afterward, on December 6, 1882, occurred a transit of Venus—a rare and much anticipated event not to be repeated for 122 years. Olin Landreth, the chair of the engineering department at Vanderbilt, and J.T. McGill, a fellow in the chemistry department, planned to observe the transit with the 6-inch Cooke refractor in the observatory that had just been built on the campus, and invited Barnard to make observations there with the 5-inch Byrne refractor. Though clouds interfered with the observation of the first and second contacts—when the limbs of Venus are just touching that of the Sun—the sky cleared in time for the third and fourth contacts, and Barnard's observations compared favorably with those of the other two. Moreover, he published descriptions of his observations in the *Sidereal Messenger* and the German journal *Astronomische Nachrichten*, which were among the best anywhere. With this fresh demonstration of Barnard's skill, Landreth decided to pitch Barnard for a role in operating the campus observatory, which had been little used since it was built. The University Chancellor and chair of the physics and astronomy departments, Landon Cabell Garland, opposed the idea, however, on the grounds Barnard had had insufficient formal education. The matter was brought to the attention of the University president, Holland Nimmons McTyeire, who ultimately had the final say. He favored Barnard's appointment, and so on March 1, 1883, Barnard was offered a special fellowship in the University "connected with Astronomy."

McTyeire was a Bishop of the Methodist Episcopal Church, South. The son of a cotton planter and slaveholder in South Carolina, he always held fanatically strong pro-slavery view, even after, or perhaps especially after, his ordination.[31] He received his bachelor's degree from Randolph-Macon College in Virginia in 1844, where Garland was serving as president at the time, and later became a leading figure in movement within the Methodist Episcopal

Church to establish an institution of higher learning ("Central University") within the Methodist Episcopal Church. After his stint at Randolph-Macon, Garland spent 20 years at the University of Alabama in Tuscaloosa, but in the same week in April 1865 that General Joseph E. Johnston's surrendered the Southern command's Confederacy forces effectively ending the War, Union cavalry forces burned the University of Alabama campus to the ground. Garland spent a year trying to rebuild, then moved on to the University of Mississippi as chair of philosophy and astronomy. McTyeire was eager to win over the celebrated academic to his higher education cause, and Garland wrote essay after essay for church publications advocating for an institution to support an "educated ministry." For a while the effort stalled because of a lack of financial resources. However, in 1873, McTyeire visited New York City for medical treatment, and while there approached Cornelius ("Commodore") Vanderbilt, the millionaire railroad and shipping magnate, whose second wife and McTyeire's wife were cousins. Vanderbilt agreed to make two $500,000 gifts (the equivalent of over $10 million today) for the establishment of a university in Nashville, on condition that McTyeire would serve as chairman of the university's Board of Trust for life. The university was to be called Vanderbilt University, with McTyeire becoming the first president and Garland the first Chancellor. It opened its doors in September 1875, occupying a seventy-five-acre site, "West of the city, beautiful for situation, easy of approach, and of the same elevation as Capitol Hill, which is in view."[32]

Garland's plan for the new University was to have four departments: Biblical Studies and Literature, Science and Philosophy, Law, and Medical. During his days at the University of Alabama, he had been concerned about the lack of discipline among the students, and tried unsuccessfully to turn it into a military institution. Ever a keen student of Scottish moral philosophy, "Old Grey" or "Old Horsehead," as he was affectionately known to Vanderbilt students, always maintained that the development of character was the central purpose of a university. He also aspired, from the first, to strive for the highest standards of scholarship—and it had been wanting to uphold this standard that had led to his initial hesitation to approve Barnard's appointment.

Initially there were but two buildings on campus, one used by the Biblical Department and the other, Main Hall (later the Kirkland Building) for everything else. The next building on campus was the observatory; ground was broken in fall 1875, and the structure completed by the following spring. The main instrument was a 6-inch equatorially mounted refractor built by Thomas Cook and Sons of York, England, but other instruments included a meridian circle, a sidereal clock, and solar and stellar spectroscopes.

As a special fellow "connected with astronomy," Barnard was to take charge of the observatory as well as provide assistance to Garland in running the astronomy and physics departments. In return he would receive instruction in any of the non-professional schools free of charge, a salary of $500 per annum (equivalent to about $15,000 today), and the use of a house, close to the observatory, rent-free. Enrolled as a special student, he was exempted from the requirements of working toward a degree. Still spry at seventy-three, Garland, despite the heavy administrative duties of the Chancellorship—and a self-imposed commitment to preaching sermons to the student body in chapel every morning—still taught the first 2 years of coursework in physics and astronomy, using an English translation of a French text, Adolph Ganot's *Eléments de Physique*. Mechanics, waves, heat, electricity, and magnetism were covered in the first year, optics and astronomy in the second, and students were often asked to recite, "and even come forward to the front of the room and illustrate their replies with the apparatus at hand" [12]. Garland was an awe-inspiring figure in every way, and Barnard—who always referred to him as the "noble Dr. Garland"—seems to have looked up to him as a kind of father-figure, making up for the one he had never known. His full-time assistant was Charles Schott, Sr.—the same man who had sold Braid the brass spyglass tube that had been used to fashion Barnard's first telescope—who cleaned and repaired the apparatus, brought it to and from the lecture table from the nearby storage room, and was present at all the class meetings. Wearing small skull caps and long laboratory coats, Garland and Schott seemed to the students like "old alchemists."[33]

Since the purpose of Barnard's special fellowship was to develop his potential for astronomical work rather than to qualify for a degree, Barnard was allowed to register for classes weeks after a semester opened, and often took reduced course roads. In addition to the coursework in physics and astronomy on which he concentrated during his first 2 years, he studied mathematics under the chair in mathematics, William J. Vaughn, who had been one of Garland's students at the University of Alabama, and English literature (and a little bit of French and German) under W. M. Baskervill, McTyeire's son-in-law and a noted scholar of Anglo-Saxon and Middle English.[34] Though based on his later continuing struggles with math Barnard does not seem to have learned very much from Vaughn, he must have benefited greatly from Baskervill's instruction, for by the time he left Vanderbilt he had become a confident and effective writer whose style resembled that of Grant's *Memoirs* as described by Mark Twain: "clarity of statement, simplicity, unpretentiousness, manifest truthfulness, fairness and justice toward friend and foe alike, soldierly candor and frankness and soldierly avoidance of flowery speech." It was a far cry from the ornate and flowery style of Percival Lowell, and far more suited for scientific exposition.

Landon C. Garland. (Credit: Public Domain via Wikipedia Commons)

Barnard's salary was actually less than he was making at the photograph gallery at the time. Nevertheless, Barnard did not hesitate; at long last he resigned the position he had held so long, where he had made so many friends, and in addition to everything else, received a thorough apprenticeship in photographic techniques that he was to parlay to great advantage in astronomy a few years hence. He, his wife, and mother moved from "Comet House" (later acquired by Ebenezer "Ebby" Calvert) into the little house on campus, and he took firm command of the observatory with its 6-inch Cooke refractor. He favored the latter instrument whenever observations requiring stability and precision, such as measurements with a filar micrometer, which made up a great deal of a nineteenth-century astronomer's routine work and which Barnard wielded with great skill (though we will say little about his work in that field here). However, in addition, he continued his comet seeking, and for that he still favored the 5-inch Byrne refractor, which he set up under the open sky as he had always done. While sweeping with this instrument in the constellation Lupus on July 16, 1884, he encountered a comet: It was the first ever discovered at the Vanderbilt University Observatory.

The observatory on the Vanderbilt campus, where Barnard, on June 16, 1884, made the first comet discovery ever made at Vanderbilt University. He made the discovery with his 5-inch Byrne refractor, set up on the lawn in front of the building, rather than with the 6-inch Cooke refractor housed under the dome. (Credit: Special Collections and University Archives, Jean and Alexander Heard Library, Vanderbilt University)

In August 1884 Barnard ventured for the first time into the wider world outside Nashville, in order to attend the American Academy for the Advancement of Science's meeting in Philadelphia. It was a very different scene for him from that in Nashville 7 years earlier, when he had been a raw and untested youngster. He was now "known," and regarded as a steadily rising figure who promised great things. En route he paid homage to the Cincinnati, Allegheny, Washington, Harvard, Albany, and Princeton observatories, and spent a day both in New York City and Boston sightseeing. It was necessary, of course, for him to practice the strictest economy, and he generally avoided hotel bills by traveling whenever possible in day coaches at night. The meeting in Philadelphia was attended by many of the great astronomers of the day. Though suffering from a neuralgia (perhaps due to the rigors of his travels), he was unable to stay through an evening address by the Irish astronomer Robert Stawell Ball. However, on another day, he was present when an American astronomer who specialized in the mathematical theory of the

Moon was holding forth for what he thought was the basis for a correct theory of its motions. In the audience was the most distinguished guest at the meeting, the Cambridge University mathematical astronomer John Couch Adams, who had achieved immortality as the mathematical co-discoverer of Neptune (with U.J.J. Le Verrier of France). What happened next Barnard described in his usual humorous manner:

> Now this same problem has been more or less Adams' hobby, and it was interesting to watch the twitching of his fingers during the progress of the paper. When it was finally through, and criticisms were in order, he quickly rose and in the most rapid manner proceeded to show that the gentleman was altogether in error, and went through the tedious formula with the ease of a master, and showed here and there where certain assumptions were erroneous, and where a false step had been made. It was quite a picture to see him. He has a peculiar affect[a]tion, causing him, when excited, to breathe through his nose with a sound like escaping steam at almost every sentence especially was this noticed when he used a word that began with "s".[35]

On his return to Nashville after the meeting, Barnard arrived "thoroughly worn out with his travels; but, after having received a most friendly welcome and recognition from the astronomers whom he had at last met in person, he could henceforth feel that he was one of the fraternity."[36] Soon afterwards, on December 1, 1884, his mother died. He must have felt the loss deeply, but it can hardly have been unexpected, and for many years she had required care. His niece and later assistant Mary R. Calvert wrote of her that her influence "must have strongly affected his whole life, but the impression I have … is that he must have been almost 'on his own' from, say, his early teens."[37]

Though much of his work at Vanderbilt involved a continuation of the search for comets, he remained a careful observer of Jupiter, and published a particularly vivid account in the *Sidereal Messenger* of an observation made on April 1, 1886, of dark projections from what is now known as the South Equatorial Belt into the Equatorial Zone:

> At 12h 45m [UT], three of the dark projections ranged from the inner edge of the belt and just south of the equator. I noticed that from the summit of each there extended for a short distance in a following direction, a dusky streak, looking like smoke. I was strongly impressed with the resemblance to what might be called a silhouette view of three volcanic peaks, ranged in a line and vomiting smoke, which a strong wind was carrying eastward.[38]

Throughout the 4 years he spent at Vanderbilt, Barnard's efforts were unceasing, and his dedication to astronomy bordered on the fanatical. He had already formed the lifelong habit of nervously staying up waiting for every fragment of the night in which there was a patch of clear sky. Landreth had called him "Enthusiastically Energetic Barnard." Bishop McTyeire, alarmed at Barnard's tendency to overdo and worried that he was at risk of breaking down from the strain, wrote a letter in August 1886 in which he strongly advocated for a vacation of 2 or 3 weeks, and even longer if he found himself away, and doing well:

> You need rest; if not now—you need rest to get strength for the future. And, certainly, you deserve it; you are entitled to it.
>
> The Stars and comets will keep on their way, and be found in the right places when you return. Forget them for awhile. Don't look up, except to say your prayers, for the next month. Rest, *rest*.[39]

A Change in His Affairs

Barnard did take off, and at the end of August he and Landreth traveled to the AAAS meeting in Buffalo where he saw (but still did to dare to approach) Newcomb and met William Robert Brooks for the first time. The meeting with his main rival in comet discovery seems to have had a stimulating effect, since soon after his return to Nashville, on October 5, 1886, he discovered another comet—his sixth. It was later to develop into a fine object, attaining to naked-eye visibility in early November, and briefly, in mid-December, became a third-magnitude object with a 15-degree tail. On its retreat from the vicinity of the Sun it moved into the southern skies, and—since its orbit is hyperbolic—headed out of the Solar System, never to return. This never-again-to-be-seen comet was the best performer of all Barnard's many comets.

Instead of savoring this triumph, as many men would do, Barnard experienced doubts about the value of what he was doing. Writing to Edward C. Pickering, the Director of the Harvard College Observatory, he complained rather bitterly of the comet-seeker's lot:

> The discovery of a comet is not so slight a matter as one might think and the discoverer very seldom gets his just dues; the discovery of a single comet represents months and sometimes years of patient searching, at the risk of health from exposure to the biting cold of winter and the damp dewey [sic] nights of summer. I have many a cold bleak winter night searched ceaselessly from sunset

until dawn, almost frozen and completely worn out and nothing to show for my labors and not able to make up for any lost sleep in the day time, before I came here [to Vanderbilt] I had to work every day, and since being here I have had to attend the classes in the day and get my lessons as any other student would.[40]

The discovery of comets had seemed the most glorious work in the world to his friends in Nashville, not least because the custom of naming them after their discoverer conferred a kind of immortality. Nevertheless, he was now finding so many of them that the work came to seem almost routine. He discovered nine by the time he left Vanderbilt, more than any living person except the even more prolific Brooks. Meanwhile he reluctantly came to realize that the enthusiasm of the unsophisticated was not shared by professional astronomers, who tended to regard the mathematical computation of a comet as far more important than the mere discovery of one. Unfortunately, Barnard, with his irregular education—which made him strong in some subjects, such as photographic techniques, but missing entirely in others—struggled with math. He had never had the foundational instruction needed to advance very far in the more complex theoretical aspects (such as those Peirce was teaching to his more gifted students at Harvard). Eventually he would manage, through sheer brute force, to compute an orbit or two, but having done it just to show that he could, he didn't care to do it again. His observations were always first-class, but his reductions of them (which included various corrections such as for atmospheric refraction, precession, etc. and required no more than basic arithmetic and tables of logarithms) were, as Edward Singleton Holden would later complain, often wrong, so that "the computers learned to check [them] before using them." He was, Holden continued, "like Sir W. Herschel cut into two parts, and only the observing faculty left."[41]

As these somewhat depressing realizations began to dawn on him, Barnard began to doubt whether the path he was treading would ever lead to the attainment of what by now become his ultimate dream of getting a position at one of the great observatories. He was approaching thirty, so no longer so very young, and worried that he was using himself up without getting any closer to his goal. Though he could always remain at Vanderbilt, he had long since absorbed everything that Garland could teach him, while the modest equipment of the Vanderbilt University Observatory was useful for teaching the courses in practical astronomy with which he was charged but never allow for more than a pedestrian program of research. Then there was his personality. He did not have an overbearing nature, but was—as Mrs. Baskervill, the wife of William his English professor and his closest neighbor on the Vanderbilt campus, would remember— "simple, unassuming, free from all self-assertion and undue self-esteem, with a disposition so kindly and lovable that the charming simplicity of his character, his unselfishness, his modesty, and his genial nature was most thoroughly appreciated by those who knew him best."[42] These characteristics might be charming in everyday life but were not easily reconciled with his almost demonic drive for astronomy. In the end, the latter always won out, and his natural reticence and tendency to self-effacement was always ruthlessly pushed aside when it came to furthering his professional goals.

Regarding that, he had a clear idea who the important figures in astronomy were that might, potentially, have the ability to assist his reach for a professional position. He still avoided Newcomb, but he developed a comfortable enough relationship with Pickering to be able to vent about the lack of recognition of comet-seekers, as above; and he also kept his eye on Edward Singleton Holden, who for better or for worse was to have more influence on his career than almost anyone else.

Holden, born in St. Louis in 1846, was a Washington University (St. Louis) graduate who married the daughter of William Chauvenet, his professor of mathematics and astronomy, furthered his education at the U.S. Military Academy at West Point, and after a brief stint in the artillery, resigned from the Army to take position at the U.S. Naval Observatory in Washington, D.C. He arrived there in 1873, just as the observatory was putting into operation its 26-inch Clark refractor, then the largest in the world. Newcomb was in charge of it, and Holden was his assistant.

Just then, plans were being made for an even larger telescope, the dream of James Lick, an eccentric San Francisco piano builder and real estate millionaire, who wanted to build as a monument to himself a telescope "superior to and more powerful than any telescope yet made." In 1874, Newcomb was approached for advice about such a telescope by Darius O. Mills, president of the Bank of California and a member of the first Lick Board of Trust, and Holden, as Newcomb's assistant, was included in these discussions. After some debate about the merits of refractors vs. reflectors, it was finally decided that a refractor would be best, and that a 36-inch was the largest feasible size to attempt at the time. The contract was awarded to the firm of Alvan Clark & Sons of Cambridgeport, Massachusetts, whose founder and head, Alvan Clark, would not live to see it through to completion. The question of a director for the future Lick Observatory also came up. Newcomb did not hesitate to recommend Holden, despite the fact that the latter was only 28 years old and had not yet had time to distinguish himself. Newcomb's word at the time was law, and the trustees seem to have assumed that Holden was the probable director from this time forward. So did Holden, who henceforth made all of his career decisions with this in mind.

During his years at the Naval Observatory, Holden's most important contribution was his *Monograph on the Central Parts of the Orion Nebula*—a bibliographic tour-de-force published in 1878 in which he presented a detailed observational history of this object in an attempt to determine whether parts of the nebula might undergo slight changes in brightness. He also attempted to make some observations with the Clark telescope but he lacked the patience and willingness to tinker at the telescope until he got it right, which is a prerequisite of all great observers. Temperamentally he was always more drawn to library work—and in 1880 gladly gave up whatever was left of his duties at the telescope in order to concentrate on the, for him, far more congenial task of reorganizing the Naval Observatory's library. (For that matter, Newcomb, who was totally absorbed in mathematical studies of the motions of the Moon and planets, also disliked work at the telescope, and at the soonest opportunity had left the Naval Observatory for the position at the Nautical Almanac Office where he could spend all this time in his calculations.) The same year Holden spent reorganizing the Naval Observatory library, the directorship of the University of Wisconsin's Washburn Observatory became vacant due to the sudden death of its previous occupant (James Craig Watson). Holden was offered the job and accepted only on receiving assurances from Captain Richard S. Floyd, a former officer with the Confederate Navy during the Civil War who had been appointed president of what was by now the third incarnation of the Lick Board of Trust, that doing so would not interfere with his acceptance of the Lick directorship. Floyd gave these reassurances, and indeed, Holden continued to receive information and to dispense advice as needed through a voluminous correspondence as progress on building the new observatory on Mt. Hamilton, a 4200-foot peak of the Coast Range, went forward apace. By 1881, a winding road had been built up the mountain, and the top of the summit blasted away to create a level surface for establishment of an observatory housing a 12-inch Clark refractor that had formerly been owned by the physician and pioneering astrophotographer Dr. Henry Draper. This telescope was to stand in until the 36-inch refractor was in readiness, and was used that November by Holden and the Chicago court reporter and double-star observer S.W. Burnham to observe a transit of Mercury. The following December 1882, Holden and David Peck Todd of the Amherst Observatory were on the mountain observing and photographing the transit of Venus with a smaller telescope Todd had brought from Amherst. Then, in 1883, Holden and several others sailed to Caroline Island, in the Pacific between Hawaii and Tahiti, for a total solar eclipse, using the precious minutes of totality to search for a possible intra-mercurial planet sought for a number of years ever since its existence had been inferred on the basis of some calculations of the French mathematical astronomer U.J.J. Le Verrier; the planet, however, failed to make an appearance.

Edward Singleton Holden. This portrait was probably taken during the time he was still at the U.S. Naval Observatory in Washington, D.C. Public Domain via Wikipedia Commons

Clearly, Holden was well on his way to great opportunities, and Barnard took advantage of a shared interest in nebulae to initiate a correspondence after Holden returned from Caroline Island. He solicited information from the author of the massive *Monograph* about a suspected band of nebulosity west of the Orion nebula which he thought he had feebly made out during his comet sweeps. In contrast to the chatty and often humorous style of some of Barnard's other correspondents, such as Swift, Holden's tone was always dignified and formal, and he suggested that the answer Barnard was looking forward was likely to be sought in the library. He suggested Barnard consult drawings by Heinrich d'Arrest and Lord Rosse. He didn't quite leave it there, however, and on September 12, 1883, wrote again, telling Barnard, "I shall be very glad to look at the nebula and tell you what I see, if it will be any advantage to you."[43] On September 21, he repeated the offer: "I will look at the nebula myself later on and let you know what I find, if you wish."[44] Presumably Barnard "wished," but there is no evidence Holden ever tried to settle the matter by direct observation. He also begged off a request from Barnard for help in sorting out new nebulae he had picked up during his comet sweeps, excusing himself on the grounds that his copy of Sir John Herschel's *General Catalogue*, which provided the definitive list of nebulae known up to that time, had been lost on Caroline Island.[45]

The correspondence lapsed for a time until, at the end of August 1885, a new star suddenly appeared in the Andromeda nebula. In hindsight, it was one of the great astronomical events of the century, as the star—which at its brightest reached a magnitude 5.85, bringing it within naked-eye visibility—is now known to have been a supernova, and remains the only one ever observed in the Andromeda nebula. Barnard as usual made a series of valuable observations of the remarkable object, which he published in the *Sidereal Messenger*, in which, in addition to providing estimates of the star's brightness as it began gradually to fade, he called attention to "a well-defined but faint nebula" that he had noticed at one end of the elliptical hazy mass, wondering whether it might be new. Holden sent a postcard to Barnard in which he characteristically suggested that the only way to decide whether it was new or not would be to go back and study old drawings, and recommended consultation of a careful engraving made by G.P. Bond of the Harvard College Observatory in 1847. This showed that what Barnard had found was not new—his "nebula" was only a "condensation" of the great nebula, and had already been catalogued by Sir John Herschel.

Being back in touch, the two men continued their correspondence about the new star until it could be followed no more, and in the course of their correspondence, Holden casually let drop that he was soon to leave Wisconsin for California in order to take charge of the University of California at Berkeley, in order to be close at hand during the final stages of the Lick Observatory's preparation, before taking charge of the Observatory itself as soon as the great 36-inch Clark refractor had been installed there. Holden mentioned the need to assemble a staff, and had already begun considering Barnard for one of the positions and had begun soliciting confidential opinions of him from other astronomers. Barnard did not know this, but he did know that this was an excellent chance to try to butter up the future director, and replied effusively by return post:

I received your card this morning, and I am very greatly delighted to hear that you were going to California to take charge of that grand observatory. I am sure no man in the country is so well suited for that great charge as you are. I'm sure I feel that the University of California deserves the highest congratulations for its success in securing one so eminently fitted for the position. I am not surprised for I have all along said you would be the man for the Lick Observatory.

You have my heartiest wishes for your welfare and I hope your health may always be good so that you can enjoy the grand instrument which under your charge I know will reveal wonders that man hath never dreamed of. Please accept my humble but most sincere congratulations, and may you live long to honor our land.[46]

Holden responded warmly and in kind: "Thank you heartily for your letter of congratulation. I appreciate it—and you are right. There is going to be a grand chance on that mountain for every man who is there—there will be four of us."[47] By saying "us," Holden could at least be thought to leave open the possibility of Barnard being among that "happy few." The correspondence continued for a while in this manner with a particularly interesting note from Barnard emphasizing his skill in photography and aspirations for celestial photography—"no doubt the Astronomy of the future," Barnard would call it[48]—which Holden shared. In addition to exchanging portraits of themselves, Barnard, after determining the photographic focus of the 6-inch Cooke refractor—about 0.17 inch outside the visual focus—obtained some tolerably sharp pictures of the Moon, which he passed along to Holden.[49]

By the spring of 1886, James E. Keeler, a pioneering astrophysicist at the Allegheny Observatory, became the first staff astronomer hired for the Lick Observatory, and Holden informed Barnard of his intention to add "some of the younger men in the country" to come to Mt. Hamilton to take up the routine work of the Observatory (such as setting up a time service), as well as to "use the leisure which I shall be able to give them, for their own special researches."[50] Barnard saw the encouragement of the phrase "some of the younger men in the country," and informed Holden in a letter of May 7, 1886, that he was planning to give up comet-seeking and devote himself to other lines of work. His letter was as close to an explicit job application that the custom of those days permitted, and included the following background information—slightly exaggerated as to his precocity in photography:

Up to my coming here I had been a photographer from the time I was about 7 or 8 years old [sic], most of that time as a photo printer and later as an operator. So you see I am somewhat prepared for the difficulties that will occur in my photo experiments here…. I wished to let you know that I am trying to do the best I can and yet give sufficient time to my studies in the University which go sometimes rather hard with me as I had never previously gone to school except about two months before I went to work in a photograph gallery....

And he added, prophetically:

It seems to me that the Lick Observatory, with the proper instruments, would be the best place in the world to do wonders in Celestial photography.[51]

A Young Man Rises to a High Place (Mount Hamilton)

In the end, Barnard's campaign of blandishments and flattery succeeded. In March 1887 he received a formal invitation from Holden to join the staff of the Lick Observatory. The offer was for $1200 a year (equivalent to about $39,000 today), with rent-free living quarters on Mt. Hamilton, duties to commence in or about October 1887. The other members of Holden's staff would be, in addition to Keeler and Barnard, Sherburne Wesley Burnham and University of Michigan astronomer John Martin Schaeberle.

It was Barnard's dream come true, and though Vanderbilt, of course, was reluctant to lose its great man, Barnard did not hesitate to follow the precept, "Go West, young man!" He immediately resigned from his position as Instructor in practical astronomy at Vanderbilt University, and without staying for his degree, he and Rhoda made short work of disposing of all their household goods and on September 14, 1887, caught the train for San Francisco. They seem to have been utterly oblivious of the fact that—as Holden had tried to communicate—the Observatory was not yet anywhere close to being ready to receive them, and they ought to stay put until further notice. Though Barnard had traveled to the East for astronomical meetings, he had never been west of the Mississippi River, and as he always did, he kept a notebook carefully recording his impressions. The route first led to the small town of Mackenzie, Tennessee, where they switched railcars for Memphis, and passed that first afternoon through a country which, he wrote, was "parched and baked with the intense heat of the Sun." Clouds of dust rose as the train passed, "stifling the nostrils and filling the eyes with an impalpable powder," and Barnard wondered how people could bear to live in such conditions. They next passed through cotton fields, white with the ripe balls of cotton, in which pickers were busily gathering in the crop, "with that intolerable sun pelting down upon them its firey [sic] beams":

Barnard carried his account only as far as Denver, and left no surviving impressions of the rest of the journey via the great transcontinental railroad through Salt Lake City and on to San Francisco. Nor does there seem to be any documentation of what happened when he first presented himself to Holden in his offices at the Lick trust. Presumably, Barnard was devastated to learn that the observatory was still months away from completion, and though he had been more than willing to work without a salary, the possibility that he would not be allowed to take up residence on the mountain had apparently never crossed his mind. Now the unpleasant reality stared him in the face. He and Rhoda would rent a room in San Francisco, and after exhausting the money from the Warner prizes, he resorted to extreme measures to make ends meet—selling the 5-inch Byrne refractor for cash to the University of California. He must have been heartbroken. Even so, the Barnards's meager resources would have soon been depleted had he not found employment copying legal documents in the law office of John R. Jarboe, one of Holden's personal friends. Though Jarboe seems to have treated Barnard well, the job did not pay very well, and Barnard continued to dip steadily into his savings. The next few months must have been dreary indeed for someone who characteristically wrote in an almost illegible hand and was also a painfully slow typist—and it also must have been almost unbearable for someone who no longer could easily satisfy his addiction to astronomical observations. Though this time was barren astronomically, Barnard perhaps gleaned something from books on astronomy, mathematics, and navigation written by Holden's father-in-law William Chauvenet, especially his *Manual of Spherical and Practical Astronomy* (1863). Barnard also borrowed for a few nights a 6.4-inch refractor belonging to the well-known astronomer and geodesist George Davidson, and used it to search for some of Lewis Swift's latest nebulae.

During the long slog in San Francisco, Barnard struggled to keep from falling prey to depression, and received only second hand and sporadic information from Holden about the latest developments on Mt. Hamilton. The Observatory was still firmly in control of Captain Floyd, who as president of the Lick trust intended to maintain his grip on the Observatory until it had been safely and legally turned over to the Regents—or, as he put it, until he had safely "launch[ed] this Observatory into the Ocean of Science."[52] Except for Keeler, no other astronomer, not even Holden, was allowed on the mountain, according to Holden on the grounds that the trustees did not want to be "pestered."[53] During all this time of prolonged waiting, only once, in early December, did Barnard—as a tourist—manage a brief visit to his future abode nearer the stars. By then, the mounting and tube of the great telescope, having been carried up the mountain in parts by teams of horses, were being assembled in the giant dome.

At last, on New Year's Eve 1887 was the 36-inch lens finally removed from the safe on Mt. Hamilton in which it had been kept for a full year and installed in the tube of the giant telescope. On hand were Captain Floyd, president of the Lick Trust, Alvan Graham Clark, a "terrible old blow and grumbler," from the Clark optical firm in Cambridgeport, Massachusetts, which had made the lens, and Ambrose Swasey, from the Cleveland, Ohio mechanical engineering firm of Warner and Swasey which had made the cast-iron mount and dome. Keeler was also on hand. They hoped that, before the last of the Old Year expired, they could take a first peek through the great telescope that, as they believed, was sure to inaugurate a new era in science. However, bad weather intervened. A "rattling South Easter" brought strong winds, drifting clouds, and freezing cold, and the dome was found to be frozen solid in its track. With great effort, workmen managed to pry open the shutter, and the great telescope was swung toward the star Aldebaran. However, the four men found that they were unable to bring the image to a focus. The source of the problem was, however, correctible. It turned out that the Clarks had misstated the focal length of the lens by six inches, with the result that the tube had been made too long. Swasey simply cut off a six-inch segment from the end and remounted the lens. By the next clear night, January 7, 1888, the telescope was ready for another test.

On this night to be remembered, Floyd, Clark, Swasey, and Keeler gathered in the dome, joined this time by Floyd's niece (and Keeler's future wife) Cora Matthews. Since the dome was still frozen in place, the observers had to wait for various objects to transit the open shutter. First across was the brilliant white star Rigel—to everyone's relief, it came to a perfect focus. Next came the Orion nebula. In the Trapezium, the famed group of hot young stars embedded in the nebulosity, Clark discovered a faint star that had never been seen before. Finally, at just past midnight, after everyone else had gone to bed, Keeler captured Saturn—"beyond doubt the greatest telescopic spectacle ever beheld by man"[54]—in the giant telescope. With a magnification of 1000 X, he detected for the first time an incredibly fine division, "like a spider's thread," [14] in the outer, A ring of Saturn. There could no longer be any doubt as to the telescope's excellence. Three days later, the Regents formally approved Holden's recommendations for the members of the Lick staff, and Holden telegraphed Keeler, Burnham, Schaeberle, and Barnard word of their official appointments: "Regents appointed you to date from day of transfer of the observatory."[55] Also receiving an appointment was Charles B. Hill, who had worked with Davidson at the U.S. Coast Survey and was hired as assistant astronomer, secretary, and librarian. In the event, the transfer would not take place until June 1, 1888.

As usual, there was still a great deal of red tape to be got through. Before turning the Observatory over to the Regents, Floyd wanted to have a complete inventory of the property done. With Barnard still chomping at the bit in San Francisco, Floyd offered him the job, "for … the same compensation that [the trustees] allow to Professor Keeler, that is $75 per month and your board and lodging here."[56] Though the situation was far from ideal, Barnard was eager to get onto Mt. Hamilton any way he could, and jumped at it. Most of the time while he was working on this, Keeler thought, Barnard looked rather glum; possibly due to the generally grey weather, or missing Rhoda, who of course remained in San Francisco, or the sheer tedium of the work. But he brightened up on at least one occasion. On the evening of April 5, 1888, Keeler, who had been observing Mars with the 36-inch refractor, glimpsed the two satellites of Mars, and woke up Barnard to show them off. "He was delighted," Keeler wrote. "It was his first view through the telescope."[57]

The inventory—worked on at breakneck pace and completed in only 2 weeks—included not only grand items such as "The Main Observatory Building" and "1 36" equatorial complete in position," but also "One single chicken house near Cottage 2," "1 3 foot step ladder," and "1 ladle, 1 skimmer, 1 egg turner." It was finished by mid-April, when Barnard returned to San Francisco to prepare his final report (which would run to 98 pages) but the night after he arrived he was taken ill—both from overwork and from disgust at the frightful conditions of his quarters on Mount Hamilton, blame for which Holden put on the Lick Trust. Barnard, he wrote "has the choice of 2 sets [of quarters]—one with a kitchen and no furniture—the other with bedrooms and no kitchen—both filled with bedbugs!—This is simply a sample of the utter disregard of the needs of the Obsy.—All the work having been put on the Main building."[58]

Holden finally "got hold" on May 1, 1888, moving into one side of the so-called Astronomers' House, a three-story brick building located near the Main Building, while the next day Barnard and Rhoda—and also Elizabeth Calvert, who had come out from Tennessee in order to provide her older sister companionship and support—set up a household in one of the small wooden cottages previously used by workmen and located in the saddle below the Astronomers' House. One of the first orders of business was to have it fumigated so as to make it fit for human occupancy. In order to make it feel more like home, the Barnards planted a small garden outside with geraniums, violets, and mignonette. Then, without more ado, Barnard got back to work, recording a first entry in the observing book on May 9: "Field of 12 in. lowest power. Found 4 new nebulae within 1° of Castor."[59] Eight dreary months of near-total telescope deprivation were finally at an end, and Barnard—an "observaholic" if ever there was one—was once more headlong into severe sleep-deprivation as he stood vigil night after night waiting to use every scrap of clear sky.

One of S.W. Burnham's moody photographs Lick Observatory, taken about 1890. The Main Building is in the background, with the 36-inch refractor dome on the left and the 12-inch refractor dome on the right. In front of it is the Astronomers' House, where Holden lived; in the middle of the photograph, just right of the windmill and water tank and on the threshold of the short road up to the 36-inch refractor dome, is the cottage where Edward and Rhoda lived from 1888 to 1894. (Credit: Essays in Astronomy (New York: D. Appleton & Company, 1900))

"Wonders That Man Hath Never Dreamed of"

Barnard's career at Lick has been documented in detail elsewhere.[60] Though when he came to Lick he expected to work with the large telescope, as the junior man on the staff, he was assigned to the 12-inch, and was given no nights at all with the 36-inch, except under unusual circumstances. Holden, as Director, reserved two nights on the great telescope, despite having at first "no regular research program apart from observing and sketching planets or nebulae, sometimes looking at satellites, sometimes trying spectroscopy with Keeler assisting him, and sometimes making long series of measurements supposedly intended to calibrate various eyepieces or micrometers…." [16, p. 95].

Of course, as director, Holden had heavy administrative duties, and so perhaps could be allowed a certain latitude whenever on a clear night he "voted it cloudy" at 10 p.m. and went to bed. He worked very hard as a director, and of course was, as he had to be, a first-rate political string-puller who was excellent at cajoling the Berkeley higher-ups such as President Horace Davis and Secretary of the Regents John H.C. Bonté. According to historian Donald E. Osterbrock, Holden had "tremendous powers for reading, assimilating, and organizing information," but "little original creative power, or even research skill. He did not produce any new ideas, and nowhere showed that he could look at what he had done critically, and modify his approach to get new information."[61] There was no getting around the fact that as an astronomer he was a complete mediocrity, and this was not likely to be overlooked by the staff, given their own exceptional abilities for research. (Only Schaeberle, apart from Holden himself, proved to be a rather pedestrian researcher.) He had hired them all, to his credit; but once they were hired, he found it impossible to control them.

There were, of course, great challenges living and working on an isolated mountaintop, which he fully appreciated. Holden wrote to Bonté, "This is rather a queer place for labor. It is very isolated and not much fun going—and the men get weary, not having either interests or variety. Then too, we are more or less like the horses of these mountains! We get 'loco,' as they say."[62] The fact that the staff were much better astronomers than he was could not be hidden for long in such a place, and exposed his limitations. One of Holden's research efforts—taken over from Burnham—involved taking photographs of the Moon and planets with the 36-inch. Barnard was assigned to develop these plates, and as Osterbrock points out, "even an assistant far less critical … could not have failed to notice that they were often out of focus, poorly centered, and otherwise far from perfect."[63] In addition, Barnard, as a skilled, eager observer—whose baggage included significant issues of personal deprivation from his difficult childhood, a tendency to be shy and awkward overcome only by sheer strength of will, a lack of confidence owing to the limitations of his education but matched with incredible and relentless drive, a complete lack of balance in his life that saw him observing on every clear night except on the rare occasions he was absent from the mountain, anxiety, chronic insomnia especially during the summer—was bound to become enraged when, assigned to the 12-inch or other smaller telescopes, he noted that Holden had shut the 36-inch down early and left the coveted instrument unused the rest of the night. Such a situation would have proved difficult for anyone to endure, but it was impossible for someone like Barnard, whom Osterbrock calls "a highly neurotic individual," who could no more give up his dream of observing with the largest telescope in the country "than commit suicide."[64] It would be Barnard who would cause Holden more trouble than anyone else.

At first, there is no indication in the Lick Observatory archives of any difficulties between the two men, and despite not having regular access to the 36-inch, Barnard was incredibly productive. He made a vast number of routine observations with the 12-inch refractor to which he was assigned, which included the discovery of many new nebulae. Also, with a 5-inch "seeker," he added new comet discoveries to the growing list, including two faint comets found in 1888 and three in 1889, of which the first was discovered on April 1, just as Lowell was setting out for Noto. These discoveries, of course, helped enhance the Observatory's reputation with the general public. In addition— and in the very shadow of the great dome itself—he repurposed a cheap old portrait lens, with an aperture of 5.9 inches and a focal length of 31 inches, referred to as the Willard lens after a photographic stock dealer in New York City[65] and temporarily stored in a rough wooden box. As Barnard knew well, these large lenses had been used for making portraits during the wet-plate period of photography, their great size having been required to collect as much light as possible and thus to shorten the exposure times for portrait settings. Following the introduction of the much more sensitive dry plates they were no longer needed, and much smaller and less expensive lenses were substituted. Holden had acquired the lens for one-time use, in order to photograph the Sun at the January 1, 1889, total eclipse (whose path of totality conveniently passed across Northern California). After the eclipse, its official use ended but Barnard, with his deep knowledge of photography, recognized its potential for use in wide-angle photography of the Milky Way and comets, and so set out on the work for which he had been born and for which he will always be remembered. He opened the leaves of the Willard like those of a flower, and each plate exposed to the sky was "receptive, budding patiently under the eye of Apollo."

The best view Barnard ever had of Venus, on May 29, 1889, using the 12-inch refractor of Lick Observatory. Usually completely bland and featureless, on this occasion the air was thick with smoke from a wildfire burning in the canyon near Mt Hamilton, which stabilized the near-ground layers of air as well as acted as a neutral-density filter. From: E.E. Barnard (1897) "Physical and Micrometrical Observations of the Planet Venus Made at the Lick Observatory with the 12-inch and 36-inch Refractors," Astrophysical Journal, vol. 5, pp. 299–303. Plate 19

He could have had few expectations as to what would be revealed in the great structures of the Milky Way star clouds, so enfeebled by distance to the eye; nor had anyone using strictly visual methods been able to make out the faint details in the tails of comets well enough to be able to follow their subtle changes in form from night to night. Photography in his skilled hands would open up hitherto inaccessible wildernesses to exploration.

Barnard had initially attempted to photograph the star clouds of the Milky Way with a 1-inch, 9-inch focal length Voigtländer lens attached to a 6 ½-inch equatorial, but this small lens, though it recorded many stars, did not allow anything more than the most feeble impressions of the clouds. At this point

Barnard turned to the old Willard lens, whose great light ratio suggested it would be more successful in such experiments. The box with the lens was strapped firmly upon the tube of the 6 ½-inch equatorial, which was used as a guiding telescope. Though the telescope was equipped with a clock drive to follow the diurnal motions of the stars, because of the additional weight of the camera, constant adjustment was required, which Barnard supplied using slow-motion rods in right ascension and declination. Barnard found the work of guiding the telescope for hours in order to register details in the clouds of the Milky Way "very tedious work," but he had been used to it ever since he had first performed such tasks as a child on the *Jupiter* camera, and was sure that "the results more than pay for the trouble" [17, p. 15]. The plates used were 8 x 10 inch gelatin bromide dry plates manufactured by the M. A. Seed Dry Plate Company of St. Louis (founded by the British immigrant Miles Anscow Seed in 1883 and acquired by Eastman Kodak in 1902). These were the same plates that Lowell was using in Japan, and were universally regarded as the finest in the world at the time. Impatient of results, Barnard typically rushed to the darkroom in the morning of the following day, and as a first step produced negatives from the plates, with great anticipation as to what might be discovered on them. Often it was a *Eureka!* moment, as something surprising and unexpected was revealed. As he had done for many years as a printer at the photograph gallery, he also produced glass positives and silver prints of the better ones.

In this photograph taken by Burnham, Barnard is observing with the Willard lens, mounted in the wooden box, in its original configuration, using the 6 ½-inch equatorial with slow motion controls to guide it during long exposures. During colder weather, he wore the Esquimaux coat made of reindeer skin shown here and heavy rubber overshoes, and frequently, on bringing his eye to the telescope, noticed the discharge of a spark causing a slight shock between the eye and the eyepiece. Needless to say, this was very annoying, and he learned to frequently touch the metal of the instrument so as to produce a discharge from his finger instead of from his eye. He later recognized the fur coat and rubber overshoes were causing his body to become electrically charged like a Leyden jar, and that on approaching the eyepiece, the electricity was discharged from the eye into the telescope. (Credit: Special Collections and University Archives, Jean and Alexander Heard Library, Vanderbilt University)

His first trial of the Willard lens on the Milky Way was made on July 28, 1889, when he exposed a plate for 1 h 17 min on the Milky Way in Scutum. Though the region is one of the Milky Way's richest—the "Gem of the Milky Way," as Barnard himself called it—the exposure was too short to show it well. Two nights later he made an exposure of 1 h 30 min on Comet Davidson, a naked-eye comet which had been discovered a week earlier from Australia. It was his first comet photograph.

His real breakthrough occurred on August 1, when Barnard made another attempt on the Milky Way—this time specifically selecting a region in Sagittarius which possessed "the most intricate and complex structure of any portion of the Milky Way above our horizon" [18, p. 243] (that is, visible in the Northern Hemisphere). Years before, while sweeping for comets in Nashville, he had discovered a "small black hole" here, and had always been curious as to what it might be. He now placed the "black hole" near the center of the field, and using a guide scope whose cross wires were maintained fixed on a following star, completed an exposure of 3 h 7 min. He later wrote:

> This remarkable picture shows the cloud-like forms like waves of spray. A … curving lane [of darkness] runs from the lower left-hand portion of the picture … and curves gracefully upwards to the place of Jupiter. It is singularly like the stem of a great leaf. At the middle of the picture it is seen to pass behind some of the clouds of stars and emerge beyond, showing us clearly which part of the Milky Way at that point is nearest to us. Imagination may aid one, but it looks as if the lines of the cloud-forms, and of the stars and vacancies, all run more or less concentric with this extensive lane…. The black hole is seen slightly to the left of the center, with the small cluster [NGC 6520] as a white spot close to the left of it [19, pp. 311–312].

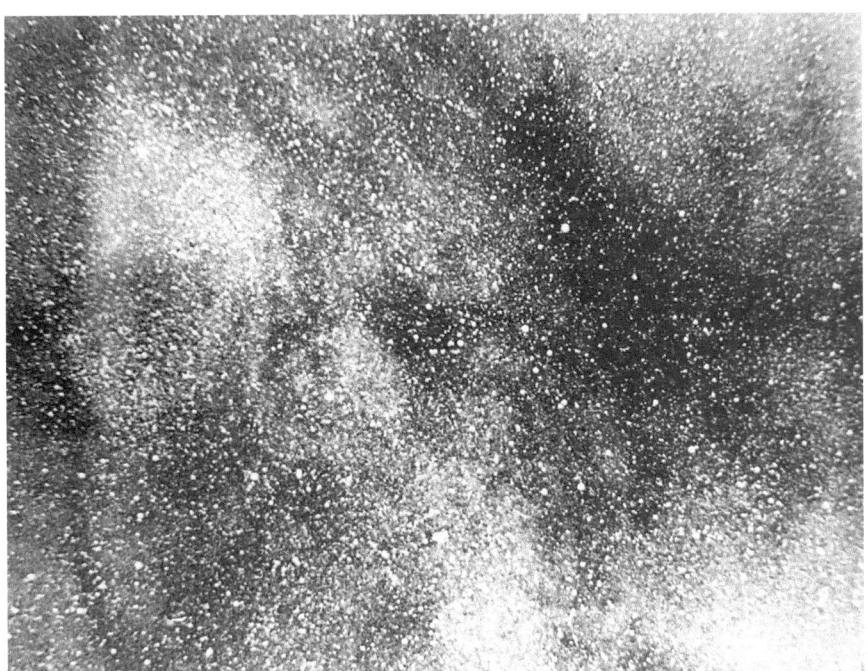

A print of Barnard's first long-exposure Milky Way photograph with the Willard lens, taken on August 1, 1889 and published in the journal Knowledge. *The plate is centered on a "black hole" Barnard had first seen while comet-sweeping in Nashville. (Credit: Historical Collections Project, Lick Observatory)*

The next night, Barnard made several more exposures on the Milky Way, centering on the open cluster M11 (also known as the "Wild Duck Cluster"), but they were not as spectacular as his plate of August 1, and meantime he became distracted by the remarkable phenomena being exhibited by the latest Comet Brooks, which his great rival—now in charge of the observatory of William Smith, a wealthy nurseryman, at Geneva, New York—had first captured in the early morning hours of July 7, 1889. The new comet proved to be moving around the Sun in an elliptical orbit with a period of about 7 years, but apart from its short orbit, it did not seem to be a very interesting object— it was rather small, and never came close to reaching naked-eye visibility. Nevertheless, Barnard picked it up in the 12-inch, and observed it carefully just as he had observed every other comet that had been visible above the horizon since the Lick Observatory had opened.

As he captured it in the eyepiece, he was surprised to find a faint nebulosity "seen close following and north of the nucleus; angle and distance measured. It is small and has a very faint nucleus. Another nebulous object [in the line between the other two] is small, round and faint."[66] The next morning Barnard returned to the telescope; he now satisfied himself that the two nebulous bodies were companion comets (which he labelled B and C), traveling in convoy with the main one (A), and took the trouble of waking up Holden to show him what he had discovered. As soon as the director took his turn at the eyepiece, he glimpsed yet a third nebulous object nearby, and before the night was over Barnard had found three more. Clearly, the comet had broken up. It was not the first time this had happened; it had occurred with Biela's Comet in 1845 and with the Great Comet of 1882, in the latter case as witnessed by Barnard himself from Nashville. As soon as a detailed history of the object had been calculated by Seth Carlo Chandler, Jr., in Boston, it was found that as far back as 1886, the comet had passed within only 0.0001 AU of Jupiter, for 2 days passing within the satellite system of Jupiter, just inside the orbit of Io. The strain of this passage had changed the comet's orbital period from 29.2 years to 7.1 years, and disrupted the internal cohesion of its structure. Thus, it appeared, as Barnard had suspected, the companions had not recently separated from the main body but had already been formed during their extreme passage by Jupiter.

For weeks, Barnard continued to follow the convoy of objects as they traveled through space, on August 4 discovering two new companions (D and E), with the 36-inch refractor. The companions disappeared after only a few days. The two main companions, B and C, separated rapidly from the main comet, until at the end of August, B became stationary with the respect to the main comet and underwent a remarkable physical change: "It enlarged rapidly, becoming extremely diffused and losing all appearance of central condensation" [20, p. 72]. Soon afterward it dissolved completely. Now only the main comet and companion C remained; Barnard's last observation of the latter was made on November 25, at which point it was no longer visible even in the 36-inch. As for the main comet, he was able to detect it with the 36-inch on November 21, 1890, with the aid of an ephemeris prepared by the noted German orbit-computer Adolf Berberich; it was then "the faintest and most difficult object that I have ever seen in the heavens" [21]. He continued to follow it in the 36-inch until January 13, 1891.

In addition to observing the comet, Barnard continued making exposures on the Milky Way, and—on the night of November 1, 1889—made one of the most celebrated observations of his career.

The Eclipse of Iapetus

As Keeler had noted on the first night of observations with the great refractor, Saturn is perhaps the most beautiful object in all the heavens. In addition to its rings, it has a numerous retinue of satellites, of which all the larger ones lie nearly in the ring-plane and so can never be eclipsed. The orbit of Iapetus, however, is inclined to the ring plane by some 15 degrees, and once every 15 years there is a several-week window of opportunity during which Iapetus may enter the rings' shadow. Such an event had been predicted for the night of November 1–2, 1889, by an alert English astronomer, Arthur Marth, who published a short notice about it in the June 1889 *Monthly Notices of the Royal Astronomical Society* and encouraged astronomers everywhere to try to observe it. Oddly enough, Marth's challenge seems to have been taken up only at Mt. Hamilton, where only the emergence of the satellite from the shadow of the globe and its subsequent passage into the shadows of the inner crape ring (the last of the main rings to be discovered, in 1850) and the bright middle Ring B were visible.

The night of November 1–2 happened to be one of those when Holden was scheduled on the 36-inch. As usual, he shut down early and went to bed and was presumably sleeping soundly when Saturn, then in the constellation Leo, rose at 12:50 a.m. Mt. Hamilton time. Barnard, however, was very wide awake in the 12-inch dome. As soon as Saturn emerged from the mists near the horizon, he saw at a glance that Iapetus was invisible, still deep in the shadow of the globe, but sketched in for later reference the positions of the other satellites—Titan, Tethys, Enceladus, Dione, and Rhea, the last on the following side of the planet. He kept a "sharp watch" on the approximate point where Iapetus's reappearance was expected, and just before 2:38 a.m., he suspected something, quite close to the positions of Tethys and Enceladus but still "faint and uncertain." Another half minute sufficed to convince him that what he was seeing was not a figment of the imagination. The satellite had reappeared from the shadow of the globe, and in the opening between the shadow of the globe and that of the crape ring was shining much more brightly than Enceladus, and indeed rivaling Tethys.

At this point Barnard had an inspiration. "The idea … occurred to me that it would be an excellent plan to test the effect of the shadow of the crape ring on the visibility of the satellite, by frequent comparisons of the light of Iapetus with that of Tethys and Enceladus."[67] In the heat of the moment he had improvised a plan to do eyeball photometry, mentally dividing the brightness-difference between Tethys (10th magnitude) and Enceladus (11 ½ magnitude) into ten equal parts, and using this scale to estimate the brightness of Iapetus as it underwent eclipse in the crape ring's shadow. The jerry-rigged method worked, and he was able to establish from a graph of his estimates—eighty in all—that "the Crape Ring is truly transparent—the sunlight sifting through it; that these particles cluster more and more thickly—or, in other words, the Crape Ring is denser as it approaches the bright rings" [22, p. 109]. Barnard wrote up accounts of his sensational observation for the *Publications of the Astronomical Pacific* (edited by Holden) and the *Monthly Notices of the Royal Astronomical Society* which Holden had submitted for publication and which included the statement, "Professor Holden requested me to observe this phenomenon with the 12-inch, as it would be out of the reach of the great telescope," [23] a statement that would continue to reverberate down through the ages in accounts of the observation by historians of the ringed planet.[68]

In fact, however, it was a fraud. Barnard later documented the true situation in a manuscript draft, "Charges against Holden," written in the late spring or early summer of 1892, which included the following statement that stood in complete and open contradiction to the above:

> Instance eclipse of Iapetus as showing how he neglects important work.—After 10 [o'clock] the big telescope lay idle the rest of the night while one of the most important obs[ervations] on record was passing unobserved except with the 12 in. [He] insisted that I state that it was out of reach of the 36 in.[69]

That same year, Barnard revealed the truth—and gave the lie to the *Monthly Notices* version—in an article for *Astronomy and Astro-Physics*, on whose editorial board he served, in which he stated unequivocally:

> For various reasons, no other observer in the world saw the eclipse of Iapetus.... [From Mt. Hamilton] the night was fine and clear and especially favorable. The planet rose at 12h 50m. At first the seeing was only ordinary, but it increased in excellence until in the latter stages of the observations it was superb. By the time the satellite had entered the shadow of the crape ring, the planet had attained a high altitude and was excellently place for observing [25, p. 121].

Saturn was not, then, "out of reach" of the 36-inch; Holden's attempted cover-up was exposed, and Barnard concluded his exposé by triumphantly—and not unjustly—claiming, "The observations of that eclipse with the 12-inch equatorial have given us more information about the crape ring of Saturn, perhaps, than could possibly have been obtained by a hundred years of ordinary observing."[70]

The Rebel

It seems that after the Iapetus eclipse, Barnard set his heart against Holden. Barnard saw himself as "a skilled, eager observer, denied the use of the great telescope by a bumbling, ineffective former Army officer, posing as a scientist! There was no justice in the world, and the young observer … began making his unhappiness known."[71]

Barnard's appeals to Holden to grant him use of the large telescope on a regular basis continued, but went unheeded, typically leading to lengthy sermons that every instrument in the observatory was for the sole object of advancing science and not for any individual's benefit or advantage. Holden seems to have been singularly lacking in psychological insight; he never glimpsed the possibility that by barring Barnard from the great telescope he was recreating the conditions of extreme deprivation of Barnard's early childhood days. In any case, the smooth, reasonable tone of his letters did nothing to appease the younger astronomer; instead they only served to provoke him, and led Barnard to embark on a campaign of fierce resistance against the director to which he eventually recruited Burnham, Charles B. Hill, and others, in which Holden would be referred to in their correspondence with such terms as "The Devil," "the Czar," "an unmitigated blackguard," "the Dictator," "Prince Holden," "the great I am," "the Great Mahatma," "that humbug," "that contemptible brute," "that immoral and incompetent man," and "our former colleague and fake."[72] This campaign eventually led to Holden's downfall, though not until after Barnard himself had left Mt. Hamilton.

Burnham was to become Barnard's chief ally in this righteous cause. He was not only Barnard's closest friend but also, 19 years his senior, probably something of a father figure. From the first, on the nights Burnham was assigned to the 36-inch and Barnard to the 12-inch, they took midnight lunches together, and would sometimes spend their days together wandering the canyons and wild landscapes around Mt. Hamilton, whose late-summer appearance Barnard would later compare to that of Mars as seen in the great telescope:

In the heat of summer everything becomes parched and dry on the mountain for the want of rain, and one feels frequently a sense of oppression in the eternal dryness and dust. Under these conditions we often went down into the canyons for relief, for a visit to these places, where spring seems never to cease, is, in the dry season, like a visit to fairy land. Here one finds the green foliage of trees, wild flowers and graceful ferns in abundance—among them the delicate maidenhair fern. Through these canyons always runs a cool crystal stream, over whose rocky margins and steep banks are spread masses of velvety green moss…. The north slopes of the canyons high above the streams, are generally covered with dense chaparral, while the south sides are more or less free of it, the bare spaces being covered with wild oats. It was easy to descend these southern slopes, say into Sulphur Creek Canyon, and one could take long running jumps of twenty feet or more on the soft soil and reach the bottom in about twenty minutes. That was the pleasure of it! But in climbing up the same slopes it sometimes took hours to return, for the wild oats and the soft soil made on slip back almost as fast as he came up. When the exhausted one finally reached the top he usually registered a vow never to descend into the canyon again. But always in a few days he was ready to go once more, for the call of the canyon was more than fascinating—it was irresistible [26, p. 318].

Vermont native Sherburne Wesley Burnham, while making his living as a court reporter in Chicago, became interested in the study of double stars and discovered 451 new pairs with a 6-inch refractor housed in a small dome (which neighbors referred to as the "Cheesebox") in the backyard of his home at 36th Street and Vincennes Avenue. This remarkable achievement led to his appointment as one of the original staff at Lick Observatory, where he continued to excel in double star observations and became E.E. Barnard's closest friend. This photograph, made from a somewhat deteriorated lantern slide, shows him at about the time the Observatory opened in 1888. (Credit:

University of Chicago Photographic Archive [apf6-04182], Hanna Holborn Gray Special Collections Research Center, University of Chicago Library)

The winters on Mt. Hamilton could be severe, and that of 1889–90 was of exceptional severity. By December the snow was piled so deep that the stage to San Jose was blocked for days at a time, and supplies and food had to be carried up the mountain on horseback; the months of January and February were even snowier, and during the blizzard of February 16 to 20, Mt. Hamilton was cut off from all communication with the outside world.

However, it was the summers and autumns that Barnard found intolerable. Nearly every night was clear, the weather hot and dry, and the mountaintop parched; he was up all night observing, but had difficulty sleeping in the daytime. (Rhoda must have required almost superhuman patience to keep him going at times.) Under these conditions Barnard's neurotic susceptibility produced a wide range of emotional responses to stress: irritability, anxiety, worry, fear, anger, and suspicion ranging to extreme paranoia. He reacted angrily to any perceived criticism from Holden, and in any statement from the director "consistently searched for the worst, and invariably interpreted anything less than fulsome, glowing praise as criticism."[73] Relations between the two men deteriorated to the point where Barnard refused to communicate at all with the director except formally, and in writing, so as to establish a documentary record of their interactions. And yet, in addition to the negative emotions of dealing with Holden, there were also positive emotions associated with making successful observations and discoveries. He must have experienced an "emotional roller coaster," and at times felt as if he were going stark-raving mad.

Barnard kept busy with observations using the 12-inch and (on occasion, by invitation) the 36-inch, and in 1890–91 made an outstanding series of observations of cloud features on Jupiter, the phenomena of the edgewise rings of Saturn, and of intriguing aspects of the satellites in various presentations. His observing reports are small masterpieces, and must have cost him enormous effort; they will always be valued by students of those objects.[74] Though fairly neglected during 1890 and 1891, the experiments with the Willard lens resumed in the spring of 1892, when Comet Swift, discovered by his old friend Lewis Swift in the first week of March, put on a fine show in the morning sky. Barnard followed it from the time of its discovery until the end of May. Perihelion passage occurred on April 6, when the comet was briefly visible with the naked eye and exhibiting extraordinary changes which were only discoverable by photography. As it turned out, Barnard had been scheduled some time before this date to give an evening lecture in San Jose, and determined to do his duty to the public, took the stage down the mountain at noon. However, at the same time he needed to do his duty to astronomy. He later recalled:

> Every morning's picture [of Swift's Comet] increased the interest and importance of the work…. I did not want to disappoint the people, and I certainly could not let the comet go by unphotographed. San Jose was nearly a mile below us in vertical height, and twenty-seven miles distant by stage road. The only possible way for me to secure the photograph and not disappoint my audience was to return to Mount Hamilton that night after the lecture. At ten o'clock I hired a horse and buggy in San Jose and drove up that lonely mountain road, the journey taking five hours, and arrived at the summit at three o'clock in the morning, in time to make the photograph of the comet…. In the main the journey was a most impressive one—alone in the mountains, with only the horse in front and my friends, the stars, above me. I doubt if my courage had not failed me entirely if the friendly stars had not encouraged me with their presence [31, pp. 283–284].[75]

As it turned out, the photograph he got that morning was, in Barnard's view, "the most successful of the series," and showed an abrupt bend at one point in the tail that had not been there the previous morning, "as if its current was deflected by some obstacle." In other places the tail reminded him of "crumpled silk" [32, pp. 387–388]. These photographs, together with others being taken at the same time by William H. Pickering from Arequipa, Peru, were the first to show the extremely rapid transformations that take place in the tails of comets, a subject which soon became yet another important area of Barnard's research.

The photographs of Comet Swift were the last taken with the Willard lens in its wooden box. After taking his pioneering Milky Way photographs in the summer of 1889, Barnard had noticed a marked deterioration in the images produced by the lens, which signified a need for it to be refigured, and for this purpose it was sent away to John A. Brashear's optical shop in Pittsburgh. On its return to Mt. Hamilton the reworked lens was to be mounted on its own equatorial and set up in a small observatory on Mt. Hamilton, using funds provided by Colonel C.F. Crocker, the vice president of the Southern Pacific Railroad Company and a member of the Board of Regents of the University of California. At Holden's request, Barnard had worked out plans for all this as early as February 1891, but much to the eager young astronomer's dismay—he later accused Holden of intentionally stalling his Milky Way work as a means of persecuting him[76]—the project was not finished until June

1892. Thus Barnard was unable to carry forward the Milky Way photography so promisingly begun for 3 years.

Barnard's design for the dome called for an unusually wide shutter needed to accommodate the telescope's 10- or 12-degree field of view and to make unnecessary too frequent episodes of rotating the dome during long exposures. It was set up near one of the observatory water tanks on the small outcrop known as "Huygens Peak," close to the cottage in which the Barnards were still living at the time (and continued to live until the summer of 1894, when they moved into a two-story brick residence on "Ptolemy Ridge"). Adjoining the small dome was a dark room, partitioned into two areas one of which was used for changing and filling the plate holders, the other for actually developing the plates.

The means of guiding was provided by a small telescope of 2.4-inch aperture fastened to the wooden box. Fine iron wires, coarse enough to be just visible, in black relief, on the dark sky, were inserted between the lenses of a negative eyepiece, and the star used for guiding thrown slightly out of focus to show behind the wires. The observer attempted to keep the illuminated quadrants behind the wire perfectly equal, something that, of course, required enormous patience and concentration during an exposure of several hours. Barnard's first plates with the Crocker telescope were trial exposures of 45 min made on June 17, 1892. They were too short to show very much, but three nights later, he obtained a 4 h 30 min exposure on the so-called small Sagittarius star cloud. It marked the beginning of a new era in his photography of the deep sky. Among the objects shown on this plate were two striking "black holes," from which ran "diverging semi-vacant lanes."[77] The more prominent of these black holes (B92 in his later catalog of these objects) was so sharply outlined on its eastern edge that Barnard thought it could hardly be more definite if drawn with a brush and India ink. On June 25, Barnard exposed a plate to the incredibly wild and dramatic landscape of the Milky Way near Theta Ophiuchi, which showed for the first time the very singular small S-shaped "vacancy" two degrees north of Theta (B72, the "Snake") and also the vast "rift" or "chasm" which runs southwest of the star. The latter, known as the "Pipe," is one of the greatest vacancies in the sky visible from the northern hemisphere. Other plates—Barnard exposed sixteen that summer, all dramatic improvements over the plates taken in the summer of 1889— brought out many other mysterious and hauntingly beautiful objects, revealed for the first time as Barnard removed them from the developing tray. On a plate exposed just inside the western edge of Sagittarius and about 8 degrees northeast of the Scorpion's tail, he discovered a sharply defined irregular black spot which reminded him of a "parrot's head," with a 10th magnitude star for the "eye" (the object was later catalogued as B97).

But he held his grandest description in reserve for plates exposed in later years to the very heart of the great Sagittarius star clouds. In describing them he achieved a height of eloquence few astronomers—or poets—have equalled before or since:

> These magnificent star clouds are the finest in the sky. They are full of splendid details. One necessarily fails in an attempt to describe this wonderful region of star masses. They are like the billowy clouds of a summer afternoon; strong on the side towards the Sun, and melting away into thin atmosphere on the other side. Forming abruptly at their western edge against a thinly star strewn space, these star clouds roll backwards into the general sky.[78]

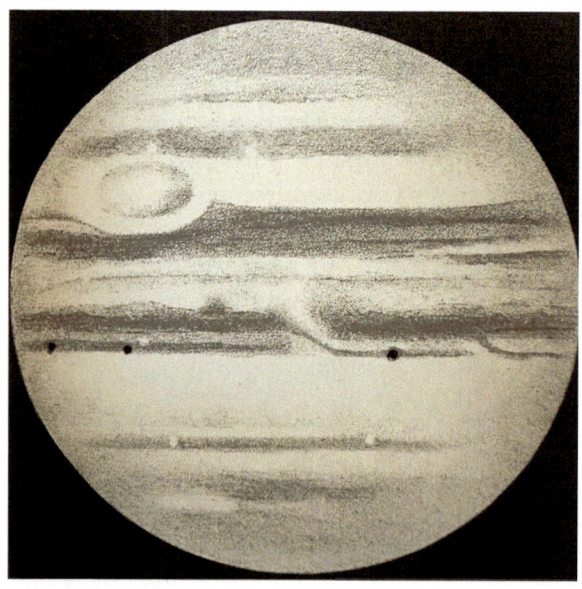

Jupiter, observed by Barnard with the Lick Observatory 12-inch refractor on July 30, 1890, showing some of the small dark spots on the northern border of the North Equatorial Belt and the white markings he called "horse-tails." Barnard's observations of Jupiter beginning in Nashville in 1879 convinced him that rather than being atmospheric in nature, the observed markings on the planet's surface attested to the surface existing in a "plastic or pasty condition," where the belts and markings were merely discolorations due to internal eruptions. (From: E.E. Barnard (1891) "Observations of the Planet Jupiter and his Satellites during 1890," Monthly Notices of the Royal Astronomical Society, vol. 51, pp. 543–558, Plate 14, fig. 1)

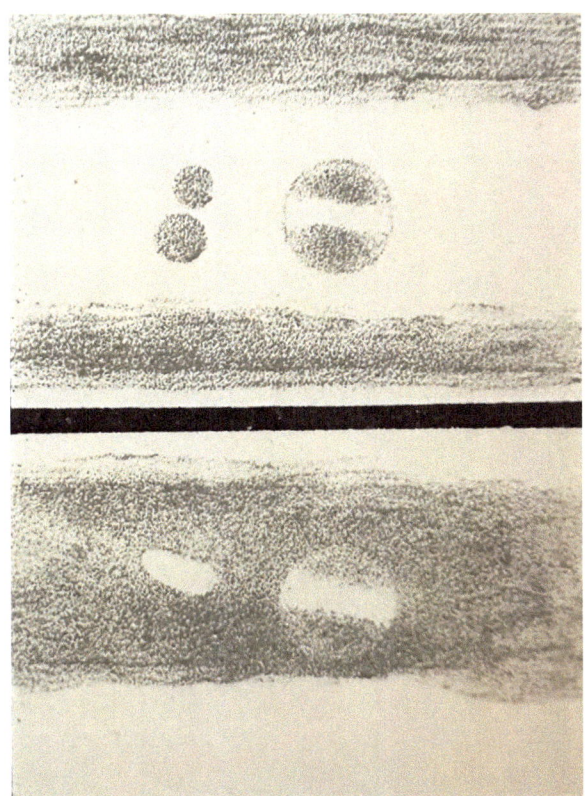

Observations of Jupiter's satellite Io, showing its appearance on September 8, 1890 (above) against a background of bright cloud and on August 3, 1891 (below) against a background of dark cloud, which led Barnard to infer (correctly, it turns out) that the satellite has a bright equatorial zone. (From: E.E. Barnard (1891) "Observations of the Planet Jupiter and his Satellites during 1890," Monthly Notices of the Royal Astronomical Society, vol. 51, pp. 543–558, Plate 14, fig. 5)

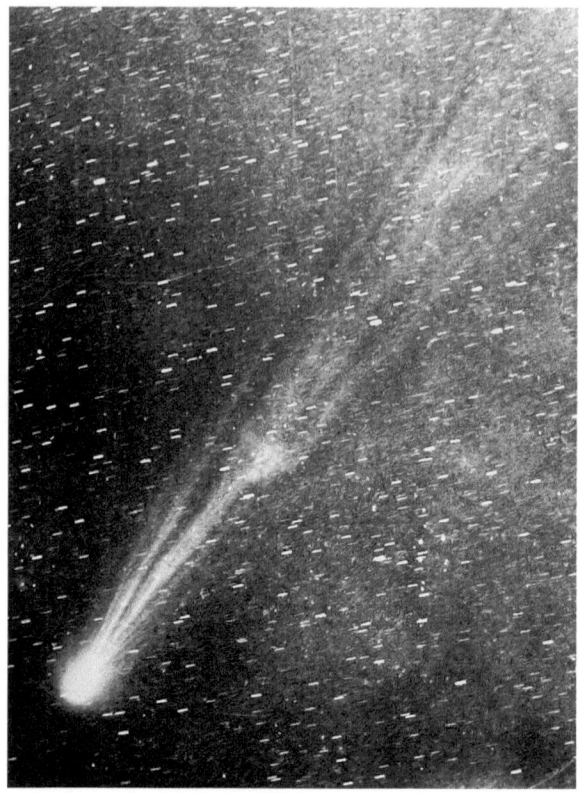

Comet Swift (1892 I), a 50-min exposure with the 6-inch Willard lens taken by Barnard on the morning of April 7, 1892, which shows the abrupt bend in the tail that made Barnard wonder whether "its current was deflected by some obstacle." (From: E.E. Barnard, "Milky Way and Comet Photographs," Plate 93)

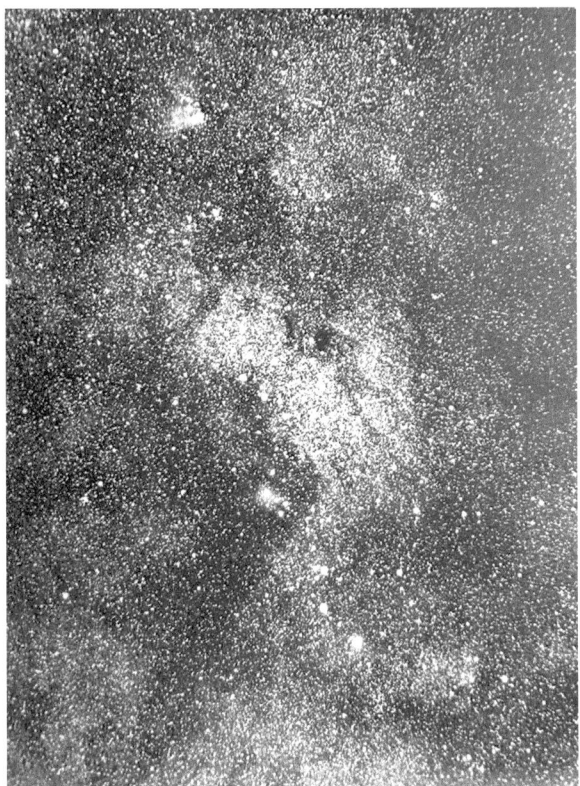

Small Star Cloud and Holes in Sagittarius. This 4 h 30 min exposure was made on June 20, 1892, on the area of sky centered on R.A. 18 h 7 mins (1855) and declination -18°.3. (From E.E. Barnard, "Milky Way and Comet Photographs," plate 55)

The Region North of Theta Ophiuchi. This exposure of 4 h 0 min was made on June 25, 1892, and shows the area of the sky centered at R.A. 17 h 12 min (1855) and declination -21°.3. (From: E.E. Barnard, "Milky Way and Comet Photographs," plate 39)

Greatness Thrust Upon Him

During this same productive summer of 1892, when Barnard was largely abandoning all pretense of getting sleep, his long-considered and thoroughly documented case against Holden regarding his use of the 36-inch refractor finally came to a head. He had decided to reach above the director's authority and appeal directly to the one body that could overrule him, the Board of Regents of the University of California. Barnard presented voluminous, if sometimes hysterical, documentation of his cause. The Regents sided with him, and so, beginning in August 1892, Barnard achieved his goal of obtaining regularly scheduled time—every Friday night—on the great telescope. Within a month, he was to be vindicated by making the most sensational, if not the most scientifically important, discovery made at Lick Observatory up to that time.

Though Mars had early been a subject of keen interest, he paid little attention to it for many years—a testament to just how difficult an object it is to observe. He made a few observations in the summer of 1892, including with the 36-inch, but the planet was unfavorably low in the sky and little was seen even in the great telescope. Early on Friday evening, September 9, 1892, he made a few routine observations including several micrometer positions of the Martian satellites, and then, as midnight approached, he swung the great refractor toward Jupiter, then in Pisces near the 5th magnitude star Mu Piscium.

As he later admitted, he was consciously searching for faint Jovian satellites. "It was simply with a desire to satisfy one's own eyes (by a personal search) that no other moon existed in the Jovian family…."[79] If such a body existed, Barnard—with the great telescope and the unrivaled sensitivity of eyesight— would be the one to find it, and so he was. Sliding the overpoweringly brilliant planet just outside of the field of the 520x eyepiece he was using, "At 12 o'clock as near as may be, to within a few minutes, I detected a tiny pint of light close following the planet and near the 3rd satellite [Ganymede] which was approaching transit. I immediately suspected it was an unknown satellite and at once began measuring is position-angle and distance from the 3rd satellite" [33, p. 749]. In his observing book that night, Barnard made a crude sketch of its position relative to Ganymede and Jupiter itself, indicating it as a tiny dot below which he wrote: "This must be a new satellite at elongation. It is apparently smaller [fainter] than either of the satellites of Mars."[80] He managed to get a few more measures and after it disappeared behind the planet waited for it to reappear, but without success; he finally gave up as the day was beginning to break.

Though sure that he had a new satellite in his grasp, Barnard nevertheless wanted to hold back any announcement until he had more measures in hand. The next night, Saturday, Schaeberle was assigned to the telescope. Though Barnard was able to successfully persuade him to let him have the telescope another night, it was visitors' night at the Observatory, and some 200 had ascended the summit in the hope of getting a glimpse of Mars near opposition. After the last of the throng had left, around 11:30 p.m., Barnard finally returned to his position at the eyepiece, and set to scanning the space on either side of the planet. There was nothing at first; but after a few minutes, he glimpsed the new satellite leaving the planet on the following side, and got a good set of measures of it at its eastern elongation. This would prove to be of critical importance in working out an orbit. The next afternoon, September 11, news of the discovery was telegraphed to the Harvard College Observatory, the Associated Press and United Press wire services, and from there went out to the rest of the world. The cable included an estimate of the satellite's distance from the center of the planet (112,400 miles), its period of revolution (erroneously given as 12 h 36 min; one of Barnard's errors of reduction that Holden loved to go on about) and the time of its observed eastern elongation. Continued measurements of the satellite's elongations—aided with a small piece of smoked mica to block out the planet's intense glare—soon led to improved calculations of its orbit.

Though Barnard was later to feel resentful that Holden failed to announce the discovery officially, it appears that Barnard himself by this time was so distrustful of the Director that he wanted the matter left in his own hands. In disregard of the usual protocol, he, not the Director, made the announcement. Regardless, the momentousness of what Barnard had done was felt immediately: for the first time since Galileo had discovered the four "Galilean" satellites in 1610, a new satellite of Jupiter had been found. The next day, Barnard might have said with Lord Byron, "I awoke one morning and found myself famous"—except Barnard, the "man who was never known to sleep," was probably already awake.

Barnard "selfie" taken with the 36-inch refractor, shortly after the discovery of the fifth satellite. The cable seen running from his hand down his sleeve was used to open the shutter of the camera lens. Note that Barnard's left leg trousers are stained, probably by chemicals used to develop photographs. (Credit: Special Collections and University Archives, Jean and Alexander Heard Library, Vanderbilt University)

Congratulations poured in from all over the world. He was honored with having a peak on the Sierra Crest named for him by W.L. Hunter, John Hunter, William Hunter and C. Mulholland, who made the first ascent a little over 2 weeks after Barnard's discovery. It has the distinction of being the highest "thirteener"—a peak between 13,000 and 13,999 feet—in the United States. He was named an Associate Fellow of the American Academy of Arts and Sciences, received the Lalande Prize of the French Academy of Sciences, and had conferred upon him an honorary D.Sc. degree from Vanderbilt, which remains the only honorary doctorate ever given by the school.

But the most touching communication came from Joseph S. Carels. "I have been getting telegrams and letters of congratulations …," he wrote, "but of all, yours brings back memories that are sad of struggles that were hard to bear sometimes and of hopes and disappointments that clustered thick about me in that newly built cottage away out on Bellmont [sic] Avenue.... I know you did your part in bringing me to public notice—in letting the people know that I was trying and wanted to do something." He listed all the friends he had had in Nashville—Albert Roberts, editor of the Nashville *American*, Anson and Fanny Nelson, Braid, the Calvert brothers, Rodney Poole, Judge John M. Lea, Bishop McTyeire, Chancellor Garland, Schott, Landreth, Vaughn, Baskervill. But it was Carels himself that moved Barnard to the greatest tribute. "Clinging to me through life has ever been the memory of [your] kind word and nod and smile of recognition to a poor sick ragged boy on his way to or from work.... This is not sentiment. It is plain and substantial reality."[81] Without forgetting those who had lent a helping hand along the way, that poor, sick, ragged boy had somehow, almost miraculously, managed to climb from a miserable existence, and now, suddenly, he stood at the pinnacle of the astronomical world.

Notes

1. Broadly speaking, this chapter follows the standard biography of Barnard: William Sheehan, *The Immortal Fire Within: the life and work of Edward Emerson Barnard* (Cambridge: Cambridge University Press).
2. E.E. Barnard, "Autobiographical Sketch"; Barnard. Edward Emerson. Papers, Special Collections and University Archives, Jean and Alexander Heard Library, Vanderbilt University.
3. His attitude was probably similar to that of Gallio, the Roman proconsul of Achaia, who in the year 51 dismissed an accusation brought against Paul the Apostle in Corinth (Acts xviii.17). As Rudyard Kipling phrase's Gallio's response in his poem "Gallio's Song," "I care for none of these things." That may well have been Barnard's attitude as well.
4. E.E. Barnard, "Autobiographical Sketch."
5. Cited in: Jake Wynn, "A Civil War Soldier Reflects on the Comet of 1861." https://emergingcivilwar.com/2017/10/25/a-civil-war-soldier-reflects-on-the-comet-of-1861/
6. Barnard, "Experiences," p. 275.
7. Barnard, "Autobiographical Sketch."
8. Barnard, "Autobiographical Sketch."
9. Ibid.
10. Ibid.
11. E.E. Barnard to Joseph S. Carels, September 21, 1892; Barnard. Edward Emerson. Papers, Special Collections and University Archives, Jean and Alexander Heard Library, Vanderbilt University.

12. Mary R. Calvert to Robert G. Aitken, November 11, 1927; Mary Lea Shane Archives of the Lick Observatory.
13. E.E. Barnard to John Trentwood Moore, Tennessee State Historical Commission, April or May 1921; Vanderbilt University Archives.
14. Barnard, "Autobiographical Sketch."
15. Ibid.
16. Barnard, "Experiences," p. 276.
17. Barnard, "Experiences," pp. 275–276.
18. E.E. Barnard, Nashville *Artisan*, August 10, 1883.
19. Braid, "First Employment," p. 9.
20. Newspaper clipping dated May 1, 1893 from the *Standard Union* of Brooklyn, New York; Vanderbilt University Archives.
21. Burnham, "Early Life," p. 194.
22. E.E. Barnard to Simon Newcomb, April 27, 1891; Barnard. Edward Emerson. Papers, Special Collections and University Archives, Jean and Alexander Heard Library, Vanderbilt University.
23. *Nashville Tennessean and the Nashville American*, April 30, 1915.
24. *Nashville Tennessean and the Nashville American*, April 30, 1915.
25. Braid Electric Co. still exists, though in 2004 it was acquired by Rexel, a global electrical equipment distributor based in France.
26. For information on the Calvert family, I am indebted to Barnard's great nephew James R. Calvert for taking the time to answer my inquiries. Also useful was an article based on recollections of Ross Calvert, son of Peter Calvert: Sara Sprott Morrow (1974) "A Portrait of the Calverts," *Nashville!* (August), pp. 40—56. After taking over Poole's and renaming it "Calvert Bros," it remained a family business for decades. All of Ebby's girls—Mary, Bertha, Zillah, and Alice—worked there, as did Peter's son, Ross, briefly, in 1913, when needing more money to marry, and left for a job compiling freight carrier distance tables for the N.C. and St. L. Railway where he stayed 37 years. Ebby's daughters—Barnard's nieces—grew up in Comet House (807 Belmont Avenue) to which Ebby had built an addition to accommodate the whole family, and when his wife, Alice, died, Elizabeth Calvert came from England to take care of the four girls. The eldest, Mary Ross Calvert, left on her twenty-first birthday, to live with the Barnards in Williams Bay, and after Rhoda's death, "looked after the professor and assisted him with his astronomical work," most notably completing the *Atlas of Selected Regions of the Milky Way*. The Calverts were devoutly religious Baptists, who never opened their studio on Sunday—indeed, Ebby wouldn't even ride on a streetcar on Sunday. Peter's son Ross recalled that when Vanderbilt's Chancellor James H. Kirkland requested that photographs of graduating classes be made on Sunday, the Calverts refused though it meant a considerable financial loss. This begs the question of what Barnard's religious opinions may have been. He went to Baptist Sunday school in his youth, and probably would have

identified as a Baptist in later years, though in all his vast correspondence and in all his publications he never says anything about his religious beliefs. He certainly followed strict habits of never smoking or drinking; but he seems seldom to have gone to church, and unlike his cousins, he didn't hesitate to work on Sundays. It seems that Astronomy was his real religion, and superseded whatever traditional Christian beliefs he had been brought up in.

27. Barnard, "Unscientific Experiences," pp. 281–282.
28. E.E. Barnard (1880). "Mars; its moons and heavens." Unpublished manuscript, Barnard. Edward Emerson. Papers, Special Collections and University Archives, Jean and Alexander Heard Library, Vanderbilt University.
29. Nowadays, of course, new comet discoveries are almost entirely the provenance of automated telescopes using charged-couple devices.
30. Barnard, "Experiences," pp. 277–278.
31. In 1859, on the eve of the Civil War, McTyeire published a book, *Duties of Christian Masters* (Nashville, TN: Southern Methodist Publishing House), in which he stated his unqalified support of slavery "as part of human nature." It was his opinion that slavery was "God's punishment and that he, as a follower of the faith, was bound to do all in his power to ensure this continued." Garland, too, was an apologist for slavery before the Civil War, and his older brother Hugh was one of the lawyers employed by the pro-slavery owner Irene Emerson in the Dred Scott case. Garland himself owned up to sixty slaves before the Civil War; the first few gifted by his parents when he married. Later he purchased others as families (out of a humane desire to keep them together, he said) and rented them out to others as house servants as a source of income. He claimed "he did not own them as property, but he instead owned their labor," a distinction he made many times as in a lecture at the University of Alabama in 1860 where he said, "The negro has, through slavery, been taken up from a condition of grossest barbarity and ignorance and made serviceable to himself and to the world, and elevated and improved, morally, intellectually, and physically." In other words, raised from the mudsill which was their natural place otherwise. It should be noted in his partial defense that in Alabama at the time, outspoken opposition to ownership slavery was a death sentence. After the Civil War, unlike McTyeire, Garland became an outspoken opponent to ownership slavery, writing many articles and sermons advocating his position in Methodist Church publications. See: Kathryn Fuselier and Robert Yee (2016) "The Legacy of Slavery at Vanderbilt: Our Forgotten Past," *Vanderbilt Historical Review* (October 17) and Jamon Smith (2006) "Slavery Marks University's Past," *Tuscaloosa News* (April 7).
32. As Bishop McTyeire described it to Vanderbilt; quoted in Edwin Mims (1946) *History of Vanderbilt University* (Nashville: Vanderbilt University Press), p.38.
33. Lagemann, *Garland Collection*, p.61.
34. Born in 1850, Baskervill belonged to a younger generation who came of age after the Civil War. In contrast to his father-in-law, he took a dim view of

slavery and was active in an organization known as the Open Letter Club which was "a loosely organized attempt to disseminate liberal propaganda concerning civil rights and education for the Negro in the South between 1887 and 1890," as noted in Essie (Wenar) Samuel (1967), *A History of Failure: The Open Letter* Club, unpublished Master's thesis, Vanderbilt University. In addition to his work in Anglo-Saxon and Middle-English, he wrote biographical and critical studies of southern writers, including Irwin Russell, Charles Egbert Craddock, Joel Chandler Harris, Maurice Thompson, and Sidney Lanier. All are forgotten now, except Harris, remembered for his Uncle Remus stories which were mostly collected directly from the slaves of the Turnwold Plantation where he had apprenticed during his teenage years, and Lanier, whose poems still turn up in anthologies now and then.

35. Quoted in E.B. Frost [13, p. 6].
36. Ibid., p. 7.
37. M.R. Calvert to Robert G. Aitken, November 11, 1927; Aitken. Robert Grant. Papers, Mary Lea Shane Archives of the Lick Observatory
38. E.E. Barnard, (1886). *Sidereal Messenger*, May 1886, p. 156.
39. H.N. McTyeire to E.E. Barnard, August 6, 1886; McTyeire. Holland Nimmons. Papers, Special Collections, Vanderbilt University Archives.
40. E.E. Barnard to E.C. Pickering, October 21, 1885; Pickering. Edward Charles. Papers, Harvard College Observatory Archives.
41. E.S. Holden to E.B. Knobel, June 16, 1893. This letter was sent by Knobel to Ernst D. Burton, president of the University of Chicago on July 14, 1923. Burton. Ernest DeWitt. Papers, Hanna Holborn Gray Special Collections Research Center, University of Chicago Library. This letter was sent by Knobel to Burton, an American biblical scholar and president of the University of Chicago, in 1923; I should never have come across it had it not been pointed out to me by my great mentor Donald E. Osterbrock, who uncovered it during his extensive research on the history of Yerkes Observatory.
42. McGill, "Edward Emerson Barnard," p. 53.
43. E.S. Holden to E.E. Barnard, September 12, 1883; Barnard. Edward Emerson. Papers, Special Collections and University Archives, Jean and Alexander Heard Library, Vanderbilt University.
44. E.S. Holden to E.E. Barnard, September 21, 1883; Barnard. Edward Emerson. Papers, Special Collections and University Archives, Jean and Alexander Heard Library, Vanderbilt University.
45. E.S. Holden to E.E. Barnard, September 26, 1883; Barnard. Edward Emerson. Papers, Special Collections and University Archives, Jean and Alexander Heard Library, Vanderbilt University.
46. E.E. Barnard to E.S. Holden, October 26, 1885; Holden. Edward Singleton. Papers, Mary Lea Shane Archives of the Lick Observatory.
47. E.S. Holden to E.E. Barnard, October 27, 1885; Holden. Edward Singleton. Papers, Mary Lea Shane Archives of the Lick Observatory.

48. E.E. Barnard to E.S. Holden, December 10, 1886; Holden. Edward Singleton. Papers, Mary Lea Shane Archives of the Lick Observatory.
49. E.E. Barnard to E.S. Holden, May 7, 1886; Holden. Edward Singleton. Papers, Mary Lea Shane Archives of the Lick Observatory.
50. E.S. Holden to E.E. Barnard, October 27, 1885; Barnard. Edward Emerson. Papers, Special Collections and University Archives, Jean and Alexander Heard Library, Vanderbilt University.
51. E.E. Barnard to E.S. Holden, May 7, 1886; Holden. Edward Singleton. Papers, Mary Lea Shane Archives of the Lick Observatory.
52. Richard S. Floyd to George C. Comstock, December 24, 1887; Comstock. George Cary. Papers, University of Wisconsin Archives.
53. E.S. Holden, diary, March 31, 1888; Holden. Edward Singleton. Papers, Mary Lea Shane Archives of the Lick Observatory.
54. James E. Keeler, San Francisco *Examiner*, January 10, 1888.
55. E.S. Holden; note. Holden. Edward Singleton. Papers, Mary Lea Shane Archives of the Lick Observatory.
56. Richard S. Floyd to E.E. Barnard, March 13, 1888; Floyd. Richard S. Papers, Mary Lea Shane Archives of the Lick Observatory.
57. J.E. Keeler to E.S. Holden, April 9, 1888; Holden. Edward Singleton. Papers, Mary Lea Shane Archives of the Lick Observatory.
58. E.S. Holden, diary, April 16, 1888; Holden. Edward Singleton. Papers, Mary Lea Shane Archives of the Lick Observatory.
59. E.E. Barnard, observing log book; Historical Collections Project, Lick Observatory.
60. The definitive biography of Barnard is William Sheehan [15].
61. Osterbrock, "Rise and Fall," p. 84.
62. E.S. Holden to J.C. Bonté, January 10, 1891; Holden. Edward Singleton. Papers, Mary Lea Shane Archives of the Lick Observatory.
63. Osterbrock, "Rise and Fall," p. 95.
64. Osterbrock, "Rise and Fall," p. 92; p. 94.
65. The lens had actually been manufactured by Charles F. Usner in New York City, who, in the early days of photography, made portrait lens for stock dealers such as Willard & Co. and others.
66. E.E. Barnard, observing log book; Historical Collections Project, Lick Observatory.
67. E.E. Barnard, observing log book; Historical Collections, Lick Observatory.
68. See, for instance, A. F. O'D. Alexander [24].
69. E.E. Barnard, "Charges Against Holden"; one of several manuscript drafts written in 1892.
70. Barnard, "Transparency," p. 121.
71. Osterbrock, "Rise and Fall," p. 95.
72. Osterbrock, "Rise and Fall" p. 81.
73. Osterbrock, "Rise and Fall," p. 95.
74. His 1890–91 planetary observations include: E.E. Barnard [27, 28, 29, 30]; and (1892) "Observations of the Reappearance of the Rings of Saturn…"

75. E.E. Barnard (1908) "A Few Unscientific Experiences of an Astronomer," *Vanderbilt Quarterly*, vol. 8, pp. 273–288:pp. 283–284.

76. E.E. Barnard, manuscript draft, "notes for Young's letter," June 5, 1892; Barnard. Edward Emerson. Papers, Mary Lea Shane Archives of the Lick Observatory.

77. E.E. Barnard (1913) "Photographs of the Milky Way and Comets", *Publications of the Lick Observatory* (Sacramento, CA: University of California), vol. 11; notes to plate 55.

78. E.E. Barnard (1913) "Photographs of the Milky Way and Comets", *Publications of the Lick Observatory* (Sacramento, CA: University of California), vol. 11, notes to plate 49.

79. E.E. Barnard to the editor of the San Francisco *Examiner*, December 25, 1893.

80. E.E. Barnard, observing log book; Historical Collections Project, Lick Observatory.

81. E.E. Barnard to Joseph S. Carels, September 21, 1892; Barnard. Edward Emerson. Papers, Special Collections, Vanderbilt University Archives.

References

1. Barnard EE (1908) A few unscientific experiences of an astronomer. Vanderbilt Univ Q 8:273–288

2. Rowling WK (1866) Cholera, as it appeared in Nashville, in 1849, 1850, 1854 and 1866. University Book and Job Office, Medical College, Nashville, pp 9–10

3. Burnham SW (1894) Early life of E.E. Barnard. Popul Astron 1:193–195

4. Braid JW (1928) First employment of Barnard; his first telescope." Edward Emerson Barnard Memorial Number. J Tenn Acad Sci 3(1):8

5. McGill JT (1928) Edward Emerson Barnard," Edward Emerson Barnard Memorial Number. J Tenn Acad Sci 3(1):32–56

6. Tenn JS (1990) Simon Newcomb: the first Bruce medalist. Mercury (Jan./Feb.):30

7. Newcomb S (1903) The reminiscences of an astronomer. Houghton-Mifflin & Co, Boston/New York, pp 191–192

8. Calvert PR (1928) Reminiscences of Barnard," Edward Emerson Barnard Memorial Number. J Tenn Acad Sci 3(1):12

9. Barnard EE (1928) Letter from Mr. Barnard regarding Mr. Braid," Edward Emerson Barnard Memorial Number. J Tenn Acad Sci 3(1):10–11

10. Howell AE (1928) Letter from Mr. Alfred E. Howell," Edward Emerson Barnard Memorial Number. J Tenn Acad Sci 3(1):8

11. Nelson FD (1883) Edward E. Barnard—fellow of Vanderbilt University. Our Day 2(2):40

12. Lagemann RT (1983) The Garland collection of classical physics apparatus at Vanderbilt University. Folio Publishers, Nashville, p 61

13. Frost EB (1923) Edward Emerson Barnard. Astrophys J 58(1):1–35

14. Keeler JE (1888) First observations of Saturn with the 36-inch equatorial of the lick observatory. Sidereal Messenger 7:79–83

15. Sheehan W (1995) The immortal fire within: the life and work of Edward Emerson Barnard. Cambridge University Press, Cambridge

16. Osterbrock DE (1984) The rise and fall of Edward S. Holden—part I. J Hist Astron 15:82–127

17. Barnard EE (1895) Astronomical photography. The Photographic Times (August):1–24

18. Barnard EE (1890) On the photographs of the milky way made at the lick observatory in 1889. Publ Astron Soc Pac 2:240–244

19. Barnard EE (1890) On some celestial photographs made with a large portrait lens at the lick observatory. Mon Not R Astron Soc 50:310–314

20. Barnard EE (1889) A very remarkable comet. Publ Astron Soc Pac 1:72–74

21. Barnard EE (1891) On the observations of comet 1889 V. Astrophys J 10:111

22. Barnard EE (1890) Observations of the eclipse of Iapetus in the shadows of the lobe, crape ring, and bright ring of Saturn, 1889 November 1. Mon Not R Astron Soc 50:107–110

23. Barnard EE (1889) Eclipse of Japetus, the VIII satellite of Saturn, on November 1, 1889. Publ Astron Soc Pac 1:126–127

24. Alexander AFO'D (1980) The planet Saturn: a history of observation, theory and discovery. Dover, reprint of 1962 ed., New York, p 234

25. Barnard EE (1892) Transparency of the crape ring of Saturn, and other peculiarities as shown by the observations of the eclipse of Japetus on November 1, 1889. Astron Astrophys 1:119–123

26. Barnard EE (1921) Sherburne Wesley Burnham. Popul Astron 29:309–324

27. Barnard EE (1891) Observations of the planet Jupiter and its satellites during 1890 with the 12-inch equatoreal of the lick observatory. Mon Not R Astron Soc 51(9):543–556

28. Barnard EE (1891) Colour changes in the markings on the surface of the planet Jupiter. Mon Not R Astron Soc 52(1):6–7

29. Barnard EE (1891) Observations of the spots and markings on the planet Jupiter, made with the twelve-inch equatoreal of the lick observatory. Mon Not R Astron Soc 52(1):7–16

30. Barnard EE (1891, 1892) On the phenomena of the transit of the first satellite of Jupiter 1890 Sept. 8, and observations of the red spots on the planet. Mon Not R Astron Soc 52(3):156–157

31. Barnard EE (1908) A few unscientific experiences of an astronomer. Vanderbilt Q 8:273–288

32. Barnard EE (1892) Observations and photographs of Swift's comet of March 6, 1892. Astron Astrophys 11:386–388

33. Barnard EE (1892) An account of the discovery of a fifth satellite of Jupiter. Astron Astrophys 11:749–750

4

Rivals of Mars

Contents

The greatest thing a human soul ever does in this world is to see something, and tell what it saw in a plain way. Hundreds of people can talk for one who can think, but thousands think for one who can see. To see clearly is poetry, prophecy and religion, all in one.
—*John Ruskin,* Modern Painters, *Vol. III, Part IV, Chapter 16, Section 28*

© The Author(s), under exclusive license to Springer Nature Switzerland AG 2024   **195**
W. Sheehan, *Parallel Lives of Astronomers*, Springer Biographies,
https://doi.org/10.1007/978-3-031-68800-3_4

Chasing the Red Planet

Somewhat surprisingly, Percival Lowell and Edward Emerson Barnard already met in December 1892, before Lowell set sail for Yokohama Bay on what would be his last visit to Japan. The evidence for this is recent. In the 1980s, the Nautical Almanac Office of the U.S. Naval Observatory disposed of its collection of volumes of the *American Ephemeris and Nautical Almanac* that were duplicates of those in the USNO Library. Scores of old volumes went into the recycle bins. Richard E. Schmidt, at the time employed as a civil service astronomer at the Observatory, managed to salvage a few. Cleaning out his office on his retirement in September 2015, he found one that had belonged to Barnard, as noted by the inscription (with date 1891) on the first page. How the volume had managed to find its way to Washington, D.C., in the first place was uncertain, but in salvaging it Schmidt had appreciated it was likely to be of considerable value, not least because Barnard had clearly had this particular volume at his side in the dome of the 36-inch refractor of Lick Observatory when he made his celebrated discovery of the fifth satellite of Jupiter in September 1892—the discovery which made him internationally famous. Schmidt generously passed the volume along to me. In addition to Barnard's inscription and a few notes including some he had entered on the night of the great discovery, the volume was found to contain something entirely unexpected: on the second page, after the one bearing Barnard's inscription and just before the title page, Percival Lowell had signed his name, in pencil, and set down a forwarding address in Tokyo above the address of his State Street office:

Percival Lowell
c/o Messrs. Walsh, Hale and Co
Yokohama, Japan.
53 State St.
Boston

Clearly, Lowell anticipated or at least hoped for further correspondence with the discoverer of the fifth satellite of Jupiter, and given the provenance of this entry, can only have entered it in person. Thus, for the first time Barnard's and Lowell's hitherto "parallel lives" had intersected, and they had taken each other's measure. It was only the beginning.

Barnard was, of course, by then internationally celebrated as the discoverer of the fifth satellite of Jupiter, known as a world-class astrophotographer, and established as a skillful observer of the planets. He had won a reputation as an inspiring figure who had risen, in typical Horatio Alger fashion, from rags to riches. Lowell, on the other hand, was a Boston aristocrat, known as a colorful writer and authority on the Far East with a smattering of interests that included polo, which he pursued with greater avidity, so far as was known, than astronomy, and had published no astronomical papers or observations. No one would then have expected that Lowell was likely to make any significant mark on the astronomical scene other than, perhaps—as was at first assumed, in fact, to be the case—in the role of philanthropic benefactor of the Harvard College Observatory.

Lowell's October 1893 prospecting with Ralph Curtis of a possible trip to Seville the following Easter suggests that at the time he had no idea either. What happened on his return to Boston that gave a sudden impulse in a new direction was the receipt of a book as a Christmas gift from his second cousin Mary (sister of James Russell Lowell, the poet). The book was a copy of the French astronomer Camille Flammarion's monumental book *La Planète Mars*, published in 1892, which effectively summarized all observations of the planet that had been made since the invention of the telescope. In addition, Flammarion was an enthusiast, who wholeheartedly embraced the notion that Mars was inhabited. As evidence, he provided painstakingly detailed summaries of the series of Mars memoirs the famous Italian astronomer Giovanni Schiaparelli had published regarding his observations at the oppositions of 1877, 1879, 1882, 1884, 1886, —1888. These summaries (in French) were invaluable to Lowell, since he did not read Italian at all.

Beginning with the map compiled at the 1877 opposition, Schiaparelli—affected by red-green color blindness, and trained as a draftsman and civil and hydraulic engineer before transferring to astronomy—had recorded a series of curious linear markings on the planet's surface—*canali*, he had called them. The term would have been best translated into "channels," which was clearly what Schiaparelli had thought them. But even before Lowell had come upon the scene, the false friend "canal," implying canals in the artifactual sense, had come into widespread use.

Giovanni Virginio Schiaparelli, as a youngish man (carte de visite photo from about 1865). He became the prince of Mars observers of the last quarter of the nineteenth century. (Credit: William Sheehan collection)

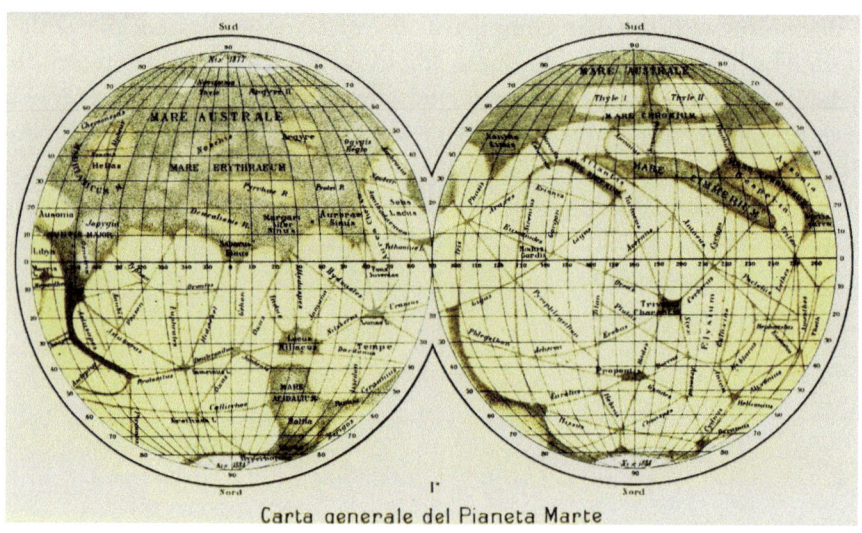

Carta generale del Pianeta Marte

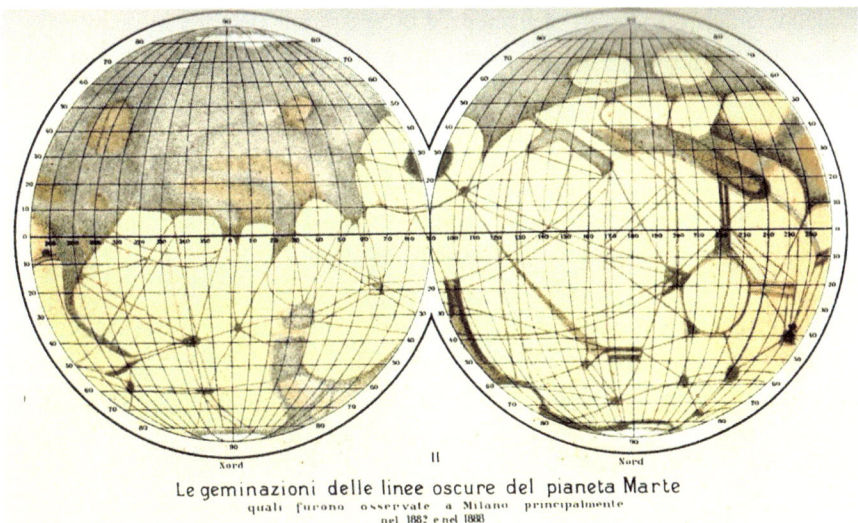

Le geminazioni delle linee oscure del pianeta Marte
quali furono osservate a Milano principalmente
nel 1882 e nel 1888

Schiaparelli's maps of Mars. Above, a general map based on observations between 1877 to 1888, which includes the nomenclature adopted for the various features. From: G.V. Schiaparelli, "La Vita sul Pianeta Marte," *Natura ed Arte*, 1895. Below, a map based on observations between 1882 and 1888 showing some of the canals affected by the geminations. From: G.V. Schiaparelli, "Il Pianeta Marte," *Natura ed Arte*, 1893. Note that in these maps the colors are as recorded by an astronomer suffering from red-green color blindness.

As Lowell had imagined within 2 weeks of arriving for the first time in Tokyo the perdurable principle of "impersonalism" as the key to the Japanese character, so now, having carefully worked his way through Flammarion's book, he seized on the possibility of "canals" on Mars in the artifactual sense—namely, irrigation canals—as striking at the heart of the great Martian mystery. The Lowell family motto was "Occasionem cognosce"—(recognize the opportunity). Percival not only recognized it, he seized it with both hands.

His mind was working at high speed, which suggests that he might have been in at least a mildly manic (hypomanic) state at the time. It is certainly a distinct possibility given the frenzy of activity that followed, which included nothing less than establishing an observatory in the space of several months in order to undertake surveillance of the planet just as it began approach within reasonable range of telescopic scrutiny. (And mood disorders did run in the family.) As described by psychologist Kay Redfield Jamison, who has written extensively about mental disorders in the Lowell family:

[Hypomanic] episodes are characterized by symptoms that are, in many ways, the opposite of those seen in depression. Thus, during hypomania and mania, mood is generally elevated and expansive (or not infrequently, paranoid and irritable); activity and energy levels are greatly increased; the need for sleep is decreased; speech [or writing] is often rapid, excitable, and intrusive; and thinking is fast, moving quickly from topic to topic. Hypomanic or manic individuals usually have an inflated self-esteem, as well as a certainty of conviction about the correctness and importance of their ideas… [1]

With no time to waste, Lowell arranged a meeting at Harvard College Observatory with William H. Pickering in January of 1894. This was done through his cousin Abbott Lawrence Rotch, an MIT graduate and founder of the Blue Hill Meteorological Observatory. Pickering, with his assistant Andrew Ellicott Douglass, had returned from a Mars observing expedition 2 years earlier at the Boyden Station of Harvard College Observatory at Arequipa, Peru, in the foothills of the stratovolcano El Misti. There, he had shown himself to be something of a publicity hound who eagerly cabled rather sensationalistic statements to the New York *Herald*, with which he had an exclusive agreement to provide reports. For instance: "Mars has two mountain ranges near the south pole… Snow fell on the two summits on August 5 and melted again on August 7." Also: "Have discovered forty small lakes on Mars." Disapprovingly, his brother Edward, the highly respected Director of Harvard College Observatory, had advised William that "In my own case I should have restricted myself more distinctly to the facts… You would have rendered yourself less liable to criticism if you had stated that your interpretations were probable instead of implying that they were certain."[1] Lowell, however, was glad to hitch his wagon to this audacious and dogmatic young astronomer's star. (Pickering was only 34 in 1892.) Seth Carlo Chandler Jr., a Boston actuary who as an amateur astronomer made calculations for the Harvard College Observatory and was thus well-acquainted with the personalities there, later told Edward Singleton Holden that Lowell had "the very lowest possible estimate of Edward's capabilities as a man of science …but, astonishing to relate, he is very much stuck on William H." Chandler appreciated some of the dangers of being "stuck" on such a man. He continued:

Lowell, I am sorry to say, has not selected the right kind of companions for his astronomical picnic. He is handicapped by them. What he needed was some young, well-equipped astro-physicist (I mean with education & the mental qualifications of a scientific man), sound and conservative, to sit on his coat-tails and keep him down to business, and prevent wild flights of fancy… [H]e was well-warned before he started … as to his unfortunate environment, and professed to know what he had to deal with in that respect, but expressed unbounded confidence in his ability to keep W.H.P. under his thumb, as he put it....[2]

Sound and conservative William was not, nor would ever be. Nor was he much of an astrophysicist. Indeed, it had been his neglect of taking spectra of stars for which his brother had sent him to Peru in favor of the observations of the Moon and planets (as well as his tendency to overrun his accounts) that had led Edward to recall him from his Southern Hemisphere assignment. Though prone to reckless statements about the interpretation of his observations—he later advocated strongly for vegetation, and even swarms of insects on the Moon—William was, however, a genius in mechanical design and also one of the first to appreciate the importance of finding the best atmospheric conditions available for astronomical observations. In 1889, he had pioneered the site at Wilson's Peak near Pasadena where the Mt. Wilson Observatory would later be built (once a deed to the land could be obtained), and reasoning from analogy to the Arequipa site, he argued that the arid desert climes found in the Arizona Territory were likely to provide conditions just as good as—and to be far more accessible than—those in Peru. On being recalled from Peru by his brother, both because of his poor financial management of the observatory and his tendency to focus on studies of Solar System bodies (especially Mars) in preference to getting the stellar spectra that were Edward's priority, Pickering remained obsessed with Mars back in Boston. He failed, however, to get Edward's support to provide Harvard funds for an expedition to observe Mars from Arizona its forthcoming opposition in 1894, nor did he at first find funding from other potential sources—not least because of the fact that Wall Street had just crashed and the country was in the midst of an economic depression.

The panic in 1893 was indeed widely felt, and put a damper on philanthropic enterprises such as the funding of new observatories. It caused fellow Bostonian Henry Adams's eldest brother John, who had made a fortune in the Union Pacific Railroad, to go "all to pieces," and indeed, wrote Henry, "The entire nervous system of Boston seemed to give way and he broke down with a whole crowd of other leading men… My own nerves went to pieces long ago" [3].[3] Though Percival spared his nerves from at least that cause of breakdown by largely leaving financial affairs to worry his cousin William Lowell Putnam, II, and other investors in the extended family, he knew that if he tried to rely entirely on his investments during his extended overseas travels such as that he had just returned from he would be bankrupt. In order to tend to his portfolio, during his visits to Boston he made regular visits to his State

Street office. In 1894 he was serving as acting treasurer of Massachusetts Cotton Mills.

In Percival Lowell, William H. Pickering found his mark. Lowell needed little fuel to flame the fire of his newfound passion for Mars, and at a dinner meeting at the beginning of the year 1894, the two men—who had once lived on the same Mount Vernon Street on Beacon Hill—entered into an exceedingly improbable collaboration which, by the end of it, would seem to involve the workings of destiny. The Lowell Observatory was a consequence that was truly written in the stars.

Lowell and Pickering needed an understudy to send West to scout potential sites, and found the perfect man for the job in Douglass. As solar astronomer John A. Eddy summarizes:

> Lowell, 39, infatuated with Mars and intrigued by the promise of mountaintop telescopes, needed a rugged advance man to take the train at once to Arizona Territory and there to seek out a suitable site…. Young Douglass, 27 and fresh back from adventures in Peru, seemed ideally suited for the task. What followed in six weeks of March and early April 1894 was a comic opera of a one-man, whirlwind site-survey—made by rail and horse-drawn wagon with a small telescope [the 6-inch Clark Lowell had taken to Japan] and a large stack of Western Union Telegraph blanks, the latter to keep his anxious employer advised, in real time, of nightly measures of the skies. After perfunctory stops at Tombstone, Tucson, Phoenix and Prescott, Douglass (with telegraphed consent from Boston) selected Flagstaff and on April 23 broke ground for the first building of the observatory that today bears Lowell's name [4, p. 70].

The mesa on which the observatory was to be built would be called "Mars Hill"—a neat and characteristically Lowellian pun combining its intended purpose to study the red planet with the name of the rock outcrop near the Acropolis in Athens where the Athenian judicial council met (the Areopagus—the Hill of Ares). In advance of the site selection, Pickering borrowed two telescopes, an 18-inch refractor by the Pittsburgh optician John A. Brashear that was to be routed to Flagstaff temporarily before being set up permanently at Philadelphia's Flower Observatory, and a 12-inch Clark from Harvard. Pickering also provided the design for a pre-fab dome that could be shipped by rail to Arizona (among Flagstaff's advantages were not only its atmosphere but the fact it was located on a rail line and also—according to what Douglass reported in advanced age—its saloons). Pickering also arranged to take a sabbatical from Harvard beginning that coming June, so that he could travel

West to establish the new observing station; Douglass, his prospecting work done, remaining in Flagstaff awaiting his arrival.

Along the way there were some misunderstandings to clear up. At first, it seems not even to be certain that Lowell was actually planning to travel west himself. A very garbled account in the Cambridge *Tribune*, published just days before Pickering and Lowell set out, reported that the observatory was under the charge of "Mr. Alexander E. Douglass," and that Pickering was due to join him out West as soon as his leave of absence from Harvard took effect. Lowell, it was said, "will probably accompany him."[4]

Even newspapers that realized that Lowell's direct participation in the affair was vouchsafed nevertheless tended to assume that his interest was principally philanthropic, and that he was going along in something of the capacity of a tourist. It was assumed that the Flagstaff campaign was an expedition of Harvard College Observatory (according to William Pickering's original plan), but Lowell rather strenuously nipped that rumor in the bud and insisted that, no, he was to be personally in charge, determined to run the thing himself. Any mention of a Harvard connection was to be rooted out, and the Observatory was henceforth to be referred to simply as the "Lowell Observatory." These developments were followed with evident amusement by the gossipy Chandler, who related to Holden that Edward Pickering had tried unsuccessfully to capture the Lowell Observatory for Harvard, "but found he had caught a Tartar in Lowell, who is a man who can see through a mill-stone, and not one it is safe to play as a sucker."[5]

Poetic License and a Talk to the Boston Scientific Society

The dome that Pickering designed for the expedition was not the only thing that was pre-fab; so were many of Lowell's ideas. As always, he approached the subject he proposed to investigate from the top down. When Flammarion's book fell into his hands, he doubtless studied it so carefully—and with such attention to detail—that the drawings and maps of earlier observers, especially Schiaparelli, which he regarded as definitive, must have been practically committed to memory.[6] Though chance favors the prepared mind," as Pasteur once said, it is also true, as art historians have long insisted, that "all paintings owe more to other paintings than they owe to direct observation."[7] Around Schiaparelli's monumental body of work, Lowell proceeded to frame his own expectations ("priors") before he had set eye to an eyepiece at all.[8]

Lowell's expectations ("priors") are documented in a poem he wrote during that busy spring and summer. From internal evidence, and the fact that the only manuscript copy exists in a notebook in Flagstaff, suggests that it was begun in early June and abandoned, in still largely unfinished state, by at latest July 1, 1894. His last serious effort in that direction, and never published, "Mars," running to 248 lines, conveys the spirit of romance with which he began the Great Attempt to prove the red planet an inhabited world:

> One voyage there is I fain would take
> While yet a man in mortal make;
> Voyage beyond the compassed bound
> Of our own Earth's returning round....
> My far-off goal draws strangely near,
> Luring imagination on,
> Beckoning body to be gone
> To ruddy-Earthed blue-oceaned Mars.
>
>
> We know just enough to long to know more
> Of that first habitable shore
> Across the ocean of the sky,
> Ocean whose aether-waves of light,
> Buoyant to naught more gross than sight,
> Tot thought alone give passage o'er
> To where the telescope's piercing eye
> Lets wondering Earthbound man descry
> ...
> A sister-planet whose sister-face,
> Featured so familiarly,
> Complete in all its rounded grace
> Mirrors much that our Earth might be
> Could we once above it rise,
> To behold it in its entirety,
> Sailing along the pathless skies.[9]

At the same time, though rather more prosaically, he set forth what he expected to find in a lecture to the Boston Scientific Society, on Tuesday, May 22, 1894. By all accounts, Lowell was a mesmerizing speaker, and his audience must have been perched at the edge of their seats when he confided to them his bold plans. He was founding his new observatory for the study the

Solar System, and looked forward, with some confidence, to the likely results. Utilizing the same all-purpose top-down Laplacian/Spencerian/pseudo-Darwinian framework with which he had found the key to the Far East, he described

> … an investigation into the condition of life in other worlds, including last but not least their habitability by beings like if unlike man. This is not the chimerical search some may suppose. On the contrary, there is strong reason to believe that we are on the eve of pretty definite discovery in the matter…. If the nebular hypothesis is correct, and there is good reason at present for believing in its general truth, then to develop life more or less distantly resembling our own must be the destiny of every member of the solar family which is not prevented by purely physical considerations, size and so forth, from doing so…[10]

The strongest evidence so far was furnished by the "so-called" canals on Mars:

> Speculation has been singularly fruitful as to what these markings on our next to nearest neighbor in space may mean…. The most self-evident explanation from the markings themselves is probably the true one; namely, that in them we are looking upon the result of the work of some sort of intelligent beings.[11]

Among those who took notice of these remarks was Holden, who in a widely quoted article in the *Publications of the Astronomical Society of the Pacific* wrote, "It seems to be the first duty of those who are writing for [the] public to be extremely cautious not to mislead; and especially to avoid overstatement. Conjectures should be carefully separated from acquired facts, and the merely possible should not be confused with the probable, still less with the absolutely certain" [6, pp. 167–168]. Here, Holden had put his finger on one of Lowell's most unfortunate and enduring characteristics—as historian William Graves Hoyt put it, "he expressed [his] conclusions dogmatically, trusting in analogy too much perhaps, and being too willing to accept the merely plausible for the actual in the absence of any immediate alternatives."[12] He was also, dangerously, in the habit of throwing his hat over the fence in describing in advance what he believed himself already to know, and then only going out and hunting up the evidence for these foregone conclusions.

Within days of giving his lecture, Lowell boarded the train for Flagstaff. He arrived on May 31, 1894, as he informed his mother with whom he was, as usual, keeping up a nearly daily correspondence. He had mailed her a copy of Holden's critique to his family. His mother wrote back, "Holden seems like a fly that only comes back to the same place to tickle and rouse you again," and signed off, "Good-bye my darling may you discover more problems on Mars!"

On first arriving in Flagstaff, Lowell, Pickering and Douglass took up temporary accommodation in the Bank Hotel, and walked along dirt streets in the still rough frontier town—whose band was still learning to play in tune—to the base of Mars Hill and on up to the top where the dome awaited. In order to maintain their vigil as continuously as possible, they sometimes slept in a tent within the dome, and Lowell later rented a house on Beaver Street just north of Aspen Avenue in town, while a four-room warm house took shape 100 feet north of the dome, consisting of a study, two bedrooms, a work room and a dark room entered via a trap door in the floor of the work room. (From this humble beginning, gradually and through rather jumbled and haphazard stages, Lowell's 25-room residence on Mars Hill the "Baronial Mansion," would take shape during his lifetime).

Harvard College Observatory astronomer William H. Pickering designed a pre-fab dome to house the 18- and 12-inch refractors borrowed for the 1894 Mars campaign. A base for the rotating dome was built on-site on Mars Hill with cedar posts and planks, and measured 34 feet in diameter. (Credit: Lowell Observatory Archives)

The structure shown in the previous image was topped with Pickering's pre-fab dome, consisting of a wooden frame covered with wire netting, and covered with stretched painted canvas. With a weight of 3.5 tons, it had been built by the Clark telescope-making firm in Cambridgeport, Massachusetts, and sent in parts for assembly in Flagstaff. (Credit: Lowell Observatory Archives)

The 12-inch refractor, with which Lowell and his staff got their first look at Mars on May 31, 1894, is in the lower part of the photograph, and the 18-inch, which they used the following evening, in the upper part. They rested on a mount, especially designed to hold both telescopes, by the Cambridgeport, Massachusetts, telescope-maker Alvan Graham Clark. (Credit: Lowell Observatory Archives)

Lowell's first observations of Mars with the 12-inch were made on June 1, and with the 18-inch on June 2. With the two telescopes mounted piggyback in the assembled pre-fab dome, it was easy to switch from the one to the other—though the 18-inch was, of course, always heavily favored.

So far as is known, Lowell had made no observations of planets whatever other than those of Mars around 1870 with his small telescope from the flat roof of Sevenels, and of Saturn in 1893 with the six-inch refractor in the garden of the house at Tokyo. Though he was, thanks to Peirce, a good mathematician, he had no experience of practical astronomy whatever, nor had he demonstrated any skill—or record of systematic effort—in draftsmanship, which was essential at a time when visual planetary observations still involved sketching at the eyepiece. Ruskin had advised, "Do not … think that you can learn drawing any more than a new language, without some hard and disagreeable labor…." [7] Lowell, so far as can be established, had never yet put forward this "hard and disagreeable labor," but—with his usual self-confidence—seems to have approached the task as rather straightforward, or at any rate as something that one could learn (as perhaps he did languages) "on the job."

For that matter, he seems not to have grasped a point well made by the philosopher of science Norwood Russell Hanson, who stressed the difference between vision (seeing), which is primarily pictorial, and statements about what is seen (seeing that) which are linguistic. "Our visual consciousness is dominated by pictures; scientific knowledge, however, is primarily linguistic," he wrote. "Seeing is, as I should almost like to say, an amalgam of the two—pictures and language" [8]. Certainly there are visual thinkers, but they are very rare; most of us are primarily verbal. Such differences might be described as "cognitive style." It was clear that Lowell was very far over on the language side (and note that mathematics, for this purpose, is also a language). Hanson coined the phrase "theory-ladenness of observation."[13] By this he meant something similar to what Einstein had in mind when he told Werner Heisenberg, "Whether you observe a thing or not depends on the theory which you use. It is the theory which decides what can be observed."[14] These ideas now seem obvious, and tend to move the perspective decisively from retina to brain, from sensation to higher-order processing. It is in the brain not in the eye where true seeing takes places. The brain is not a passive filter of sensations that come to it from the external world, says the University of London neuro-scientist Karl Friston, but a "phantastic organ" (from the Greek word *phantastikos*), "a statistical organ that generates hypotheses or fantasies that are tested against sensory evidence" [10, p. 148].

Though Lowell would not have cared for the term, he was then and always would be regarded by professional astronomers as an amateur astronomer. A Grand Amateur, perhaps, in the Victorian tradition of men of means able to afford the specialized equipment needed to pursue their individual interests, but an amateur nevertheless. One might, perhaps, have expected at least a certain degree of circumspectness, at least at the beginning, as he began his apprenticeship (or to use Chandler's term, his "astronomical picnic") under the tutelage, such as it was, of William H. Pickering, something of the caution Barnard had recommended long before in "Mars; His Moons and His Heavens," and attempted to "fetter the wings of fancy and restrain its heights."

One would, however, expect wrong.

"One Watcher Alone with the Dawn"

Lowell's notebooks are interesting, in showing his rapid progress in ability to "see." His first sketches show dark areas around the south polar cap, including Syrtis Major, which he always saw as richly colored—vivid blue-green, or even robin's egg blue—in contrast to Pickering who always found it greyish. In part, the difference had to do with the fact that the observers divided their time at the telescope into different watches of the night; Douglass and Pickering had the earlier watches, and Lowell had the time right around sunrise, and at this stage, got his best views of Mars in the blue sky after sunrise, with the background of the daytime sky undoubtedly giving its own tinge (due to simultaneous contrast) to the colors reigning on the Martian disk. In any case, color perception is notoriously subjective, and it is at least worth mentioning that the greatest Mars observer of the era, Schiaparelli, suffered from the classic red-green color-blindness, which limited his ability to make out subtle differences of color on the planet and perhaps led to a tendency to overemphasize boundaries between different regions and to show them as hard, sharp lines.

As intriguing as Lowell's sketches are, even more intriguing is his commentary upon them. His strong linguistic tendency (as opposed to visual tendency, which is rather weak) is evident from the first. Already in these first views he is standing tip-toe in his perceptions, with a great deal of "theory" to what he is seeing. Thus, regarding the dark areas, he would state, "We have at present every reason to believe [them] to be water… A part of them may be vegetation… but a part can hardly be anything but water and perhaps all may be… The Hour-glass Sea [Syrtis Major] seems to be the center of the oceanic system of the planet; it is with the exception of the temporary polar sea [along the edge of the melting polar cap [the darkest and therefore presumably the deepest body of water there…."[15]

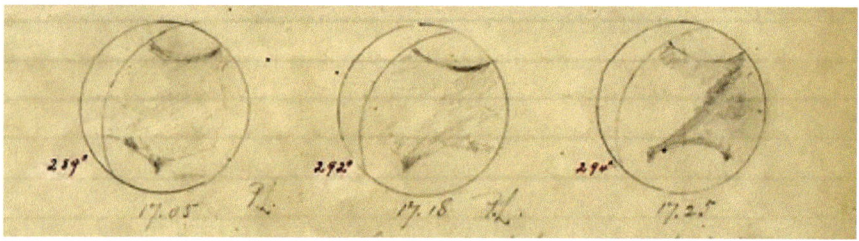

Percival Lowell's first sketches of Mars from Flagstaff, made with the 12-inch refractor on June 1, 1894. (Credit: Lowell Observatory Archives)

He was, of course, impatient of seeing the canals, and since they who seek often find, gratification was not long delayed—in spite of Mars's great distance from the Earth and small telescopic size. He first glimpsed canals on June 7, when he was still—by every reckoning—a naïf in astronomical observation. With dawn approaching, he entered a note, "Seeing growing dim owing to the growing day," only to realize moments later, "Ha! Ha! Not day but dome in the way!" That morning, "one watcher alone with the dawn" saw what he believed to be flashes of sunlight from the South Polar Cap (and though recorded only as a single line of text in his observing logbook would later be worked up into some of the purplest of purple prose in his book *Mars*), and then his very first canal, seen again on successive occasions:

> The first one I saw was on June 7th, the Cerberus in all probability from its position. On June 9th it appeared persistently. I noted it as looking very broad. When on the meridian I glimpsed it double. I had only one glimpse of it as such and though this was definite enough, I put no very great faith in it. I mention its duplicity for future verification. On the 9th and more certainly on the 10th, another canal showed debouching into the same bay but turning more to the right, as it went northward, the Galaxias perhaps… During the next week I made out Gigas, Titan, Gorgon, Sirenius and Eumenides…. Up to this time the canals were of great difficulty, and were neither sharp nor phenomenally straight; fugitive glimpses [of the canals], many of which flitted before me but which I have recorded only the more certain ones. But when the region of the Lake of the Sun began to come round the canals in its neighborhood came out in much better shape; Phasis, Eumenides and Agathodaemon.[16]

In addition to these prose descriptions, Lowell also recorded his impressions with far greater emotion in the poem, composed mostly, it seems, in those early days of June. These passages recount his observations of the glints of sunlight from the polar cap, and the dark areas (supposed seas) and the light areas (supposed lands). The features appear rather vague and uncertain (as they always do to novice observers—even Schiaparelli recounted this when he first observed the planet in late August and early September 1877); then, in glimpses lasting a fraction of a second—"revelation peeps," as he was later to call them—the markings represented on the maps appear well-defined and without doubt. In his elaborations of what he is seeing, Lowell's "phantastic organ" reveals itself to be working overtime:

> On that lucent disk, afire opal
> That flashes in the finder's flare,
> About the top, in the great glass, glows

A circular spot, bowed to oval,
Girt about by a steel-blue band,
So dazzlingly white on its tilted throne,
It looks larger than life in irradiate glare;
The south polar ice-cap it is in glisten,
Flashing its news of a frozen zone…
Till you almost stop to listen
For the crash as a berg newborn
Startling the sea of an arctic morn
Plunging scatters the polar spray,
Dips and turns and then floats away
Into the swoon of a tropic day.

Then other markings slowly emerge
From the first blinding dazzle of sunlit sheen.
Some orange and red, some ultramarine,
Markings that make its disk appear
To the mind's eye so neighborly near
…
Mottling the globe to its very verge
In tremulous pulsings that waver and surge,
Baffling the eye to tell them till
All of a sudden they stand stock still,
For a single instant revealing clear

Listed features in one swift bo-peep…

Three Weeks and Gone

Lowell's first Martian observing campaign lasted only from May 31 to June 24, when he decamped from Flagstaff.[17] He wrote up his initial findings in an article signed as from "Lowell Observatory, July 1894," but which was actually written in Boston—and at frantic speed, to judge from the rapidity with which 20 single-spaced printed pages appeared in the journal *Astronomy and Astro-Physics*. It was only the first instalment of many to come, and already gave rather strong hints of the shape that Lowell's "Theory of Intelligent Life on Mars" would take.

Lowell tried to give the impression that the observations of Mars at Flagstaff had been nothing if not thorough. In fact, his own work was hardly extensive.

In his absence, Pickering and Douglass muled on, pushing the research in some independent directions. Douglass made a determined effort to record any deformations of the limb and terminator that might suggest elevations and depressions. Pickering, turning astrophysicist, deployed an instrument called an Arago polariscope to find evidence of water in the so-called seas. The limb deformations—some of which were also observed at Lick, and which were generally believed to be mountains or clouds—were argued away by Lowell with some rather elaborate (and confusing) statistics. Lowell insisted they were nothing of the sort, and ran on to his conclusion—which he would never abandon—that Mars's skies are almost always cloudless and its surface almost uniformly flat. He did, however, accept the polariscope finding and also the detection by Douglass of canals criss-crossing the dark areas in concluding as evidence that the seas were not seas after all, as he had still believed when he wrote his poem, but rather vast tracts of vegetation as called for in his later theory.[18]

After his initial flurry of observational activity during June, Lowell was only back in Flagstaff again from August 19 to 29, and then—for the main show, just before and after the October 20 opposition—-from October 9 until November 22.

Three weeks of peering through the telescope at the still-tiny gibbous disk still four long months away from opposition had been sufficient to impress upon Lowell just how much of a factor the Flagstaffian atmospheric conditions were in showing the more delicate detail on the planet:

> How little this matter of atmosphere is duly appreciated is evidenced by the way in which a large instrument is assumed to be necessarily superior to a small one, quite irrespective of what it is that is to be observed. Now the fact is that there are two quite different classes of celestial phenomena; those dependent on quantity of light and those dependent on quality of definition for their visibility, and the two means to these ends go anything but hand in hand. For the one, the illumination, the size of the instrument is the prime requisite; for the other, the definition, the atmosphere is the first essential. As an object-lesson in this it is worth noticing that the biggest instruments have not always given the best views of Mars. In matters of Martian detail it is amply evident from the results that observer, atmosphere, instrument is the order of weight to be given as the factors of an observation.[19]

He proceeded to lead the reader along with him up an ascending ladder of seeing in which, under bad seeing, Mars appeared much like it does in the experimental (and very blurred) photographs Pickering had taken in 1890, up through the "portrait stage" captured by Nathaniel Green, and finally to the superior vantage point attested in Schiaparelli's visions. Only at this stage—revealed only in remarkable moments of stillness of the Flagstaff air—did the planet reveal all the magic of the canals. He ended the ascent of the ladder of seeing here: Schiaparelli's visions (and their further additions from Flagstaff) represented the *ne plus ultra* of observational prowess. Lowell never seems to have considered the possibility that still better views of the planet—which would be to Schiaparelli's views as Schiaparelli's to those of photographs—might one day show a quite different Martian reality. But the British solar astronomer E Walter Maunder had already cautioned from his observations of tiny sunspots, sometimes appearing close together and merging into an apparent line, that "We easily fall into the mistake of supposing that the most delicate details which we see really form the ultimate structure of the surface; but it is not possible that they can do so. The finest granule, the smallest pore, as we see it, is only the integration of a vast aggregation of details far too delicate for us to detect; and the minute speck of brighter or duller material may, and probably does, contain within itself a wide range of brilliancy, not to speak of varieties of temperature, of pressure, of motion, and of chemical constitution" [13]. If this were too indirect, Maunder himself would apply the lesson to the case of the canals of Mars, asserting, "We have no right to assume, and yet we do habitually assume, that our telescopes reveal to us the ultimate structure of the surface of the planet" [14, p. 251].

Though a whole school would develop embracing Maunder's skepticism, at the moment the canal literalists led by Lowell were completely untroubled by the distinct possibility that, as Pickering put it, "What the public generally does not understand … is that while the drawings may look thoroughly artificial, and may be most carefully made, yet that the planet itself, if sufficiently well seen, might not look artificial at all."[20]

In the Library at Sevenels

It was not at the observatory at Flagstaff but in the library at Sevenels that the meaning of the Martian phenomena directly accessed at the telescope during the first 3 weeks of June seemed to leap off the page and blare their secrets to the blandishments of his analysis:

> The phenomena seem to point to a vast vernal freshet, now in process upon the planet. In the first place we see the great polar snow-cap vanishing under our eyes. Now this might take place either by evaporation or by melting. Doubtless

it does take place in both ways, but what is of interest is that we have direct evidence of the latter process. Instead of simply growing beautifully less, the snow-cap appeared persistently fringed by a dark line which steadily kept pace with its retreat. This can hardly be anything but water although it may be water in a very different condition from what we know it, neither fresh nor salt, for example, but something unlike either. The character of the water is to certain extent a vital question, as irrigation with salt water does not commend itself to our notions.....

Here we have a *raison d'être* for the canals. In the absence of spring rains, a system of irrigation seems an absolute necessity for Mars if the planet is to support any life upon its great continental areas.[21]

Pioneering Lowell biographer William Graves Hoyt emphasized that though Lowell "… had already arrived at some positive and quite sensational conclusions about life on Mars in particular and extraterrestrial life in general before he ever looked through a telescope from Flagstaff, … on his own testimony at least, he did not formalize his thinking into what he considered to be a full-blown scientific theory until late in July of 1894."[22] That is, after thinking hard on what had been, up till then, 3 weeks of observing the planet during a season (Martian spring) when the canals were not yet their most visible, by an inexperienced amateur, with it still very far from Earth.

Lowell did not need to add more and more observations. He did not reach his conclusions by induction—i.e., bottom-up reasoning from specific instances (facts) to general principles. As a mathematician at heart, he believed, with Peirce who had said in his 1870 textbook *Linear Associative Algebra*, "Mathematics is the science which draws necessary conclusions," that it was better, surer, neater to start at the top with premises, then work down to the mudsill level—the untidy and disorderly shambles of observation—by logic and logic alone. But this is a too passionless assessment of his method. There was an emotional side to it as well—better reflected in his poem "Mars" than in his *Astronomy and Astro-Physics* articles; for Lowell was subject to what we would perhaps call a manic excitement but what the psychologist William James described, contemporaneously, in the *Varieties of Religious Experience*, as a "psychopathic temperament"—one that brought with it "ardour and excitability," and for whom "conceptions tend[ed] to pass immediately into belief and action." According to James, men of this temperament were "not mere critics and understanders with their intellect. Their ideas possess them, they inflict them, for better or worse, upon their companions or their age" [15].

The logical structure of Lowell's "Theory" was simple and straightforward enough to be perfectly understandable to an intelligent child of about ten (which is one of the reasons it appealed so immediately to the general public). In a letter to Douglass, he boiled it down to the following:

> Roughly speaking the evidence seems to be that Mars has (1) some but not much atmosphere; (2) is an aged world with no water to speak of except what makes the polar caps; (3) is provided with an elaborate system of line markings which are best explained by artificial construction… (4) shows what seem to be artificially produced oases as the termini of the canals—what we see and call canals being merely strips of vegetation watered by the canals, the canals themselves being too narrow to be seen.[23]

This, in a nutshell, was the "Theory" that, as Hoyt observes, "for the remaining twenty-two years of his life, Lowell would direct his own energies and the work of his observatory to the accumulation of an incredibly massive body of data to bolster."[24] In other words, one might say he placed the theoretical cart ahead of the observational horse.

Following his first shot across the bow in *Astronomy and Astro-Physics*, Lowell proceeded to publish at short intervals a flurry of articles in journals and newspapers. Many of its supposed discoveries were announced through press releases, which produced headlines such as the San Francisco *Examiner's*

THEY CANNOT ESCAPE NOW

If There Are People Living on Mars We will Soon Know It

Under most rigid scrutiny

Lowell's second article in *Astronomy and Astro-Physics*, again written in Boston, followed his 2-week return visit to Flagstaff at the end of August and first part of September.[25] Even more than the earlier one, it sounds as if written in a state of frenzy:

> During the past three months Mars has been observed here every night with but few exceptions; and although it is still (Sept. 10th) a month and a half to opposition the results already obtained are very encouraging, amply confirming the importance of choosing as good air as possible for an Observatory site.
>
> In this preliminary account of some of them I may with a certain propriety begin, so to speak, at the flood, inasmuch as the prediction which I ventured to make in my last paper with regard to the Martain [sic] vast spring freshet has already been fulfilled—although whether it be a surface freshet or an aerial one still remains in a degree doubtful. But the fact that in the planet's southern

hemisphere at this season (from two months after vernal equinox to the summer solstice) a wholesale transference of water takes place from the pole to the equator, is practically beyond question. Whether what we see be the water itself or only the effects of it is more uncertain…

… Most suggestive of all Martian phenomena are the canals. Were they more generally observable, the world would have been spared much skepticism and more theory. They may, of course, not be artificial but observations here indicate that they are; as will I think, appear from the drawings. For it is one thing to see two or three canals and quite another to have the planet's surface mapped with them upon a most elaborate system of triangulation.

In the first place they are at this season bluish-green, of the same color as the seas into which the longer ones all eventually debouch. In the next place, they are almost without exception geodetically straight, supernaturally so, and this in spite of their leading in every possible direction. Then they are of apparently nearly uniform width throughout their length. What they are is another matter. Their mere aspect, however, is enough to cause all theories about glaciation fissures or surface cracks to die an instant and natural death.

… It is their singular arrangement that is most suggestively impressive. They have every appearance of having been laid out on a definite and highly economic plan. They cut up the surface of the planet into a net-work of triangles instantly suggestive of design. What is more at each of the junctions there is apparently a dark spot… The larger of these appear on Schiaparelli's chart as lakes. A short half-hundred of them were seen at Arequipa in 1892 and others have recently been detected here….[26]

As in a stage play, the main characters have now been assembled: the melting polar caps, the problem of how water was to be transported from one pole to the other given that it cannot proceed (as has been demonstrated) by means of vapor or gas, the existence of marshes and lakes (which he soon began to refer to as "oases"), the canals. The only thing left was for the Martians to don their buskins and make their grand appearance in the tragedy of their doomed and dying desert world. They were already waiting in the wings.[27]

"Try to Get the First Place at Lick Observatory"

With almost solipsistic self-absorption, Lowell seems, at times, to have believed himself to have the planet all to himself that summer. He alone was able to penetrate the murk that had surrounded the mysterious planet to see the sun of truth. Though thoroughly informed (thanks to Flammarion) of the history of observations of the planet, and as noted, hardly yet in possession of more than a handful of observations of his own, he seems to have had little interest in what others were doing at observatories elsewhere.

But he was, of course, far from being alone in taking advantage of the opportunity to observe Mars during its October 20, 1894, apparition, when it was reasonably well placed for study (for the last time in the nineteenth century). Above all there was Barnard, who had always regarded Mars as the "most interesting planet" though, owing to the difficulty of its observation, also the most frustrating. Small telescopes showed very little, but even with the 12-inch and 36-inch refractors at Mt. Hamilton, during the very close opposition of August 1892, it had been a disappointing object owing to its location far south of the celestial equator.

The Lick telescope, shown in this drawing as it appeared in the year after it began operation on Mount Hamilton. (Credit: Public domain, Library of Congress's Prints and Photographs division)

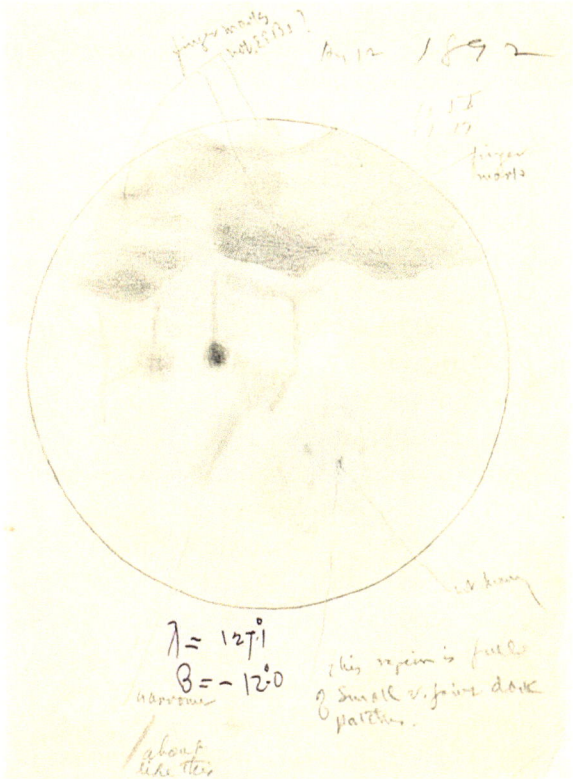

A sketch of Mars by Barnard with the 36-inch refractor, August 12, 1892, one of his first regularly scheduled nights on the telescope. It shows a few dark spots and stubby "canals" as well as a region "full of small v. faint dark patches" in Tharsis. Just under a month after this sketch was made, he used the telescope to discover the fifth satellite of Jupiter. (Credit: Edward Emerson Barnard Papers, University of Chicago Library, Hanna Holborn Gray Special Collections Research Center)

Even without Mars, Barnard had kept himself very busy since the discovery of the fifth satellite of Jupiter and his brief meeting with Lowell in San Francisco, just before the latter set sail for the last time for Yokohama Bay. By then, he was feeling tired—his relentless self-demand to utilize every scrap of clear skies to observe would have worn anyone down—but in addition the battle with Holden had depleted him. Within 2 weeks of the discovery of the fifth satellite, the San Francisco newspapers were already reporting that Barnard had "resolved to resign his position under Holden so soon as an opening presented itself elsewhere" and crying for Holden's head. According to an editorial in the San Francisco *Examiner*:

Candidly, the people of California are sick of Mr. Holden and his overbearing methods. Were there any intrinsic worth hidden beneath this thick and unpleasant coating of selfish egotism, we could afford to disregard the storm of popular criticism.... Even if Holden is incapable of intelligent research himself, it is only fair that he should be restrained from antagonizing those at the Observatory who are.[28]

Having won over both the Regents and the newspapers, Barnard was now seen as a potential loose cannon, especially after Burnham leaked to the newspapers—apparently without Barnard's knowledge—that he was seriously considering leaving Lick for the University of Chicago's new Yerkes Observatory, which was planning to build a 40-inch telescope in order to, in Sir William Huggins's phrase, "lick the Lick."[29] Predictably, the *Examiner* reacted to the rumor with hysterical outrage. Yet the fear of Barnard leaving California was exaggerated, if only because there was not, at this stage, any position to lure Barnard to. Burnham, who had led the effort to draw Barnard away, admitted as much. "Now it will probably be at least a year before anything can be done practically," he confided, "and at least two years before the 40-inch telescope would be ready" (an estimate that itself proved to be overly optimistic by 3 years). In the meantime, Burnham urged his friend, "... you can act with entire independence with this for the future. If you can get the first place at L[ick] O[bservatory], had you not better take it? Leaving the future as to change [can be] considered when the time comes."[30]

Barnard, of course, had no interest in climbing the slippery pole and trying to get himself installed as Director of the Observatory, as Burnham seemed to hope, but he was, as the year 1892 ended, clearly in desperate need of a break. He had continued to work like a demon even after discovering the fifth satellite, and among other things, had published in the *Astronomical Journal* an extensive set of micrometric measures of the faint object made with the 36-inch, to which he added a comment showing some of the obstacles faced in doing such work: "The greatest enemy ... to accurate measures is the wind..... Very often, in the exposed position that Mt. Hamilton occupies, there are winds prevailing night after night [and] when it is necessary to face these winds, as has been the case in many of the observations of this new satellite ... the great portion of the telescope exposed to them causes a constant vibration of the tube" [16, p. 161]. He suggested that in the construction of domes for large refractors, a protecting screen could be introduced in the observing slit to cut off the direct wind from the tube, and for this purpose proposed using a canvas curtain which could be raised from a roll at the base of the tube. In response to these comments, Holden later published a note in *Astronomy and Astro-Physics*, a new journal edited by George Ellery Hale and W.W. Payne, which explained that, in response to similar complaints by Burnham, he had hired a machinist 3 years earlier to put up two rods inside the dome, one on each side of the slit, together with a series of loose sliding rings to each rod. Holden elaborated, "We never went so far as to attach canvas to these rings, for Mr. Burnham finally decided that he did not wish it to be done on his account. The rods and rings have been in place for several years, and any of the experiments suggested can be tried at any time in half an hour by anyone interested who knows of the existence of the rods, etc." [17] After wasting countless hours trying to make sensitive measures while the great telescope was shaking like the proverbial reed in the wind, Barnard was furious to learn that such a device had already existed but that he had never been informed about it.

As a year packed with both agonies and ecstasies drew to a close, Barnard was invited to give a lecture in Nashville. He was of course eager to return home again, but on giving the matter some thought, decided to try for something more extensive. Thus, on New Year's Eve, 1892, he submitted to the Regents a request for a 2 months' leave of absence, in which to undertake an extensive lecture tour back East. Then, a few days later, he learned that the French Academy of Sciences had voted him its prestigious Lalande Prize for the discovery of the fifth satellite, and so sought from the Regents a further extension of four more months in order to travel overseas and receive the prize in person, as well as undertake a grand tour of the leading astronomers and observatories of Europe. The Regents were only too eager to accommodate, and granted his request—with full salary— which, as the San Francisco *Bulletin* did not fail to observe, was "a perquisite that many Berkeley men have often sighed for, but seldom receive."[31]

Accompanied by Rhoda, Barnard left Oakland by train on the evening of February 11, 1892. After giving lectures accompanied with slide presentations of his comet and Milky Way photographs in Butte, Montana, and Northfield, Minnesota, where the journals *Popular Astronomy* and *Astronomy and Astro-Physics* were published, he arrived in Chicago on February 23, on a "raw wet cold" afternoon, and stayed several days with Burnham. He then continued south to Nashville, which neither he nor Rhoda had seen since their departure for the Lick Observatory five years earlier. Received by a large and enthusiastic audience at the Vendome theater on March 6, he was introduced before his lecture, "The Astronomer and His World," by Chancellor Garland, 83 years old, who announced the wish of the now-famous astronomer's alma mater "to anticipate the result of the world in bestowing … honors"[32] by conferring upon him the honorary degree of Doctor of Science. Barnard's lecture was so well received that he was prevailed upon to repeat it to an equally packed house several nights later.

After leaving Nashville, Barnard lectured in St. Louis and in Chicago, then visited Swasey in Cleveland, Keeler at the Allegheny Observatory in Pittsburgh, and Newcomb in Washington, D.C., where at last, in contrast to their humiliating meeting in Nashville in 1877, he could now stand on equal ground and toe to toe with the doyen of American mathematical astronomy. In Rochester, New York, he was introduced on the lecture platform by his old friend and fellow comet discoverer Lewis Swift, before heading on to Hartford, Connecticut, where he received kind receptions from the famous American authors Harriet Beecher Stowe and Mark Twain,[33] and New York City, from where he and Rhoda were to embark on the *Aurania*, a steamer of the Cunard line. They left on May 20, 1893. Their fellow passengers—and this shows just how far up in the world Barnard had come—included some British nobility, Lord and Lady Aberdeen.

After landing in Liverpool, the Barnards traveled by rail to Bradford, and from there continued another 7 miles east to Morley. Rhoda had lived there at Gillroyd Mills (a textile mill) before emigrating to the United States. Rhoda had always been intensely uncomfortable in social situations with astronomers, and especially their wives, most of whom came from a much higher social class, and so rather than accompany her husband on the ground tour, she stayed with her own humble people in Yorkshire. He set out for London, Greenwich, Cambridge (where he gave his first lecture in England, at the behest of Sir Robert Ball, the recently appointed director of the Cambridge University Observatory and future star Victorian popularizer of astronomy), and was there joined by Arthur Cowper Ranyard. Though called to the bar (Lincoln's Inn) in 1871, Ranyard devoted as much of his time and means as he could to his avocational interest, astronomy, and in 1889, upon the death of Richard Anthony Proctor, succeeded the latter as editor of the journal *Knowledge*. One of his innovations, the introduction of high-quality collotype reproductions, made it possible to reproduce Barnard's Milky Way photographs with some success. Ranyard, in fact, had been the first one to suggest that the dark markings recorded in them might consist of interstellar dust clouds rather than be vacancies as generally believed, including by Barnard at the time. Together the two men set out across the Channel. Barnard's itinerary on the Continent was carefully planned to take him to all the great observatories, and he was hosted—or rather feted—by many of the world's leading astronomers at every stop.

He and Ranyard went to Paris, to Juvisy (Flammarion's observatory; the latter couldn't believe that Barnard never drank), and to Meudon. The view of Paris to the east of the Meudon park was magnificent. At this point, Ranyard, who was unwell, turned back, and Barnard traveled on alone to Marseilles, Nice (another magnificent view), Genoa, and Milan, where he "conversed" with Schiaparelli in writing, since Barnard did not speak Italian at all and Schiaparelli could not speak English but could read and write it. He then crossed over the St. Gothard Pass, past Lake Lucerne (magnificent as seen in storm), to Munich, Vienna, Salzburg, back to Vienna again, thence to Dresden, Leipzig, Berlin (where he paid homage to the telescope with which

Neptune was discovered), Potsdam, Kiel on the Baltic, Cologne, Bonn, up the Rhine to Coblenz, to Bingen, across a splendid open country to Mainz, to Heidelberg, Strasbourg, Brussels (with a short sightseeing trip to Waterloo). Then back to England, up to Edinburgh, across to Belfast, to Dublin, to Parsonstown (where he visited Birr Castle, as guest of Lawrence Parsons, the Fourth Earl), and to Cork, where the extreme poverty of the people impressed him. Finally, he returned to Bradford to rejoin Rhoda. There, on July 11, he gave his second lecture in England (after the one in Cambridge), on "visual and photographic astronomy," which included some of his Lick Observatory photographs illuminated by limelight.[34] They then set out together to Liverpool for the trans-Atlantic crossing to New York, and returned by train to Lick Observatory where they arrived on August 24, 1893, which was some 3 months before Percival Lowell arrived back in Boston from Japan.[35]

E.E. Barnard, "cabinet card" (successor to the carte de visite) inscribed to Camille Flammarion, whom Barnard visited during his European tour in 1893. On the back, Barnard indicated the date of the photograph as March 17, 1893, so it must have been taken while he was in Nashville. (Credit: William Sheehan collection)

From the moment he again set foot on the mountain, Barnard made a point of not reporting for duty to Holden. The Director later complained bitterly to the Regents of this neglect of official courtesies, and also of the fact that Barnard failed subsequently to inform him of absences from the mountain for the purpose of delivering lectures; which required them to pen a letter informing Barnard, in as nonconfrontational a manner as possible, that "whatever the personal relations of men may be, official courtesies should never be ignored or overlooked."[36] Henceforth Barnard did comply at least with the letter of the law on this; he would inform the director of the occasions when he needed to be away from Mt. Hamilton by means of terse written notes. It was clear that Barnard's leave of absence from the observatory—which Holden had enjoyed almost as if he were on vacation—had resolved nothing at all between them.

Though the California newspapers, especially the San Francisco *Examiner*, took Barnard's side, and were relentless in their attacks on the director, astronomers elsewhere tried to take a more even-handed view. Even before meeting Barnard on his European grand tour, Sir William Huggins had confided to Hale, "We at a distance are rather puzzled what to think. The general view seems to be that the Director is a little difficult to get on with."[37] Newcomb, who had doubtless heard Barnard's complaints aired freely during the latter's visit to Washington, D.C., later felt obliged to push back a bit on the side of his former protégé, writing to Barnard and at length confidentially and at length about "how things look from this side of the continent":

> To the outer world, the regime of the Lick Observatory seems of the most liberal kind. You and the rest of the staff seem to make what observations you can, and to publish them at pleasure without supervision, whereas the general rule is, at all observatories of a private character, that the director alone has the right to publish work done at the institution. The theory is that the work belongs to the obs[ervatory] and not to the worker, so the former alone has the right to it, and in publication the director represents the institution.
>
> Now I would like you to think over another point you mention. Why should H[olden] want to get rid of men like yourself and Burnham? He was the latter's most devoted friend, and supporter, and when the observatory was started, invited him to it because he believed him to be the best man in the world to do work of a certain kind. Why such a change came over the spirit of his dream? Not jealousy of the work he did or of the reputation he was making, because he himself knew all about those before he sent for him…

Similar remarks apply to your own case. What has seemed most questionable in his conduct toward you was his failure to publish officially the discovery of the 5th satellite of Jupiter. But it has also been rumored that he left the announcement to you because you wished it. I asked you about this at my house, but you evaded the subject…[38]

Barnard, however, could not see any good in Holden, and wrote back, "If I ever have the opportunity to see you again, I shall expose to you the rascality and humbugery [sic] of the man who has made my life here one of misery."[39] Newcomb, nonetheless, remained unconvinced, and as far as possible tried his best to remain on good terms with both men. Keeler refused to join the attacks on his former director, while Lewis Boss, the Director of the Dudley Observatory in Albany, New York, who saw Barnard's behaviors in an unfavorable light, suggested they were partly to be excused as "due to the former influence of another" (i.e., Burnham).[40] Perhaps the least charitable assessment was, however, that of the gossip-mongering Chandler, who had been an ally of Barnard's during his Nashville days but now regarded him as little more than a spoiled *prima donna*. He wrote to Holden, "If not trenching on too delicate a matter, I have for some time desired to say, with reference to certain public references to the trouble which thin-skinned and big-headed assistants of the Lick Obs. have been stirring up, that my sympathies are with you." He continued,

…. I fear that Barnard has been under bad influences. He ought to have developed into a good man. But it looks as if he had been spoiled by adulation, and over-estimate of his own abilities and standing…. I know too that other astronomers feel much the same way. It would have been far better for him if he had seen the need for instruction and training; but I suspect that he thinks he knows it all, because he has a good eye….[41]

The Milky Way and Comets

At night, in the solitude of the small Crocker dome or the 36-inch refractor (for Barnard always preferred to observe alone), Barnard was largely able to forget these troubles by doing as he always did, and throwing himself into his work. He resumed the photography of the Milky Way with the Crocker telescope that had been interrupted during his absence from the observatory, but from his return that late summer until the end of his tenure on Mt. Hamilton

2 years later, he carried it on almost continuously, except when prevented by clouds or moonlight—or by the appearance of a celestial phenomenon too remarkable to be neglected. This was the case in October 1893, when he followed, mesmerized, the breathtaking changes in the tail of a small comet (1893 IV) recently discovered by Brooks. On the morning of October 17, the comet was low in the sky, much dimmed by haze and mist; on the morning of October 21, in a clearer sky, the main tail was straight and presented "a rather graceful appearance." But nothing could have prepared him for what he would find the following day when the plate was removed from its bath of chemicals in the developing tray:

> To say the least the resulting picture was astonishing. It presented the comet's tail as no comet's tail was ever seen before. The graceful symmetry was destroyed; the tail was shattered. It was bent, distorted and deflected, while the larger part of it was broken up into knots and masses of nebulosity, the whole appearance giving the idea of a torch flickering and streaming irregularly in the wind [19, pp. 146–147].

These photographs recorded for the first time what is now known as a *tail disconnection event*. Barnard strongly suspected that the comet's tail during its flight through space had suddenly been shattered on encountering "some kind of resisting medium—a cosmical cloud—a swarm of meteors—certainly a region of resistance in some form" [20, p. 790]. In fact, as he later came to appreciate, they are related to electromagnetic disturbances from the Sun. What happens in a tail-disconnection event is that the comet's ion tail, being driven away by the solar wind, encounters a sudden gust; the ion tail is violently torn away from the coma while a new tail forms behind it. Nothing like this had ever been seen before, and Barnard, utterly captivated by the drama unfolding in the heavens, awaited with "considerable anxiety" the developments of the following morning. The pre-dawn sky was cloudy but, he wrote, "the clouds were breaking and flying in the face of almost a hurricane.... The little observatory [in which the Crocker telescope] was housed rocked in the wind and the dome threatened every moment to fly away in the direction of San Francisco." Despite the horrific conditions, Barnard obtained a useful exposure in the moments when the flying clouds permitted the image of the comet to fall on the plate, in which the disturbance of the previous morning was confirmed.

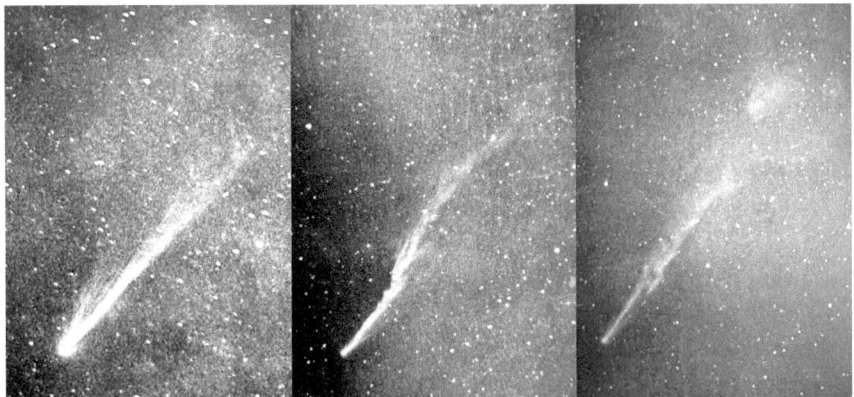

Three photographs of Comet Brooks (1893 IV), taken by Barnard with the Crocker Telescope on the mornings of October 20, 21, and 22. The first two were exposures of 35 min and the third, taken in adverse conditions (high wind and scudding clouds), of 45 min. These photographs document for the first time a tail disconnection event, in which the comet's ion tail encounters a gust in the blustery solar wind that tears the tail away (the third image shows the ghostly remnant drifting off) even as a new tail forms. (From: "Comet and Milky Way Photographs" (1913), plates 107, 108, and 109)

Thereafter clouds and moonlight interfered until the morning of November 2, at which point Barnard obtained yet another remarkable plate. Now the tail appeared strongly concave in the direction of motion, as if it were "beating against a current of resistance." The end was suddenly found to be bent backward at almost a right angle.

Barnard kept up his series of photographs of this remarkable comet through November 19, and later regarded them as some of the most valuable records he ever obtained. In his view—and in this he was quite correct—they initiated a new era in the study of comets. Whereas only a few years before a small comet like this would have been seen as an utterly unremarkable affair—indeed, Brooks's comet never attained to naked-eye visibility—yet with the aid of photography, it revealed the existence of an interplanetary medium, and pointed the way toward the much later use of comet tails as windsocks for tracking the space weather produced by the blustery solar wind.

Ranyard Prods Barnard on the Dark Markings on the Milky Way

Arthur Cowper Ranyard, Barnard's companion earlier in the year in making the passage from London to Paris, was the editor of the journal *Knowledge*, where he was always keen to publish high-quality collotype reproductions of Barnard's photographs. Soon after taking over as editor, he had published Barnard's first image of the Milky Way, taken with the Willard lens, on August 1, 1889, and he had just published Barnard's photographs of Comet Brooks, regarding which he had agreed that the only possible explanation for the changes in its tail was the encounter with some kind of resisting medium. He was predisposed to believe that space was far from empty, but filled with streams of opaque nebulous matter, heated but not luminous and ejected like the gases in solar prominences across vast distances of space. He further believed that this diffuse and all-pervading dark matter might, in its outer portions, condense into stars or into masses of luminous nebulosity such as that visible in the Orion Nebula [21]. Ranyard believed that apparent holes, chasms and other apparent vacancies revealed in Barnard's photographs were actually clouds of obscuring matter, and offered the evidence of the collotype reproductions of his photographs as evidence of this belief. That evidence was, perhaps, to be criticized for "blurring the boundaries between the published collotype, the source negative and the astronomical phenomena themselves" [22], which created confusion as to what exactly was under discussion. Barnard himself was well aware of the problem of substituting reproductions for source negatives—much less for the phenomena themselves—and later, as we shall see, worried himself almost to death trying to get satisfactory reproductions of his plates. The collotype would seem promising, but in the end disappoint his hopes.

Ranyard's chief interest at this time was the structure of the Milky Way, and so he found most intriguing a plate Barnard had exposed to a region in the constellation Cepheus on October 13, 1893—a week before Comet Brooks began to corner his attention. This plate shows a prominent series of irregular dark lanes ramifying in various directions. These features Barnard found completely baffling. As he wrote to Ranyard on November 14:

> When viewed at a distance of a few feet the effect is enhanced; it is then seen that the sky (or Milky Way) is broken up into numerous black cracks or crevices. Looking at these peculiar features, I cannot well see how one can avoid the conclusion that they are necessarily real vacancies in the Milky Way, through which we look out into the blackness of space. I am aware that you are opposed

to this view, and I would like to have your opinion of the real nature of these apparent crevices in the Milky Way, as shown on this particular plate.[42]

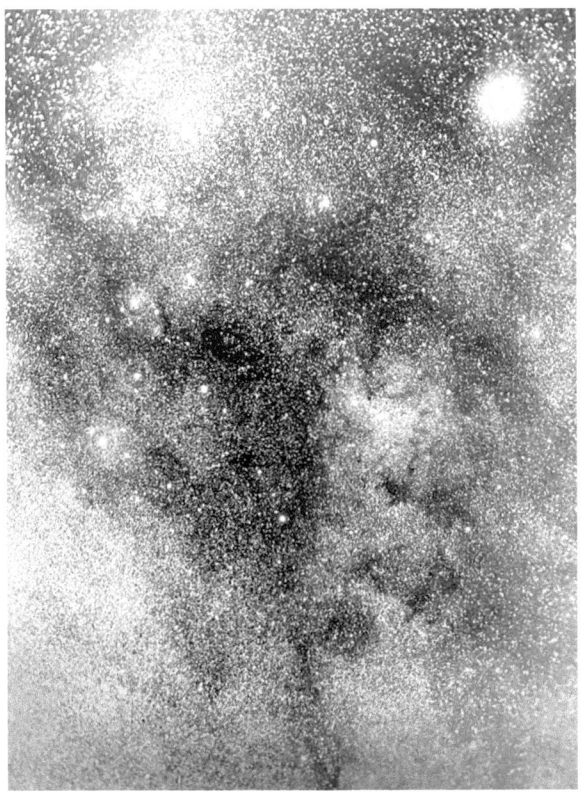

A 7-h exposure on an area in Cepheus, taken by Barnard with the Crocker telescope, October 13, 1893. The photograph is centered on R.A. 21 h 40 min. and declination + 58°. (From: E.E. Barnard, "Comet and Milky Way Photographs" (1913), plate 82)

Ranyard was convinced that the apparent crevices were not vacancies but obscuring matter of some kind, and W.H. Wesley, Savilian astronomer at Oxford University, assistant secretary of the Royal Astronomical Society, and a respected authority on astronomical photographs, tended to agree with him. "That the dark spaces in the Milky Way are caused by dark … matter, lying between us and the stars," he wrote, "appears a reasonable supposition, and in many cases it seems difficult to account for the appearance observed in any other manner" [23, p. 182].

Ranyard published Wesley's comments in *Knowledge*, and they inspired Barnard to return once again, on July 6, 1894, to the wild regions of the

Milky Way near Theta Ophiuchi where he had previously (on June 25, 1892) recorded dramatic structures which seemed to him strongly suggestive of true vacancies and chasms. He sent this plate to Ranyard, with the comments: "It is essentially a region of vacancies. There is a great chasm here in the Milky Way [the "Pipe," described earlier].... It will be noticed that in many of these vacancies there are 'deeper depths' yet, which almost suggest that the appearance of diffused nebulosity over the region is real nebulosity, and that these dark and black places in it are thin places and actual holes" [24].

The "deeper depths yet" could be appreciated on the original negative but were lost in the saturated black of the collotype reproduction—something which must be kept in mind when interpreting Ranyard's comments on what might be visible in this image. Ranyard declared:

> The dark vacant areas or channels running north and south, in the ... picture referred to by Prof. Barnard, seem to me to be undoubtedly dark structures, or absorbing masses in space, which cut out the light from a nebulous or stellar region behind them.... It is comparatively easy to conceive of a narrow stream of dark nebulosity or foggy matter cutting out the light of a uniform background; while if the narrow dark regions correspond to thin places or holes in the nebulosity, they must be holes or thin places extending in a direction away from the earth. The probabilities against such a radial arrangement with respect to the earth's place in space seem to my mind to conclusively prove that the narrow dark spaces are due to streams of absorbing matter, rather than to holes or thin regions in bright nebulosity.[43]

This stimulating discussion ended, alas, without conclusion, as Ranyard passed away from cancer on December 14, 1894, at the age of only forty-nine, leaving Barnard to wrestle—largely alone—with the interpretation of the dark markings of the Milky Way. In the meantime, he would lament to Wesley the loss of his friend, who was taken all too soon:

> My wife and I were sincerely grieved when finally we received notice of the sad end. We came more in contact with Mr. Ranyard by far than with any other person cross the ocean. In every way as a scientific man and as a perfect gentleman, we learned to admire and to love him. It is not only a great loss to English science, the death of a man like him but it must be felt by the entire scientific world. Such eminently honest and able men are so few that the world is in every way a loser when they die....[44]

"To Save My Soul"

Despite the distractions of the Holden situation, Barnard pressed on relentlessly with several lines of important scientific research during the years 1893–94, and in addition to publishing numerous articles both in scientific journals and more popular venues, he regularly left the mountain to lecture in San Jose or San Francisco. On the other hand, he had considerable freedom to pursue his own research. He had no teaching responsibilities, and never had students or assistants (at least until later in life when his niece Mary Calvert assumed that role).

For all his busyness with the comet and Milky Way photographs, he did not fail to notice during the spring and summer of 1894 the increasing prominence of a commanding reddish object in the night sky—the planet Mars.

That year, he and Rhoda had, after 5 years of living in a cottage, taken up occupancy in a two-story brick house on "Ptolemy Ridge," just across the Mt. Hamilton Road from the Main Building. It was hardly luxurious, but it was far more comfortable (and easier to keep warm) than the cottage had been, and the Barnards became happily settled there. In addition, he had now command of the 36-inch refractor not only on Friday but also on Saturday nights, and when Mars came to its most opportune situation for viewing he devoted both nights, from dark until dawn, exclusively to the Red Planet.

Barnard first scrutinized the planet when it was still very far (some 65 million miles) from the Earth, on May 21. (This was 10 days before Lowell arrived in Flagstaff.) It was already brighter than all but a handful of the stars in the night sky, but the disk was small, not reaching 9 arcseconds in apparent diameter until early June. Nevertheless, he managed to make out a dusky patch in the midst of the south polar cap, which had not yet begun the season of its rapid melting. His next drawings of Mars were made toward in the latter part of July, when the planet's diameter had increased to almost 14 arcseconds, it began to show breathtaking detail—in part a testament to the great power of the Lick refractor but also to the excellent quality of the "seeing" conditions that generally prevailed during the long dry summers on Mt. Hamilton. On July 23 and 30, with the Mare Sirenum region in view, he recorded several small dusky spots which appeared "very feeble and faint when near the middle of [the] disk" but grew almost black as they drew toward the terminator. As was typical of him, Barnard simply set down what he saw, and did not waste time speculating about what these spots might be—though as we now know, three of these spots lie in the positions of great Martian shield volcanoes, Ascraeus Mons, Arsia Mons and Olympus Mons, which appear dark when they are not hooded with clouds.

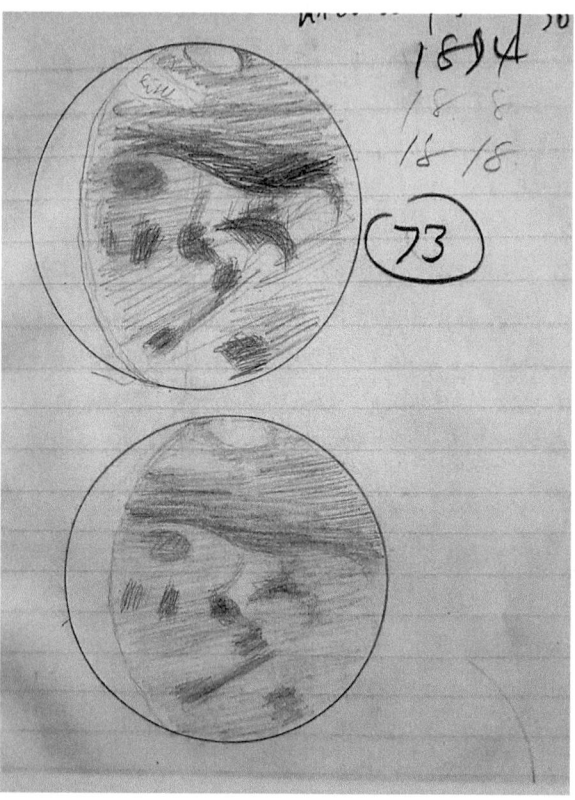

Barnard sketches of Mars with the 36-inch refractor, July 30, 1894. The disk was only 13.5 arcseconds across—about the same as at an aphelic opposition—and yet the planet was already showing remarkable detail. The dark spots connected by a dark diagonal are, as we now know, the shield volcanoes Ascraeus Mons and Pavonis Mons; the dark patch in near the bottom of the disk is Olympus Mons. (Credit: Historical Collections Project, Lick Observatory)

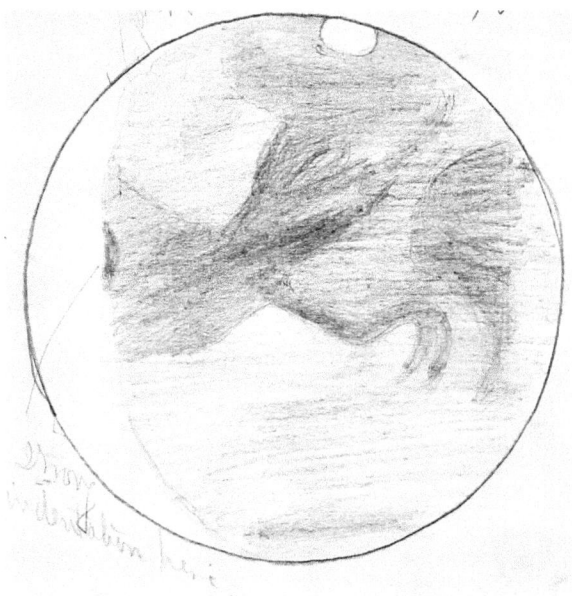

Another early Barnard drawing, August 12, 1894, 17 h 10 min to 17 h 30 min. Two months before opposition, the apparent diameter was already 199 arcseconds. (Credit: Edward Emerson Barnard Papers, University of Chicago Library, Hanna Holborn Gray Special Collections Research Center)

The Martian disk expanded in size—from over 17 arcseconds at the beginning of September, to over 19 arcseconds by mid-month, and 21 arcseconds by the first of October. Barnard attempted to keep pace with the increasingly detailed revelations by sketching on an ever more generous scale—from 1½ inches to the diameter of the planet in July, to not quite 3 inches in early August, to 5 inches in mid-August, September, and October. (By comparison, Lowell and his associates never departed from the scale of 2 inches to the diameter of the planet. In comparison with Barnard's sketches, theirs appear rather quaint.)

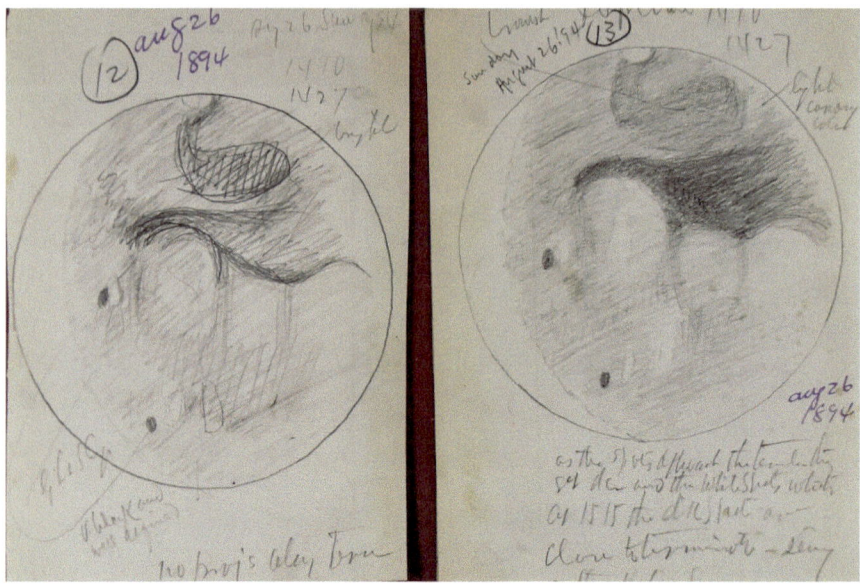

Mars according to Barnard on August 26, 1894, 14 h 10 min to 14 h 27 min. The one on the right is a somewhat more finished version of the one on the left. The "canals" shown here, extending from Mare Sirenum, actually exist, and are thin trails of wind-blown dust. They have been prominent in recent years and appear much as Barnard shows them. The dark spots near the terminator are in the positions of the shield vol-canoes Arsia Mons (upper) and Olympus Mons, whose nature was first revealed in Mariner 9 photographs in 1971–72. The behavior of Olympus Mons is quite puzzling: it is sometimes dark and sometimes hardly visible against the desert, depending on the presence or absence of thin white cloud (especially in the morning or evening) sur-rounding the lower parts of the caldera. Here the upper slopes of the caldera stand out as dark by contrast. (Credit: Edward Emerson Barnard Papers, University of Chicago Library, Hanna Holborn Gray Special Collections Research Center)

Small vs Large Telescope Controversy

Barnard's customary practice was to enter sketches of various celestial objects directly into the pages of his observing books, and so those of Mars appear among pages and pages of arithmetic representing micrometer measurements, reductions, and notes (written in his rather sloppy but usually fairly legible hand). However, by early September he was seeing such an incredible wealth of detail on the planet that he felt compelled to depart from his usual practice and began to sketch the planet on separate sheets of higher quality paper, in order to have the best chance of capturing all that was there. Looking at these records and then at those of Lowell, one has the impression of comparing the manuscript scores of Beethoven and Mozart. Lowell's sketches are beautiful in their way, rather delicate and graceful, but Barnard's are full of titanic energy. One can hardly believe that these two men were observing the same planet.

But then their observing techniques and level of experience were dramatically different, and they also stood on opposite sides of the small versus large telescope controversy which had been waged since the mid-1880s. One of the clearest statements of the small-telescope position was given by the English amateur astronomer William Frederick Denning, who did all of his own planetary work with a 10-inch reflector and had argued in 1891:

> Our atmosphere is always in a state of unrest. Its condition is subject to many variations. Heat, radiated or evolved from terrestrial objects, rises in waves and floats along with the wind. These vapours exercise a property of refraction, with the result that, as they pass in front of celestial objects, the latter at once become subject to a rapid series of contortions in detail. Their outlines appear tremulous, and all the features are involved in a rippling effect that seriously compromises the definition. Delicate markings are quite effaced on a disk which is thus in a state of ebullition; and on such occasions are rarely able to attain their ends… In large instruments these disturbances are very troublesome, as they increase proportionately with aperture…. The observer becomes conscious that what he has gained in light has been lost in definition. At times—perhaps on one occasion in fifty—this experience is different; the atmosphere has apparently assumed a state of quiescence, and objects are seen in a great telescope with the same clearness of detail as in smaller ones. It is then the observer fully realizes that his instrument, though generally ineffective, is not itself in fault, and that it would do valuable work were the normal condition of the air suitable to the exercise of its capacity [25, p. 33].

In general, amateurs, including Lowell, defended this ground. Indeed, many of the most prolific observers of the Martian canals used relatively small telescopes—Schiaparelli, for instance, had discovered them with only an 8.6-inch refractor, while the Serbian-Austrian amateur astronomer Leo Brenner (real name Spiridon Gopcevic), wielding a 7-inch refractor from his private observatory at Lussinpiccolo on the shores of the Adriatic Sea, would record no fewer than 164 canals (as well as absurdly claim to have worked out the perennially uncertain rotation period of Venus to within 1/10,000th of a second!). Brenner would claim that "comparing the drawings of all observers in Flammarion's 'Mars,' we found that, excepted the two refractors of Schiaparelli and the 30-in. of Nice, no other telescope in the world (not even the Lick 36-in.) has shown the surface of Mars with such a distinctness as we always see the planet" [26, p. 167]. Enjoying the unrivalled seeing in Flagstaff, Lowell threw his hat into the ring, and proclaimed with his usual air of dogmatic certainty:

> No amateur need despair of getting interesting observations because of the relative smallness of his object-glass…. In matters of planetary detail size of aperture is not the all-essential thing it is tacitly taken to be…. A large glass in poor air will not show what a small glass will in good air [27].

Though of course not showing nearly as much detail as the 18-inch, Lowell insisted that even in the six-inch refractor he had brought with him to Japan, "Lake Tithonus and the Agathodaemon canal" were evident as a dark line, while the half-tone areas of Argyre and Pyrrhae came out with "singular prominence—the dark and light markings being more contrasted … than in the 18-inch."[45] He also did not find high magnifications useful, arguing that "simply increasing the power will not do, for atmosphere, owing to its inevitable lack of homogeneity, prescribes a limit. With the 18-inch at Flagstaff, a power of 420 was about this limit on Mars…. The greater number of my drawings were made with 370."[46] (This last comment recalls to mind somewhat Pickering later admitted, that at Arequipa, where he and Douglass used a 13-inch refractor to observe Mars in 1892, "We used a low power of 350 much of the time, because we wanted to see the canals. Had we used a higher power, we should perhaps have seen what was really there" [28, p. 412].

Barnard, who had started his career with small telescopes, took a keen interest in this debate. He tried different apertures on the planets, but in the end decided to stick with the full aperture of the 36-inch. Experience was his teacher. He found that at least when the seeing was good, as it often was at Mt. Hamilton in the late summer and early autumn of 1894, the larger aperture afforded a decisive advantage. A soft laminar airflow often blew from the ocean up to the summit where the Observatory occupied the first high ground the flow encountered, conditions which were conducive to superb seeing.[47] He maintained a vigil on the planet throughout the night—and it could be physically as well as mentally demanding work. On the night of Sunday September 2 and the morning of September 3, when the seeing became splendid after sunrise and he enjoyed perhaps the most sublime revelations about Mars ever granted to man up to that time, there was no water in the engines that wound the clock drive mechanism or turned the dome, so he had to do all this by hand—"dreadfully hard and exhausting work," as he noted in his logbook.[48] (In contrast to Lowell, Barnard observed alone, and had no one to assist him.) His observations on that magical night have already been mentioned (in Chap. 1), but we need to reintroduce them here within the wider context. His drawings speak for themselves, but in addition, he recorded in his observing log book:

> … [W]hile making drawings I have examined Mars most thoroughly under good conditions. The region of the [L]ake of the Sun [Lacus Solis or, as now usually known, Solis Lacus] has been under review. There is a vast amount of detail. … I however have failed to see anything of Schiaparelli's canals as straight narrow lines. In the regions of some of the canals near the Lacus Solis there are details—some of a streaky nature, but they are broad diffused and irregular and under the best conditions could never be taken for the so-called canals….
>
> The Lacus Solis seems to be very much transformed from what it was the first of this past month. It is irregularly joined to the Sea preceding it. It is broken up into several masses.[49]

The views of September 2–3 produced a *Eureka!* moment for Barnard. Though he did not have the pre-set ideas about the planet that Lowell did, it seems, from what he says here, that he nevertheless was well aware of the standard view of the character of the Martian surface based on Schiaparelli's maps. However, here his direct perception stood against expectation; his brain was as always busy generating predictions (prior beliefs) and comparing them with the sensory signals that were arriving at his eyeball, and was here dramatically confronted with a whopping prediction error—the difference between predicted and actual sensory signals. The result was adaptation to a new reality depicted in the image, leading to a subsequent belief that minimized prediction errors by concluding that there are no actual canals on Mars.

His next opportunity on the 36-inch was September 9–10. Again, he watched Mars throughout the night, though because there was no water in the engines that turned the dome and wound the clock drive, he had to do this himself by hand, finding it "dreadfully hard and exhausting work." (This was one occasion on which he must have regretted not having an assistant.) But the magnificent views he was getting of Mars made it unthinkable to sacrifice the opportunity and wait for the engines could be repaired, so he did as he had on countless prior occasions and pushed through the fatigue. The magnificent Hourglass Sea—Schiaparelli's Syrtis Major—was coming into view, and a letter to Newcomb penned on September 11 described all that he was seeing:

I have been watching and drawing the surface of Mars. It is wonderfully full of detail. There is certainly no question about there being mountains and large greatly elevated plateaus. To save my soul I can't believe in the canals as Schiaparelli draws them. I see details where he has drawn none. I see details where some of his canals are, but they are not straight lines at all. When best seen these details are very irregular and broken up—that is, some of the regions of his canals; I verily believe—for all the verifications—that the canals as depicted by Schiaparelli are a fallacy and that they will so be proved before many favorable oppositions are past…. It is impossible to adequately draw all that can be seen….[50]

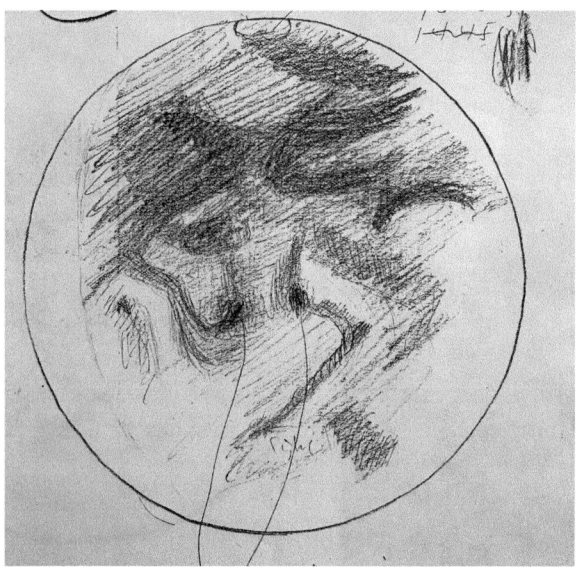

Barnard's best view of the Solis Lacus and Tharsis region, September 3, 1894. This inspired his comment "There is a vast amount of detail…. I however have failed to see anything of Schiaparelli's canals as straight narrow lines." The diagonal sash across the lower part of the disk connects two dark patches now known to be shield volcanoes on Mars—Ascraeus Mons and Arsia Mons. When Barnard made this sketch, Lowell was on the train back to Boston, and would not return to Flagstaff for over a month. (Credit: Edward Emerson Barnard Papers, University of Chicago Library, Hanna Holborn Gray Special Collections Research Center)

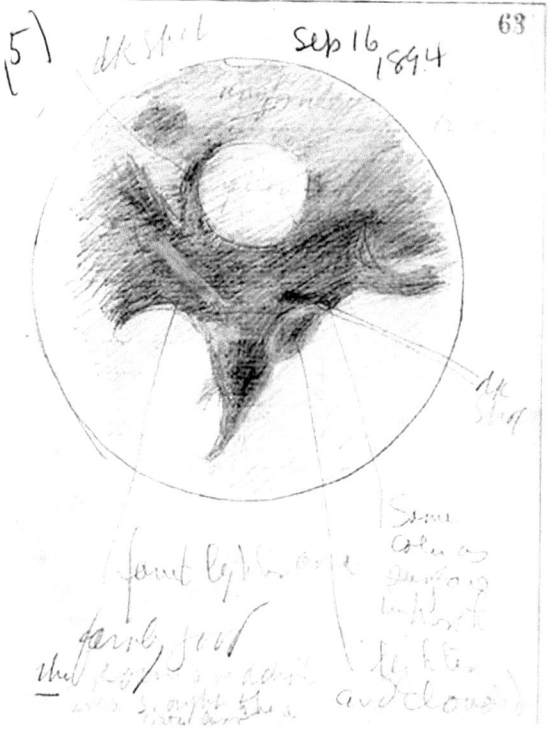

(5) dk spot   Sep 16 1894   63

dk
spot

faint light one

faintly seen
the _____

Some
colour
part of
which
littles
and clouds

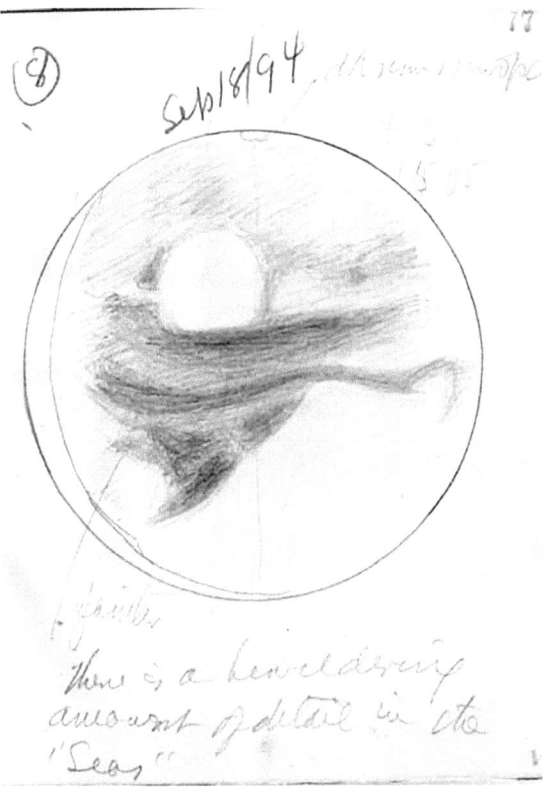

(6) Sep 18/94 dk seen on spot

faintly

there is a bewildering
amount of detail in the
"Seas"

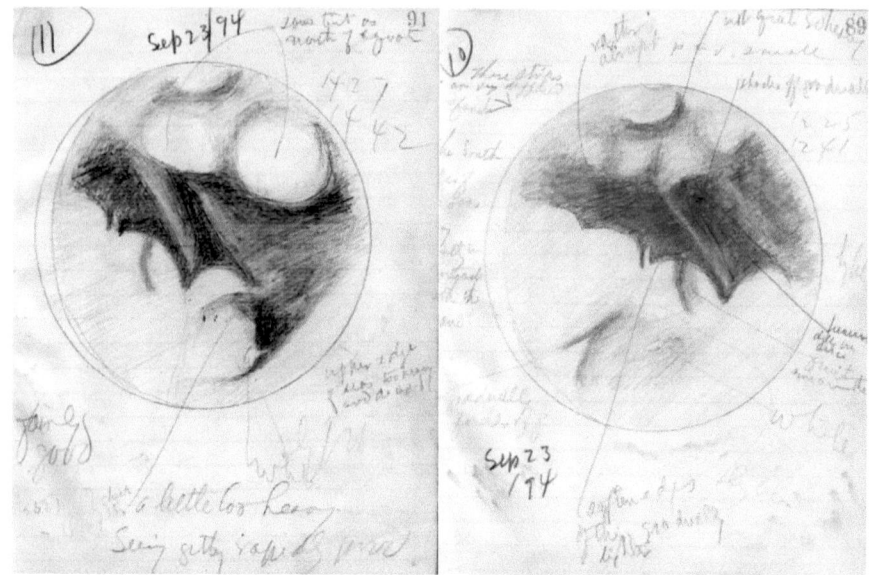

Barnard sketches of Mars, made on September 15 (upper), 18 (middle), and 23 (two, lower), 1894. The Syrtis Major is rotating off the disk and the Mare Tyrrhenum region, which Barnard compared to a bat wing in appearance, is rotating into view. Under his September 18 drawing, Barnard has entered the note: "There is a bewildering amount of detail in the 'seas'." (Credit: Historical Collections Project, Lick Observatory)

Throughout the time that Barnard was enjoying these unprecedented views of Martian surface detail, Pickering and Douglass continued to work, but Lowell did not join them until October 9, when he resumed his observing campaign just before the planet's closest approach (October 13) and opposition (October 20). For the first time, his observations and Barnard's overlapped.

Though Barnard had already told Newcomb, "There is certainly no question about there being mountains and large greatly elevated plateaus," Lowell (unaware of course of any of this) was convinced—and would remain so for the rest of his life—that Mars was in fact remarkably flat, rather resembling the Noto peninsula as he had described it several years earlier, "mappy at best." He insisted on the basis of his calculations that there were no elevations higher on Mars higher than about 3000 feet. He also believed that Mars's atmosphere was both thin and as clear as any desert sky on Earth, and it was unlikely that any moisture could possibly be transported across the vast desert regions by wind-driven clouds and rain. Of the absence of clouds, he would write in his book *Mars*:

The first and most conspicuous of its characteristics is cloudlessness. A cloud is an event on Mars, a rare and unusual phenomenon, which should make it more fittingly appreciated there than Ruskin lamented was the case on Earth, for it is almost perpetually fine weather on our neighbor in space. From the day's beginning to its close, and from one end of the year to the other, nothing appears to veil the greater part of the planet's surface.[51]

In fact, Mars was so intensely desert that not only was it unlike the Earth with water to spare, but its annual water allotment was entirely confined to what could be unlocked from the polar snows. "Upon the melting of its polar cap, and the transference of the water thus annually set free to go its rounds," Lowell continued, "seem to depend all the seasonal phenomena on the surface of the planet."[52] This included the seasonal changes, in which the "temporary sea" around the melting polar cap and blue-green areas in June (the Martian southern hemisphere's spring) had turned yellow and apparently sere in August (early summer) and late October (late summer), as if these vast areas, once containing seas or marshes, had dried up.

These assertions were based on the evidence of Lowell's drawings—one each from the three short observing windows during which he had had the planet under surveillance. He simply asserted, without evidence, that the phenomena represented were continuous across the intervals which went unobserved.

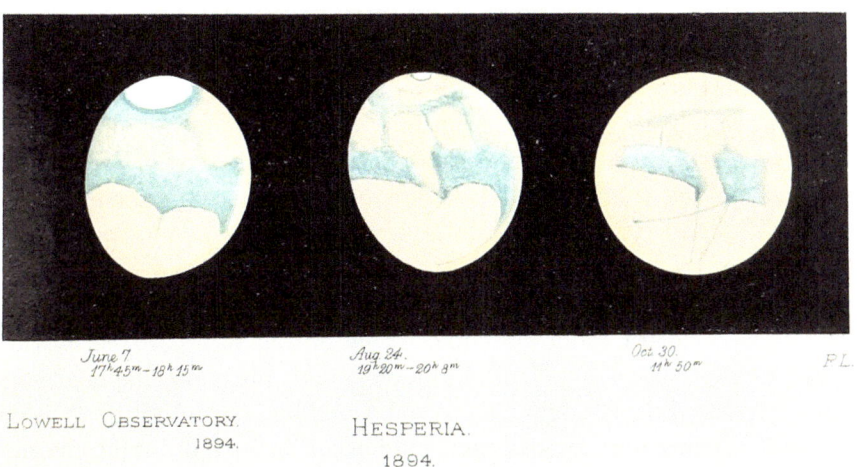

Lowell's color drawings of the Hesperia region of Mars. One each has been taken from the three periods in which Lowell was at Flagstaff: the one at left, labeled "early spring," was made on June 7, the one in the middle, "early summer," on August 24, and the one on the right, "late summer," on October 30, 1894. (From: Annals of the Lowell Observatory, *vol. 1, plate III)*

What Lowell was now convinced were not seas but vast irrigated tracts blooming with Martian vegetation, showed the changes to be expected of vegetation passing with the approaching autumn into the sere. Thus, he declared, "Mars looked more Martian than he had in June."[53] He later detailed the vast changes sweeping across the planet in his book *Mars*:

> Toward the end of October, a strange, and, for observational purposes, a distressing phenomenon took place. What remained of the more southern dark regions showed a desire to vanish, so completely did those regions proceed to fade in tint throughout. This was first noticeable in the Cimmerian Sea, then in the Sea of the Sirens, and in November in the Mare Erythraeum about the Lake of the Sun. The fading steadily progressed until it had advanced so far that in poor seeing the markings were almost imperceptible, and the planet presented a nearly uniform ochre disk.
>
> This was not a case of obscuration; for in the first place it was general, and in the second place the coast-lines were not obliterated. The change, therefore, was not due to clouds or mist.[54]

In fact, the "distressing phenomenon" did, in fact, allow an alternative explanation to the one he asserted. The alternative explanation was precisely the one he dismissed. Because he was away from the Observatory from late August to October/November, he failed to appreciate that an obscuration on a vast scale—what we now know to have been a large regional dust storm—was taking place. It appears to have commenced the day before Lowell resumed his place in the observing chair. According to British Astronomical Association Mars Section Director Richard McKim, the dust storm began about October 10 in either Libya or Northern Ausonia, covered Mare Cimmerium, the southern desert of Ausonia through to Phaethontis and part of Hellas, then extended at least thinly over Mare Sirenum, Mare Australe, and Aonius Sinus.[55] The event even obscured the South Polar Cap, which disappeared for the first time in the observational history of the planet on October 13, as Douglass noted that night and Barnard the night following. The dust seems nearly to have cleared about November 1, but unusually, it gave rise to widespread so-called anomalous (and temporary) darkenings in Hesperia, Mare Chronium, Xanthus, Promethei Sinus and Amenthes. (Not until 1960s was it realized that temporary dark patches could occur in the wake of dust storms, so that for decades the appearance of these patches mystified successive generations of observers and encouraged the vegetation idea.) Barnard was front-row witness to it all, and his pre- and post-dust storm views of the Syrtis Major region, in particular, show that a whole Martian hemisphere was, for weeks, covered in dust.

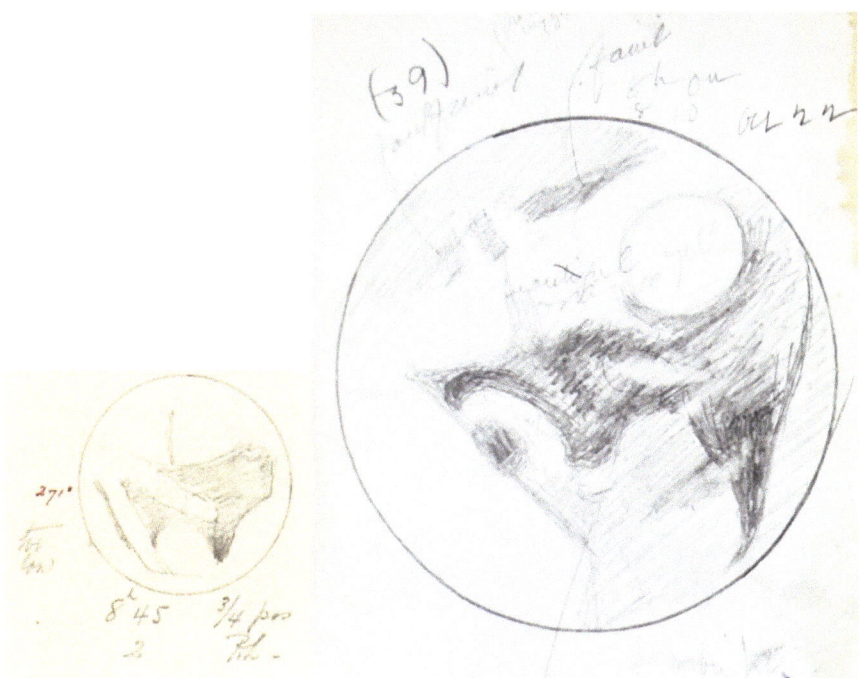

Views of Mars by Lowell (left), October 21, 1894, in poor seeing conditions (seeing reported as 2/7), and by Barnard (right) October 22. The dark line rendered by Lowell as a canal (Amenthes) is an anomalous dark streak, one of several associated with the large regional dust storm that developed over Mare Cimmerium a week or so before these observations were made. Note that the region to the left (east) of the anomalous steak in Barnard's sketch is completely blank—this area has apparently been completely obscured by dust. South is at the top. (Credits: Lowell Observatory Archives and Edward Emerson Barnard Papers, University of Chicago Library, Hanna Holborn Gray Special Collections Research Center)

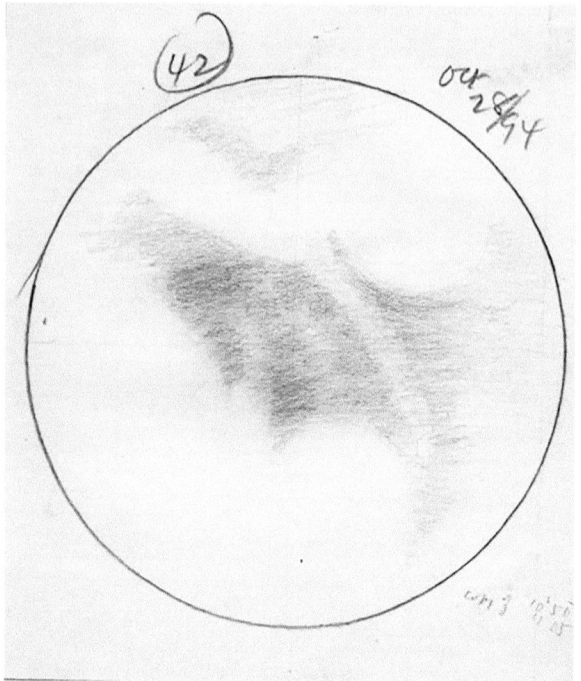

Barnard's sketch of October 28, 1894 shows almost the same identical suite of markings as in the previous drawing, but the whole region centered on Hesperia and Syrtis Minor are now greatly faded, with dust obscuring also Syrtis Major to the right (west). (Credit: Edward Emerson Barnard Papers, University of Chicago Library, Hanna Holborn Gray Special Collections Research Center)

Such storms are now well-known to students of the planet, but Lowell—firm in his "priors"—would never admit that obscurations on such a scale could take place on Mars. And yet, ironically, Lowell had also witnessed the obscuration in progress, only to argue the evidence away by claiming it to be nothing more than the fading out of the vegetation tracts as called for by his theory. He saw the thing, but was blind to its meaning.

As for the hitherto unprecedented appearance of the anomalous dark markings—which appear, according to modern views about the planet, to have been produced by the removal by winds of a cover of fine light dust from darker areas, and were recorded accurately in Barnard's sketches—Lowell saw them as canals. Indeed, during the period when the anomalous dark areas were becoming prominent, between October 29 and November 2, Lowell's drawings show a profusion of canals. His observing log book contains entries such as these:

October 29, 9h 50m. Suspicions of many canals to the east of Cerberus.
October 30, Canals not so faint as at previous presentations, other markings
   more so. Important this as regards water circulation on the planet.
Nov. 1, 10h. Canals as conspicuous (!) as the seas.

Nov. 2 9h 12m. The canals come out in distinct flashes, as well as being, the
more conspicuous of them, always dully there.

-------- 9h 26 m. This is the first time the canals have really appeared strikingly
straight and in profusion.

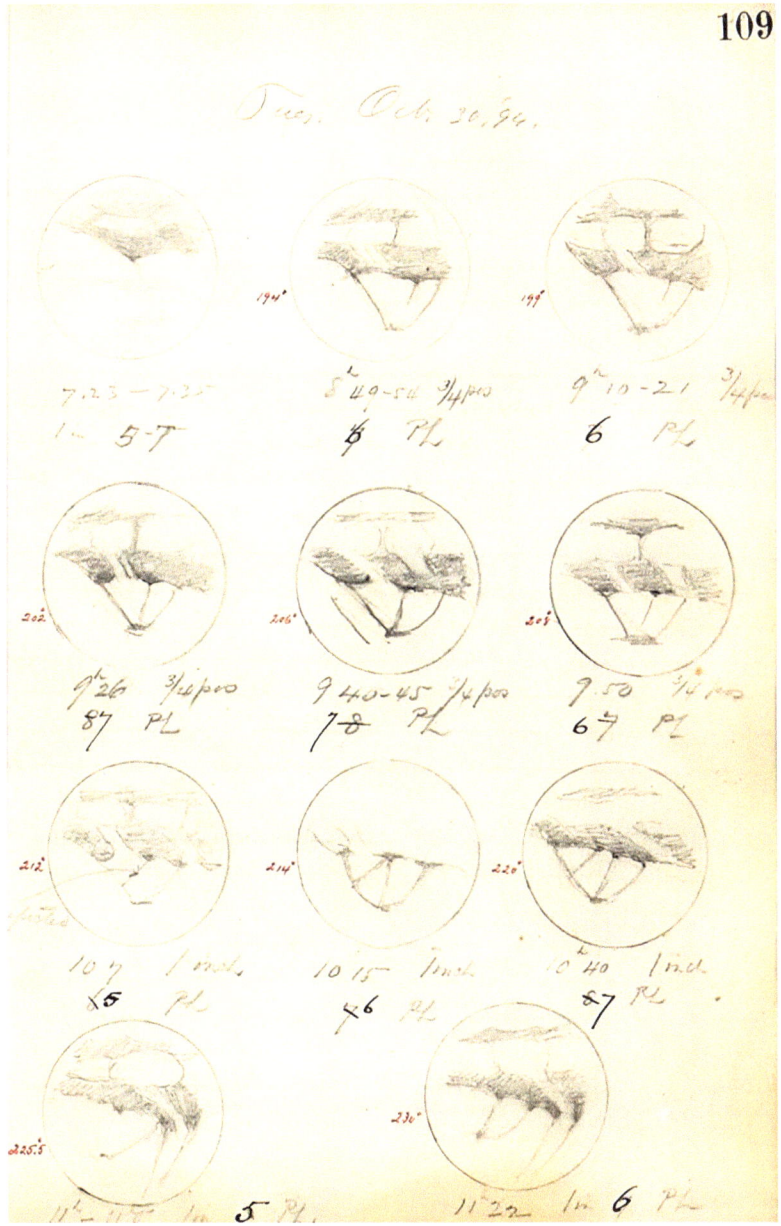

As the large regional dust storm subsides, the canals play begin to hide-and-seek in
flashes of fine seeing, as shown in this series of sketches by Lowell on October 30, 1894.
(Credit: Lowell Observatory Archives)

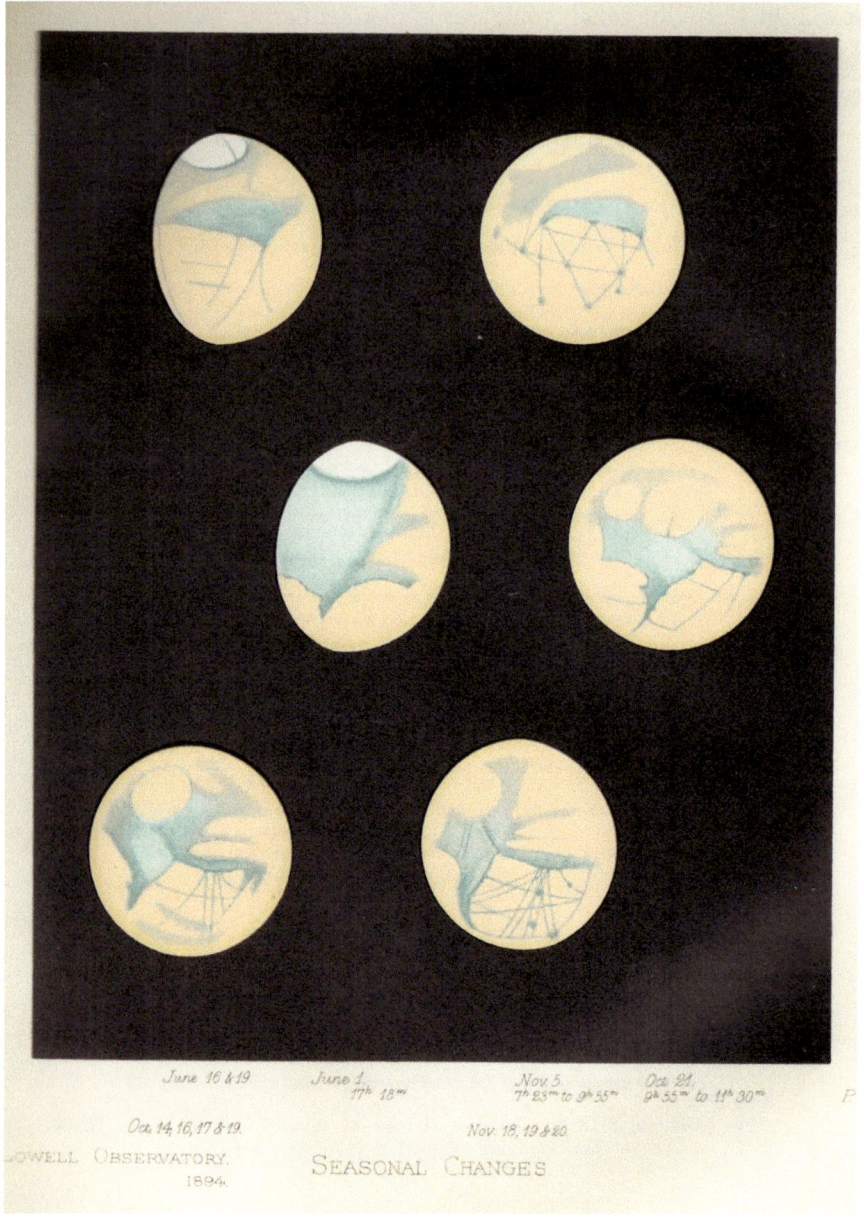

Lowell's drawings of Mars, purportedly showing seasonal changes. (From: Annals of the Lowell Observatory, *vol. 1, Plate IV)*

This also fitted with Lowell's "Theory." As the vegetation faded with the advanced season, the Martians opened up the locks of their canals and let the water, carefully husbanded from the melting polar cap, debouche across the arid landscape. On Mars, as on Earth, if winter comes, can spring be far behind? The now desiccated yellow regions would bloom again, just as Lowell had envisaged in June, which must now have seemed ages ago, in an unpublished poem "Mars":

> Tracings which Schiaparelli has spied,
> So called canals, in parallel pairs
> That for a thousand miles run straight nor swerve,
> Save as parts of great circles curve...
> From certain centres stretching out
> To band the whole great globe about,
> And changing in hue with the changing year
> From a dark green tint to an ochre clear
> With the varying mantle that verdure wears
> When fields of grain from lush grow sere,
> Colors that elsewhere come and go
> O'er the surface of Mars as its seasons grow.

So ended the great opposition of 1894, the last favorable opportunity to observe the planet of the century. The observations having been made, it remained only for the astronomers to interpret them and to publish their definitive results.

As often, the one who spoke first would speak loudest.

The hemisphere drawn by Barnard on September 2–3, 1894, as imaged by the Indian Space Research Organization (ISRO) Mars Orbiter Mission on October 17, 2014. (Credit: ISRO)

Notes

1. Edward C. Pickering to William H. Pickering, August 24, 1892. Quoted in: Bessie Zaban Jones and Lyle Gifford Boyd [2].
2. Seth Carlo Chandler, Jr. to Edward Singleton Holden, September 4, 1894; Holden. Edward Singleton. Papers, Mary Lea Shane Archives of the Lick Observatory.
3. Barbara W. Tuchman, *The Guns of August* and *The Proud Tower*, Margaret Macmillan (ed.) (2012) (New York: Literary Classics of America), p. 698.

4. Cambridge (MA) *Tribune*, May 19, 1894.
5. Chandler to Holden, September 4, 1894; op. cit.
6. Rather as his ancestor Francis Cabot Lowell had committed the blueprints of plans he had seen in England to memory, and so brought them home with him unsuspected of theft.
7. E.H. Gombrich (1961) *Art and Illusion: a study in the psychology of pictorial representation*. The A.W. Mellon Lectures in the Fine Arts, 1956. Princeton: Bollingen Series, 2nd ed., p. 317.
8. See: Jennifer Putnam and William Sheehan [5].
9. I first discovered a notebook containing a version of this poem while doing research among Lowell's manuscripts at Lowell Observatory in the summer of 1982. At the time, they had not been fully inventoried, and many of them were still stored in the observatory's old vault in the basement of the Administration Building. Despite its being only in the rough form of a heavily revised draft, and of perhaps not of the highest literary merit, I recognized its importance in conveying the emotional excitement and preconceptions with which Lowell set out for his Mars Hill redoubt. The poem may have been started even before he arrived in Boston, though the effort clearly continued after he started observing; he includes descriptions that can only have been inspired by firsthand experience at the telescope, including alluding to a flash observed on the South Polar Cap on June 7 1894, which was later expanded into his set-piece "on a hilltop alone with the dawn," in P. Lowell (1895) *Mars* (Boston: Houghton, Mifflin and Co.), pp. 86–87. On the basis of internal evidence, it would seem that the draft was abandoned before he left for Boston in July, since it refers to the blue areas of the planet as "seas" and uses the old adjective "Martial" instead of "Martian" which he adopted that summer. However, though I thought that the poem was probably abandoned once he got more fully underway with his observing campaign, this doesn't appear to be the case. While doing research in the Lowell Archives, science writer David Baron discovered another notebook which contains a revised draft, including passages about color changes of the dark areas attributable to vegetation. This follows Lowell's later thinking after he had concluded the dark areas were vegetation tracts rather than seas. Thus, he writes: "This at least we seem to know/That plants of some sort there survive… For …patches distinctly show/Changes of color as the seasons grow,/Greens and yellows that come and go." He also goes beyond Schiaparellian descriptions to explicitly the "canals" as "huge waterways/At which we wonderingly gaze." However, even this draft is quite imperfect and would probably have required extensive efforts to make suitable for publication. Sometime, probably in the late summer of 1894, and possibly recalling the remarks of his friend W.B. Mason of his poem abut the Japanese mountain Azuma-yama—"I should not have thought it possible to commit literary suicide with such thoroughness and dispatch"—he gave up. For Mason's comment (and Basil Hall Chamberlain's concurrence), see Strauss, *Percival Lowell*, p.78.

10. Lowell, "Lowell Observatory: Percival Lowell before the Boston Scientific Society, May 22, 1894." Boston *Commonwealth*, May 26, 1894.

11. Lowell, "Lowell Observatory".

12. Hoyt, *Lowell and Mars*, p. 311.

13. Hanson, *Patterns*, p. 19.

14. Werner Heisenberg quoted in: A. Salom et al. [9].

15. P. Lowell [11, pp. 548–549]. This article was signed "Lowell Observatory, July 1894," but was actually written in Boston rather than Flagstaff.

16. Ibid., p. 549.

17. A somewhat more personal perspective on the astronomers' activities during the first month in Flagstaff suggesting they weren't all work and no play is furnished by Hoyt, *Lowell and Mars*, p. 67: "Soon after Lowell's return to Boston after only 1 month of observing, Douglass wrote him that 'I am earnestly [sic] requested by the young ladies to state that in our parties and social gatherings your absence is deeply felt.' Lowell replied: 'Pray convey my continued regrets to the fairer, the picnic half of our world,' and later urged: 'Keep the girls till the end of the month if prayers can avail.' Again, in noting Douglass' terminator work, he declared: 'My compliments on your "irregularities" of the terminator. Do you connect them in any way with the (wild) oats which alone we observed planted by man in the neighborhood of Flagstaff?'"

18. Many of the sketches by Pickering and Douglass suggest the effects either of irradiation—the apparent bleeding of bright areas into the dark space beyond the limb or terminator—or even mere effects of tremors in the atmosphere. Lick Observatory astronomer W.W. Campbell, in his review of Lowell's *Mars*, points out that the observation of bright projections at the terminator had originally begun at Mt. Hamilton in 1890, and continued in 1892 and 1894. Of those observations, he says, "there are some arguments in favor of there being clouds, but many more in favor of their being mountains." Regarding the 736 irregularities observed at Flagstaff, Lowell notes (says Campbell) that "694 were not only recorded but measured, and that of these 403 were depression. However, he rejects 346 out of 403 depressions as not real, since they lay on the dark areas of the planet and were due to the smaller irradiation at those places. He holds that the remaining 57 depressions were due to clouds within the terminator, and the 291 projections were clouds outside the terminator; because if they were mountains, the number of depressions should equal the number of projections. To my mind the argument is not convincing.... When we consider that these clouds or mountains (or something else) are immersed in an illuminated atmosphere, we cannot expect the projections and depressions to be equal in number." He concludes somewhat scathingly: "I confess my inability to unravel Mr. Lowell's discussions of Mr. Douglass's observations.... Mr. Lowell writes of the 'long and low' irregularities, that the projections averaged 0".136 in height; the depressions, 0".125 in depth.

These are the distances from the approximately elliptic arc that would have formed the apparent terminator if the irregularities had not existed. Thus we have the heights of the irregularities from a curve that did not exist given to three decimals of a second of arc! And there is nothing to show that the varying distances of the planet were taken into account, either. Every practical astronomer knows that the *first* decimal place is uncertain; the systematic error in such cases can easily and generally do exceed a tenth of a second. To say that the results are accurate because they are the mean of a large number of observations, is to say that if a stranger to Colorado's clear atmosphere should waken unexpectedly on Pike's Peak and guess the distances to several hundred neighboring peaks, the mean of all the guesses would be very near the average distance." See: W.W. Campbell [12].

19. P. Lowell (1894a), "Mars," pp. 539–540.
20. William H. Pickering (1908) "Different Explanations of the Canals of Mars," *Harper's Monthly*; reprinted in: William H. Pickering (1921), *Mars* (Boston: Richard G. Badger), p. 151.
21. P. Lowell (1894a) "Mars," pp. 549–550.
22. Hoyt, *Lowell and Mars*, p. 68.
23. P. Lowell to A.E. Douglass, March 25, 1895; Lowell Observatory Archives.
24. Hoyt, *Lowell and Mars*, p. 69.
25. Lowell was in Flagstaff only from August 20 to September 4, and so left just as Mars was getting really interesting. At least some of the time with his crony William Edward Story, a mathematics professor at Johns Hopkins and Clark University to whom Lowell dedicated his book *Mars*.
26. P. Lowell (1894b) "Mars," pp. 649–650: p. 649.
27. In his poem "Mars," Lowell had already suggested what he expected of the "Martials" (as he called the putative inhabitants of the planet until the summer of 1894 when the term "Martians" was first introduced):

"If beings still be there after our kind,
They surpass us in body, excel us in mind…
More godlike than any Grecian god
Who ever on Olympus trod,
Whose mental faculties transcend ours
As our own do the poor brute's powers?"

28. "Holden and the Regents," San Francisco *Examiner*, September 25, 1892.
29. Sir William Huggins to George Ellery Hale, December 10, 1892. Hale. George Ellery. Papers, Hanna Holborn Gray Special Collections Research Center, University of Chicago Library.
30. S.W. Burnham to E.E. Barnard, October 14, 1892. Barnard. Edward Emerson. Papers, Special Collections and University Archives, Jean and Alexander Heard Library, Vanderbilt University.

31. San Francisco *Bulletin*, February 24, 1893.

32. Nashville *Daily American*, March 6, 1893.

33. It would be interesting to know how they hit it off, as both shared a dry sense of humor. Barnard probably did not know that Twain had made mischievous use of some aspects of his career as a comet discoverer in a talk he gave at Yale, after receiving an honorary M.A. degree in June 1888. Returning later in the year when he the degree, he addressed the students pretending that receipt of the degree gave him full powers over the faculty and curriculum: "I found the astronomer of the university gadding around after comets and other such odds and ends—tramps and derelicts of the skies. I told him pretty plainly that we couldn't have that. I told him it was no economy to go on piling up and piling up raw material in the way of new stars and comets and asteroids that we couldn't ever have any use for till we had worked off the old stock. At bottom I don't really mind comets so much, but somehow I have always been down on asteroids. There is nothing mature about them. I wouldn't sit up nights the way that man does if I could get a basketful of them. He said it was the best line of goods he had. He said he could trade them to Rochester for comets, and trade the comets to Harvard for nebulae, and trade the nebulae to the Smithsonian for flint hatchets. I felt obliged to stop this thing on the spot. I said we couldn't have the university turned into an astronomical junk-yard. And while I was at it I thought I might as well make the reform complete. The astronomer is extraordinarily mutinous, and so, with your approval, I will transfer him to the Law Department and put one of the law students in his place. A boy will be more biddable, more tractable, also cheaper. It is true he cannot be intrusted with the important work at first, but he can comb the skies for nebulae till he get his hand in." From: Mark Twain (1923) *Mark Twain's Speeches*, with an introduction by Albert Bigelow Paine and an appreciation by William Dean Howells (New York: Harper & Brothers), pp. 142–143.

34. "Dr. Barnard at Morley," Leeds *Mercury*, July 15, 1893. In a letter to Charles Whitmell (September 5, 1913) Barnard gave some details of his heartfelt connection (through his wife) to Morley. "My wife is a Yorkshire woman," he wrote. "She has two nieces there. Miss Rosa Federer (who is teaching art at some other town) and Miss Bertha Federer who I think lives at 8 Hallfield Road [Manningham Lane, Bradford]. You see then I'm pretty near Yorkshireman myself, though I was born in Nashville, Tennessee." Quoted in David Sellers [18].

35. Among the leading astronomical personages Barnard met during his 3 months in Europe (the list is not complete) were John Hartnup, Jr., director of the Liverpool Observatory, Captain William Noble, an amateur astronomer who had a private observatory at Maresfield, Sussex and was a leader of the Royal Astronomical Society at the time, Agnes M. Clerke, the astronomical historian, Dr. and Lady Huggins, Sir Robert Ball, G.H. Darwin, Miss Klümpke and the Henry Brothers in Paris, Jules Janssen (who spoke very litttle English)

at Meudon, Flammarion (who did not speak English) at Juvisy, Henri Perrotin, a noted planetary observer and Auguste Charlois who discovered 99 asteroids at Nice (neither spoke English), Schiaparelli at Milan, Hugo von Seeliger at Munich (whom he noted was a fellow teetotaler), Edmund Weiss (who seems to have been the one who referred to asteroids as "vermin of the sky") and Johann Palisa (who discovered 122 of them) in Vienna.

36. T.G. Phelps, C.F. Crocker and H.S. Foote to E.E. Barnard, November 21, 1893. Barnard. Edward Emerson. Papers, Mary Shane Archives of the Lick Observatory.

37. Sir William Huggins to George Ellery Hale, December 10, 1892; Hale. George Ellery. Papers, Hanna Holborn Gray Special Collections Research Center, University of Chicago Library.

38. Simon Newcomb to E.E. Barnard, August 19, 1894. Barnard. Edward Emerson. Barnard. Edward Emerson. Papers, Special Collections and University Archives, Jean and Alexander Heard Library, Vanderbilt University.

39. E.E. Barnard to Simon Newcomb, September 11 and September 18, 1894. Newcomb. Simon. Papers, Library of Congress.

40. Lewis Boss to Edward Singleton Holden, August 13, 1895. Holden. Edward Singleton. Papers, Mary Lea Shane Archives of the Lick Observatory.

41. Seth Carlo Chandler, Jr. to Edward Singleton Holden, April 16, 1893. Holden. Edward Singleton. Papers, Mary Lea Shane Archives of the Lick Observatory.

42. E.E. Barnard to A.C. Ranyard, November 14, 1893; published in (1894) *Knowledge*, vol. 17, p. 17.

43. Ibid.

44. E.E. Barnard to W. H. Wesley, February 2, 1895; Library of the Royal Astronomical Society

45. Lowell (1894a) "Mars," p. 539. That the Tithonius Lacus and Agathodaemon canal should be visible as a clear dark line in even a six-inch refractor is hardly surprising; I remember well seeing them easily with only a 4 ¼-inch reflector in August 1971, shortly before Mariner 9 revealed their true character as dark-floored components of the vast Valles Marineris canyon system.

46. Lowell, "Mars," p. 539. This last comment recalls to mind something Pickering later admitted: At Arequipa, where he and Douglass used a 13-inch refractor to observe Mars in 1892, "We used a low power of 350 much of the time, because we wanted to see the canals. Had we used a higher power, we should perhaps have seen what was really there" [28].

47. Aspects of the small versus large telescope discussion are not well understood even now. It's certainly true, as Lowell emphasized, that there's an advantage in using a telescope aperture that matches the size of the "seeing patches," typically around 16 inches, in the pupil plane of the telescope. However, there is a need for numerous qualifications to such a statement. In general, much larger differences in the refractive index of air from one point to another tend to exist near the ground, because different objects and areas on the surface

have unlike temperatures. So the nature of seeing depends on the given location. Also—though some seem to imagine that the atmosphere has blobs of some fixed size that limit resolution—in fact the structure of turbulence, and hence the quality of images, changes with time. All of this is well explained in the classic article: Andrew T. Young [29, p. 150].

With a large telescope, there's a chance of seeing fine detail occasionally, but those fine details are never seen in the small telescope. This is something that the greatest visual planetary observers such as Barnard and Lowell realized. According to Andy Young, "In addition, there are complications due to the nature of the human visual system. The larger telescope gathers more light; so, even with the same point-spread function in the optical system, the signal/noise ratio is better with the bigger telescope. That allows the observer to 'perceive' more of the detail, even with the same point-spread function.

Remember that the critical flicker frequency of the eye is higher at higher intensity levels. The eye and brain adjust the time resolution of the system to keep the photon noise just beyond the level of perception, in these low-light situations. Also, as the visual observer is limited in color sensitivity in the mesopic range, the one with the bigger telescope can see more gradations of color than the one with the small telescope. Color variations are generally rather subtle, both on Mars and on Jupiter, so this kind of difference really means the bigger telescope is better in similar seeing conditions.

Remember, too, that the frequencies at which the image varies due to seeing are mostly well above the eye's critical flicker frequency. The observer with the smaller telescope is already stuck with much of the blurring that the one at the bigger telescope sees.

In my planetary work, I've had a few occasions on which I enjoyed very good seeing for a few minutes with both reflectors and refractors. I remember two times in particular, on which the seeing was very good for a few minutes, and I cold see an amazing amount of fine detail and color contrast on Mars: once with the 15-inch refractor at Harvard, and once with the 36-inch reflector at McDonald Observatory. It is those moments of unusually good seeing that can make the full resolution of the optical system available to the visual observer." Andrew T. Young to William Sheehan, personal communication, February 21, 2024),

48. E.E. Barnard, observing log book; Historical Collections, Lick Observatory.
49. Ibid.
50. E.E. Barnard to Simon Newcomb, September 11, 1894; Newcomb. Simon. Papers, Library of Congress.
51. Lowell (1895) *Mars* (Boston: Houghton, Mifflin and Co.), p. 45.
52. Lowell, *Mars*, p. 113.
53. Lowell, *Mars*, p. 121.
54. Lowell, *Mars*, pp. 120–121.
55. On the 1894 regional dust storm, see: Richard J. McKim [30].

References

1. Jamison KR (1993) Touched with fire: manic-depressive illness and the artistic temperament. The Free Press, New York, p 13
2. Jones BZ, Boyd LG (1971) The Harvard college observatory: the first fur directorships, 1839–1919. The Belknap Press of Harvard University Press, Cambridge, MA, p 307
3. Tuchman BW (2012) In: Macmillan M (ed) The guns of August and the proud tower. Literary Classics of America, New York, p 698
4. Eddy JA (1986) Review of George E. Webb, tree rings and telescopes: the scientific career of A.E. Douglass. J Hist Astron 17:69–71
5. Putnam J, Sheehan W (2021) A complicated relationship: an introduction to the correspondence between Percival Lowell and Giovanni Virginio Schiaparelli. J Astron Hist Heritage 24(1):170–227
6. Holden ES (1894) The Lowell observatory in Arizona. Publ Astron Soc Pac 6:160–169
7. Ruskin J (1971) The elements of drawing. Dover Publications reprint of 1904 edition, New York, p 26
8. Hanson NR (1958) Patterns of discovery: an enquiry into the conceptual foundations of science. Cambridge University Press, London/New York, p 25
9. Salom A et al (1989) A life of physics. World Scientific Publishing, Singapore, pp 31–35
10. Friston KJ et al (2018) Computational psychiatry: the brain as a Phantastic organ. Lancet Psychiatry 1:148–158
11. Lowell P (1894a) Mars. Astron Astro-Physics 13:538–553
12. Campbell WW (1896) Mars, by Percival Lowell. Publ Astron Soc Pac 8(51):207–200
13. Walter Maunder E (1894) The Tenuity of the Sun's surroundings. Knowl A Mon Rec Sci (March 1):49–50
14. Maunder EW (1894) The canals of Mars. Knowledge (November 1):249–252
15. James W (1902) The varieties of religious experience. Henry Holt & Co, New York, pp 23–24
16. Barnard EE (1893) Micrometer measures of the fifth satellite of Jupiter. Astron J 12:161–174, 161
17. Holden ES (1893) Screens to protect telescopes from wind tremors. Astron Astrophys 12:471
18. Sellers D (2019) The early history of Leeds astronomical society—1859–1918. Leeds Astronomical Society, 2019, Leeds, p 27
19. Barnard EE (1893) Photographs of brooks' comet. Pop Astron 1:145–147, 146-147
20. Barnard EE (1894) Photographs of a remarkable comet. Astron Astrophys 13:789–791, 790
21. Ranyard AC (1893) What is a nebula? Knowledge 16:10–12

22. Mussell J (2009) Arthur Cowper Ranyard, *knowledge*, and the reproduction of astronomical photographs in the late nineteenth century periodical press. Br J Hist Sci 42:345–380
23. Wesley WH (1894) On the distribution of stars in the milky way. Knowledge 17:179–182, 182
24. Barnard EE, Ranyard AC (1894) Structure of the milky way. Knowledge 17:253
25. Denning WF (1891) Telescopic work for starlight evenings. Taylor and Francis, London, pp 29–30, 33
26. Brenner L (1895) Large versus small telescopes. J Br Astron Assoc 5(3):166–167, 167
27. Lowell P (1894) Mars. Pop Astron 2:1–2
28. Pickering WH (1914) Monthly report on Mars—no. 6. Pop Astron 22:407–422, 412
29. Young AT (1971) Seeing and scintillation. Sky Telescope 42:139–141
30. McKim RJ (1999) Telescopic Martian dust storms. Mem Br Astron Assoc 44:28–31

5

A Tale of Two Telescopes

Contents

> *Beyond the path of the outmost sun through utter darkness hurled—*
> *Further than ever comet flared or vagrant star-dust swirled—*
> *Live such as fought and sailed and ruled and loved and made our world...*
> *'Tis theirs to sweep through the ringing deep where Azrael's outposts are,*
> *Or buffet a path through the Pit's red wrath when God goes out to war,*
> *Or hang with the reckless seraphim on the rein of a red-maned star.*
> *—Rudyard Kipling, Dedication,* Barrack-Room Ballads

Hanging on the Rein of a Red-Maned Star

And so the memorable opposition of 1894 ended. At the end of November, Lowell and Pickering returned to Boston. Pickering returned to Harvard from his sabbatical. Henceforth he and Lowell's paths would diverge.[1] Rather than return to Sevenels, Lowell took occupancy of a small high house at No. 11 West Cedar Street for the winter.[2] He may have wished to strike a new note of

© The Author(s), under exclusive license to Springer Nature Switzerland AG 2024   **257**
W. Sheehan, *Parallel Lives of Astronomers*, Springer Biographies,
https://doi.org/10.1007/978-3-031-68800-3_5

confidence and independence—but he may also have wanted to escape witnessing the final stages of his mother's long and painful illness. An invalid for many years, she had been diagnosed with Bright's disease, consisting (according to the medical lights of the time) of a triad of dropsy (edema), albumin in the urine, and kidney disease. Since the family need not worry about expense, she had doubtless taken her turns with such treatments as warm baths, bloodletting, squill, digitalis, mercury compounds, opium, diuretics and laxatives, and various diets, but ultimately all in vain. For the last year or so her health had taken a serious turn for the worse—and this may help to explain why her "Darling Percy" was so often back in Boston during the summer of the great Mars campaign. Though we do not know the details of Elizabeth Lowell's particular case, her daughter and Percival's sister Bessie—later to become a campaigner for better prenatal care—became convinced that her Bright's disease had been brought on by childbirth and complications related to eclampsia [2]. Whatever the case, Katharine probably lingered for a while before death, mercifully, came at last in April 1895.

Percival dealt with his grief by throwing himself all the harder into his work. In addition to writing up articles about Mars, he sometimes made the rounds of his publishers—as Ferris Greenslet says of a somewhat later time, "usually two or three times each winter he would look in at the publishing offices on Park Street … to inquire into the continuing sale of his Japanese books or discuss the possibility of a new edition."[3] Greenslet came to Boston from New York in 1901 and served as an editor for the *Atlantic Monthly* and later a literary advisor and director of the Houghton Mifflin Co. He remembered vividly meeting Lowell on some of the occasions when he came around, and wrote:

> This reporter has met many of the so-called great men of his time, but none with a more potent personal quality than Percival Lowell. He agrees with another witness that one felt it even before, or almost before, he entered the room. It was as if one had been suddenly deposited in a powerful magnetic field.[4]

Quoting an unidentified member of the staff at Flagstaff who had called Lowell "buoyant with strength, ambition, love, sincerity of purpose, in fact all that was highest in life," the editor found but one exception that seemed to prove the rule; the report of an "unimpeachable witness" who "once saw him personally eject his butler down the steps of his house on West Cedar Street and hurl a steamer trunk after him."[5] It was not the only or last time the Great Man would act with violence toward a subordinate. Doubtless, under the circumstances, Lowell believed the man had only received the treatment he deserved.

Above all, Lowell kept his eagle eye trained on developments at the Observatory in Flagstaff. Unlike Pickering who had left, Douglass remained in Lowell's employ, and stayed on through an abysmally dreadful winter. As Mars shrunk from its maximum size in October 1894 to a mere paring of its erstwhile self (4 arcseconds across) the following April, Douglass did his best to keep it under constant surveillance. But most of the time the effort was futile, and Douglass could not hide from Lowell the fact that the vaunted Flagstaff seeing was generally awful. He must have received in alarm Lowell's pronouncements in which he expressed in increasingly damning terms his disappointment. In January—when Flagstaff received 126 inches of snow for the month which, though an unofficial reading, has yet to be surpassed— Lowell wrote, "Flagstaff in the winter is a fraud, apparently."[6] In March, he added, "The seeing seems to be so perpetually poor now that I see little use in keeping up the observatory any longer."[7]

Since Flagstaff in the winter was obviously hopeless, and the next opposition of Mars was due to occur on December 1, 1896, Lowell decided to try to observe it in a more favorable climate. Thus, in the spring of 1895, Lowell dispatched Douglass to Mexico to scout sites between Chihuahua and Mexico City. Though Douglass found no site obviously better than Flagstaff in summer (based on the experience of 1894), in winter Mexico was obviously superior. In the meantime, though the work of observing Mars was over, Lowell's and his Observatory's efforts to interpret and publicize the 1894 observations went on at breakneck pace.

Lowell's literary output was extraordinary. He churned out article after article, adding to the three *Astronomy and Astro-Physics* articles written in July, September and November 1894, five more in *Popular Astronomy* between September 1894 and March 1895 ("Mars," "The Polar Snows," "Spring Phenomena," "Atmosphere," "The Canals I," and "Oases"), four more in *The Atlantic Monthly* between January and August 1895 ("Atmosphere," "The Water Problem," "Canals" and "Oases"), and two, of a somewhat more technical cast ("On Martian Longitudes," May 1895, and "Evidence of a Twilight Arc Upon the Planet Mars," August 1895) which were published, with more reluctance than enthusiasm, in the first volume of the *Astrophysical Journal*, edited by George Ellery Hale and James Keeler. (This journal was intended to replace and to establish a higher standard than its predecessor *Astronomy and Astro-Physics*, which Hale had co-edited with W.W. Payne of Goodsell Observatory in Northfield, Minnesota; Payne continued to edit *Popular Astronomy* as before.) In addition, in February 1895 Lowell delivered a series of evening lectures before the Lowell Institute (Augustus was still trustee) in Huntington Hall on the MIT campus—the first of a number of lecture series he gave over the years, to which the public response would always be enthusiastic.

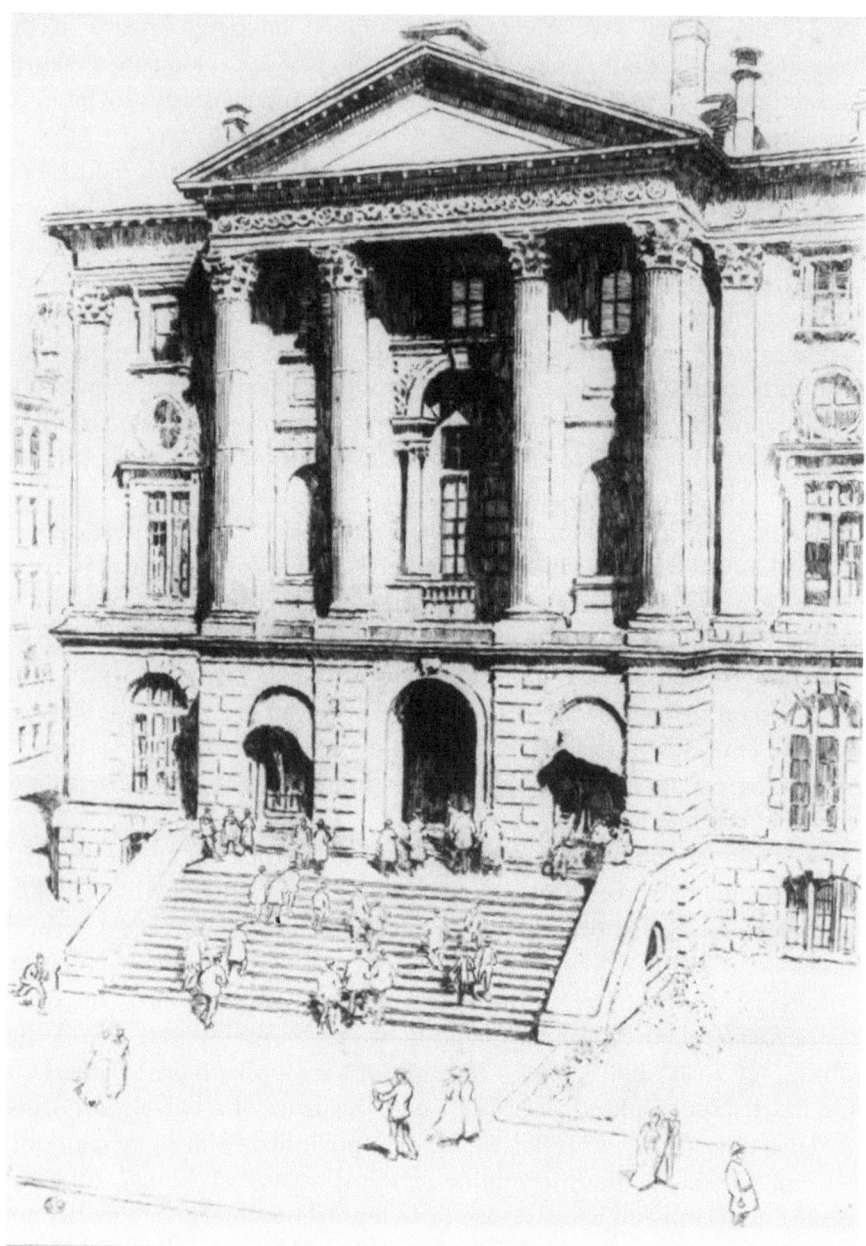

Huntington Hall on the MIT campus, circa 1895, when Percival Lowell gave the Lowell Institute lectures there. (Credit: Public Domain courtesy of MIT Museum)

Lowell in an undated photo, but likely February 1895 when he gave the Lowell Institute lectures. (Credit: Lowell Observatory Archives)

Among those in attendance was author and Unitarian minister Edward Everett Hale (nephew of the celebrated orator Edward Everett), who published the following vivid account:

Mr. Lowell's four lectures on the planet Mars were heard by crowded audiences of people who filled every seat and all the standing-room in Huntington Hall. For once, we got the very latest advices from that planet. The observatory in Flagstaff, as our readers know, was established by Mr. Lowell himself, and the position of Mars in the last summer gave him opportunity to make such observations as have never been made before, and to reveal to us what are marvels indeed. The result, as our readers know, is the firm conviction in his mind that intelligent beings occupy the planet Mars, who know how to work in the common good, who have contrived public works of vastly larger extent than we of the Earth have dreamed of, and have carried out their contrivances with a precision and strength wholly unknown in mundane affairs....

There are not more than twenty people in this Earth who have seen what he has seen. Even some of the great observatories of the world are so situated that they have not noted the marvels which the Flagstaff observatory has revealed to us. But truth is truth, and it matters but little whether at this moment it have twenty apostles or two thousand. It is certain that the revelations which the Flagstaff observatory has made from its signal station to the world are revelations which shall be accepted....[8]

From this veritable raft of material, articles, newspaper accounts, and lectures—many of course involving repetitions or variants on the same themes—he would select the best bits, carefully rework them, and give them a final literary polish in *Mars*, his best-selling book, the manuscript of which was finished in the West Cedar Street house *en garçon* during November 1895. The scene of this literary activity was captured by Greenslet when, on a later occasion, as a young editor passed with a bag of manuscripts to his own modest establishment in the next block, he "used to observe Lowell every weekday at five-thirty. His handsome head was to be seen *vis-à-vis* the *Boston Evening Transcript* beneath a life-sized plaster Venus similar to those that infest the Athenaeum. Visibility was perfect, for the shade was always raised to the very top of the window as if to admit no impediment to a message from Mars."[9]

Lowell was, at least at first, highly successful in gaining widespread support for this theory of intelligent life on Mars—an achievement that was equally indebted to his active publication strategies, to his claims of having unrivaled seeing at Flagstaff, and to the visual authority of his maps (discussed below). He has had few rivals before or since in his ability to cross disciplinary boundaries and make complex subjects accessible to the public. The reaction of Lafcadio Hearn, who had lived in Japan since 1891, was fairly typical: "It is strange that Lowell should have written the very best book in the English language on the very old Japanese life and character, and the most startling *astronomical* book of the period, —'Mars'—more interesting than any romance."[10]

Lowell was first and foremost a literary man, and though as he had done as a Harvard undergraduate, he kept a foot in both the scientific and literary cultures, at least at this stage of his career he put decidedly more weight on the literary foot. At least among his more rigorously scientific peers (including Barnard), he never became a proper scientist at all. Lowell's biographer David Strauss emphasizes that his appeal to readers lay in

> … the dramatic and personal way in which he presented his work on Mars and East Asia. No doubt, his enthusiasm for the literary romances of Robert Louis Stevenson, Edward Bulwer-Lytton, and Frederic Stimson played an important role in Lowell's conceiving of science as a story that could, as Hearn pointed out, rival the adventures recounted in novels. Lowell's books thus appealed to an audience of men and boys who sought in literature, as in life, male heroes with whom they could identify. The romance provided the perfect vehicle to deliver these stories to the public in an attractive form.[11]

"The romance worked for Lowell's readers," Strauss adds, "because they experienced a vicarious sense of participation through identification with the narrator or hero."[12] Lowell gave to his accounts a strong quality of eyewitness reporting, often enlisting personal friends (accomplished in law or zoology or letters but not in astronomy) such as Judge Edward M. Doe and Edward Sylvester Morse to the telescope to stand in for "everyman" readers in verifying the detail that he saw. So powerful was this technique that readers often came away as convinced as if they had seen for themselves. In addition, he encouraged readers to take sides by allowing them into his confidence, asking them to form an alliance with the author as he presented himself as an underdog waging a lonely battle against prevailing scientific opinion to establish his own views. Certainly, "… this emphasis on the personal adventure in doing science was well received by the educated public, Lowell's astronomy colleagues were clearly disturbed by his emotionally charged rhetoric and egocentric focus."[13]

An example of how he worked up sometimes rather meager data into a personalized adventure story appears in what on its face appears to be a striking but rather ordinary observation made early in the Flagstaff campaign. Just after sunrise on June 7, 1894, he was observing alone with the 18-inch when he saw—or thought he saw; probably it was just a trick of the seeing—a pair of "dazzling white specks" on the south polar cap. The contemporaneous record of the observation occupies only a single line in his observing log book! However, in his book *Mars*, it is expanded into a composition full of the most glorious purple prose of which his ornate and flowery style was capable.

Meanwhile an interesting phenomenon occurred in the cap on June 7. On that morning, at about a quarter of six …, as I was watching the planet, I saw suddenly two points like stars flash out in the midst of the polar cap. Dazzlingly bright upon the duller white background of the snow, these stars shone for a few moments and then slowly disappeared. The seeing at the time was very good. It is at once evident what the other-world apparitions were,--not the fabled signal-lights of Martian folk, but the glint of ice-slopes flashing for a moment earthward as the rotation of the planet turned the slope to the proper angle just as, in sailing by some glass-windowed house near set of sun, you shall for a moment or two catch a dazzling glint of glory from its panes, which then vanishes as it came. But though no intelligence lay behind the action of these lights, they were none the less startling for being Nature's own flash-lights across one hundred millions of miles of space. It had taken them nine minutes to make the journey; nine minutes before they reached Earth they had ceased to be on Mars, and, after their travel of one hundred millions of miles, found to note them but one watcher, alone on a hill-top with the dawn [3].

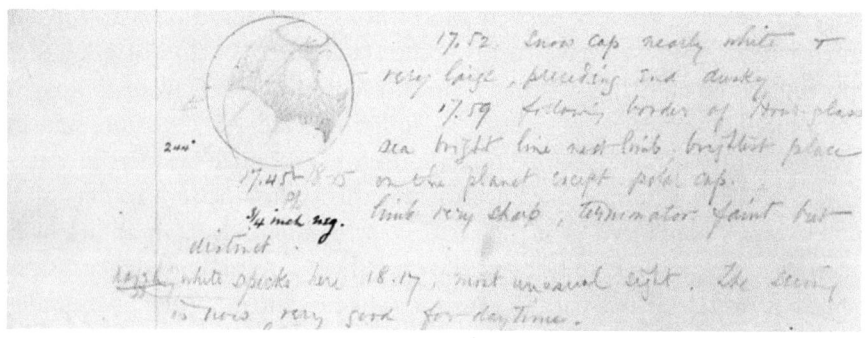

Lowell's sketch of June 7, 1894, in which he indicates the position on the South Polar Cap of "dazzling white specks … a most unusual sight." He identified their position in the midst of the cap with the "Mitchel Mountains," seen as an area detached from the rest of the cap by O.M. Mitchel in 1845 and by Nathaniel Green in 1877. (Credit: Lowell Observatory Archives)

Another characteristic of his writing is that it was marked by a strong tendency to dogmatic assertion. In part, this was owing to his deductive style of reasoning, where the conclusions followed, rather syllogistically, from the premises. The latter had to be far from certain in order to set the chain of reasoning going; they had merely to be plausible. And so he could never resist the temptation to present possibilities as probabilities, probabilities as certainties—a style of presentation which, at least in the court room, sways juries. According to experiments on testimony by psychologist Elizabeth Loftus, jurors will "trust almost totally the witness's report and largely disregard objective information." Also, they are more likely, as noted by linguist George Lakoff, to respond positively to "power language," typically more characteristic of male than female speech, in which hedges and qualifiers are absent and instead statements are expressed directly and emphatically and without hesitation. (It should be kept in mind, by the way, that women were not even allowed on juries in Lowell's day, and most of his readers were undoubtedly men.[14]) The last pages of his book *Mars* read like the closing argument for the prosecution delivered by an extremely able lawyer, and would have found few jurors who would not have found the Martians "guilty as charged"; guilty, that is, of inhabiting a dying planet, and covering it with a vast network of artificial canals. Thus, Lowell ran through his arguments one by one, with each gaining plausibility from the last:

> To review, now, the chain of reasoning by which we have been led to regard it probable that upon the surface of Mars we see the effects of local intelligence. We find, in the first place, that the broad physical conditions of the planet are not antagonistic to some form of life; secondly, that there is an apparent dearth of water upon the planet's surface, and therefore, if beings of sufficient intelligence inhabited it, they would have to resort to irrigation to support life; thirdly, that there turns out to be a network of markings covering the disk precisely counterparting what a system of irrigation would look like; and, lastly, that there is a set of spots placed where we should expect to find the lands thus artificially fertilized, and behaving as such constructed oases should. All this, of course, may be a set of coincidences, signifying nothing; but the probability points the other way....[15]

As significant as Lowell's rhetorical devices were his drawings and maps. It was the case, as Lowell himself would likely have quipped, that a picture is worth a thousand words. As the historian and geographer K. Maria D. Lane has argued in her highly original book *Geographies of Mars*, "maps are typically viewed as objective representations of reality.... Cartographic representation has therefore been extremely powerful in its ability to present certain claims as inevitable or incontestable" [4]. In the history of Martian cartography, Schiaparelli, whose map of 1877-78 showed "a greater level of detail with a definitive style of marking," won out over the naturalistic map produced at the same opposition by Nathaniel Green, whose "hazy and indistinct colorations ... faced the impossible challenge of demonstrating *more* authority by presenting *less* detail."[16] Thus the Schiaparellian tradition established itself as authoritative, and a series of successive observers attempted to outcompete one another by adding more and more canals to the map, attesting to another feature of cartography emphasized by Lane, that "given the authority and nature of the cartographic data-recording format, it was nearly impossible to erase canals that had been mapped by a credible astronomer. Just as was true for many of the terrestrial expeditions of the day, prestige inhered in putting things *on* the map, not taking them off."[17] Not only did the craze to add more and more canals threaten the integrity of Mars maps by introduced more and more uncertain details, but it invited the unscrupulous—observers like the Serbian-Austrian astronomer Leo Brenner—and created the embarrassing situation where the canals indicated on the various maps often did not agree in position or orientation. In addition, mapping canals as Schiaparelli did required skill in draftsmanship but no outstanding artistic ability, and so observers of limited artistic skill, who included Pickering, Douglass, and Lowell—succeeded in passing themselves off as virtuosos wielding the pencil, rule and compass, whereas they would have shown very badly with paint and a brush. Thus, the drawings and maps they produced looked highly artificial and unrealistic, produced with a knowledge of what was to be arrived at, and notably, Lowell's Mars resembled his Noto, about which he had said, "panoramic views are painfully plain. They must needs be mappy at best, for your own elevation flattens all below to one topographic level." For Mars, like Noto, "you [had] might as well lay any good atlas on the floor and survey it from the lofty height of a footstool…. The geometric lines stood in ludicrous insistence," and "any child could have drawn the thing …mechanically."[18]

Readers viewing Lowell's Mars map of 1894 might, for a moment, have been puzzled by what it purported to show. They likely would not have considered that it had been produced by a man of evident literary ability, extensive Far Eastern travels, but very limited experience in astronomical observations and unknown, but probably fairly limited, artistic skill. What they did know—and Lowell repeated it over and over again until it seemed to leave no doubt as to its truth—was that he had a keen eye and observed at Flagstaff where the air was good as was to be had anywhere on the surface of the Earth. His map showed more detail than any hitherto published—confirming not only all but two of Schiaparelli's canals but adding 116 more to make a grand total of 184. In addition, he recorded 64 of the dark spots which Schiaparelli had called "lakes" but he rechristened "oases," which were often found at the intersections of several canals. Moreover, his canals were of the exceeding narrowness—"fine lines and little gossamer filaments, cobwebbing the face of the Martian disk, but threads to draw one's mind after them across the millions of miles of intervening void" [5], as he would put it later; the oases were no more than tiny dots. Admittedly, "when the air was not steady more difficult details, such as the canals, showed as broadish streaks smooching the disk." However, "in the best air they contracted in width, standing revealed as narrower lines. The same thing was true of the round dots at their intersections" [6]. All of these numerous features were given "a local habitation and a name," ordered and catalogued and legitimized and turned into solid substantial objects by acquiring impressive- sounding Latinate names, ultimately based on the Schiaparellian scheme of nomenclature but adroitly expanded to meet the growing need. Lowell had once said that "naming a thing is man's nearest approach to creating it."[19] Then he stood very near to the Creation itself.

One look at this map and the reader would have immediately concluded that, if Mars looked anything like it, Lowell's theory must be true, for the appearance was thoroughly "unnatural" and "artificial." Moreover, regardless of what other astronomers thought of his theory, Lowell generally was accorded respect for his maps. Even Simon Newcomb, a noted critic of Lowell, wanting a good map with which to illustrate an article in *Encyclopaedia Britannica* he was writing, appealed to Lowell, saying, "I know no better source than the publications of your observatory."[20]

Note that there was nothing in Lowell's approach like that of an artist—a Claude Monet, for instance, who had once said, "When you go out to paint, try to forget what object you have before you, a tree, a house, a field, or whatever. Merely think, here is a little square of blue, here an oblong of pink, here a streak of yellow, and paint it just as it looks to you, the exact color and shape, until it give you your own naïve impression of the scene before you."[21] This whole approach is sometimes referred to as the "innocent eye," a term introduced by John Ruskin, who said that "the whole technical power of painting depends on our recovery of … a sort of childish perception of these flat stains of color, merely as such, without consciousness of what they signify—as a blind man would see them if suddenly gifted with sight."[22] Though ultimately innocence of the eye is an unattainable ideal, it does get at something of the same thing we have tried to capture in the metaphor of the flower and the bee. To be sure, there was no "innocence" in Lowell's eye; to a greater extent, there was in Barnard's. Nor was there anything in Lowell's approach resembling Ruskin's theory about unclarity, or "mystery," in nature. As described by Leonard Campbell, the artist "who represents what he sees before him is able to see either not at all or only in part … providing he can rid himself—as much as is humanly possible—of the pictorial conventions through which he, and to a great extent, all of us who have been exposed to paintings, illustrations or photographs, have learnt, since infancy, to grasp nature conceptually."[23] Applied in the present case, this meant that he ought not to study and rehearse and memorize in every detail Schiaparelli's map before turning to the eyepiece to look at the planet directly. Ruskin did not, of course, discuss attempting to draw an image in a telescope, but instead suggested trying to draw a bank of grass, with all its blades, or a bush, with all its leaves. But the lesson would have been the same: "you will soon begin to understand," he said, "under what a universal law of obscurity we live, and perceive that all *distinct* drawing must be *bad* drawing, and that nothing can be right, till it is unintelligible."[24] In Ruskin's terms, almost every single drawing of Mars made during the heyday of the so-called "Mars furor," the 1890s, was a bad drawing.

In fact, it would have been impossible, presented with an ambiguous and inconstant stimulus like Mars in the eyepiece, to approach it with an innocent eye, as the ability to make out anything at all required looking *for* rather than simply *at* the planet. An observer like Schiaparelli or Lowell—bent on cartography—existed for the purpose of making the surface of the distant planet intelligible, even impossibly so. As a result, the observer had to make do not with fully realized perceptions but fragments of perception, built up little by little into a coherent—though possibly misleading—whole. The casual reader or consumer of astronomical maps would not have appreciated that all of the detail shown in Lowell's maps could not possibly have been visible at once, much less seen (or "glimpsed"), stored in short-term memory, then transcribed, with as much speed as the artist could command, to a sketchpad. It was very unlike what many readers who had never attempted to observe through a telescope imagined—namely, that sketching a planet somehow resembled drawing a still life like a vase-full of flowers or a bowl of fruit. Because of the relentless churning of the air, the abundant details came and went fleetingly; the system of canals—or lack of system of whatever the surface markings might actually consist of—revealed itself in piecemeal fashion, a few canals at a time, and then only for a fraction of a second, in flashes, like a dance of the seven veils. The process of learning how to see them is attested in Lowell's observing log book:

> June 3. "Glimpse of pig-tail turning to the east was the beginning of the Nilosyrtis."
>
> June 7. "The marking on the meridian north of the centre of the disk is probably the Lethes…. In the glimpses in which it was caught, sometimes thought it more inclined to the north. This was probably the Amenthes failing into it."
>
> June 9. 16h 35m. "[Trivium Charontis] Glimpsed very thick canal along parallel."
>
> --"Lethes and Amenthes both out; the former ending in a black hole [the western end of the Propontis] very near the limb."

June 10. 17h 50m. "Canal so persistently visible is Cyclops probably."

June 13. "Canals once (about 16h 25m) under higher power (1/4 or 3/8 inch eyepiece) and good seeing appeared as narrow lines."

June 15. 17h 23m. "Canals were seen only imperfectly, and all needed more or less confirmation. The darkest are most certain."

June 16. At 17h 20m. "If this was no cheating vision, it was the Eumenides."

June 19. 17h 20m. "With the best will in the world, I can certainly see no canals."

-- Fifteen minutes later the canals came out distinctly and four good drawings were made of them.[25]

The canals were fugitive visions, gone as soon as they came. Glimpse by glimpse, canal by canal, the observer gradually acquired confidence in what was seen, in a process that he would never describe better than in the following passage in *Mars and Its Canals* (1906):

> When a fairly acute-eyed observer sets himself to scan the telescopic disk of the planet in steady air, he will, after noting the dazzling contour of the white polar cap and the sharp outlines of the blue-green seas, of a sudden be made aware of a vision as of thread stretched somewhere from the blue-green across the orange areas of the disk. Gone as quickly as it came, he will instinctively doubt his own eyesight, and credit to illusion what can so unaccountably disappear. Gaze as hard as he will, no power of his can recall it, when, with the same startling abruptness, the thing stands before his eyes again.
>
> By persistent watch ... for the best instants of definition, backed by knowledge of what he is to see, he will find its comings more frequent, more certain and more detailed. At last some particularly propitious moment will disclose its relation to well known points and its position be assured. First one such thread and then another will make its presence evident; and then he will note that each always appears in place. Repetition in situ will convince him that these strange visitants are as real as the main markings, and as permanent as they.[26]

Imagine trying to make a map of the Earth by adding in a continent, seen for a fraction of a second then dissolving, or a sea, or an island, or a river—following each apparition successively, sketching as quickly as one can before it fades from memory, then overlaying each impression one on top of the other at the end. Or imagine sketching clouds as they appear and dissolve—and treating them all like solid substantial masses—that, in a sense, describes Lowell's map of Mars.

For the observer or reader willing to to engage the "suspension of disbelief," these partly illusory emanations became solid facts. His honeybee-like "buzzing here and there impatiently from a knowledge of what is to be arrived at" at last alights and comes to rest in a kind of dreamland. There, as he once wrote in *The Soul of the Far East*, "Only the superficial never changes its expression; the appearance of the solid varies with the standpoint of the observer. In dreamland alone does everything seem plain, and there all is unsubstantial."[27]

A Not-So-Innocent Abroad

With *Mars* off his hands, Lowell set sail for Europe, taking along with him 28-year-old Wrexie Louise Leonard, a native of Troy, Pennsylvania, whom he had met in Boston in 1893 and hired as his personal secretary, a role which was actually far more demanding than the word "secretary" suggests (today we might say "executive assistant") and in which she would remain until his death. She was intelligent, infallibly loyal (as all of Lowell's employees had to be), and supported him in ways that went beyond the office to even include taking turns at the telescope to make observations of the planets that were strikingly similar to his own. Though her role was completely subordinate, it did allow her to go about as far as a woman from her background and experience could hope to go during those late-Victorian times, when women were as stifled in career opportunities as in their dress—and even Lowell's mother had been so—and not even allowed to vote. (Percival, by the way, a staunch conservative in things political, would oppose women's suffrage until his dying day.) She was clearly deeply in love with him—indeed, he had a magnetic effect on women—and it appears likely that they had a physical relationship during the early years of their association, though there is no direct evidence and, as Lowell himself might have said, it is impossible to prove a negative. Later, his great nephew (and Sole Trustee of the Observatory) William Lowell Putnam III told me that in later years the Lowell staff could never understand why Percival never married Miss Leonard, but explained that it would have been no more reasonable for someone of his social class to marry his secretary than for him to have taken "a fair Jap partner" during his Far Eastern period.[28]

Wrexie Louise Leonard, observing at the 24-inch Clark refractor, probably about 1897. Though Percival's secretary, she played many roles for him, including regularly taking turns at the telescope. She was completely loyal, and became the most consistent member of his staff in confirming the controversial spoke system he saw on Venus. (Credit: Lowell Observatory)

Leonard, in typical late Victorian costume, strikes an attitude perhaps consciously recalling that of the girl in Jean Honore Fragonard's famous painting, "Young Girl Reading." Photo probably taken about 1900. (Credit: Lowell Observatory Archives)

In Boston, Lowell was feted by the Boston *Globe* as a departing hero, "a scholar by instinct, and an astronomer by choice." He survived a mishap in the early morning hours of December 19, 1895, as his steamship approached the port of Southampton—the pilot, "aged 73 and bordering on imbecility," Lowell remarked, veered to starboard rather than toward port and ended up lodged on a reef, requiring the service of a tug to make it to port. A few days later, in Paris, he headed to Flammarion's flat in the top floor of a five-story apartment block at 16 Rue Cassini, across the Street from the Paris Observatory. Though Flammarion spent his summers at Juvisy, where Barnard had met him, he wintered in town, and though Lowell and Flammarion had corresponded they had never met. Flammarion, however, was a great admirer of Lowell's work, and honored him with dinner in his astronomically themed apartment. "There were fourteen of us," Lowell wrote, "and all that could sat on chairs of the Zodiac under a ceiling of pale blue sky, appropriately dotted with fleecy clouds, and indeed most prettily painted."[29] No doubt the fourteen included Flammarion's wife Sylvie, and may also have included Flammarion's assistants at Juvisy, Ferdinand Quénisset and Eugène Michel Antoniadi—though if he met the latter, who lived not at Juvisy but in Paris, he cannot have possibly imagined what a crucial role the 26-year-old astronomer from a Greek family in Constantinople would later play in the Martian canal debate. At the time Antoniadi was about to add to his role at Juvisy directing the Mars Section of the British Astronomical Association, and to sort out the reports of members of the Section which invariably included numerous canals. Indeed, Antoniadi's own drawings and maps at the time were heavily influenced by Schiaparelli's drawing style. Several maps from 1896—by Antoniadi, Vincenzo Cerulli at his private observatory at Teramo, and by Leo Brenner—all rival Lowell's own in the number and thinness of their canals.

In Paris, he visited the Paris glassmaking firm of optician Mantois, which had provided the glass used in the Lick 36-inch refractor about procuring a 24-inch lens for a refractor to replace the borrowed 18-inch used in 1894. He then secured, through the Société Astronomique de France, which Flammarion had founded, the assistance of Charles Trèpied, director of the Algiers Observatory, to test astronomical sites in the Sahara. Together with Trèpied, Leonard, and Ralph Curtis, he visited Boghari and Biskara and even sites on the northern fringe of the Sahara Desert, but in the end, none trumped Flagstaff. Recrossing the Mediterranean, he continued his travels, setting his sights above all on Milan where he hoped to meet his "Martian Master," Schiaparelli, at Brera. Deteriorating conditions in the increasingly industrialized city of Milan and growing problems with Schiaparelli's eyesight—his right eye was congenitally poor, and his left eye (perhaps, as he believed, damaged by observing Mercury too close to the Sun) was useful for ordinary purposes but was so near-sighted that he had to read with a magnifying lens. Lowell recorded his impressions of their interview; when he described to the latter the small spots at the intersections of the canals—his "oases" — Schiaparelli's eyes lit up: "I suspected them myself," he said, "but could never see them well enough."[30] Actually, Schiaparelli's views of Lowell at this stage (they were to evolve over the years) were rather guarded; he worried to his old friend Otto Struve at the Pulkovo Observatory about the profusion of canals on recent maps (including Lowell's), and expressed his hope that Lowell and Cerulli, another prolific recorder of canals, "will not see too much ... but will see well."[31] Lowell, of course, did not know anything about Schiaparelli's skepticism as he headed off to Lussinpiccolo, Istrien, in the Adriatic, to meet Brenner, who had (as noted above) with only a 7-inch refractor nearly out-Lowelled Lowell with the numerousness and intricacy of his canals. His quarrelsome and self-destructive nature led, however, almost immediately to an irreparable rupture between them, and Brenner often enjoyed disparaging Lowell, often in highly personal terms, in later years.

Lowell was back in Boston in March 1896, awaiting the verdict of the reviewers. There is no reason to think he was in the least apprehensive. He assured Douglass, with one of his typically atrocious puns, that regarding the canal network, "no explanation yet advanced [apart from that of their artificiality] will ... hold water. The system is what would occur on an old waterless planet if inhabited.... *Voila*, quite unconvincing until you go into the details when the argument proves very strong."[32]

The book was certainly widely noted, and generally well received by the general public and even a few professional astronomers. Journalist Garrett P. Serviss, who wrote on science topics for the Hearst newspapers, approved. "[H]is theory accounts very well for what is seen. At any rate, it is the most complete theory that has yet been advanced." Lick Observatory astronomer William Wallace Campbell—hired in 1891 to replace Keeler in taking charge of the Observatory's spectroscopic work—noted that Lowell had taken "the popular side of the most popular scientific question about,"[33] and that his observational results had agreed with his statements in the lecture hall before he put eye to telescope at all. And yet he found a great deal to like about the book, and commended Lowell's "lively and entertaining style," his allocation of a good part of his considerable private means to establish and conduct an observatory, and his search after the best seeing conditions. He did insist that Lowell's views were not as novel as might seem, and that many of them were actually based on old ideas of W.H. Pickering and even Schiaparelli. He also thought that the whole concept of a planet-wide irrigation system such as that Lowell envisaged made no sense from a hydraulic engineering standpoint:

> We are asked to believe that the equatorial region of Mars, forming a strip at least seventy degrees wide, can be and is irrigated from the north and south poles; the "canals" in the two cases of opposite flow being identical! The contemporary problem on Earth would be to irrigate San Francisco, Chicago, New York, Rome, Tokyo from snow melting at the South Pole; and to irrigate Valparaiso, Cape of Good Hope, Australia, from the snow melting at our North Pole; all the irrigated land lying between New York, etc., on the north and the Cape of Good Hope, etc., on the south, to be irrigated alike from the North and South Poles.[34]

Similar objections were raised with even greater directness by the respected historian Agnes M. Clerke, who in the *Edinburgh Review* opined:

> We can hardly imagine so shrewd a people as the irrigators of Thule and Hellas wasting labour, and the life-giving fluid, after so unprofitable a fashion. There is every reason to believe that the Martian snow-caps are quite flimsy structures. Their material might be called snow *soufflé*, since, owing to the small power of gravity on Mars, snow is almost three times lighter there than here. Consequently, its own weight can have very little effect in rendering it compact....
>
>   No attempt has yet been made to estimate the quantity of water derivable from the melting of one of these formations; yet the experiment is worth trying as a help towards defining ideas.... [We find that] only one-seventh of a foot of water, ... could possibly be made available for their fertilization, supposing them to get the entire advantage of the spring freshet. Upon a stint of less than two inches of water these fertile lands are expected to flourish and bear abundant crops; and since they completely enclose the polar area they are necessarily served first. The great emissaries for carrying off the surpluses of their aqueous riches would then appear to be superfluous constructions, nor is it likely that the share in those riches due to the canals and oases, intricately dividing up the wide, dry, continental plains, can ever be realised.... Further objections might be taken to Mr. Lowell's irrigation scheme, but enough has been said to show that it is hopelessly unworkable.[35]

Behind the scenes, the professionals generally took umbrage at Lowell's presumption in setting himself up as an authority, despite being little more than a neophyte observer. The strongest objections were stated by the generally even-tempered Keeler. In addition to his pioneering work as an astrophysicist, he was an excellent visual observer and artist, and wrote to Hale, "I dislike [Lowell's] style… [I]t is dogmatic and amateurish. One would think he was the first man to use a telescope on Mars, and that he was entitled to decide offhand questions relating to the efficiency of instruments; and he draws no line between what he sees and what he infers."[36] Both men agreed that manuscripts Lowell had submitted to the *Astrophysical Journal* had been editorial headaches as well as falling short of the high standards which they hoped the new publication would reach, and banished Lowell from their journal. In practical terms, he wasn't much affected, since other outlets remained to him, including *Popular Astronomy* (still edited by Payne), *Atlantic Monthly*, *Monthly Notices of the Royal Astronomical Society*, *Astronomische Nachrichten* and— when all else failed—his own outlets. Three volumes of the *Annals of the Lowell Observatory* would be published at Lowell's expense between 1898 and 1905, followed by a long series of Lowell Observatory *Bulletins*. For Lowell, money was speech.

In addition, the dissent—and even disgust—of many professional astronomers made little difference to the public—or indeed to Lowell's long-lasting legacy in shaping human thought about Mars. He offered what was ostensibly a "theory," in the rigorous sense of being "a supposition or system of ideas intended to explain something, especially one based on general principles independent of the thing to be named."[37] Here, the evolutionary entropic sequence deriving from the nebular hypothesis provided the principles. It was a theory that depended on a vast amount of circumstantial evidence (much of it already provided by Schiaparelli, Pickering and others) and could neither, in the final analysis, be either proved or disproved (falsified). But the idea of Lowell's theory as being a rigorous scientific formulation involved, in fact, a categorical error; it was more tale than theory, and one which was matched to the moment. During the *fin-de-siècle* epoch when things on Earth sometimes seemed to be tottering toward the abyss (and another 20 years hence indeed would do so), it proved simply irresistible to vast swathes of readers, as a majestic, colossal tale of an extraterrestrial civilization struggling admirably and courageously and with pathos to avert—or at least postpone for a few centuries—the annihilation they knew would soon overtake them.

Lowell had smuggled a story—a work of the imagination, a literary creation—into the public mind as a scientific investigation, and it was never really that at all. Literary critic Robert Crossley writes in *Imagining Mars* that Lowell did not so much initiate a revolution in the popular perception of Mars as give it shape and definiteness and conviction…. The Mars that Lowell championed was big, exciting, romantic, and richly textual. Lowell's vision was extraordinarily artful and literary in presentation. Nevertheless, it was more creation than discovery…. Lowell saw himself as the successor of Schiaparelli, but, in strictly scientific terms, he did little to advance the study of Mars beyond where Schiaparelli had taken it. If anything, the effect of Lowell's telescopic surveys of Mars … was to institutionalize a few features of Schiaparelli's Mars—above all, of course, the canali—and to assemble a vast speculative theory [story] about Mars and Martians that rested on, reinforced, and overreached the Schiaparelli observations…. [10]

Similarly, Robert Markley, in his book *Dying Planet* notes that one of the legacies of Lowell's writings about Mars has been "the persistence of [an] image of a dying planet in science fiction" ever since, a persistence which "reveals something of the dark underside of modern myths of technological and social progress" [11]. He places Lowell's achievement squarely in this broader context—a context which still has strong ecological echoes in the current moment in which planetary catastrophe presents an increasingly plausible outcome, and not in geological but historical time, for our own Earth:

Lowell's conviction that Earth was beginning its inevitable decline to a Mars-like desiccation underwrites his analogies between the two planets. In his mind the "lambent saffron" of the deserts of Northern Arizona are reminiscent of "the telescopic tints of the Martian globe." Such analogies are emotive rather than intellectual. Lowell does not so much investigate the Martian ecologies he imagines as invoke a version of the scientific sublime, an admixture of terror and awe that cannot be represented by the familiar language of terrestrial experience… [H]is fascination with the "unspeakable death-grip" of the nebular hypothesis is crucial to understanding its popularity. His sublime, even tragic, depiction of a dying world suggested that astronomy involved more than mere data gathering, and despite the glaring lapses that he made … audiences … were swayed by the aesthetic and philosophical implications of his insistence that "the earth …is going the way of Mars."[38]

The "sublime, even tragic" dimension of Lowell's vision of Mars was, as we shall see, to largely follow the contours of his own career, which ended, to some extent, in a kind of "nameless horror" similar to that he envisaged in the deserts of Mars.

Meanwhile, at Mt. Hamilton: Transitions

Meanwhile, what of Barnard, who had seen Mars in 1894 better than anyone had before, and so ought to have been able to speak with greater authority than anyone else on the planet?

For him, the timing could hardly have been worse. He did give several benefit lectures in San Francisco at which he mentioned his work on Mars. In contrast to Lowell, whose informal remarks were always given in the same rather otiose style as that which characterized his writing, Barnard was inclined to speak in a more self-effacing and homespun manner. One lecture, given at Metropolitan Hall on December 8, 1894, to raise funds for the Teachers' Mutual Aid Society, was hampered by a terrific downpour and the blockading of several car lines by flooding, which "pruned the attendance severely." As an audience slowly began to filter in, he addressed the question of whether the planets might be inhabited using an interesting analogy to an orchard, "where all the fruit would never be ripe at the same time." He discussed as among the most important developments of the 1894 opposition his observations of the obscuration affecting Mare Cimmerium, and showed a series of lantern slides of his drawings of September 23, October 21 and 22, and October 28. According to the reporter's account:

> At the beginning of the recent opposition the "Features" of Mars were very clear and distinct; later they were … obscured.… In September the Professor was sketching Mars when the Cimmerian Sea became entirely obscured, and it was not until the 28th of October that it was again clear enough for him to recognize the markings with which astronomers of the day are familiar…[39]

In another talk, given in February 1895 at the Y.M.C.A. Hall, he began by affirming his commitment to making astronomical knowledge accessible to the wider public:

> I believe … that people who ordinarily are supposed to only appreciate such reading matter in the daily papers as pertains to prize-fights and horse races, will be found to be deeply interested in astronomical subjects, if the same are only placed before the in an easily intelligible form.[40]

He then launched into a masterful summary of the history of Mars observations up to those he had made on Mt. Hamilton in 1894:

The earliest observations with telescopes showed darkish regions upon the surface of Mars that resembled great oceans. The rest of the surface was of a uniform reddish color. As repeated observations showed that these markings were apparently permanent features of the planet's surface, astronomers concluded the dusky regions were oceans and that the general reddish surface was land…

Continuous observations with sufficiently powerful telescopes have shown that considerable temporary changes take place in the extent of some of these dusky regions. There is a very great difference in the distinctness of outline in some of these markings. There is a very great difference in depth of shade or darkness. Some portions are very diffuse and indefinite in outline and extent. These in general are the lighter and more hazy portions of the ocean regions. Sometimes certain of these regions seem to cease to be seas; that is, they assume the same general reddish tint of the continents. The change is more or less periodic and depends upon the Martian seasons, and especially upon the melting of the great snowcaps. Schiaparelli's explanation … [is that] annual floods occur of great extent… Certain low-lying regions will, therefore, sometimes be dry land and at other times temporary seas…

From measurements of the oceans and continents upon Mars it is found that about two-thirds of its surface is occupied by the great continental regions, differing very much in this respect from the earth. Nearly all the seas are found in the southern hemisphere, the continents occupying almost the whole of the northern hemisphere. These great [brick]-red regions of the southern part of the planet were found by Schiaparelli in 1877 to be covered with a network of fine dark lines crossing in all directions, but apparently connected with the great southern ocean system…[41]

Apparently minimally (if at all) aware of Lowell's work, he denied that the Schiaparellian network could consist of canals in the artifactual sense:

As we are dependent upon Schiaparelli for a knowledge of their existence, we also are indebted to him for a plausible explanation of their utility in the economy of nature in her dealings with the world Mars. From peculiarities developed during the seasonal changes of the planet, Schiaparelli concludes that the canals are a natural water system for the irrigation of the great arid regions of the northern hemisphere of Mars.… No one believes that they are the artificial production of living beings that may exist upon the planet.[42]

Finally, he made it clear that the planet was without a doubt subject to significant obscurations:

… Visual observations point strongly and conclusively to the presence of an atmosphere. Indeed, the opposition just past gave the best evidence yet obtained of an atmosphere surrounding the … planet.

It is true that clouds are seldom seen covering any great area of Mars, but clouds do exist there. In October of 1894 great changes were noticed over the region known as the Cimmerian sea. A large whitish obscuring mass blotted out the well-known details of this region for several days and then passed away… These must have been clouds. Immediately afterwards the entire atmosphere of the planet seemed to become filled with haze through which the seas and continents were seen with the utmost difficulty. This occurred at a time when Mars was nearest to the earth, and therefore ought to be best seen. Details that were clearly and strongly seen when the planet was 60,000,000 miles distant were now, at a distance of some 38,000,000 miles, invisible or seen with the utmost difficulty. This could only be explained by the supposition of a dense, hazy condition of an atmosphere surrounding the planet.…[43]

These remarks, though interesting, were heard by only a few members of the public, and reported in the perishable medium of the newspapers. His tendency, moreover, to prefer simple, direct and somewhat understated communications perhaps made less impact on the public that Lowell's tendency to dogmatic statements and penchant for ornate and flowery prose. There is no reason to think, however, that Barnard intended his remarks in these lectures to be anything more than preliminaries, previews of some more carefully considered and definitive publication in a professional journal. That publication came, eventually, though not for another year, during which time Lowell was able to attain to a virtual monopoly in matters Martian.

Barnard's reasons for holding back, despite the intense public interest in Mars, had simply to do with the pressure of other work, for as always, Barnard had his hands full with several active lines of investigation. Thus, though during the summer of 1894, Barnard made Mars his priority on the nights he was assigned to the 36-inch refractor, he was also busy pushing forward the Milky Way photography with the Crocker telescope, and obtained some outstanding results, including the deep plates on the region near Theta Ophiuchi, earlier discussed. In addition, the personal difficulties with Holden were continuing, and he was being pressured to leave Mt. Hamilton for Yerkes

Observatory. Despite the fact that the University of Chicago was operating on a tight budget at the time, after the "panic" of 1893, the relentless Yerkes director George Ellery Hale wanted another $15,000 for astronomer's salaries in order to hire both Barnard and Keeler immediately to the staff, whereupon University president William Rainey Harper reminded him that "astronomy is *one* department of the university."[44] In the end, Harper agreed, in December 1894, to a scaled-back budget, with Barnard the only addition to the staff.

Barnard, however, hung fire. He wanted to see if he could persevere at Lick at least through the summer in order to take advantage of the prime observing season, in which to push forward the work with the Crocker telescope. This remained the plan even though Burnham, knowing Barnard's difficulties with Holden and perennial complaints about his health—largely psychosomatic, as Barnard himself realized—urged him to take the Yerkes position at once. "If you are wearing down, getting worse, or no better, it may result in your complete physical collapse, or something worse, and then your work will be ended," Burnham advised him in February 1895. "You will see then how short sighted it was to sacrifice health and years of astronomical work for the sake of continuing there a few months longer."[45]

By early March 1895, the barrage of entreaties from Burnham, Hale, and Harper were beginning to have their wanted effect, and Barnard was on the verge of accepting the Yerkes position. But a new commitment, that he could not refuse, had meanwhile come to the fore. He was making every effort to collect in one volume the comet and Milky Way photographs taken with the Willard lens, which, he urged, were "of the highest scientific value and would, I am sure … be a great credit to the Observatory."[46] He knew that the cost of reproducing sixty to a hundred of his finest plates would be in the range of $1500 to $2500, not a trivial sum given that Barnard's annual salary at the time was only $2400." Unfortunately, the Observatory was also strapped for funds, and Holden suggested that, as Barnard had just a few months earlier submitted a number of these photographs to the Royal Astronomical Society for publication in their *Monthly Notices*, it would be better to wait "until we know whether the R.A.S. will accept or refuse your offer. If the Society declines to print your negatives I should like to know on what grounds they do so, before I make any recommendation to the Regents."[47] Naturally, Barnard was infuriated by the Director's offer, and at once penned a turbulent letter to Timothy Guy Phelps, the chairman of the Lick Observatory committee of the Board of Regents, in which he urged the Regents to intervene, as they had so often done, in his behalf. He reminded them that his sixteen photographs of Brooks's Comet of 1893 were "the most wonderful and important set of photographs of a comet in existence," and that his Milky Way photographs were equally important—and infinitely more so than Holden's photographs of the Moon which, he reminded Phelps, had been published by the Lick Observatory.[48]

While engaged in these struggles, Barnard realized that if anything were to leak out about the Chicago offer, it would ruin his chances with the Regents of getting his photographs published, and at the moment, that was the overriding priority. In the end, Barnard's friends and supporters raised a significant amount of money for him to publish the photographs, but progress was delayed by shortcomings in the reproduction process, and Barnard was nothing if not a perfectionist. Collotype, such as Ranyard had used for the photographs published in *Knowledge*, photogravure, and halftone were the available methods, and Barnard found fault with each of them. In the end, he could never satisfy himself and would leave Mt. Hamilton with the project in suspension and a sense of obligation to return the money to his subscribers; the photographs were not to be published for almost 20 years, and then by the photogravure method he had earlier rejected. His Milky Way photographs taken during 1895 were, meanwhile, the best of the lot. Thus, on the night of June 21-22, 1895, he made a 7 ½ h exposure on the region of Rho Ophiuchi, and the following night another, of 8 ½ h. Here William Herschel long before had noticed a "hole in the heavens." Barnard wrote of this region:

> I do not think there is any other region of the heavens so extraordinary as this, about, and in the immediate vicinity of, Rho Ophiuchi. One hesitates at any attempt to describe it. Perhaps even ore remarkable than the nebulosities are the vacant lanes that run eastward from the great nebula, and those in the upper part of the plate. The considerable dark space 2 degrees northeast of Tau Scorpii is unreal.[49]

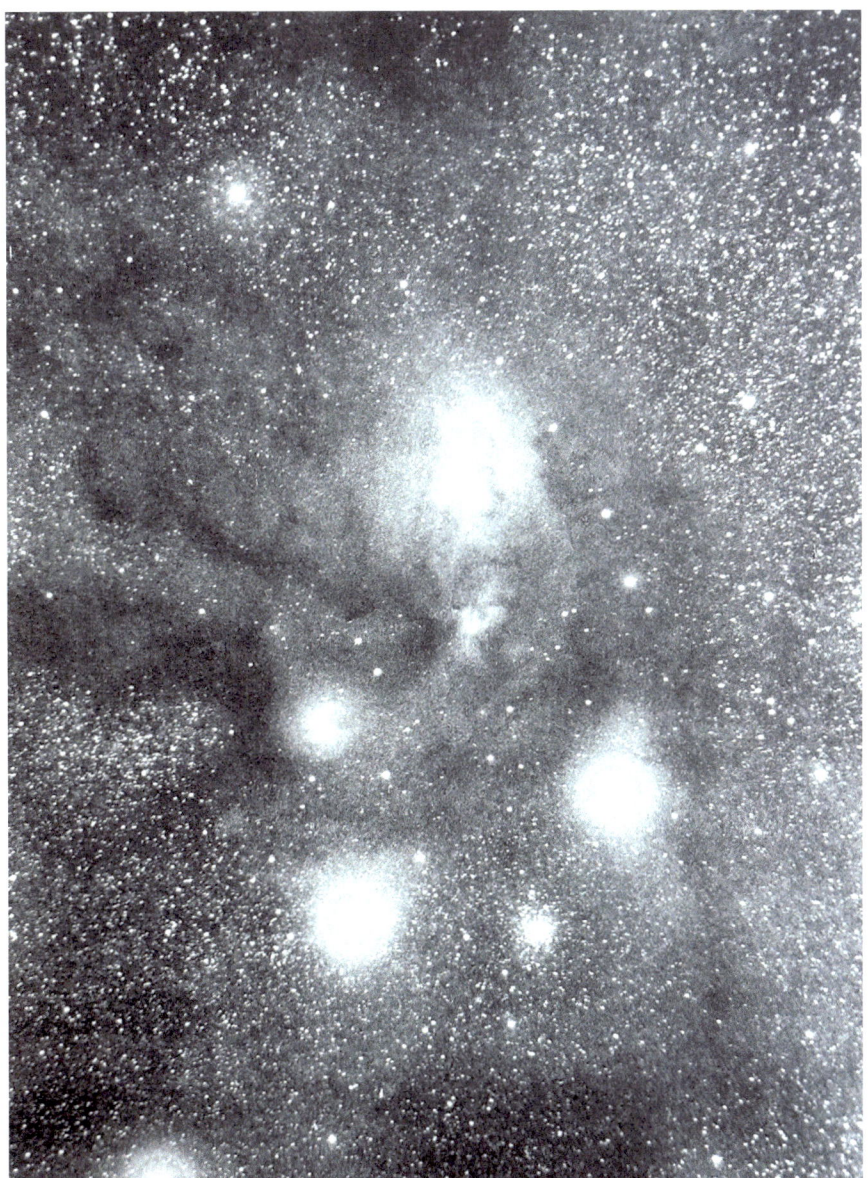

An 8 ½-h exposure with the 6-inch Willard lens on the great nebula of Rho Ophiuchi, June 22-23, 1895. The photograph is centered on R.A. 16 h 20 m and declination -23°. It is similar to but on an enlarged scale to Plate 35, taken the night before, in "Comet and Milky Way Photographs"

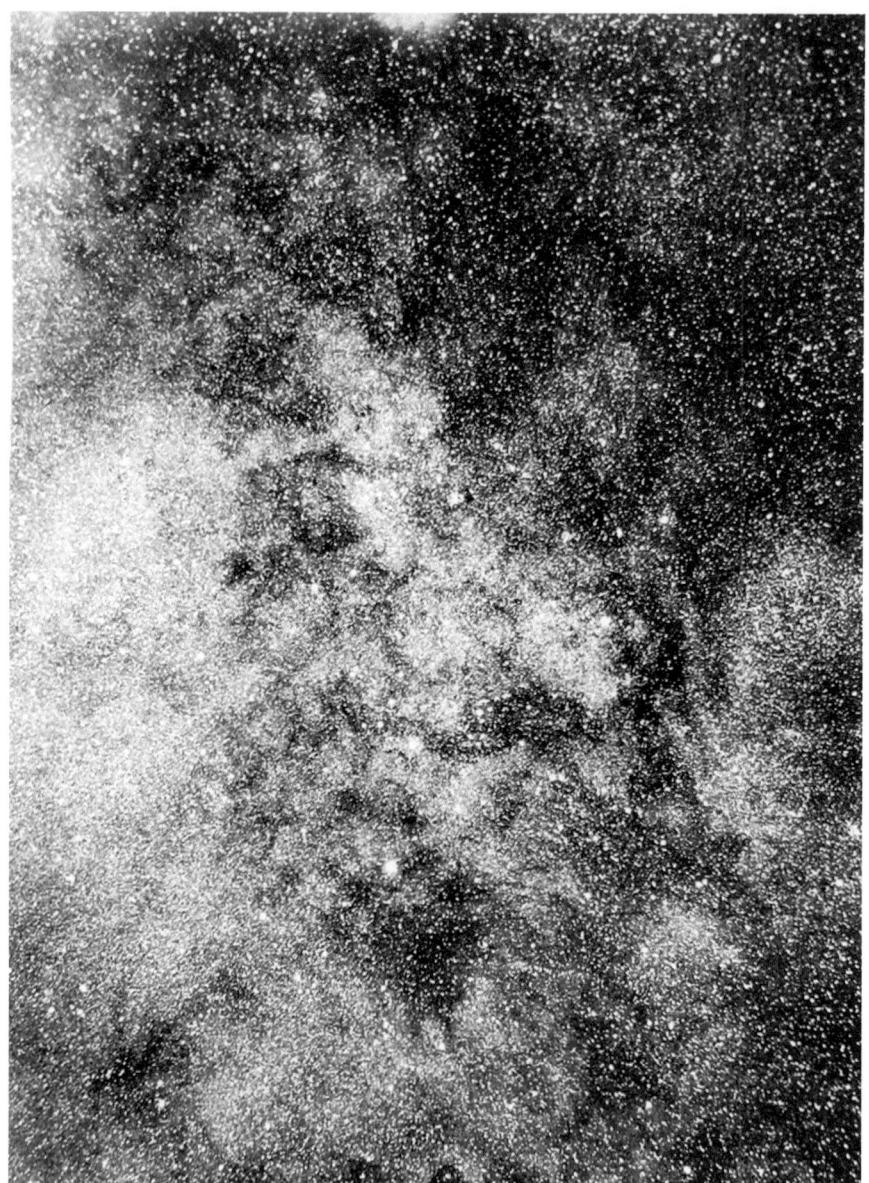

The Great Star Clouds in Sagittarius, captured in an exposure of 3 h with the Willard lens on August 13, 1895—one of the last Barnard took before leaving Mt. Hamilton. The image is centered on R.A. 17 h 52 min, declination -28°. It appears as Plate 50 in the "Comet and Milky Way Photographs" (1913)

In the end, Barnard's negotiations to leave Mount Hamilton for Yerkes could not be kept secret indefinitely, and in June, the matter finally came out in the open—his name was listed in a University of Chicago catalog of its summer courses; Barnard was listed on the first page as a professor of astronomy in its department of mathematics and astronomy. Holden took this as "public news." Barnard had continued to assure Phelps that he would be the first to learn of any imminent departure—he must have felt betrayed on first learning of the Chicago catalog announcement from a newspaper reporter. So the deed was done, it could not be undone. That same day, Barnard's colleague J.M. Schaeberle, who to the end of his life was convinced that Barnard had never really intended to leave Mt. Hamilton at all but had been outmaneuvered by others, would recall long afterwards, "A most pathetic figure, poor broken-hearted Barnard came to me, rested his hands and head on my shoulder, broke down and cried convulsively like a child."[50] Barnard's resignation of the Lick Observatory was officially to take effect on October 1, 1895, and in the months that remained he worked himself harder than ever trying to finish up whatever work he could. A week before he was due to leave, he wrote to Hale, "I am rushed to death and tired out."[51] He observed on the very last night on Mt. Hamilton, making a series of routine micrometer measures of double stars and exposing a 20-min plate on the Pleiades with the Crocker telescope. The following day, carrying along observing books and Milky Way and comet plates, he and Rhoda left Mt. Hamilton, and set out for Chicago.

His great nemesis Holden, who remained embattled with his staff even after Barnard had gone, followed the same road out some 2 years later, the event being noted by staff member W.J. Hussey in his diary: "… the Observatory conveyance is seen driving down the Mountain Road. It is Edward S. Holden leaving Mt. Hamilton forever."

Small Versus Large Telescopes: Portrait Lens Versus Large Reflector

When he arrived in Chicago, the Yerkes Observatory was still in the dreaming stage. A site, on the north shore of Lake Geneva, near its western end, about a mile from the center of the small town of Williams Bay, had been acquired, and construction of the buildings had been underway since April 1895. The 40-inch refractor was still being worked on in the Clark optical workshop in Cambridgeport, Massachusetts, and would not be delivered until May 1897.

Barnard took up temporary residence in Chicago not far from Hale's mansion in Kenwood, where the only telescope available to him was the 12-inch refractor in Hale's personal observatory. This meant an unwelcome hiatus in Barnard's observational activity, but he did have time to work on the large backlog of reports on his Mt. Hamilton observations, including writing notes on his Comet and Milky Way Photographs.

At this time he was also provoked into a rather unpleasant controversy with a British "grand amateur," Isaac Roberts, who after making a fortune in the Liverpool building trade devoted himself in retirement entirely to astronomy, setting up a 20-inch reflector for photography first at Maghull, near Liverpool, then (from 1890) in the clearer skies of Crowborough, Sussex. Roberts used this large reflector to photograph nebulae and star clusters, and in the course of this work, decide to attack Barnard's photographs taken both with the Willard lens and a smaller 1 ½-inch lens. Roberts insisted that extended diffuse nebulosity Barnard recorded around the star 15 Monocerotis did not exist at all or else consisted of the light of countless unresolved stars. These were, however, resolved in photographs with his large reflector.[52]

Roberts first attacked in 1895-6, and did so again in 1898 and 1903. The disagreement ended only with Roberts's death in 1904. In retrospect, Barnard proved to be mostly right and Roberts mostly wrong. Barnard readily agreed that Robert's larger aperture would always win out when it came to recording faint stars but, as he realized through practical experience rather than theory, a lens's focal ratio rather than aperture was what mattered in photographing extended objects. Both the Willard and Roberts's reflector had focal ratios of f/5, but the Willard with a wider field of view and used under clearer, more transparent skies, was able to record extended nebulous masses, not only around 15 Monocerotis but also in the Pleiades and in Orion ("Barnard's Loop") far more quickly and effectively than Roberts with his large reflector used under the less favorable conditions at Maghull and Crowborough.[53]

The debate began during a time when Barnard was under a great deal of stress. Under the circumstances, he managed, usually, to maintain a reasonably polite demeanor against the often prickly adversary who had initiated the attack against him. Roberts, according to historian Lee T. McDonald, seems to have had "a tendency to overreact when crossed, and an intransigence in the face of evidence that he was in the wrong, even to the extent of not bothering to check his facts."[54] In this debate, Barnard was thrown into taking up the cause of small telescopes against large ones, the exact opposite of position in the case of planetary observations.

Hoisting one half of a shutter for the 40-inch refractor dome, September 17, 1896. Credit: University of Chicago Photographic Archive [apf6-00060], Hanna Holborn Gray Special Collections Research Center, University of Chicago Library

House of Edward Emerson Barnard and his wife Rhoda Calvert Barnard, taken by Barnard from the horse-shoe drive with Yerkes Observatory visible in the background. Barnard acquired the site and began building even before the Observatory was finished, wanting to be just a short walk from the Observatory so he could get to it at short notice whenever the sky cleared. The photograph was taken on April 17, 1904. (Credit: University of Chicago Photographic Archive [apf6-00676], Hanna Holborn Gray Special Collections Research Center, University of Chicago Library)

Large Versus Small Telescopes: Lick Refractor Against the World

Just before Christmas 1895, Barnard wrote a preliminary summary of his most recent planetary observations at Lick, including, at long last, those of Mars in 1894. In contrast to Lowell's writings, which always aimed to appeal to the largest possible audience, Barnard's Mars work was mentioned, almost as an afterthought, in an article about Saturn for the *Monthly Notices of the Royal Astronomical Society*; the ponderous title—"Micrometrical Measures of the Ball and Ring System of the Planet Saturn, and Measures of the Diameter of his Satellite Titan, made with the 36-inch Refractor of the Lick

Observatory in the year 1895. With some Remarks on Large and Small Telescopes"—didn't mention Mars at all. From a public relations point of view, the title was a disaster. Barnard finished writing it at the Kenwood Observatory on Christmas Eve 1895, and it was published with all possible speed in the January 10, 1896, issue of the *Monthly Notices*. Not a single one of Barnard's Mars drawings was included.

Nevertheless, for any one who happened to read it through, the paper was full of interesting things. After describing his observations of Saturn, Barnard turned to address the small-versus-large telescope controversy which had raged especially in the 1880s, and described the results of some experiments he carried out on Venus and Saturn on July 14-15, 1895. "These experiments," he said, "were carefully made to satisfy the desire that seems to exist in the minds of a number of astronomers to know if a reduction of the aperture of this great telescope would make it show planetary markings that were not visible with the full aperture." The small telescope might have an advantage, he thought, "If the object is very bright, or the air unsteady." However, even then, Barnard preferred to reduce the light by a cap with a small hole placed over the eyepiece. In general, reducing the aperture did not seem helpful in the detection of delicate planetary markings. Instead, he concluded, "It has always appeared to me, when I have heard large telescopes decried in this connection, that if these same observers could look at Saturn or Jupiter with a great telescope, under first-class conditions, they would themselves be astonished at the difference, and would at once decide for the larger aperture."[55]

These preliminaries out of the way, he addressed the Mars work of 1894, which he emphasized had been done with the great telescope under first-class conditions, with the planet on the meridian, thus having a high altitude in the sky and being extremely favorably placed for observation. The best conditions were experienced shortly after sunrise, at which time the "surface with the great telescope has shown a wonderful clearness and amount of detail." There could be no question of the planet being as shown in what he calls "the average drawings of recent years." Instead, with the great telescope on Mt. Hamilton:

> … This detail … was so intricate, small, and abundant, that it baffled all attempts to properly delineate it. Though much detail was shown on the bright "continental" regions, the greater amount was visible on the so-called "seas." Under the best conditions these dark regions, which are always shown with smaller telescopes as of nearly uniform shade, broke up into a vast amount of very fine details. I hardly know how to describe the appearance of these "seas"

under these conditions. To those, however, who have looked down upon a mountainous country from a considerable elevation, perhaps some conception of the appearance presented by these dark regions may be had. From what I know of the appearance of the country about Mount Hamilton as seen from the observatory, I can imagine that, as viewed from a very great elevation, this region, broken by canyon and slope and ridge, would look just like the surface of these Martian "seas." During these observations the impression seemed to force itself upon me that I was actually looking down from a great altitude upon just such a surface as that in which our observatory was placed. At these times there was no suggestion that the view was one of far-away seas and oceans, but exactly the reverse. Especially was I struck with this appearance in the great "ocean" region of the Hour-glass Sea, and especially in the equatorial portion of this region. These views were extremely suggestive and impressive. I have not seen these small and delicate details described elsewhere, and I feel confident they would scarcely be shown in a much smaller telescope. The details shown on the "continental" regions were usually irregular features, principally delicate differences of shade. No straight hard sharp lines were seen on these surfaces, such as have been shown in the average drawings of recent years. I would mention specially the region of the Solis Lacus and following it. Some short diffused hazy lines—rather irregular—were also seen here, running between several of the small very black spots that abound in this region. On several dates—principally about September 30—two long hazy parallel streamers were seen running from the preceding end of the "Cimmerian Sea" toward the north following.[56]

So ended the remarks on Mars, followed only by Barnard's promise that more details would be forthcoming later. "[D]etails … will be treated of specially when my work upon Mars is published." They never were. He did allow four finished drawings to appear (on a very small scale that did not do justice to the originals) in the 1895 edition of Chambers's *Story of the Solar System*, and one of these drawings (that of September 3, 1894) served as a frontispiece for an article on his observations of the South Polar Cap in the *Astrophysical Journal* in 1903. In the latter article, he had noted that "however much the general surface features of Mars may be misrepresented, and however much the canals, single and double, may be illusory, the polar caps seem to have escaped the general deluge of uncertainty and misinterpretation" [15, p. 249]. That, however, would be all the world would see of his 1894 work on Mars until the 1980s and 1990s, when his drawings—long thought lost, or even destroyed by Barnard himself—were rediscovered. By then, they were of academic interest only, as the debate over the canals of Mars had long since been resolved.

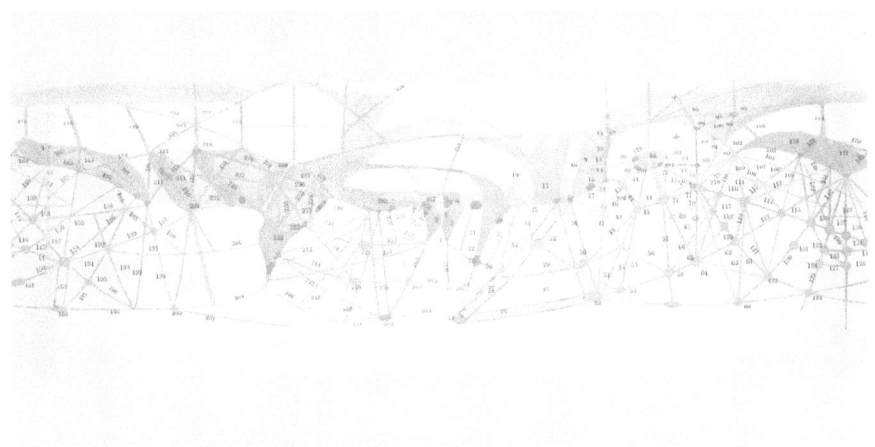

Lowell's map of Mars from 1894, published in Mars (1895), Plate XXIV. A new projec-tion by Joel Hagen, for comparison with the Barnard map below

A map of Mars compiled on the basis of Barnard's unpublished drawings from 1894, produced by astronomer-artist Joel Hagen. The projection has been chosen to match the map of Lowell on p. 227, so as to emphasize the striking differences. (Credit: Joel Hagen)

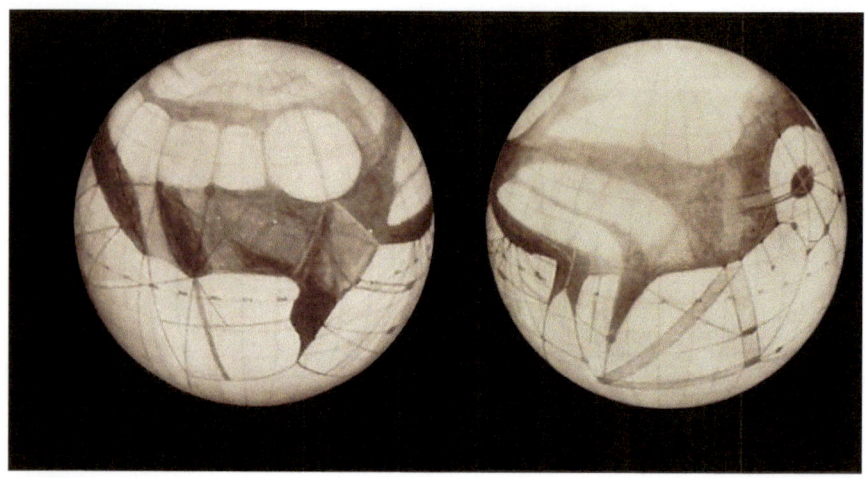

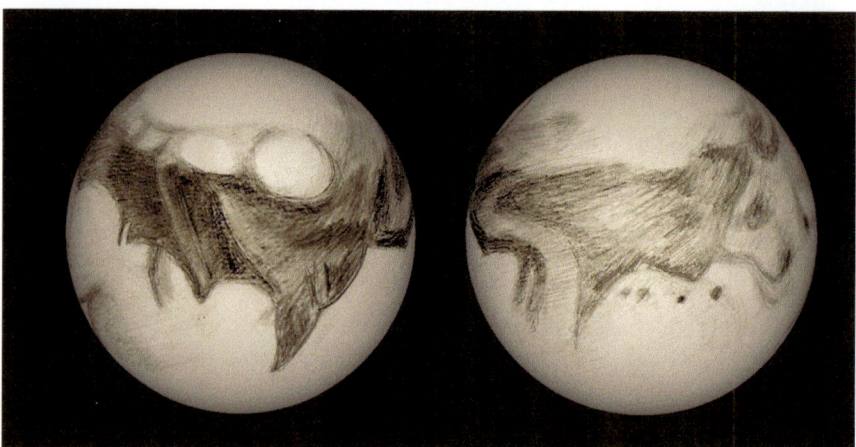

Lowell's and Barnard's views of Mars in 1894, compared. Above, two views of Lowell's globe of Mars, from his book Mars *(1895). Below, similar views of a globe produced by astronomical artist Joel Hagen based on some of Barnard's unpublished drawings. (Credit: Joel Hagen)*

Notes

1. Regarding Pickering's colorful and somewhat sad later career, see Howard Plotkin [1].
2. Noted in Boston *Herald*, December 16, 1894.
3. Greenslet, *The Lowells*, p. 366. His publisher at the time was Houghton, Mifflin, and Co., officed at no. 4 Park Street, just down the street from his grandparents'

old mansions and the Massachusetts State House. For his last three books—*Mars and Its Canals* (1906), *Mars as the Abode of Life* (1908), and *The Evolution of Worlds* (1909)—he published instead with the New York firm Macmillan.

4. Greenslet, *The Lowells*, p. 366.
5. Greenslet, *The Lowells*, pp. 366–367.
6. P. Lowell to A.E. Douglass, January 5, 1895; Lowell Observatory Archives.
7. P. Lowell to A.E. Douglass, March 5, 1895; Lowell Observatory Archives
8. Edward Everett Hale (1895) "Latest News from Mars," *Publications of the Astronomical Society of the Pacific*, no. 41, pp. 116–118. (Hale's article had previously appeared in the Boston *Commonwealth* and *Scientific American*.)
9. Greenslet, *The Lowells*, p. 366.
10. Quoted in Strauss, *Percival Lowell*, p. 74.
11. Strauss, *Percival Lowell*, p. 74.
12. Ibid., p. 75.
13. Ibid.
14. Elizabeth Loftus, *Eyewitness Testimony*, pp. 15–16.
15. Lowell, *Mars*, p. 201.
16. Ibid.
17. Ibid., p. 45.
18. Lowell, *Noto*, pp. 103–104.
19. Lowell, *Mars*, p. 141.
20. Simon Newcomb to Percival Lowell, October 30, 1905. Newcomb. Simon. Papers, Library of Congress.
21. Claude Monet, quoted in Lawrence Campbell [7].
22. Ruskin, *Elements of Drawing*, p. 27n.
23. Campbell, Introduction, p. xi.
24. Ruskin, *Elements of Drawing*, p. 120.
25. Lowell, "Observations," p. 186.
26. Lowell, *Mars and Its Canals*, pp. 174–175.
27. Lowell, *The Soul of the Far East*, p. 5.
28. William Lowell Putnam III to William Sheehan, personal correspondence, June 2012.
29. Greenslet, *Lowells*, p. 361. Flammarion must have been flattered that Lowell's interest in Mars owed much to his book *La Planète Mars*, and was impressed with Lowell's having spared nothing in the quest to set up an observing station with the best possible air for planetary work. However, he seems to have confused W.H. Pickering's expedition to Arequipa with Lowell's to Mars Hill. Thus, summarizing a lecture Lowell gave in French to the Société Astronomique de France, Flammarion wrote: "The speaker … took care to search for the purest air that was in the United States and found it in the Southwest region, in Arizona, on Mount Arequipa." C. Flammarion [8].
30. P. Lowell to Simon Newcomb, March 15, 1903; Lowell Observatory Archives.

31. G. Schiaparelli to O. Struve, October 6, 1896; G.V. Schiaparelli (1963), *Corrispondenza su Marte*, Volume 1 (Pisa: Domus Galileana). The Lowell/ Schiaparelli correspondence was conducted in French, and has been published side-by-side in French and in English translation by Jennifer Putnam and Wiliam Sheehan (2021) "A Complicated Relationship: An Introduction to the Correspondence Between Percival Lowell and Giovanni Virginio Schiaparelli." *Journal of Astronomical History and Heritage*, vol. 24, no. 1, pp. 170–227.

32. P. Lowell to A.E. Douglass, March 25, 1895; Lowell Observatory Archives.

33. Campbell, "*Mars*, by Percival Lowell," p.209.

34. Ibid.

35. Agnes M. Clerke [9]. Clerke's argument here largely anticipated the line of attack in Alfred Russel Wallace (1907), *Is Mars Habitable?* (London: Macmillan).

36. James Keeler to George Ellery Hale, December 27, 1894. Hale. George Ellery. Papers, University of Chicago Library, Hanna Holborn Gray Special Collections Research Center.

37. The definition is from Oxford Languages and Google.

38. Markley, *Dying Planet*, pp. 69–70.

39. "Mars from Mount Hamilton." San Francisco *Examiner* Dec. 8, 1894.

40. "Mars as Seen by Barnard: A Lick Professor Tells an Interesting Story About our Neighbor Planet." San Francisco *Examiner*, Feb. 10, 1895.

41. Ibid.

42. Ibid.

43. Ibid.

44. William Rainey Harper to George Ellery Hale, December 6, 1894; Harper. William Rainey, Papers, University of Chicago Library, Hanna Holborn Gray Special Collections Research Center.

45. S.W. Burnham to E.E. Barnard, February 6, 1895; Barnard. Edward Emerson. Papers, Special Collections and University Archives, Jean and Alexander Heard Library, Vanderbilt University.

46. E.E. Barnard to E.S. Holden, April 10, 1895; Holden. Edward Singleton. Papers, Mary Lea Shane Archives of the Lick Observatory.

47. E.S. Holden to E.E. Barnard, April 11, 1895; Holden. Edward Singleton. Papers, Mary Lea Shane Archives of the Lick Observatory.

48. E.E. Barnard to Timothy Guy Phelps, April 22, 1895; Barnard. Edward Emerson. Papers, Mary Lea Shane Archives of the Lick Observatory.

49. Barnard, "Comet and Milky Way Photographs," notes on Plate 36.

50. J.M. Schaeberle to Edwin Brant Frost, March 14, 1923; Barnard, Edward Emerson Papers, University of Chicago Library, Hanna Holborn Gray Special Collections Research Center.

51. E.E. Barnard to G.E. Hale, September 24, 1895; Hale George Ellery, University of Chicago Library, Hanna Holborn Gray Special Collections Research Center.

52. Roberts began photographing nebulae in 1885. On October 1, 1888, he obtained a 3-hour exposure on the Andromeda Nebula, which clearly showed for

the first time that it was a spiral nebula highly inclined to the Earth. Roberts was convinced that the photograph was a "confirmation, if not demonstration," of Laplace's nebular hypothesis. See Lee T. McDonald [12, pp. 242–243]. Roberts's first attack on Barnard, involving Barnard's photograph of the nebulosity around 15 Monocerotis, is described in "Meeting of the Royal Astronomical Society, Friday, December 13, 1895," *The Observatory* (1896), vol. 19, pp. 35–42:p.37; Barnard's response in E.E. Barnard [13, p. 14].

53. Barnard pointed out that the exterior nebulosities to the Pleiades had been "amply verified (if such a verification were at all necessary) by two other astronomers, H.C. Wilson of Goodsell Observatory and S.I. Bailey of Harvard. Roberts in 1896 began a systematic survey of 52 regions where William Herschel had reported faint nebulosity, and in 1902 he reported that in 48 of them there was "no trace of extensive nebulosity." Two of the four regions where Roberts found nebulosity were in Orion, and two in Cygnus. However, three were free from nebulosity. In one of the latter Max Wolf, in 1901, exposing a plate for 6 ¼ hours with the 16-inch Bruce photographic telescope at the Königstuhl-Heidelberg Observatory, found what he described a "great [nebulous] snake which lies around Orion, with [zeta] Orionis as center." See McDonald, "Isaac Roberts, E.E. Barnard and the Nebulae," pp. 250–251. Wolf's nebulous snake had previously been discovered by E.C. Pickering. However, because of Barnard's spectacular images of it, it became known as "Barnard's Loop," and among Barnard's comets, dark nebulae and dwarf galaxy, it is, with Barnard's Star, the most famous of his eponymous objects.

54. McDonald, "Isaac Roberts, E.E. Barnard and the Nebulae," p. 255.

55. E.E. Barnard [14, p. 165]. He does concede "if the seeing, however, is bad or very indifferent, I would prefer the smaller telescope."

56. Barnard, "Micrometrical Measures," pp. 165–167.

References

1. Plotkin H (1993) William H. Pickering in Jamaica: the founding of Woodlawn and studies of Mars. J Hist Astron 24:102–122
2. Sankovitch N (2017) The Lowells of Massachusetts. St. Martin's Press, New York, p 292
3. Lowell P (1895) Mars. Houghton, Mifflin and Co., Boston, pp 86–87
4. Maria K, Lane D (2011) Geographies of Mars: seeing and knowing the red planet in an imperial age. Chicago, University of Chicago Press, p 39
5. Lowell P (1908) Mars as the abode of life. New York, Macmillan, p 146
6. Lowell P (1898) Observations of the planet Mars during the opposition of 1894-5, made at Flagstaff, Arizona. In: Annals of the Lowell observatory, vol 1. Boston/New York, Houghton, Mifflin and Co., p 5

7. Campbell L (1971) Introduction to John Ruskin. In: Elements of drawing. Dover reprint of first edition, 1857, New York, p ix
8. Flammarion C (1909) La Planète Mars et ses Conditions D'Habitabilité, vol 2. Gauthier-Villars, Paris, p 109
9. Clerke AM (1896) New Views about Mars. Edinburgh Rev 184:368–385
10. Crossley R (2011) Imagining Mars: a literary history. Wesleyan University Press, Middleton, p 782
11. Markley R (2005) Dying planet: Mars in science and the imagination. Duke University Press, Durham/London, p 22
12. McDonald LT (2010) Isaac Roberts, E.E. Barnard and the Nebulae. J Hist Astron 41:239–257
13. Barnard EE (1896) On comparison of reflector and portrait lens photographs. Mon Not R Astron Soc 57:10–16
14. Barnard EE (1896) Micrometrical measures of the Ball and Ring system of the planet Saturn, and measures of the diameter of his satellite Titan, made with the 36-inch refractor of the lick observatory in the year 1895. With some remarks on large and small telescopes. Mon Not R Astron Soc 56(4):163–172
15. Barnard EE (1903) The south polar cap of Mars. Astrophys J 17(4):249–254

6

The Spokes of Venus

Contents

> *The few experienced observers of Venus could only laugh at the "canals" Lowell*
> *said he had seen on both Venus and Mercury, but the remaining astronomers*
> *[without their experience] like the general public bought into the fact that Lowell*
> *had observed with a giant (24-inch) telescope from an elevation of 2300 meters,*
> *so news of Lowell's wonders quickly made the rounds of the European press.*
> —Leo Brenner, Spaziergänge durch das Himmelszeit *(Leipzig: Eduard Heinrich*
> *Mayer, 1898), p. 125. (Translation W. Sheehan)*

The Clark Refractor Cometh

In March of 1896, Lowell returned from his European adventures, determined to make what had been a temporary expedition in 1894 permanent. Having returned the borrowed 18-inch to Brashear, he needed a replacement, and ordered from his friend, the optician Alvan Graham Clark, a 24-inch refractor, for the sum of $20,000 (equivalent to about $650,000 today). At this point, after the horrible winter conditions experienced (by Douglass) in 1894–95, he was far from sold on Flagstaff as a suitable site for an Observatory,

© The Author(s), under exclusive license to Springer Nature Switzerland AG 2024
W. Sheehan, *Parallel Lives of Astronomers*, Springer Biographies,
https://doi.org/10.1007/978-3-031-68800-3_6

and had already decided to observe the December 1, 1896, opposition of Mars from Tacubaya, then a suburb of Mexico City close to the site of the National Observatory, where winter conditions were guaranteed to be better. In the interests of economy, he planned to reuse the same pre-fab dome from 1894, and so specified a shorter focal length for the new telescope to make it fit. This would prove to be a mistake, as even before the new telescope was delivered, the old dome had been discarded in favor of a new one Lowell was having built by Godfrey and Stanley Sykes, proprietors of a local bicycle and fix-it shop in Flagstaff. It was set up temporarily in Flagstaff, with a canvas covering, before being sent to Tacubaya for the opposition of Mars.

As a result of trying to accommodate the telescope to the previous setting, the focal length (15 meters) made the telescope rather short, and thus subject, in significant degree, to chromatic aberration. This was a problem not only for the Lowell refractor but for other large telescopes; in all of them, though the lenses were "achromatic" (consisting of a compound of flint and plate glass components), the color correction was far from perfect; light of different wavelengths is brought to a slightly different focus, and blurred blue and red images are superimposed on an in-focus yellow image causing the image to appear swathed in a magenta haze of unfocused light blurring details.[1] Some of this haze can be eliminated by means of a filter—in 1894, Lowell found that a thin piece of ochre glass placed in front of the eye-piece was, as a rule, conducive to the detection of detail; a neutral-density filter was also tried, sometimes, to reduce the quantity of light. Perhaps less satisfactory, but more often employed during the early years, was to stop down the aperture. In the case of the Lowell refractor, this was initially accomplished by putting on and taking off diaphragms of various sizes that were placed in front of the objective lens; eventually, a large iris diaphragm was used for the purpose—though Douglass preferred to do as Barnard had done with the Lick 36-inch refractor, and use a small diaphragm over the eyepiece to cut down the pencil of light entering the eye which had the effect of reducing the effective area of the objective. In general, as is clear from examination of Lowell's log books, the aperture of the 24-inch glass was stopped to 16 inches, and very rarely was the full aperture of 24 inches used. Somewhat idiosyncratically, Lowell strongly favored using relatively low powers—310 or 370 X for Mars and 130 X to 170 X for Mercury and Venus. Pickering and Douglass, like Barnard, tended to use higher powers, in the range of 370 to 750 X for Mars. With the low powers, Lowell told Douglass, "he saw the detail on Mars either exceedingly well or not at all" [1, p. 82]. In addition, lower powers, because of an effect known as irradiation—the tendency of bright areas to bleed over into a darker area—made broad stripes appear as narrow lines, which may in part account for the fact that Lowell always represented detail harder and sharper than by others.[2]

Had Lowell more carefully canvassed the literature regarding the experience of other observers, he might possibly have found enlightening some comments from the Princeton University astronomer Charles Augustus Young, who, though best known as a solar astronomer and spectroscopist, had made numerous planetary observations with the 23-inch and 9-inch refractors of the Halsted Observatory in 1885–86 with various diaphragms, in different conditions of seeing, in an attempt to contribute to the small vs. large telescope controversy that was raging at the time. Based on his experience, he noted that "most commonly … when I have failed to see with the large instrument anything I supposed I saw with the smaller, it has turned out on examination that the larger instrument was right, and that imagination had constructed a story that was not true by building up faintly visible details and hazy suggestions furnished by the smaller lens" [2]. Though many of his observations had involved Jupiter and Saturn, he immediately saw the applicability to the case of Mars:

> The discordance between the different maps of Mars indicates that the best and most keen-eye observers unconsciously supplement what they really see, with details which they only think they see; so that in the finished drawing fact and fancy are inextricably mingled. The later observer with larger telescope and higher power naturally fails to recognize many features, and some, he has to repudiate [3, p. 2].

In other words, small telescopes and low powers were conducive to illusions.

Lowell, Clark, Clark's daughter, and Miss Leonard arrived by train in Flagstaff on July 22, 1896, with the Clark lens in tow. As soon as the Clark lens was installed in the tube, Lowell and Douglass began making observations of Mars—still months from opposition and far from the Earth. At first only the main markings, such as the Syrtis Major, appear in the sketches, but within only a week a few of the canals were declaring themselves, as usual, in flashes.

Mars remained under scrutiny—not only by Lowell and Douglass but by several new assistants, who included two young college graduates, Wilbur A. Cogshall and Daniel A. Drew. Meanwhile, Lowell, to bide the time, also decided also to test the Clark on the inner planets, Mercury and Venus, which lay quite close together in the daylight sky. In addition to Lowell himself, Drew and Leonard participated in these observations.

The practice of observing the inner planets during daylight had been pioneered by Schiaparelli. From his studies of surface markings on these planets—more or less definite in the case of Mercury, vague and uncertain in that of Venus—the great Italian astronomer announced that both planets rotated in the same period as they revolved about the Sun, in 88 days and 225 days, respectively. Lowell, emulating the master, found as soon as he set eye to the eyepiece on August 21, 1896, "a pretty little moon nearly lost in the vast blue sky," on which he immediately noticed a striking set of "narrow irregular lines … very dark," on the surface. "They were not in the least like the markings on Mars," he took pains to emphasize. "Here were no large patches of shade on the one hand, nor fine, regular pencilling on the other. Its lines were fairly straight, but broken and of varying width. 'Cracks' best explains their appearance, and probably their nature" [4].

The fact that the positions of these lines were unchanged even after as much as 5 hours' interval proved at once that the planet must have a long rotation, and in continuing to observe it on following days, he thought he could make out that they were gradually shifting in position until they passed over the edge of the terminator. He interpreted this as due to the swaying of the planet back and forth known as libration, familiar in the case of the Moon, and due to the eccentricity of a planet's orbit. Schiaparelli had been completely vindicated—or so it seemed. But though he claimed he independently arrived at the same tidal locking conclusion as Schiaparelli, his drawings (see those between March 5 and 16, 1897) show the same linear features ("canals") day after day. This despite the fact that, as we now know, the planet had rotated through 50 degrees of longitude. We cannot escape the conclusion that he was simply seeing what he was expecting. Expectation created illusion.

MERCURY.

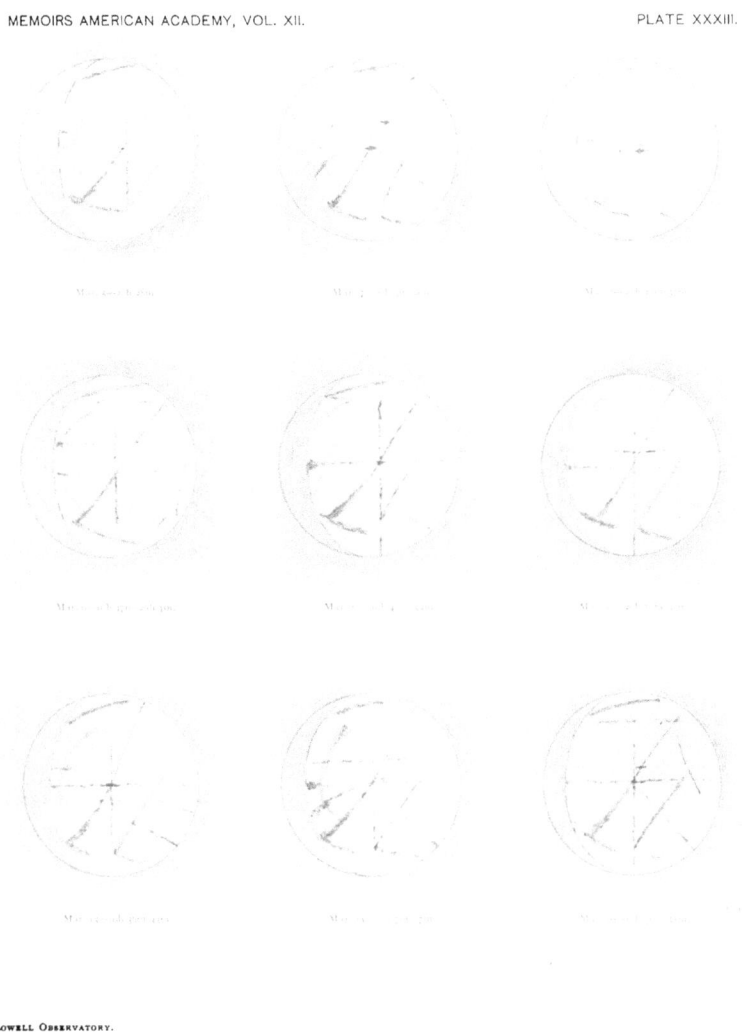

LOWELL OBSERVATORY.
MEXICO, 1897.

Some of Percival Lowell's sketches of Mercury, made with the 24-inch refractor from Tacubaya, Mexico, early March 1897. (From: New Observations of the Planet Mercury, American Academy of Arts and Sciences, vol. 12 (1898)*)*

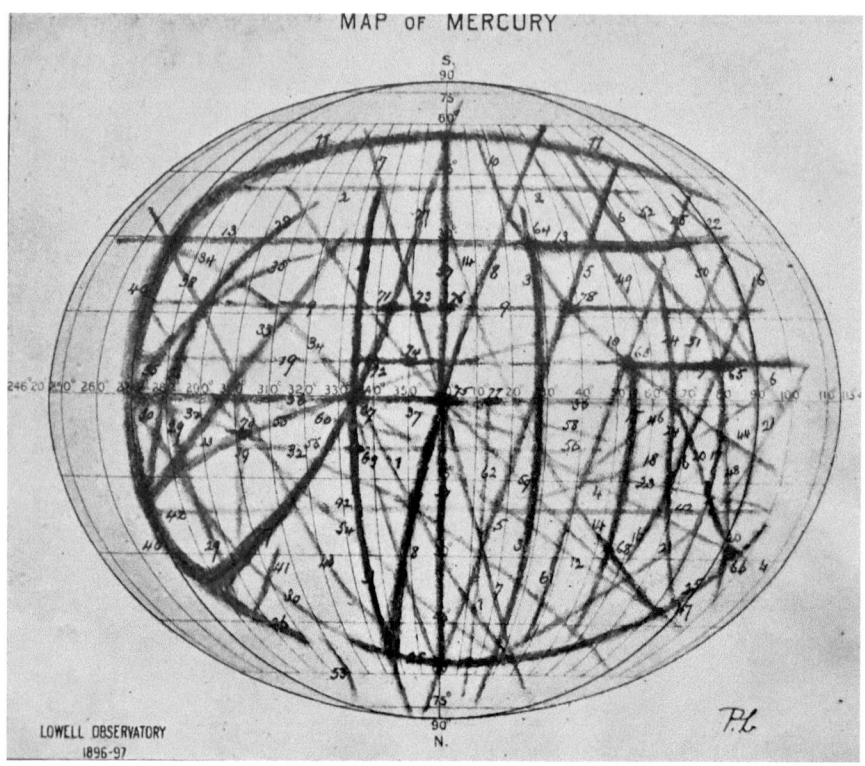

Lowell's rather bizarre map of Mercury, from "New Observations of the Planet Mercury" (1898), Plate XXVIII. The numbers on the map refer to 78 named features

He would continue to observe Mercury at intervals in October and November, 1896 and again in February and March, 1897. Eventually, he wrote up his results in "New Observations of the Planet Mercury," a sumptuously produced memoir for the American Academy of Arts and Sciences.[3] Of this it is fair to say that never have marginal perceptual data been more beaten to death with analysis. Mercury having been disposed of, he turned to Venus, close to the Sun in a brilliant daylight sky as it emerged from superior conjunction.

His first recorded observation of that was made on August 21. Within only 2 days, using what he called his "comet-seeker" eyepiece (a 3-inch focal-length eyepiece giving a magnifying power of 130 X), he succeeded in glimpsing "many markings" on the nearly full disk; a dark spot near the center being particularly apparent. He recorded the following entries in his observing log:

Aug. 24. 2h 52m. Plenty of markings on disk but none sure.

– 3h 13m. There seem to be certain markings—one black spot in special situated thus.

– 5 hrs. Apparently many markings but illusive.

Aug. 30. 3h 41m. Thought to see same spot that I saw on the 23rd with many less dark markings.

– 4h 10m. There is no doubt, I think, of markings but they are too elusive to map.

– 4h 40m. Spot again, *the* spot.

By early October 1896, the markings were sorting themselves into a spoke-and-hub pattern, which remained completely stationary from hour to hour and day to day. Douglass, Cogshall, Thomas Jefferson Jackson See (who was then a quite famous binary-star observer hired away from the University of Chicago), and Leonard all took their turns sketching the planet at the telescope, and even the occasional visitor, such as Lowell's close friend Judge Edward M. Doe, had a try. Most of the others saw very little, but Leonard's sketches often show the spoke-like pattern with a boldness approaching Lowell's own.

These features were quite unlike anything that had been seen before. Lowell attributed their visibility at Flagstaff to the steadiness of the seeing. Apparently he had no reservations whatever about their reality, and announced his findings about both inner planets in an "official dispatch…for distribution to astronomers" to the Boston *Evening Transcript*, which read:

> Venus and Mercury rotate around their axis in the same time in which they go around the Sun; Venus has an atmosphere, but no clouds, Mercury has no atmosphere.[4]

Back in Boston, where he went to regroup before leaving for Mexico, Lowell gave a lecture to the Boston Scientific Society—also duly reported in the *Transcript*—in which he added further tantalizing details about the latest discoveries. The markings on the planet were, he said, "surprisingly distinct; in matter of contrast, as accentuated, in good seeing, as the markings on the Moon and owing to their character much easier to draw…. A large number of them … radiate like spokes from a common center."[5] The surface of the planet, he declared, was a design in black and white over which was thrown a brilliant straw-colored veil. As the markings had the appearance of "ground or rock," and were not only permanent but permanently visible, he assumed they were surface markings seen through an extensive but transparent Venusian atmosphere. The spoke-system appeared to be phase-locked—thus his conclusion for the 225-day "day," equal to the length of the year—and as he had done for Mercury, he produced a chart, on which he introduced mythological names for the various features (Eros, Adonis regio, Aeneas regio, etc.)[6] Though his views about the planet overthrew many existing opinions about it, to some extent here too expectation was involved, since for Venus as for Mercury, Schiaparelli had affirmed a synchronous rotation. The master could not have been wrong.

Lowell blitzed the astronomical journals with a rapid succession of articles on both Mercury and Venus: *Popular Astronomy*, *Astronomische Nachrichten*, and *Monthly Notices of the Royal Astronomical Society* all carried detailed reports of his observations, lavishly illustrated with drawings and maps.[7] In addition, he worked up a much longer and more literary on Venus for *The Atlantic Monthly* [10]. His strategy [11]. He must have thought that by stating the same arguments over and over in much the same terms he was somehow making them more convincing. In fact, however, he was sleep-walking into disaster.

There was no avoiding the fact that the markings he described were simply bizarre. Moreover, as Lowell seems to have failed to appreciate, he was now placing himself in what historian David Sutton Dolan calls "double jeopardy." His Mars results had been one thing; they had been based on observations of markings first noticed by the great Schiaparelli, and recognized from the first as apparently unique to the surface of that planet; since they were artificial in appearance, the irrigation hypothesis seemed not only the best available but almost irresistible. So, says Dolan, "the entire edifice was built not just on the reality of marginally observable phenomena, but on specific aspects of those disputed phenomena."[8] In foisting upon the astronomical world the "cracks" on Mercury and the "spokes" on Venus, he had—no matter how hard he tried to emphasize the differences with the canals on Mars—loaded the astronomical world with too many strange phenomena, and in doing so risked straining credulity to the limit.

But even as bizarre as the Mercury observations were, the Venus observations provoked an even greater reaction. Nothing like what Lowell was reporting had been seen by astronomers elsewhere in centuries of observation. Not even Schiaparelli's keen eye had made out anything like it, while generations of astronomers going back to the early eighteenth century had all acknowledged that the planet's dazzling but almost blank disk could only be "a shell of clouds," on which "the eye … finds sure anchorage nowhere," as noted astronomy historian Agnes M. Clerke wrote" [13, p. 22]. Typical were Barnard's results. He had had only one good view of surface details on Venus, when the air around Mt. Hamilton lay heavy with smoke from wildfires on May 29, 1889. Otherwise, surface markings on Venus were, he wrote in 1897, "nearly always present, but they were always very elusive, and at no time could a satisfactory drawing be secured…. I am confident that the faint elusive spots … were real, but whether they were of a permanent nature it was impossible to tell, for the same spots were not recognized with certainty at different observations. The impression, however, was that they were not permanent" [14]. And yet here was Lowell, a neophyte observer of the planet and, to judge from his published drawings, a clumsy artist, using small apertures and low powers, claiming matter-of-factly that all of these other astronomers were wrong.

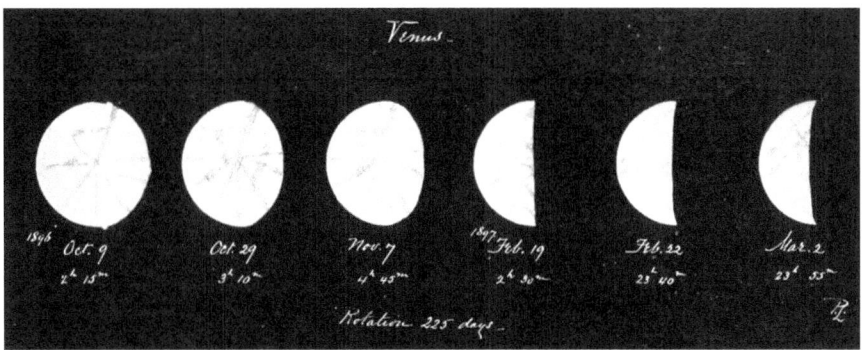

The spoke system of Venus, somewhat enhanced to improve legibility and worked over to tidy them up from the original log book sketches, according to Lowell, October 9, 1896 to March 3, 1897. Though supposed to conclusively demonstrate a 225-day isochronous rotation of the planet around the Sun, E.M. Antoniadi later noted "at a glance" that these drawings implied that Venus presented always the same face not to the Sun "but to Flagstaff"! In 1902, even Lowell himself briefly admitted they might be caused by some illusion, but he later reaffirmed their reality. (From: Percival Lowell, The Evolution of Worlds *(1910))*

Monthly Notices of Royal Astronomical Society.   Vol.LVII. Plate 6

CHART of VENUS.

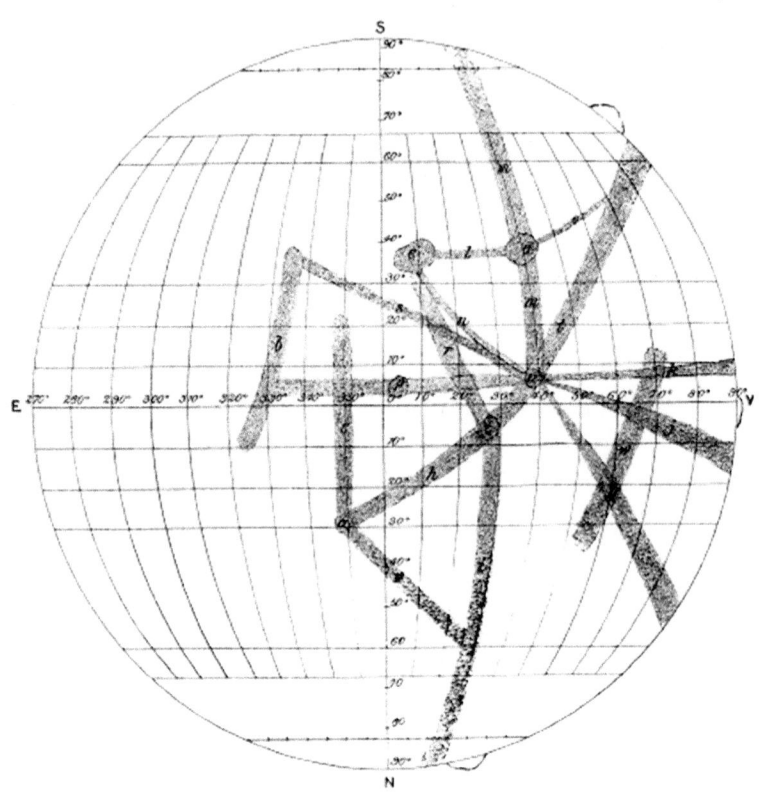

Percival Lowell del.

Lowell's chart of Venus, based on his sketches of October 15, 1896, as first published in the Monthly Notices of the Royal Astronomical Society *in March 1897*

A few days after his talk to the Boston Scientific Society, and on the last day of November 1896, Lowell closed on the house at 11 West Cedar Street on Beacon Hill which he had rented since the winter of 1894–95. The agent was Constance Savage Keith, a woman from a solidly middle-class background—her father worked in insurance—who was becoming established as an enterprising interior decorator and real estate entrepreneur in the Back Bay and Beacon Hill neighborhoods of Boston. She would hardly merit mention here except for the fact that 12 years later, she became Percival's wife. Meanwhile, the Sykes brothers' 42-foot wooden cylindrical dome and the 24-inch Clark telescope (minus the lens) were on their way to Tacubaya, where everything was to be reassembled by local workmen under Douglass's supervision in time for Mars's December 1 opposition. As usual, Lowell was a bundle of nerves, regularly telegraphing his staff to "hurry things please," though in fact his fretting was in vain—installation of the telescope and dome took longer than expected, mainly because of difficulty adjusting the double track upon which Sykes's dome would revolve. Lowell himself, bearing personally the precious lens, did not arrive in Mexico until the end of December, by which time Mars was already retreating far from the Earth. His first Mars observation was not recorded until New Year's Eve, 1897.

The Clark refractor, mounted in the Sykes dome (with canvas cover) in Tacubaya, Mexico, for the December 1, 1896, Mars opposition. (Credit: Lowell Observatory Archives)

With Mars offering no great revelations, Lowell and his staff—and especially Leonard—rejoined the campaign on Venus, and obtained a series of observations as the planet's phase, a fat gibbous the previous October, dwindled with decreasing distance from the Earth. The spoke-like markings and central hub were evident in drawing after drawing—those by Lowell and Leonard often standing, one beside the other, like planetary love knots. After 3 months in Mexico, Lowell returned to Boston at the end of March 1897, though Douglass remained behind, having applied for, and been granted, several weeks' time-off in order to climb the Mexican stratovolcano Popocatépetl. In the meantime a momentous decision had been taken: Lowell decided that it would perhaps be better to keep the Observatory in Flagstaff—a decision hailed with great joy by the Flagstaff *Sun-Democrat*, which announced the news on April 22 under the headline "Skylight City Beats Mexico."

The 24-inch Clark reinstalled on Mars Hill in its original canvas-covered dome, built by Godfrey and Stanley Sykes, with the observer's house—later to be subject to a rather haphazard development until it had become the 18-room "Baronial Mansion"—in the background. This photograph was taken in 1897. (Credit: Lowell Observatory Archives)

But the skies were far from entirely blue in Lowell's world. For some time clouds had steadily gathered, and suddenly a torrential storm of criticism burst.

Commenting on Lowell's map of Venus at a meeting of the Royal Astronomical Society, Captain William Noble could not help wondering whether "Mr Lowell has been looking at Mars until he has got Mars on the brain, and by some transference … ascribed the markings to Venus."[9] Camille Flammarion, who had been on friendly terms with Lowell when they met at Juvisy the spring before and was at least broadly sympathetic to Lowell's work on Mars, remarked that Lowell's observations of Venus were "entirely at variance with all that has gone before"; while his brilliant assistant E. M. Antoniadi, who would later emerge as Lowell's *bête noire* on Mars, wrote: "It is to be hoped … [the] canals of Venus, though negatively advancing our scanty Aphroditographical knowledge, will advance optical science in a positive manner, and enable us, in a near future, to have a clearer grasp of the canaliform illusion which so violently agitated of late the public mind" [15]. Even Leo Brenner, who had just revised his own improbably precise estimate of the rotation period of Venus from 23 h 57 m 36.2396 s to 23 h 57 m 36.27728 s, piled on.[10] Reporting in April 1897 that he too had seen the spoke-like markings, but only in "boiling" seeing, Brenner dismissed them as nothing more than an effect of unsteady air [16]. As for Lowell's drawings he found them "unbelievably grotesque," and rather similar to certain inscriptions on Chinese coins [17].

Apparently, the onslaught of criticism took Lowell completely by surprise. In utter perplexity, he noted that Barnard had written to him "that he has been unable to make out the markings on Venus discovered … last autumn in Flagstaff and seen there and in Mexico through three glasses and by every member of the staff, six persons in all, to say nothing of several outsiders." Obviously, the six observers and three glasses could hardly all be wrong. The problem, therefore, must lie with Barnard, who was now handicapped in having to use the world's largest refractor in poor conditions—in "a case of scientific criminal neglect," Lowell quipped, Yerkes had "buried" its large glass in southern Wisconsin.[11] Lowell had emphasized that the spoke-like markings were usually just as clearly defined in the 6-inch refractor as in the 24-inch, and Douglass, who served as a go-between for Lowell in his battle against his critics, claimed that he never saw them well except with small apertures, between 1.6 and 5 inches, when "Certain lines were unmistakable."[12] The case for small apertures—paralleling the small vs. large telescope controversy of earlier years—was made by Douglass who attempted to buttress the argument on the basis of his studies of the way that atmospheric turbulence affected telescopic images. A telescope of small aperture accesses only a small portion of the wave front, and if the small-sized wrinkles in the wave front have amplitudes much smaller than the wavelength of light, the image appears sharp and well-defined; the larger aperture integrates over a larger area of the wave front, so unless the wave front happens to be nearly plane (very good seeing) the image is blurred. Summing up some of his experiments, Douglass wrote:

> Such is the effect of using different apertures. As a matter of fact, we rarely have such simple conditions in actual experience. We have a given telescope and usually three series of air waves which may be all of different sizes. By a big diaphragm we can get rid of the blurring effect of the largest set. By medium and small diaphragms we can improve successively the bad effect of the other series but in doing so the light is enormously decreased. We may summarize this matter of aperture by saying that the smaller the aperture the more bodily motion and less confusion of detail; the larger the aperture, the less bodily motion and the more confusion of detail. This leads us directly to the aperture required for

certain classes of work. For seeing planetary detail we should use a small aperture unless the seeing is at its very best [18, p. 78].

Lowell later added yet another rather ad hoc explanation for Barnard's failure to see the markings that were detected at Flagstaff: He simply didn't have the right kind of eyesight. There was no doubt that he had a *sensitive* eye, that is, one well suited to the detection of faint stars or satellites or nebulae to be seen, but what was needed for the observation of planetary detail, such as the spokes of Venus or canals of Mars was an *acute* eye, such as Lowell himself, presumably, had to an exceptional degree. Moreover, he asserted—completely without proof—"the two quantities do not go together."[13]

In general it must be admitted that, as Lowell biographer William Graves Hoyt has said,

> While Lowell could claim that other observers had seen the canals of Mars, no other astronomer had, or indeed, has, ever seen anything like the "surprisingly distinct" features he described on Venus. Not even his assistants saw them as he did, although Miss Leonard's drawings … approach his own in boldness and detail….[14]

In Leonard's case, loyalty to her employer and the power of suggestion probably suffice to explain her results.

For what it's worth, the author has used the Clark refractor, under various conditions, and never succeeded in confirming anything like the hub-and-spoke system. Perhaps Lowell was simply recording shadows of the vessels in his own retina?[15]

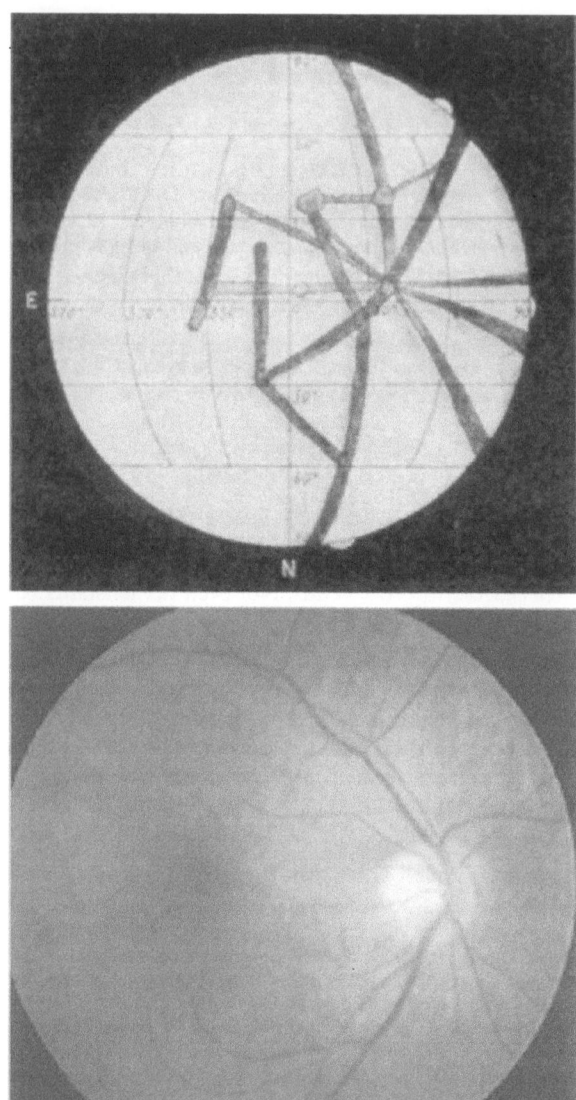

Lowell's chart of the surface of Venus, compared with a typical ophthalmoscopic photograph showing the characteristic pattern of blood vessels ramifying from the optic cup. (From: William Sheehan and Thomas Dobbins, "The Spokes of Venus: An Illusion Explained" (2003))

A Crisis in His Affairs

Though Lowell clearly had every intention of mounting a fierce defense of his Venus observations, in April 1897 the effort was suddenly suspended by the onset of an apparent case of nervous exhaustion (or "neurasthenia," to give it the rather ill-defined designation fashionable at the time, a condition to which brain-workers were supposed to be especially susceptible). Already suffering from anxiety, irritability and fatigue while still in Mexico, he had briefly attempted to return to Flagstaff from Boston in order to continue his work, but made it only as far as Chicago before having to turn back. For at least 3 years he had existed in a state of high nervous energy and excitement; it seemed that now "the machine" had simply broken down. That, at any rate, was the view his brother Lawrence would take:

> Although he could stand observing day and night without sufficient sleep while stimulated by the quest, the long strain proved too much, and he came back to Boston nervously shattered. Such condition is not infrequent with scholars who work at high speed, and although the diagnosis is simple the treatment is uncertain. The physician put him in bed for a month at his father's house in Brookline [i.e., Sevenels].[16]

He must almost at once have been given the diagnosis of "cerebral neurasthenia," a disorder which was regarded as a peculiarly American one—the psychologist William James, himself a sufferer, called it "Americanitis." It was the downside to the superiority over Eastern peoples that Lowell had conjured up in the *Soul of the Far East*, and was rife among the scholastic professions and brain-workers, with their competitive and ambitious characters and inability to relax; they had gotten dangerously far out front of the rest of the world, and the effects of modernity and the hurried pace required to keep up put them at risk of using up their limited allotment of energy and breaking down like an overheated machine. (The sufferers were, of course, members of the white, Protestant, Northern privileged classes; it followed that less evolved groups, such as Blacks, Native Americans, Southerners, and members of the laboring class, were relatively immune, simply because they didn't have the mental capacities to overuse in the first place.)[17]

Neurasthenia was such a grab-bag and all-encompassing diagnosis that it no doubt included a number of conditions. In Lowell's case, there can be no doubt that he had kept a breakneck pace for the previous 3 years, finishing *Occult Japan*, establishing the observatory, writing *Mars*, churning out papers on Mercury and Venus. It is easy enough to believe that he had simply run himself into the ground from exhaustion. The condition was commonly referred to as "brain fag," and the usual symptoms were described by the well-known physician Thomas Clifford Albutt:

> One of the most characteristics of cerebral neurasthenia is a weary brain. The sensation is familiar enough to any fagged man, especially if he fall short of sleep. Impressions seem to go half into one's head and there sink into a wooly [sic.] bed and die. Voices sound far off, the lines of a book run into one another and the meaning of them passes unperceived. Doors bang and windows rattle as they never did before; if a shoestring breaks, an imprecation is upon the lips. Business matters are in a conspiracy to go wrong. Letters [and reviews!] are left unopened partly from want of will, partly from a senseless dread lest they contain bad news. At night the patient tosses on his bed possessed by all the cares which blacken the darkness. Headache is common, loss of memory is distressing, and in severe cases, it is wider and deeper than mere inattention can explain. There is often the torture of acute hearing, or an inability to suppress attention; the hater of clocks and crowing cocks is a neurasthenic [20].

Much of this dovetails with what we know of Lowell's symptoms, and fits with what we would now probably diagnose as anxious depression. As stressed by Lowell's biographer David Strauss, "it appears that instabilities of mood and nerve tended to complicate Lowell's life at important junctures,"[18] with the collapse on the return from Mexico paralleling the earlier one in his late twenties in which Lowell had scandalously broken off his engagement, abruptly retired from business, and tried to find refuge by escaping halfway around the world. Strauss emphasizes that the crisis of 1897 took place in the context of the attacks on his Venus work, which sowed doubts as to "the validity of his observations" and threw into question his "future prospects in astronomy."[19] Of course, it is dangerous to attempt a firm diagnosis of someone no longer available for a psychiatric interview. However, it does appear that Lowell, having hitherto enjoyed the "inflated self-esteem, as well as a certainty of conviction about the correctness and importance of his ideas of a hypomanic state," now began to experience something of "the morbidity and flatness of mood along with a slowing down of virtually all aspects of human thought, feeling, and behavior that are most personally meaningful" that many poets and artists experience in their downcast moods.[20] It was a condition, by the way, familiar to many of his relatives, including his second cousin the poet James Russell Lowell.[21]

Though he had always seemed self-confident and assured, perhaps to a certain extent, it had all been a great bluff, behind which lurked something of the same dynamic later claimed for his youngest sister Amy, in whom it would be claimed that "her extroversion was no more than a calculated escape from a troubled psyche, an inner self that repudiated repose. Her egotism, though infinitely elastic, was a fragile skin enclosing a gigantic inferiority complex. At any moment the skin might burst."[22] The criticism—veering into ridicule—of his Venus observations seems to have made the skin burst.

He tried to find refuge in his West Cedar Street home. At least at first, his symptoms did not seem too severe, and a prompt recovery seems to have been expected. He presented "New Observations of the Planet Mercury" to the American Academy of Arts and Sciences on May 12, and continued working, at least intermittently, on a companion memoir on Venus. He staggered forward a bit, then stopped. In the end—though he continued to hope to return to it—he never finished, which is itself, perhaps, rather telling. He suffered a setback when his friend Alvan Graham Clark died, unexpectedly, in June, but by mid-summer, he could assure Douglass, "I get better myself slowly. My quantity increases and they tell me that my quality will come after it."[23] The rally, however, was not sustained; from mid-September until mid-December, realizing the need for more support, he moved back into Sevenels with his father and youngest sister Amy. A doctor was consulted, who gave orders for him to have "absolute rest," with no visitors and no work permitted—and no involvement with astronomy for at least a year. In October, See found Lowell worse, and wrote to Douglass (then in California) "I feel awfully sorry for Lowell—the poor man wanted so badly to continue his work, and must give it up for a considerable time."[24] Meanwhile, Lowell's brother Lawrence added his counsel, telling Percival to "take very little mental exertion, live in the open air and take a considerable amount of exercise," adding, jocularly, that the easiest way to approximate the needed conditions would be to become a mule on a cargo ship sailing from New Orleans to the Cape of Good Hope.[25] It is unlikely that Percival, in his present state, found this at all amusing.

Lawrence's advice, however facetiously framed, did hew to common views about the treatment of neurasthenia at the time. Since the cause was supposed to be the inability of the human body (and especially the brain) to hold up to the strains of modern life, an alternative to the rest cure (for men at any rate) was a whole-hearted embrace of rugged life on the Western frontier was often recommended—as in the case of Theodore Roosevelt. He had been regarded as a dandy (like Percival), known as "Young Squirt" and "Punkin-Lily" when he was a New York state assemblyman but tried to reinvent himself in the West as a rugged rancher, hunter, and outdoorsman—as a man, as he later put it, possessed of "few of the emasculated, milk-and-water moralities admired by the pseudo-philanthropists; but he does possess, to a very high degree, the stern, manly qualities that are invaluable to a nation" [21]. The image of the future "Rough Rider" and Bullmoose was born.

Lowell also came to despise the rest cure; its only effect seemed to be to further weaken his already damaged nervous system and delay his recovery. Though in earlier days he had identified masculinity with the vigorous activities of polo and mountain climbing, he now eschewed the highly masculinized extreme of treatment of a Teddy Roosevelt and instead, upon turning over the supervision of all of his affairs, including those related to the running of the Observatory, to William Lowell Putnam, II, he embarked with a young Boston physician, Dr. Alfred Lindstrom, on a healthful, and only moderate, regimen of travel, light exercise, and abundant sunshine. How much of this Lindstrom participated in is not recorded—he died in Boston in May 1900 at the age of thirty-two. In addition to taking a farmhouse in Chocorua, New Hampshire, where Lowell studied trees and shrubs and sent specimens to the director of Harvard University's Arnold Arboretum, Charles Sprague Sargent, he spent time in Georgia, at his sister Katharine Bowlker's summer house at Mount Desert Island, Maine, and, when the urge to foreign travel returned, in Bermuda.

One of his most notable forays was to Amherst, Massachusetts, home of David Peck Todd, director of the Amherst College Observatory, and his extraordinary wife Mabel Loomis Todd, the daughter of an elderly assistant in the Nautical Almanac Office in Washington, D.C., whom Todd, then an assistant astronomer at the U.S. Naval Observatory, had met in June of the summer in which Asaph Hall had discovered the two moons of Mars. In 1881 the Todds had moved to Amherst, the little college town nestled between the Pelham and Mount Holyoke ranges, where David had accepted a position as professor of astronomy over Mabel's misgivings. It was then poorly equipped with astronomy, boasting only the small observatory he had used as a student there in timing eclipses of Jupiter's satellites; David, too, whose real talent lay in mechanical engineering, would eventually see it as a mistake. Mabel at first was miserable in what seemed to be a backwater, but her view of the situation improved when she met the Dickinsons. Austin, a lawyer, a solemn-looking man with penetrating eyes and cutting a fine figure on horseback, was the leading man of the town, and lived with his wife in the "Evergreens," next to the Dickinson homestead where lived his sisters Lavinia and Emily—the reclusive poet whom Mabel described as "the rare, mysterious Emily, who listened silently in the dark."[26] The Todds moved into a house across Dickinson Meadow at the "Dell," and Austin and Mabel began to see one another regularly. In 1882, when David was off to Mt. Hamilton to photograph the transit of Venus, they crossed their Rubicon and began a 12-year affair which ended only with Austin's death. (David was strangely tolerant—even encouraging—of the relationship, perhaps because it gave him latitude to pursue his own affairs; later he would whistle "Martha" at night whenever he was preparing to enter the house, to give the lovers a chance to regroup.) Despite—or perhaps because of—the unusually open nature of their relationship, David and Mabel remained on good terms with one another, and in 1887 traveled to Japan for a total eclipse, the first of many they were to undertake together as pioneering "eclipse chasers."

In 1897, Todd published a new textbook on astronomy, which not only defended Lowell's Mars theory—"[T]he explanation of the canals themselves … seems very plausible," he wrote; "of course, acceptance of this theory implies that Mars in ages past, has been, and may be still, peopled by intelligent beings"—but even his observations of Venus, which Todd called "interesting work" [24, p. 360 and p. 354]. Lowell was immediately taken with an astronomer who was "not a fossil but belongs to a living species,"[27] and in November 1899 he and Miss Leonard spent several months with the Todds in Amherst. As many women did, Mabel found Lowell "charming," and presented him with a copy of her 1894 book, *Total Eclipses of the Sun*. Lowell, still not entirely recovered from his depression, responded appreciatively, "I can only hope that the next time I have the pleasure of seeing you I shall not even be under a partial eclipse."[28]

Perhaps the highlight of the Amherst visit was a series of Venus observations made with the 7 1/4-inch refractor in the College's old Octagon Observatory. Lowell and Leonard found the best views when diaphragmed to 3 or 4 inches, and the spoke system showed up with its usual bold definition. The Todds also took turns at the eyepiece and were apparently completely persuaded of the markings' reality. Many years later, after Lowell's death, David looked back more circumspectly. He wrote to E.S. Morse, "Of course you know what a (perhaps unconscious) hypnotist Lowell was. Autumn of 1899 he came out to Amherst—stayed several months—made me observe Venus—and by Jove! I was so dead sure I saw the Venusian markings that he did…. when—along came his published admission [in 1902] that he found, to his dismay, that he 'must have been deceived some way.' I have always though he might (very honestly) have been self-deceived about many of the Martian canals also."[29]

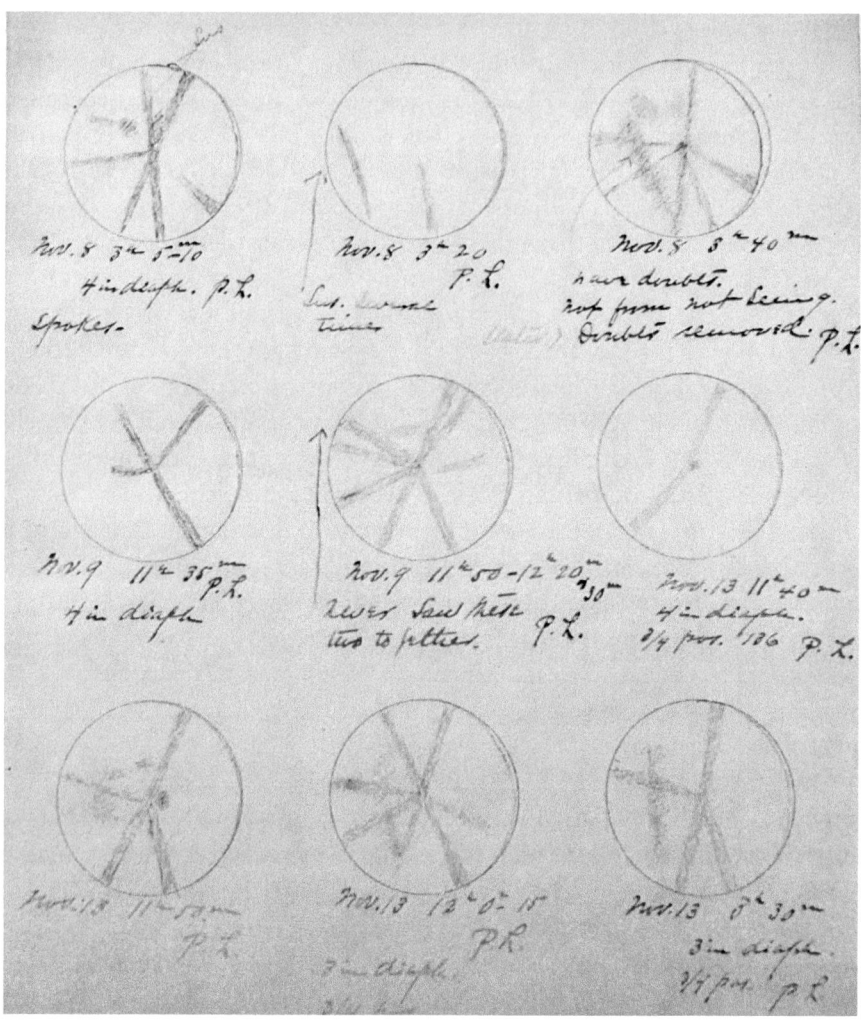

Lowell's drawings of Mars with the 7¼-inch refractor at Amherst College, November 9 to 13, 1899. The best views were often obtained with the objective diaphragmed to 40 or 3-inches. The drawing at the upper right, made on November 8, contains the comment, "have doubts," about one of the markings, followed by another "Doubts removed." (Credit: Lowell Observatory Archives)

By now, Lowell's recuperation was gathering steam. He traveled to England, to the French Riviera (where he met psychologist, philosopher, and fellow neurasthenic William James and presumably shared with James some of the 400 Temptation Key West cigars he had ordered sent to him) and Tripoli, where the Todds, with funding from Lowell, caught up with him and hoped to view their next total eclipse on May 28, 1900.[30] Lowell arrived 4 days before the event, and Mabel found him "quite jolly." The eclipse, lasting only 51 seconds, offered little more than a fleeting glimpse of the prominences and corona. Mabel found Lowell much better than he had been in Amherst, though occasionally, she noted in her journal, "there are irresponsible bursts that show a diseased mind."[31] (What she meant by "irresponsible bursts" is impossible now to say; probably only a series of petty tantrums when something didn't go Lowell's way.)

A month after the eclipse, Augustus Lowell died. Though the green shoots of his recovery had already been evident when he had visited the Todds in Amherst, and he was even further along when they saw him in Algiers, he now recovered well enough to return to contemplate returning to the Observatory. Though his emotional response to his father had always been complex, his father's passing allowed some kind of resolution. Moreover, he and his siblings received a considerable inheritance; Amy was able to use her share to buy Sevenels from the rest, while Percival now had the financial means to devote himself to upgrading the equipment at the Observatory—where, he found, things had deteriorated considerably over the nearly 4 years of his absence. (Also, significantly, when the estate was divided among the children, Percival took all of his mother's belongings with him.)

Setting up equipment during the Lowell Observatory expedition to Tripoli. (From: W.L. Leonard, Percival Lowell: An Afterglow (Boston, 1921))

Lowell Observatory Goes "Wild West"

While Putnam took over management of the Observatory's financial affairs, Douglass was placed in charge of the continuing scientific work. His salary was hardly munificent—$800 a year out of a total Observatory budget of $6000, or marginally more than the $665 average annual salary of a factory worker at the time (and much less than astronomers at Lick were paid; Barnard had made $2000 a year in 1893). He certainly earned every penny he made. His first major task was to reinstall the telescope on Mars Hill in the Sykes dome. No sooner had he overseen this, however, than he suffered a severe attack of rheumatism—perhaps related to his recent strenuous ascent of

Mount Popocatépetl—and left for several months' recuperation with an aunt in San Diego. This left See in charge of the Observatory.

Though See seemed to have more impressive credentials than anyone else associated with the Observatory, including a Ph.D. from the University of Berlin, 3 years'experience as an instructor of astronomy at the University of Chicago, and a rising reputation as a keen-sighted observer of double stars, unfortunately he suffered from a highly disordered personality, and managed, in short order, to alienate every member of the staff, leading to a marked exodus of assistants over the next year. Drew said of him that he "generally makes a good impression on slight acquaintance [but] he does not bear intimacy. The more you see of him the worse you like him. He makes trouble wherever he goes. I do not want to associate with him, if for no other reasons, simply because he is ever talking about and seeking his own worth, and I should be pleased never to see him again."[32] Drew left the Lowell Observatory in June 1897 in order to take up a position as principal of schools in Baraboo, Wisconsin, though he would temporarily return for several months the following year. Cogshall followed in September, after a fire ("possibly of incendiary origin," claimed See) razed the Grand Canyon Hotel where See and Cogshall had shared a room together; the fire destroyed their personal possessions. Soon afterward, Cogshall returned to Indiana University. As his replacement, See hired Samuel Boothroyd, one of his former students at the University of Chicago, who also lasted about a year. At the same time George A. Waterbury was hired as an assistant, and remained at the Observatory for 2 years. Records in the Observatory archives suggests that staff regarded See as arrogant, dishonest, and sloppy in his work, and also lacking in personal morality, that these attributes contributed to the frequent comings and goings of staff. Since, however, See is not a central character, we need not go into this further here.[33]

There was a temporary reprieve in November, when See requested and received permission for an extended period of leave back East, where he hoped to prepare his double star observations—which he would always claim to be the most important work done at the Lowell Observatory—for publication. With news of See's departure, Putnam able to persuade Douglass to return from California and take charge of the observatory again. At once Douglass threw himself into a great deal of work that had been put on hold since first Lowell's and then his own illnesses. In addition to editing the first volume of the observatory's *Annals*, most of the observations which had already been arranged and interpreted by Lowell himself at the time of Lowell's collapse so that he had only had to correct page proofs, prepare the index, and see the work to the printer (it was published in 1898), he began work on a second volume, which Lowell, always impatient, urged him to complete within only 4 months—though since Douglass had to write most of it, and was a far slower writer than Lowell, it only appeared in late 1900. Attempting to keep the Observatory active in pursuing the lines of research in which Lowell was interested, Douglass resumed observations of the satellites of Jupiter and loyally (if somewhat clumsily) rose to the defense of Lowell's embattled Venus work. Though Douglass had personally made few observations of Venus, he wrote vigorously, though perhaps not very effectually, in April 1898:

> No matter how difficult to obtain, a just hearing is our right. No one is entitled to cry out against us until he can show that his atmosphere is approximately as good as the one through which Mr. Lowell discovered these markings. Let our dubious friends ... devote a portion of their valuable time to work at the telescope under better atmospheric conditions, and no one will misinterpret the silence which will follow [25, p. 320].

Douglass's challenge to astronomers to observe Venus from Lowell Observatory and experience its superior atmospheric conditions had only one taker: E.E. Barnard. Barnard, of course, had been following developments in Flagstaff and Mexico City, with ever-mounting skepticism as to the reality of the markings being recorded there—or at the very least, in the implausible way in which they were depicted.

Barnard's own observing had been very limited since leaving Mt. Hamilton—the 40-inch lens had not arrived in Williams Bay and been installed in the giant tube until the end of May 1897. At last it seemed that its active career would get underway. On the night of May 28, Hale observed on the telescope till 12:45 a.m., then yielded the instrument to Barnard and Ferdinand Ellerman. At a little before 3 a.m., in order to observe the "Swan" nebula (M17) in Sagittarius, they raised the elevator floor (modeled on that at Lick) to within six inches of its highest point, and continued to observe until day began to break and they went home. For the convenience of some workmen who were scheduled to do some work on the tube in the morning, they left the floor in this position. Within an hour or so of the observers' departure, at 6:43 a.m., one of the workmen coming to the Observatory heard a loud crash, and on investigating found that the cables suspending the floor on its south side had given way, and the floor on that side had fallen forty-five feet, leaving a mass of splintered wood and twisted iron. It was a miracle that no one was injured; as Barnard reflected, "A couple of hours either way, and death in all probability would have come to one or the other of us. Only a few nights before this accident the President of the University of Chicago and some thirty or more trustees and prominent men of the university had seen through the telescope, and the floor had been up and down with them on it. … It was providential, then, that the floor fell when it did, for the fault in the attachment of the cables made it certain that it must soon have fallen" [26, p. 286]. (News of the collapse of the floor seems to have greatly stressed the optician responsible for crafting the giant lens, Alvan Graham Clark: he died suddenly and unexpectedly of a stroke only a week afterwards.)

Obviously the collapse of the floor meant another postponement of work with the telescope, and so Barnard made the most of the time available to continue writing up the backlog of observing reports. With all the controversy about Venus's markings and other planetary details that had broken out at the time, he reviewed the work he had done on that planet at Mt. Hamilton. He wrote, "Venus has been examined on a number of occasions with the 36-inch, when the planet was beautifully defined…. Nothing was seen of the singular system of dark narrow lines shown in recent years by observers to cover the surface of the planet. Every effort was made to show them, by reduction of the aperture and by the use of solar screens and various magnifying powers. They were also looked for with the 4-inch finder. Previous attempts with the 12-inch here also failed." In addition to doubts about the Venus markings, he also took exception to the linear markings Douglass was reporting on the third and fourth satellites of Jupiter. Drawing on his personal experience he wrote: "The markings I have seen have always appeared to be large and more or less diffused—except in the case of certain white polar spots…," and added some Ruskin-like strictures against the general style of drawing adopted by the Flagstaff observers:

> … It has always appeared to me that when representing any planetary detail, it should be drawn as nearly accurate, as to appearance, as possible. If the marking is vague and uncertain it should be made so on the drawing and no definite outlines should be given where none exist. Some observers seem to be in the habit of giving a definite boundary to markings whether such are really seen or not. This is very misleading and no object should be shown with a definite outline unless it really has such.
>
> This diagram method of drawing is specially noticeable in a good many drawings of Mars, and it is perhaps through the study of these and not by any inspection of the planet itself that so many queer and unnatural ideas have ben propagated concerning the physical appearance of Mars. Many of the regions shown on the drawings and maps of Mars as definitely bounded are in reality very diffused and uncertain in their outlines.[34]

Despite these public criticisms, Barnard remained on friendly terms with Douglass, Lowell, and even See. On May 5, 1898, he wrote to Douglass expressing a keen desire to visit Flagstaff so that he could, if possible, observe these details for himself. "I do not have any ill feelings—why should I have?", he wrote. "It is simply to satisfy my own eyesight that I want to come. Dr. See and Mr. Lowell have always been friendly towards me and if for nothing else, I should want to come so that I shall cease to do them an injustice, for I doubt not that my failure at L[ick] O[bservatory] to see these things so easily seen from Flagstaff may have had some sort of influence against the observations."[35]

Come to Flagstaff Barnard did, arriving on May 30, 1898 at a season when See had predicted the best seeing. See was still there, though at just this moment the campaign against him among the staff—with Douglass now joining the effort—was coming to fruition. Barnard undoubtedly sensed that something was up, but he is unlikely to have found out any of the details. They would not have surprised him since, needless to say, he knew all about the strained interpersonal relations among staff at observatories, given his experiences on Mt. Hamilton; but See remained his main contact at the Observatory and, with assistant George A. Waterbury and likely others, took turns with Barnard at the telescope.

George A. Waterbury, left, and Andrew Ellicott Douglass, on the steps of the Clark refractor dome, 1898. (Credit: Lowell Observatory Archives)

Observing Venus was Barnard's main objective, though Mercury, Jupiter, Saturn, and Uranus were also observed, as Barnard recorded in his observing log book:

May 30. High wind blowing—more or less threatening clouds. Observing Jupiter power 750. Especially attention was paid to the 1st satellite. The disc was seen once in a while, but poorly. Sometimes the image would separate into two alike. They would slowly recede from each other … and then join again.

Uranus. The image was poor… I could see nothing on its surface but Dr. See and several others saw a belt on it.

Saturn. The image so bad that I could only faintly see the [Cassini] division in the ring.

May 31. 10:30 a.m. No clouds; considerable S.W. wind… Examined Mercury… The image occasionally was fairly well defined; but it was constantly jumping so that it was not possible to see the planet steadily…. Spent about one hour on this watch. There were no markings seen….

May 31. Venus… The jumping was constant so that nothing was seen.

June 2nd. Venus. Image jumping excessively. High S.W. wind. The seeing continued very variable, running from a blurred mass to a fairly well defined limb, but the image was jumping badly…. Dr. See saw a dark line from … limb to the center of the planet…. I saw nothing—not even the old vague markings which I had seen in former years.

June 3rd. 4:30 p.m. Examined Venus up to 5:30…. The image was quite steady; I saw no markings; Mr. Waterbury, however, saw the usual Flagstaff markings.

June 4th. Fearful high N.E. wind. 6:00 p.m. Venus. Mr. Waterbury … saw the usual Flagstaff markings easily—I could not see them.[36]

Barnard never published an account of his visit to Flagstaff—mainly as a courtesy to Lowell, who had not been present—but he did offer a summary to W.W. Campbell at Lick. "My going to Flagstaff was purely a desire to verify Mr. Lowell's observations of Venus and Mercury—Mars was out of reach," he wrote. "I had almost no prejudice at this time and had a sincere hope that I could give a favorable report of my visit."[37] He regretted the fact that conditions were so bad when he was there; even the regular observers admitted that they were unusually poor for that season. However, on June 4, his last day in Flagstaff, he and Douglass climbed to the summit of 12,340-foot Agassiz peak of the nearby San Francisco Peaks. (One would like to know all they discussed; we do know that Douglass confided to Barnard the results of some experiments he was beginning to perform with artificial planet disks, which were leading him to suspect that many of the Lowellian markings were likely illusory—a venture into apostasy that would in due course prove fatal to his Flagstaff career). From Mt. Agassiz, Barnard enjoyed a panoramic view of the situation of the observatory, a view which led him to suspect that, while seeing conditions at the time of his visit may have been atypical he could easily guess what they were likely to be at other times:

> If I were to judge of the conditions from a mere inspection of the surroundings of the Lowell Observatory, I would not suppose that it was a very suitable site.... It is on the edge of a Mesa or bluff some hundreds of feet high. On one side (to the East) is a broken country and the hot desert; on the other a vast forest of pines—the observatory being on the edge of the forest. Very near—to the North—are the San Francisco peaks some 12,000 ft. High, which have snow on them most of the year.[38]

Despite having taken every opportunity over several years to defend the myth of Flagstaff seeing insisted upon by his employer, Douglass, quite independently, had already reached very similar conclusions to Barnard. In his classic paper "Atmosphere, Telescope and Observer," written in Mexico in April 1897, he had reported on his own experiments into how atmospheric currents at different levels in the atmosphere affect seeing. He wrote:

> One of the most striking instances of the use of these observations, was the discovery of the reason why some of the east winds at Flagstaff gave good seeing and others bad. When the seeing was good the currents seen through the telescope came also from the east but when the seeing was bad they did not do so at all. Instead, they came from the north or northeast and the mountain range extending from ten miles due north to about six miles east, northeast was shown to be responsible both for the change of direction in the surface movement and the very bad quality of the stream which was passing by at considerable altitude overhead. It seems probable from this that neighboring mountain ranges are not good.[39]

Much of what Douglass and Barnard noticed about the Flagstaff seeing has been borne out since. Certainly, as Douglass first noted during the winter of 1894–95, winter conditions are often poor. From the experience of several decades, long-time Lowell astronomer Brian Skiff notes that a northeast wind right after a snowstorm, when the cold air drains down to lower elevations from higher up, can cause some of the worst seeing imaginable—something like an arc minute, so that even Jupiter twinkles as seen with the naked eye! Here, the passage of a low-pressure front is followed by conditions in which the wind at the surface (say up to the top of the San Francisco Peaks) veers around to the northeast; however, higher aloft it remains from the northwest, often at high velocity (the jet stream). At points of contact where the two run up against each other, they usually give rise to extreme turbulence. On windless nights in winter, a cool flow off Observatory Mesa into the town of Flagstaff tends to make the seeing at the Clark dome poor, though right at sunset, before that starts to happen, and also later in the night the seeing can be very stable. In the summer, the monsoons, which begin from about the beginning of July to the end of September, disrupt astronomical work.

Clyde Tombaugh, a veteran of many years of observing with the Clark, used to say that he never saw the seeing worse nor better than he experienced in Flagstaff.[40] Barnard was unfortunate in happening to visit during one of the times when the seeing was very poor, and left Flagstaff with a very low opinion of the work being done there.

Hiatus: The Douglass Era

In Lowell's absence—and after the departure of See and the other assistants—Douglass remained at the Observatory alone. From a distance, Lowell attempted to direct the work, and urged his assistant to make micrometric measures of the ball and rings of Saturn (in an attempt, no doubt, to show that he could do as well at Flagstaff as Barnard had done at Lick), observations of the planet's satellites, and scrutiny of Venus and Jupiter for markings and to determine the rotation period of Jupiter's belts. The main priority, however, was to observe the next Mars opposition of February 1901, which Douglass did between August 11, 1900, and March 31, 1901.

Much of this work was of a completely routine nature. However, on the night of December 7–8, 1900, Douglass, who had paid close attention to projections and irregularities of the Martian terminator since his days at Arequipa, noted a particularly pronounced terminator projection. As usual he telegraphed a report to Lowell: "Last night projection north edge Icarium Mare lasted seventy minutes." Lowell in turn passed the announcement along to astronomers elsewhere, and before long it was circulating in the press, with a syndicate based in Philadelphia writing to Douglass that it was "very much interested in recent announcements concerning the message which it is reported you received from the planet Mars sometime since."[41] Though Douglass never claimed to have seen anything other than a high-altitude cloud, it was not until February 1901 that interest in the report finally began to wane. The experience seemed to have a sobering effect on Douglass, for at the same time interest in the "message from Mars" was reaching its peak intensity, Douglass began corresponding with astronomers elsewhere about his increasing disillusionment with the work of the Lowell Observatory.

Indeed, during Lowell's long absence, Douglass had developed a strong independent streak. Here his association with Barnard in 1898, though brief, had played a critical role, and had sown seeds of doubt about the seeing at Flagstaff which had always been Lowell's ace in the hole and produced almost clairvoyant views of planetary detail that could not be approached by observers elsewhere. In August 1898, he presented a paper, at Barnard's request, at the Boston meeting of the American Association for the Advancement of Science, in which he presented details about Lowell's tendency to diaphragm even small objectives and also to use very low powers. Thus, he said, Lowell "saw the detail on Mars either exceedingly well or not at all," which was "the characteristic effect of using low power." Further, according to Douglass, "although the detail in the lower power is far more easily seen, it is, I believe, not so well seen. For to me it sometimes appears distorted, wide at places, narrow elsewhere, and difficult to locate on the planet and in general less reliable than in the higher power."[42] He added that if the detail being observed was complex, it might be necessary to increase the magnifying power. Douglass also presented the results of his extensive research into the effects of atmospheric currents, scintillation and the like, and concluded that the claims for the superiority of the seeing at Flagstaff in revealing planetary details was rather circular. "The explanation that is usually offered for successful effort is that the seeing is better. But such statements in regard to the seeing are very loose because we judge the seeing by the number of things seen. The sentence 'We see more at a certain place because the seeing is better,' if translated into non-astronomical English, becomes 'We see more at a certain place because we see more.'…".[43]

In February 1899, he and Boothroyd carried out tests at various other northern Arizona sites than Flagstaff using a 12-inch mirror that had been supplied by Clark and Sons for the purpose; again, Flagstaff did not stand out, and he suggested—rather prophetically— that perhaps more isolated sites might be considered for future facilities [27, pp. 363–64]. Though he tried his best to defend Lowell's work, especially on Venus, he also came to have serious doubts about such details—including even the canals of Mars. Writing several years later, he admitted that inconsistencies in the positions of the canals recorded by various observers seemed problematic:

> It is right and natural that we should first regard these faintest of markings as realities upon the planet. The writer can certify to their apparent genuineness, for he has pictured numbers of them in half a dozen favorable oppositions since

1892. To him they were real until time proved that in the faintest markings astronomers failed of satisfactory agreement. In the larger markings, and even in the larger canals, conflicts of evidence do occur, but are never troublesome. One may confidently say that such realities do exist. But with the very faint canals whose numbers reach occasionally well into the hundreds, discordance reigns supreme, and it is frequently found that different drawings by the same artist antagonize each other across the page [28, p. 76].

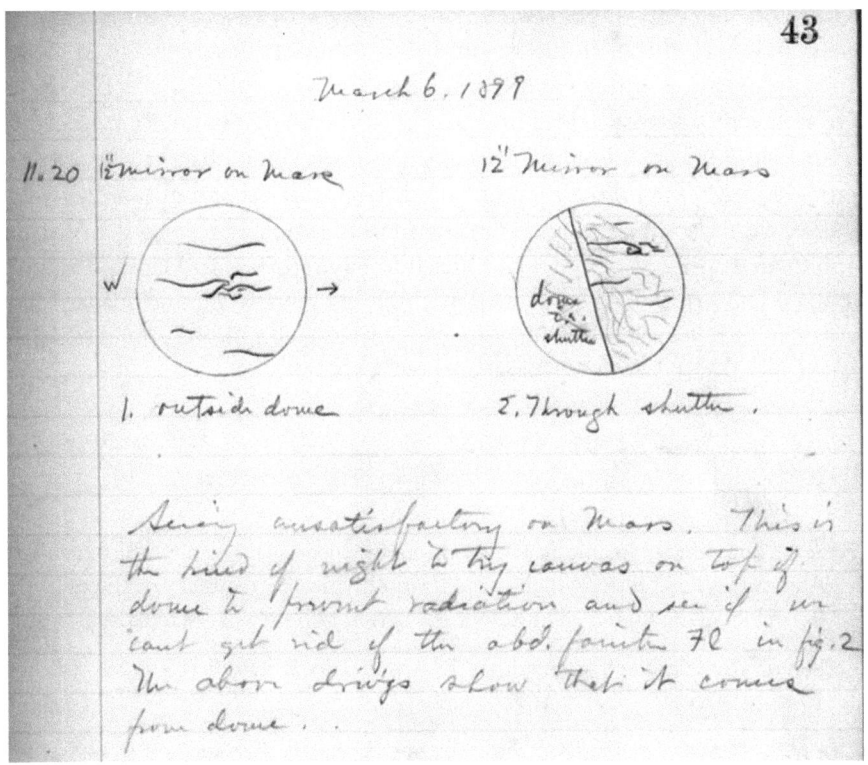

Sketches by Douglass March 6, 1899, on a night of unsatisfactory Mars images, showing atmospheric currents observed with his 12-inch mirror, (left) outside the dome and (right) through the shutter from inside the dome. On the basis of these observations, Douglass notes that the bad seeing is coming from the dome, and suggests possibly trying canvas on top of the dome to prevent heat radiation from inside the dome affecting the seeing. (Credit: Lowell Observatory Archives)

Lowellian and indeed Schiaparellian orthodoxy insisted on the reality of the doubling of canals (geminations), which contributed significantly to the artificial cast of the planet's features in drawings and maps and which Lowell explained as probably due to the Martians' attempting to relieve strain on individual canals during seasons of high water flow through the system. However, not everyone saw the geminations; Pickering and Douglass never did. Douglass found that in addition to seeing effects related to the local topography, views in the 24-inch refractor were badly affected by thermal currents from inside the dome passing through the shutter. Consequently, details such as the double canals which had been "laboriously memorized" were only recognized when (as he later put it) "for an instant, Heaven vouchsafes him a brief view."[44] These elusive flashes of detail hardly served to inspire confidence, and in 1899, Douglass embarked on a series of experiments to see whether similar effects could be produced using artificial planets in which either no details or diffuse and irregular details were substituted. The results were sufficiently robust to lead Lowell to shut the work down because, according to Douglass, "it was evident it cast doubt on some observatory publications"—though the following year Lowell reconsidered and allowed them to resume, with a special emphasis on the markings on Venus and the Martian double canals [29]. When Lowell himself finally returned to Flagstaff, in time for the Mars opposition of February 1901, he gamely offered himself as a subject to this experiment. He promptly drew a double line where only a single line existed!

Notes

1. A measure of the chromatic aberration in an objective is given by the Chromatic Aberration (CA) ratio, which is the f/ratio divided by diameter of the objective (in inches). The higher the number the better. For a few large telescopes:

   Yerkes 40" Clark refractor with a 19.4 meter focal length (f/19) has a CA of 19/40 = .475

   Lick 36" Clark with a 15 meter focal length (f/16.5) has a CA of .536

   Meudon 32.7" Henry Brothers with a 16.2 meter focal length (f 19.5) has a CA of .596

   Lowell 24" Clark with a 9.75 meter focal length (f/16.3) has a CA of .679

   Brashear 18" with an 8 meter focal length (f/17.5) has a CA of .972.

   Nice 30" with a 17.9m focal length (f/23) has a CA of .77.

The problem is serious in all these large telescopes; the Brashear used by Lowell in 1894 is best, but even the Nice refractor, which was built with an extremely long focal length specifically to optimize its use for planets, does not do very well. To give a comparable example involving smaller apertures: a CA of 1.0 is equivalent to that in a 6-inch refractor with a focal ratio of 6, which no amateur would choose as the instrument of choice for use on the planets, while a CA of 0.5 is equivalent to a 6-inch refractor with a focal ratio of 3. Only with a CA in the range of 1.5 to 3 is the level of chromatic aberration considerable filterable, and only above 3 is it considered negligible. That level corresponds to the situation in a 6-inch refractor with a focal ratio of 18—a very long telescope indeed.

2. Clyde Tombaugh, who as a member of the Lowell staff was a veteran of many years of observing with the Clark refractor, wrote (personal correspondence to William Sheehan, December 5, 1986): "For best definition the air cells of uniform refraction in front of the telescope should be nearly as large as the aperture used, and the air cells should be about 15 inches across. This was one of the reasons Lowell used 16 inches most of the time. The other was the serious secondary or residual chromatic aberration for an f/16 refractor. For large refractors one can improve the view by using yellow filters to knock out the blue-violet. Lowell's views ere deceiving because he used too low a magnifying power of 310 to 400, and irradition bled in on the dark stripes, etc., making them appear more narrow than they really were…. When I used the same telescope parameters as Lowell used, I saw the canals much as he drew them!"

3. P. Lowell [5]. Though Schiaparelli saw some of the drawings of Mercury Lowell published in 1896–97, he did not see Lowell's memoir until 1909. It came as something of a shock to the aging astronomer of Milan. "I am infinitely indebted to your … for your great Memoir on Mercury, which had remained entirely unknown to me up to this day," Schiaparelli wrote. "I had seen, I don't remember anymore where, some drawings of Mercury that were said to have resulted from your observations. These drawings seemed to me so different at first sight from what I had seen myself that I believed it was unnecessary to attempt a comparison before having seen your definitive publication. Now that I have it… It terrifies me, that is the word! Would Mercury, then, have the approximate structure of a polyhedron, regular and symmetrical, like a faceted diamond?…" It was clear that he could not entirely reconcile himself to Lowell's bizarre drawings and maps. G.V. Schiaparelli to P. Lowell, September 2, 1909; Lowell Observatory Archives.

4. Hoyt, *Lowell and Mars*, p. 108.

5. J. Ritchie, Jr., Boston *Evening Transcript*, November 28, 1896.

6. To show just how *outré* Lowell's observations were, only one astronomer before him had believed himself to be able to see the surface of Venus directly, and attempted to draw a map: the Italian priest-astronomer Francesco Bianchini in 1726–27.

7. Among them: P. Lowell [6–9].

8. David Sutton Dolan [12]. Published at http://ro.uow.edu.au/theses/1712

9. "Report of the Meeting of the Royal Astronomical Society" (1896), *The Observatory*, vol. 19, p. 420.

10. Brenner was not the first to estimate the rotation of Venus to the ten-thousandths of second. The Jesuit astronomer Francesco de Vico, at the Collegio Romano, had published a value of 23 hrs 21 mins. 21.934 secs. in 1841. See Patrick Moore (1961) *The Planet Venus* (London: Faber and Faber), p. 133.

11. P. Lowell (1897) "Atmosphere: in its effect on Astronomical Research." Unpublished manuscript; Lowell Observatory Archives.

12. Douglass, "Summary," p. 75.

13. P. Lowell (1905), "Means, Methods and Mistakes in the Study of Planetary Evolution." This manuscript, dated April 1905, was submitted to W.H. Wesley, the assistant secretary of the Royal Astronomical Society, for publication in the *Monthly Notices*, but possibly because of its rather contentious tone, was rejected.

14. Hoyt, *Lowell and Mars*, pp. 111–112.

15. Douglass said that the spokes came out best when the aperture was stopped to only 1.6- to 5 inches. With this set up, the telescope's exit pupil in front of his eye would have been shrunk to a pinhole of diameter less than 0.5 millimeter, effectively turning the telescope into a gigantic ophthalmoscope. Such a small exit pupil often casts shadows of blood vessels on the retina, making them visible. According to Andrew T. Young of San Diego State University, "As the eye pupil moves with respect to the small telescope pinhole, the angle at which rays arrive at the retina changes, so that features that are within the eye but a little removed from the retina, like blood vessels, may cast shifting shadows and be seen." Andrew T. Young to William Sheehan, personal communication, February 2, 2021. See also: William Sheehan and Thomas Dobbins [19].

16. A. Lawrence Lowell, *Biography*, p. 98.

17. On neurasthenia, see: Julie Beck, "'Americanitis': the disease of living too fast.'" *The Atlantic*, March 11, 2016.

18. Strauss, *Percival Lowell*, p. 173.

19. Ibid.

20. Jamison, *Touched by Fire*, p. 18.

21. According to Heyman, *American Aristocracy*, p. 78, James Russell Lowell's symptoms included "… bouts of wallowing and melancholy … accompanied by mental torpor, paranoia, unexplainable fears that he was going mad…. The tumultuous conflicts, guilts, and fears that possessed him seemed to be the result of the incongruities between his personal longings and the actualities of his existence—the eternal conflict in his soul between infantile intensity of passion and profound self-control, between Brahman conservative roots and romantic impulse. But there was no clinical explanation, only the florid symptoms…."

22. Heymann, *American Aristocracy*, p. 79.

23. P. Lowell to A.E. Douglass, June 15, 1897; Lowell Observatory Archives.

24. T.J.J. See to A.E. Douglass, Oct 7, 1897; Lowell Observatory Archives.

25. Strauss, *Percival Lowell*, p. 22.

26. Polly Longsworth [22]. Mabel's greatest accomplishment was in recognizing the value of Emily's achievement, and in editing and publishing them long after the poet's death. See: Julie Dubrow [23].

27. Percival Lowell to Andrew Ellicott Douglass, August 9, 1897; Lowell Observatory Archives.

28. Percival Lowell to Mabel Loomis Todd, letter undated but likely May 16, 1897. Yale University Library, Mabel Loomis Todd Papers, Box 17.

29. David Peck Todd to E.S. Morse, April 25, 1924. Peabody Essex Museum, Edward S. Morse papers. Perhaps the effect of Lowell's strong, dogmatic personality on those around him can be compared to the shared psychotic disorder referred to as *folie à deux*, in which the stronger personality imposes a delusional belief upon another person or persons who would not be delusional if left to their own devices. This is true not only in the case of weaker and more submissive associates such as Leonard and E.C. Slipher, but arguably of generations of (especially younger) readers who came under his spell (*folie imposée*) at an unformed and suggestible stage of development. Those younger readers included the present author who was utterly convinced by Lowell's arguments at age 10.

30. William James had been fascinated by Lowell's work on Shinto trances, and cited *Occult Japan* (p. 96) in his "Exceptional Mental States Lectures," given at the Lowell Institute in 1896. The latter was a sketch for what became the celebrated *The Varieties of Religious Experience*, based on his 1901–02 Gifford Lectures on natural theology given at the University of Edinburgh. Lowell's hypnotic effect on his Japanese subjects was to be repeated among collaborators of his planetary work such as Wrexie Leonard, E.S. Morse, E.C. Slipher and others; see note 45 above.

    Regarding Lindstrom, just how much of Lowell's travels he shared is unclear; probably only the first part. An Alfred H. Lindstrom, M.D., probably the same person as Lowell's companion, is recorded having died in Boston in May 1900, at age thirty-two.

31. Mabel Loomis Todd journal, June 13, 1900. Quoted in Benfey, *The Great Wave*, p. 206.

32. D.A. Drew to "To whom it may concern"; undated but ca. June 1898. Lowell Observatory Archives.

33. For a long time this correspondence was kept from public view, but it was finally published, in part, by the Lowell Observatory's Sole Trustee, William Lowell Putnam, III, in 1994. See William Lowell Putnam, III and others (1994) *The Explorers of Mars Hill: More Than a Century of History at Lowell Observatory* (West Kennebunkport, Maine: Phoenix Publishing), pp. pp. 27–31. On reviewing the litany of complaints against See, William Lowell Putnam, II fired him.

34. E.E. Barnard (1897) "On the Third and Fourth Satellite of Jupiter," *Astronomische Nachrichten* No. 3453, columns 321–330: column 321.

35. E.E. Barnard to A.E. Douglass, May 5, 1898; Lowell Observatory Archives.

36. E.E. Barnard, observing log book; Barnard. Edward Emerson. Papers, University of Chicago Library, Hanna Holborn Gray Special Collections Research Center.
37. E.E. Barnard to W.W. Campbell (1898) "Observations at the Lowell Observatory in May and June 1898"; manuscript, Campbell. William Wallace. Papers, Mary Lea Shane Archives of the Lick Observatory.
38. Ibid.
39. Douglass, "Atmosphere, Telescope and Observer," p. 67.
40. Brian Skiff to William Sheehan, personal communication, December 1, 2023. Much of what is known about the factors affecting the seeing in Flagstaff has been determined by the release twice-daily weather balloons, sometimes attaining to altitudes of 35 miles.
41. Hoyt, *Lowell and Mars*, p. 125.
42. A.E. Douglass [1, p. 82]. This paper by Douglass was prepared at the request of Barnard, who was then program director for the astronomical section of the American Association for the Advancement of Science, and presented at the Boston meeting of the A.A.A.S. in August 1898.
43. Ibid. 79.
44. Douglass, "Illusions of Vision," p. 78.

References

1. Douglass AE (1899) A summary of planetary work at the Lowell observatory and the conditions under which it has been performed. Popul Astron 7:74–84
2. Young CA (1885) Observation of the red spot of Jupiter. Observatory 8:172–174
3. Young CA (1886) Small telescopes vs. large. Sidereal Messenger 5:1–5
4. Lowell P (1903) The solar system. Houghton, Mifflin and Co, Boston, pp 29–30
5. Lowell P (1898) New observations of the planet Mercury. Mem Am Acad Arts Sci 12:433–477
6. Lowell P (1896) Mercury. Popul Astron 4:360–363
7. Lowell P (1896) Detection of Venus' rotation period and fundamental physical features of the planet's surface. Popul Astron 4:281–285
8. Lowell P (1897) Determination of rotation period and surface character of the planet Venus. Mon Not R Astron Soc 57:148–149
9. Lowell P (1897) The rotation period of Venus. Astron Nachr 3406:361–364
10. Lowell P (1897) Venus in the light of recent discoveries. Atl Mon 5:327–343
11. Holloway J (1953) The Victorian sage: studies in argument. Macmillan, New York, p 8
12. Dolan S (1992) Percival Lowell: the sage as astronomer," Doctor of Philosophy thesis,. University of Wollongong, pp 208–209
13. Clerke AM (1890) The rotation periods of Mercury and Venus. J Br Astron Assoc 1:20–25

14. Barnard EE (1897) Physical and micrometrical observations of Venus. Astrophys J 5:299–304
15. Antoniadi EM (1898) English Mechanic 67:474
16. Brenner L (1897) The Observatory 20:260
17. Brenner L (1898) Spaziergänge durch das Himmelszelt. Astronomische Plaudereien mit besonderer Berücksichtigung der Entdecker der letzten Jahren. Mayer, Leipzig, pp 125–126
18. Douglass AE (1897) Atmosphere, telescope and observer. Popul Astron 5:64–84
19. Sheehan W, Dobbins T (2003) The spokes of Venus: an illusion explained. J Hist Astron 34:53–63
20. Albutt TC (1900) A system of medicine, by many writers, vol 8. Macmillan, New York, p 138
21. Roosevelt T (1902) Ranch life and the Hunting Trail. Century, New York, pp 55–56
22. Longsworth P (1999) Austin and Mabel: The Amherst affair and the love letters of Austin Dickinson and Mabel Loomis Todd. University of Massachusetts Press), Amherst, p 64
23. Dubrow J (2018) After Emily: two remarkable women, and the legacy of—America's greatest poet. W.W. Norton, New York/London
24. Todd DP (1897) A new astronomy. American Book Company, New York
25. Douglass AE (1898) The markings of Venus. Mon Not R Astron Soc 58:313–320
26. Barnard EE (1908) A few unscientific experiences of an astronomer. Vanderbilt University Quarterly 8:273–288
27. Douglass AE (1899) The effect of mountains on the quality of the atmosphere. Popul Astron 7:355–356
28. Douglass AE (1907) Illusions of vision and the canals of Mars: a study of Martian canals by experimental methods. Scientific Am Suppl 64(1648):76–78
29. Webb GE (1983) Tree rings and telescopes: the scientific career of A.E. Douglass. University of Arizona Press, Tucson, p 48

7

Adventures with the Bruce

Contents

> *[T]here are [in Ophiuchus and Scorpio] vast regions almost entirely free from stars, in a surrounding region thick with small stars. These regions seem veiled over with some sort of material in which occur blacker spaces, as if all this part of the sky were involved in a thin faint nebulous sub-stratum which partly veils the blackness of space beyond. In this, apparently, occur rifts and openings giving a clearer view of space.*
> —E.E. Barnard, "On the Vacant Regions of the Sky" (1906)

Barnard in Eclipse…

The aphelic opposition of January 18, 1899, was the first of what might be called the H.G. Wells era of Mars, coinciding with the publication of the struggling-school-teacher-turned-writer whose novel *The War of the Worlds* as a magazine serial in 1897 and between covers in 1898. Lowell's literary genius struck a sympathetic chord in Wells, though in place of the intelligent, benign

© The Author(s), under exclusive license to Springer Nature Switzerland AG 2024
W. Sheehan, *Parallel Lives of Astronomers*, Springer Biographies,
https://doi.org/10.1007/978-3-031-68800-3_7

and peace-loving folk, struggling for existence on their doomed world by building canals as Lowell had proclaimed, Wells's Martians were given a more menacing cast. Gazing across space from their dying world, and possessed of technology far beyond that of the canal-builders that even included means of travel across interplanetary space, they had cast their eyes on the lush and verdant world inward toward the Sun:

> No one would have believed in the last years of the nineteenth century that this world was being watched keenly and closely by intelligences greater than man's and yet as mortal as his own: that as men busied themselves about their various concerns they were scrutinized and studied, perhaps almost as narrowly as a man with a microscope might scrutinize the transient creatures that swarm and multiply in a drop of water. With infinite complacency men went to and fro over this globe about their little affairs, serene in their assurance of their empire over matter. It is possible that the infusoria under the microscope do the same. No one gave a thought to the older worlds of space as sources of human danger… Yet across the gulf of space, minds that are to our minds as ours are to those of the beasts that perish, intellects vast and cool and unsympathetic, regarded this earth with envious eyes, and slowly and surely drew their plans against us… [1].

H.G. Wells's Lowell-inspired novel War of the Worlds *(1898) was written when Wells and his wife were living at Woking, south of London. The first shell shot with a giant cannon across interplanetary space—depicted here unscrewing to reveal a Martian confronting the astronomer Ogilvie, the first victim of the Martians' heat ray, in a 1906 illustration by Brazilian illustrator Alvim Corrêa—landed in the sandpit just behind Wells's house on Horsell Common. A visitor to the area, equipped with a late 1890s version of Ordnance Survey maps, can still follow exactly where the Martians went, and even which houses they destroyed. Later, the Martians mount towering tripod machines and use their heat ray to lay waste to the English countryside and London. (Credit: Public domain via Wikipedia Commons)*

The opposition was also the last of the nineteenth century, and the first accessible to E.E. Barnard with the 40-inch Yerkes refractor.

Since his arrival in Chicago in 1895, Barnard had discovered just how brutal the winters in the Midwest of the United States could be. As always he worked himself relentlessly, in all kinds of weather, and from its dedication in October 1897, devoted all the time on the great telescope available to him in the attempt to push forward work—largely involving micrometric observations of faint satellites, the central stars of planetary nebulae, and star clusters—begun with the 36-inch refractor on Mt. Hamilton. Observing during winter took a toll on him however, and on Christmas Day 1898 he wrote to the pioneering English spectroscopist Sir William Huggins, "I am writing this propped up in my sick bed. I have been sick in bed for some time and I do not know when I shall be able to get up again."[1] The following May he wrote to an old friend from California (where he doubtless wished he could still be), George Davidson, recently appointed professor and chair in the department of geography at the University of California, Berkeley:

> I have been unwell much of the past winter. It has been a dreadful winter—something like 30° below zero [Fahrenheit] several times, and below zero a great many times.... It was the worst winter they have had here for many years—and I hope the worst they will ever have.[2]

It was almost impossible to do anything useful under these circumstances, but ironically, Barnard found that most of the bad seeing seemed to be generated within the dome of the 40-inch telescope itself, leading him to refer to it as "dome seeing." It was due to thermal turbulence created by weakly mixed convection inside the dome and especially around the slit, which caused the air passing in front of the telescope objective to appear like a stream of running water. Barnard tried everything he could to improve the definition, even resorting to using a diaphragm with the great telescope—the very practice he had decried at Mt. Hamilton—often stopping it down to 24- to 30- inches, but it was to little avail. His observing log book contains entries such as:

> Dec. 13, 1898. Can make out no details on [Mars]. Have lowered the curtain to the base of the slit—this has improved the seeing some—but it is still bad. Bitterly cold wind from the west. Temp. outside -4° [Fahrenheit]
>
> Feb. 6, 1899. The seeing is excessively bad.... Pretty high cold north wind. Mars ... can making nothing of it—the seeing is so bad.

Feb. 11. Temp. in observing chair = -14°; outside -19°. The seeing is too bad to do anything with it.—[Mars] is a great blur of light.

Mar. 29. The image of Mars is excessively bad—fearful blurring—dome [e]ffect.[3]

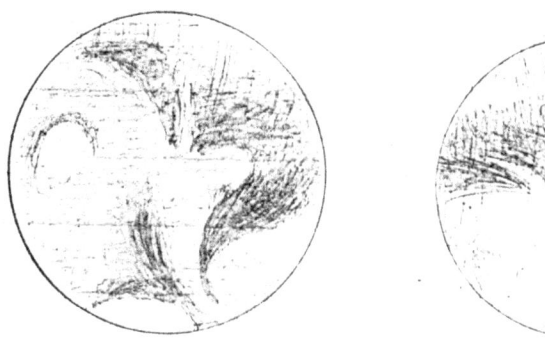

Barnard's best sketches of Mars with the Yerkes 40-inch refractor, 1899. (Left) February 5, 1899, made through a sky wet with haze and clouds; the temperature in the dome was only 7.5 degrees F. and outside 6.5 degrees, and a heavy wind was blowing from the South. (Right) February 13. (Credit: Barnard observing log book, Barnard. Edward Emerson. Papers, University of Chicago Library, Hanna Holborn Gray Special Collections Research Center)

Barnard realized that in such a climate he would never again having stunning views of Mars like those he had enjoyed at Mt. Hamilton during the magic summer and fall of 1894. With the exception of an observation of Mercury showing three dark smudges like prints of an inked thumb against a piece of paper which he compared to the naked-eye markings on the Moon, which he noted in unusually good daytime conditions on August 31, 1900, and a once-in-a generation great white spot on Saturn he discovered in June 1903, he did very little significant work on the planets. Indeed, in contrast to his record of publication in the period up until the time he left Mt. Hamilton, in which his work on the planet accounts for more than half of all of his productivity, he hardly published any papers on the planets henceforth that did not draw on the work at Mt. Hamilton. The main exceptions were papers on the phenomena of the edgewise rings in 1907–08, and about his attempts to photograph Mars in 1909, which will be mentioned in due course. Lowell and his colleagues, with the superior seeing conditions they enjoyed for such work at Flagstaff, thus largely commandeered the field for a number of years.

Though Barnard did no significant work on Mars at the oppositions of 1901, 1903, and 1905, he was busy with other things. He observed the total eclipse of May 28, 1900—the same which Lowell observed with the Todds from Tripoli—with a Yerkes Observatory expedition to Wadesboro, North Carolina. The observations were exclusively photographic, and to take maximum advantage of the brief period of totality—only a little more than a minute from Wadesboro—he and his colleague George Willis Ritchey had to deny themselves a direct view of totality, since in order to assure the accuracy of their exposures, they placed themselves in a long, shaded horizontal tunnel to minimize thermal air currents. To the extent they glimpsed the solar corona, it was only as it was projected onto their photographic plates.

The Yerkes eclipse camp at Wadesboro, North Carolina, May 28, 1900. Left to right: Mr. Wolff, William Brasington, George W. Ritchey, Ferdinand Ellerman, Albert A. Nijland (Utrecht Observatory director, seated), George Ellery Hale (standing), J. H. Wilterdijk (Leiden Observatory, seated), George S. Isham, Edwin Brant Frost, and Barnard. (Credit: University of Chicago Photographic Archive [apf6-00774], Hanna Holborn Gray Special Collections Research Center, University of Chicago Library)

Photograph taken by Barnard and Ritchey on the Yerkes Observatory expedition to Wadesboro, North Carolina, May 28, 1900. University of Chicago Photographic Archive [apf6-02118], Hanna Holborn Gray Special Collections Research Center, University of Chicago Library

In addition to Barnard and Ritchey, the Yerkes expedition included Hale, Edwin Brant Frost, and Ferdinand Ellerman, while a group from the British Astronomical Association was also on hand, and accompanied Barnard to Williams Bay where they stayed for 2 days as guests of the Barnards. One of them, the Rev. J.M. Bacon, later recalled that visit:

Here is an ideal retreat, for one whose whole life is given up to his work;—a luxuriant garden of Natures own planting, where the sumac is the undergrowth, and flowers, prized in English gardens, grow as weeds. On the one side the wide reach of Big Foot Prairie, on the other, far below, like the blue waters of Lake Geneva. I cannot recall the scene without picturing a calm, clear evening with the light of the after-glow already fading in the west, and in the distance the retreating figure of a man, nearing middle life, yet hurrying with all the activity of vigorous youth across the grass to his long night's labour [2].

The Barnards, Edward and Rhoda, at home at Yerkes Observatory, as photographed by Gertrude Bacon, 1900. (Credit: Letter Archives, British Astronomical Association. Used with permission of the British Astronomical Association)

Though excellent plates were obtained, the Wadesboro eclipse was too short to be very satisfying. By contrast, the forthcoming eclipse of May 18, 1901, would be the event of a lifetime, with totality lasting six-and-a-half minutes on the central line—almost the maximum possible. However, it was accessible only from the very Far East, the shadow of the Moon first touching the Earth at sunrise near the south end of the island of Madagascar, then moving eastward over the island of Mauritius, across the long stretch of the Indian Ocean, and onto land again at noon on the west coast of Sumatra. It then continued on its course over the China Sea, across the islands of Borneo and Celebes, and on to the coast of New Guinea where it left the Earth at sunset.

To take advantage of this exceptional opportunity, Barnard signed on with a U.S. Naval Observatory expedition to Sumatra led by Professor Aaron Nichols Skinner. Before sailing out from San Francisco Bay, Barnard paid a brief visit to Mt. Hamilton, which he had not seen since his departure in August 1895. A former colleague, Richard Tucker, described him as "nervous and unwell, with no strength, and generally a disappointed man I think."[4] Tucker was probably right. Since the early days of deprivation in which he had by sheer dint of will and relentless application risen to become one of the world's leading astronomers, Barnard had always been restless and dissatisfied—always in pursuit of unachievable perfection, and unhappy with anything which (as inevitably it did) fell short. He must have been aware, too, that compared to the spectacularly productive (though tumultuous) years at Lick, he so far had comparatively little to show for 6 years of work at Yerkes, largely because southern Wisconsin was, as he put it, "a mirey climate for a great telescope."[5]

Barnard and other members of the U.S. Naval Observatory expedition to Sumatra, standing on deck of the U.S. Army Transport Sheridan *awaiting departure from San Francisco Bay to Manila, February 1901. (Credit: Special Collections and University Archives, Jean and Alexander Heard Library, Vanderbilt University)*

The trip, lasting from February to July 1901, took him a third of the way around the globe. Yet, in the end, Barnard had nothing to show for it other than the greatest nightmare in any astronomer's life: being clouded out at the critical moment of totality. The observing station selected was at Solok, in the western part of Sumatra, where totality was calculated to last for 5 mins 52 s. For several days, around noon (the time of the eclipses), Barnard checked the weather, and things remained promising until May 17, the day before the eclipse when he laconically noted in his diary, "Raining heavily." At dawn on the day of totality, Barnard rose to find the sky more or less full of clouds, and though as the day advanced there were tantalizing hints of clearing, things changed for the worse—at noon the sky was hopelessly covered with thick mackerel clouds. Having traveled so far for this, Barnard could do nothing but go through the motions of exposing plates in the hope of registering some trace of the corona, visible feebly through the clouds, but nothing of any use was obtained. The almost 6 min of totality passed quickly, and Barnard could only note bitterly in his notebook, "The best preparations ever made for the observations of a total eclipse of the Sun had come to naught" [3, p. 540]. Another member of the expedition, Samuel Alfred Mitchell, who had spent a year as a research assistant at Yerkes and was now at the Naval Observatory, who shared a bed with Barnard in the hotel Orange in Padang, lamented afterwards, "Never will I forget the despair and dejection of an astronomer almost heart broken." On the trip over, Barnard had been in great spirits—the "life of the whole party, always telling jokes, always interested in the landscape and scenery and strange peoples we saw, and always thrilled with the prospect of doing interesting and valuable astronomical work." In this upbeat mood, he had taken hundreds of photographs of the scenery and people he encountered in the Far East. By contrast, "the trip back to America was one of deep gloom for Barnard; he was not interested in anybody or anything" [4].

The Bruce Photographic Telescope

Gradually, after returning to Williams Bay, Barnard got over his keen disappointment and was able to admit more philosophically:

> Though the scientific results were meager, the opportunity given the astronomer to see one of the most interesting and beautiful countries in the world was of some consolation—not to Science itself however but to the astronomer who had made the long journey and who had faithfully performed his duty through it did result only in failure.[6]

The "philosophic mind" took a while to take hold, however, and immediately on his return to Yerkes, Barnard was interested in nothing and nobody, and in particular he was not interested in revisiting the by now long overdue project of publishing the Milky Way and comet photographs he had taken with the Willard lens at Lick, for which he had raised funds by subscription to publish, only to find no method of reproducing them that would satisfy his very high standards. On first moving to the Midwest, he had tried to publish them with a firm in Chicago specializing in the collotype and photogravure processes, but was unable to satisfy himself with the results. Later efforts, which included encouragements and offers of help from George Ellery Hale and William Wallace Campbell, also went nowhere; the project was not to be completed until 1913.

Instead of taking up again the photographs made with the Willard lens which, important as they were in their time, were no longer state-of-the-art, he preferred to set his sights on the future. He was now eager to take the wide-angle photography of the Milky Way to a new level with a telescope specifically designed for the purpose. Had Barnard been Lowell, he could have easily funded the telescope himself, but he depended on the kindness of strangers, and found a patron in Miss Catherine Wolfe Bruce of New York City, the heiress of a type-setting fortune. She was a highly cultured woman who spoke several languages and had once studied painting in a serious way, and became a generous benefactor of astronomy in her later years. In the summer of 1897—just as the 40-inch was about to begin work—she granted Barnard $7,000 ($250,000 in current value) for a photographic telescope. The grant was given with the stipulation that "all plans, specifications and contracts for the instruments and building are to be approved by me [Barnard] in writing and that all work [was] to be done under my supervision and to my entire satisfaction."[7] Barnard's "entire satisfaction" was not so easily achieved, however; he was a ruthless perfectionist, and wanted his telescope mounted so as to allow long exposures to be carried through the meridian, which had been impossible with the Willard lens which had sat on an ordinary equatorial. He personally suggested the idea of making a mount with a bent pier, which would allow the telescope to swing freely under it in all positions of the instrument. But the mount, fashioned on Barnard's plan by the Cleveland, Ohio, engineering firm of Warner and Swasey proved to be the easy part. The long delay came in making the lens.

Barnard had a good deal of practical knowledge of opticsbut of theoretical knowledge hardly any at all. He therefore had unrealistic expectations for a large, aberration-free telescope, with a lens 10-inches in diameter, having "a wide angle and flat field with as short a focus as was consistent with these two qualities." He put the matter to his close friend, the Pittsburgh optician John Brashear, and the latter made several good lenses of 4-inches diameter and upward, but none was ever good enough. Undaunted, Brashear, on his own initiative, went ahead and in March 1899 obtained glass disks and began figuring them into a 10-inch Petzval doublet (i.e., based on the optical design invented by the German-Hungarian optician Joseph Petzval). Barnard, in the meantime, had learned that an English firm, Thomas Cooke and Sons, had a triplet lens which sounded as if it might meet his requirements, and wrote to Brashear, "It is impossible to go ahead with the lens without looking into the matter.... I need hardly tell you that it has been my earnest desire all along that you should make the lens of the Bruce telescope ... mainly ... on account of the friendship and esteem in which I have always held you. You, yourself, however, must see that my duty not only to Miss Bruce and to the Yerkes Observatory, but to the Science of Astronomy as well, demands that if this lens is what is claimed for it, it is necessary to get it from the Bruce telescope."[8] Thus, Barnard spent from December 1899 to March 1900 in England and France. In the end, the Cooke lens did not satisfy him either. After seeing what England had to offer, Barnard crossed the Channel and visited the Palais de l'Optique at the Paris Exposition Universelle (1900), walking with Camille Flammarion and the latter's assistant Eugène Michel Antoniadi through the 197-foot-long tube of the rather infamous 49.2-inch refractor which was then on display. The tube in this unusual telescope was fixed, and starlight was directed through it by a 79-inch "siderostat" mirror, which tracked the stars as they moved across the sky. Unfortunately, it proved to be a great disappointment—the only scientific results ever obtained with it being Antoniadi's drawings of M57 in Lyra, NGC 7009 in Aquarius, and a few other planetary nebulae—and after the Exposition ended the tube was broken up and the optical components placed in storage at the Paris Observatory (where they remain today). Carrying with him the fresh impression of this unsuccessful telescope, which owed so much to overreaching, Barnard at last acknowledged that his own ideal for the Bruce telescope had been "too high and one not attainable with optical skill." He finally decided that Brashear's doublet would have to do after all [5, p. 37]. With Barnard's go ahead, Brashear rapidly finished work on the lens, one of his masterpieces. It was ready by September 1900. The focal length was only 50 inches, giving a 9-degree field, and though images were fairly poor at the edges the definition was sharp in the central 7 degrees. A 6 ¼-inch Voigtländer lens having a focal length of 31 inches was mounted on the same bent pier mounting. The two telescopes were chosen because together they gave a very desirable range of scale, and Barnard, almost without exception, made simultaneous exposures with them. A 5-inch visual refractor was employed as a guide telescope. Later, two additional cameras, a 3.4-inch Clark lens of 20-inches focal length and a so-called "lantern" lens of 1.6-inch aperture and 6.3-inch focal length, were also employed.

The 10-inch Bruce astrograph, in its dome at Yerkes, with the smaller 6 ¼-inch and 3.4-inch photographic telescopes and 5-inch visual telescope for guiding, mounted together on the bent pier mount designed by Barnard himself. The twin tapering boxes contain two smaller cameras. The circular screen is the shutter for the Bruce camera, while the rectangular screen at top pivots to serve as the shutter for the two smaller cameras. The objective adjacent to and at the left of the Bruce lens is the guide scope. The rod carrying the shutters could be rotated with a terminal hand knob located near the eyepiece of the guide telescope to begin an exposure with all three cameras. Careful study of the photograph reveals that turning the rod by 90 degrees will cover all three cameras. The fourth small astrograph visible just above and to the left of the declination setting circle has a separate shutter. (Credit: University of Chicago Photographic Archive [apf6-00576r], Hanna Holborn Gray Special Collections Research Center, University of Chicago Library)

"It Is Cold and Disagreeable Here"

So Barnard had his telescope; now the question was where to put it.

Despite the five degrees of more northerly latitude at Yerkes compared to Mt. Hamilton and the rigors of a continental rather than an oceanic climate, Barnard had nevertheless hoped that the transparency of the skies on which success in deep sky photography would depend would be reasonably good. This proved, however, not to be the case. In addition, the skies at Yerkes were often cloudy.

Clearly, Yerkes might not be the best place in which to deploy the Bruce telescope after all. Though, as noted above, the 10-inch doublet lens was ready by September 1900, the rest of the telescope, including the unusual bent pier mounting, was not delivered until 1904. In the meantime, Barnard began to meditate on a new plan for the Bruce, thanks to the efforts of the irrepressible Yerkes Director, George Ellery Hale.[9]

In June 1903—just as Barnard was discovering with the 40-inch refractor a rare great white spot on Saturn, which he used to determine the planet's rotation period at the latitude of the spot (10 h 39 min)—Hale himself set out for Pasadena, then a settlement of only 15,000 inhabitants, in order to prospect Mt. Wilson (then still known as Wilson's Peak) as the possible site for a planned high-altitude observatory. He and W.W. Campbell, the Lick Observatory Director, rode burros along an old Indian trail leading from Sierra Madre up the canyon of Little Anita stream. At the summit Hussey awaited them with a 9-inch telescope. Hale used it to study sunspots and solar granulation with a clarity he had rarely experienced at Yerkes. At the time, Campbell and Hussey both expressed concerns with the amount of dust that was stirred up from the desert near the mountain, but Hale—another probable manic-depressive—was then in one of his sunny moods, and brushed the objections aside.[10] By the time he returned to Chicago he had made up his mind that this is where he was going to build his solar observatory.

Unfortunately, the Carnegie Institution rejected his first attempt to get funding for the plan, and Hale hung fire at Yerkes for the next few months. However, unwilling to wait on Carnegie for funds, he and a Dartmouth-trained spectroscopist, Edwin Brant Frost, who had joined the Yerkes staff in 1898, returned to California in December 1903, and carried out some additional tests on Mt. Wilson. At this point Hale was paying practically all the expenses out of his own pocket. In addition, Barnard, following Hale's reports of the excellent conditions on Mt. Wilson, began lobbying him to provide funds to take the Bruce there for the wide-angle Milky Way photography. At the moment, Hale did not have the money, and though Frost tried to secure $500 (about $16,000 in today's currency) from University of Chicago president William Rainey Harper to cover the cost of doing so, Harper turned him down.

At this point, it appeared that Yerkes would have to do, and in anticipation of the imminent delivery of the mount and tube from the Warner and Swasey Company in Cleveland, Barnard constructed of a small building for the Bruce on the grounds between the Observatory's main building and Barnard's house on the shore of Lake Geneva. It had not yet been completed at Christmas 1903, when Barnard groused to Hale,

"It is very cold and disagreeable here. It has been as low as 20° below zero. It remains cold all the time with tremendous winds…. We got the tinner to come and finish the shutters on the Bruce dome, but he has yet to come and do all the soldering. I hope to get him here in the January thaw. It is too cold now. The tin needs painting, but it cannot be painted until the soldering is done. The plastering is all dry but the final coat cannot be put on because of the cold."[11]

Proof that Barnard did not exaggerate when he told Hale "It is Cold and Disagreeable Here." This photograph shows him shoveling snow in January 1910, when he was busy searching, both photographically with the Bruce and visually with the 40-inch refractor, the area of the sky where Halley's comet was believed to be lurking. In later years, Barnard always wore the tam-o'shanter and white kerchief he is wearing here in the 40-inch dome on cold nights. (Credit: University of Chicago Photographic Archive [apf6-04460r], Hanna Holborn Gray Special Collections Research Center, University of Chicago Library)

Not until February 1904 did the remaining parts needed to assemble the telescope arrive from Warner and Swasey, but Barnard remained reluctant to set it up permanently on a brick pier in the now-finished dome; instead he assembled it temporarily in the main corridor of the observatory's main building. In the end, his procrastination paid off. Hale finally came through with a grant from Joseph Hooker, a hardware- and steel-pipe millionaire with an interest in astronomy, for $1000 to bring Barnard and the Bruce telescope out west. The game was afoot.

The Bruce Observatory on the grounds of Yerkes Observatory, with Lake Geneva in the background. This view shows it from the northwest; it was taken taken by Barnard on April 16, 1904, shortly after construction was completed. (Credit: University of Chicago Photographic Archive [apf6-02718], Hanna Holborn Gray Special Collections Research Center, University of Chicago Library)

Annus Mirabilis—Mt. Wilson

Though Hale had finally come through with the funds, he took pains to dampen Barnard's tendency to be overly optimistic about the fate that awaited him. Hale pointed out that there was at the time only a rugged trail up Mt. Wilson, and that Barnard might be better off trying to observe from nearby Mt. Lowe, which could be reached by a railway extending from Altadena up Rubio Canyon to Echo Mountain, and then four miles further into the San Gabriel range to reach the mountain itself.

Astronomical observations had already been carried out at Echo Mountain by Lewis Swift, who had been one of Barnard's closest friends during the latter's comet-seeking days in Nashville. Swift, who had long directed H.H. Warner's observatory in Rochester, New York, until the latter's bankruptcy in 1893, had come to Echo Mountain with the Warner observatory's 16-inch refractor at the invitation of the many-sided entrepreneur Thaddeus Lowe, whose careers had included being a Civil War balloonist, dry ice manufacturer, and California real estate speculator. From Echo Mountain, Swift continued to add to his record of comet discoveries, finding three, of which the last—1899 I—was discovered when Swift was almost seventy. Finally, in 1900, he returned to Marathon, New York, where blind, deaf, and alone he died in 1913.

Immediately on receiving Hale's suggestion, Barnard was all-in, and planned to ship the Bruce telescope to Mt. Lowe at once, and to arrive himself by early May 1904. Hale, however, counselled him rather "to come out here as late as possible, since the weather is frequently not settled until after the middle of May…. I advise you not to come before June 1, if this will serve your purpose."[12] There was another reason besides the weather to put Barnard off. Hale was still funding the expedition out of his own resources. He operated on an increasingly slender budget after bringing Yerkes staff astronomer Ferdinand Ellerman out to California, and also wanted to bring out a promising young spectroscopist, Walter Sydney Adams. However, he now found himself in the embarrassing situation of not even having funds to pay the astronomers' salaries. Without additional funding, he would have to scale back plans for the great solar observatory and reflecting telescope. The last thing he needed was the additional burden of having Barnard on his hands.

When Barnard received Hale's latest communications, he was, as almost always by the end of winter in Williams Bay after observing in all kinds of weather, sick in bed with a bad case of bronchitis. Knowing now of Hale's financial difficulties, Barnard tried his best to commiserate. He urged the director not to allow his own eagerness to interfere with Hale's own plans. However, he could not resist adding an afterthought:

> It is needless to say that I am extremely anxious to get a chance at the southern part of the Milky Way from Mt. Lowe. It would be a wonderful chance for the Bruce, and I believe that we should get something extremely valuable. I looked out the other morning about 3 o'clock while sick, at the Milky Way and that glorious region in Scorpio and Sagittarius was coming up in the low southeast, and it made me feel that that it would be a great thing to get at it with the Bruce from Mt. Lowe where the altitude would be 9° or more greater, and the atmosphere clear and transparent with the further fact that one could get at it every night when the moon did not interfere.
>
> I am satisfied that you will do all in your power to get the Bruce out there and I hope you will succeed, but as I say if it interferes with your plans for your other work, why let it go.[13]

Under the circumstances, with the chances for an expedition to Mt. Lowe fading, Barnard finally moved the Bruce telescope from its temporary placement in the corridor of the main building and mounted it on the brick pier in the small building built for it. The latter was crowned by a 15-foot dome with an unusually wide shutter (almost 100 degrees) in order to accommodate the telescope's exceptionally wide field. Unfortunately, though predictably, the telescope proved more difficult to get into working order than expected. As of June 1904, Barnard had managed to get only a few exposures. "When the moon is away, it is cloudy all the time here," he complained to Hale. But Hale at the time was encountering his own difficulties. The dust stirred up from the desert to the east which Hale had initially ignored was again causing serious problems, and the sky was often covered with a milky haze. "It seems ... that would hamper the work [with the Bruce] badly—so that it has been decided not to take the Bruce telescope there," Barnard wrote sadly to Campbell. "I am very much disappointed but it is something that cannot be helped." Fortunately, the problem with the dust was only an intermittent one, and by fall Hale found conditions on Mt. Wilson had greatly improved, and also a new trail had been opened up to the summit. Mt. Wilson was now no less accessible than Mt. Lowe. Anticipating another dreadful Midwestern winter, Barnard could not wait to escape, and hung on Hale's go-ahead. It finally came on December 4, 1904. That day, Barnard shipped the telescope ahead—all except the 10-inch lens, which Barnard would carry with him personally as "hand luggage." There was only one more delay. Rhoda—who had planned to travel to England to spend the time while Barnard was away on Mt. Wilson with relatives there—suddenly fell seriously ill. She had been suffering from heart problems, and not until January 2, 1905, did she improve enough to set out from Chicago to New York, where she would stay with a friend until shipping out for England on the evening of January 6. While nursing Rhoda back to health, Barnard heard the tremendous news that at long last the Executive Committee of the Carnegie Institution had come through with funding for Hale's plans, providing a grant of $150,000 (equivalent to $3.6 million today) for each of the next 2 years and had also approved his plans for the Mt. Wilson Solar Observatory.

The Committee also announced its intention that the new solar observatory should be a separate institution from the University of Chicago. Thereupon, Hale immediately announced his intention to resign from the Yerkes directorship so that he could devote all of his energy to setting up the Mt. Wilson establishment. Barnard was somewhat disoriented by the blow. Though he expressed his confidence that Frost, Hale's successor designate at Yerkes, would "deal justly with everyone here," he also admitted to feeling a regret at Hale's departure. "I for one shall miss you greatly here," he wrote, "for it was always a great pleasure and satisfaction to feel the uplifting and cheering influence that you always exerted over me."[14] Such was his mood as he saw Rhoda off and, with the 10-inch lens in hand, caught the train from Chicago to the West Coast to begin what astronomically would be the most productive year of his career.

Go Thou to the Mountain

Barnard reached San Francisco on January 6, 1905. After a brief visit to Lick, he continued by train to Pasadena where on the afternoon of January 10 he disembarked, and started the trek up Mt. Wilson on the so-called "new" trail. At the summit was the living quarters which Hale, inspired by his reading about the monasteries of the Levant, "perched on rocky promontories, looking out on distant peaks,"[15] had called the "Monastery." The name also expressed his resolve, forged from hard-earned experience with the personality conflicts among the women of the isolated community at Williams Bay, "who had no stars to watch, no figures to conjure with, no spectra to measure, and no other absorbing occupations." He had vowed that if he ever founded another observatory the astronomers and their families would be barred from ever living on the observatory grounds.[16] Barnard reached the solitary eyrie via a steep and precipitous path about nine miles long, after a climb lasting about 5 h, mostly on foot, with "the 10 in. lenses on one side of a horse or mule and balanced on the other side."[17] The final stages took place in fog and rain.

By this time, Hale had already recruited several members of the Yerkes staff—Walter S. Adams, Ferdinand Ellerman, and Ritchey—as well as another transplant from Yerkes, not human but instrumental: the Snow solar telescope, a horizontal telescope of coelostat design that Hale and Ritchey had set up at Yerkes in the autumn of 1903 and had since disassembled and shipped to Mt. Wilson. Others who came out to Mt. Wilson on a temporary basis during Barnard's sojourn would be Charles Greeley Abbot of the Smithsonian Institution; his assistant, Leonard Ross Ingersoll; and Henry Gale, a former football player turned University of Chicago physicist who specialized in sunspot spectra. Adams would later recall fondly those early days at the Monastery:

Hale had introduced Abbot to the Oriental stories … on the monasteries of the Levant, and our evenings usually started off with a dramatic rendering by Abbot of the tale of the Jew of Constantinople and Solomon's seal which he knew by heart. Occasionally the Smithsonian challenged all comers to a game of duplicate whist, but more often the group would gather around the fireplace for discussions of plans of work or of the state of the world in general. Hale's amazing breadth of interests, his great personal charm, and his stories of important figures in science and international affairs make these evenings stand out in memory [6, pp. 99–100].

Mt. Wilson group 1905. Seated in front is F. Ellerman; from left to right are Construction Superintendent H.L. Miller, C.G. Abbot, G.E. Hale, L.R. Ingersoll, W.S. Adams, E.E. Barnard, and C. Backus. (From: W.S. Adams, "Early Days at Mount Wilson," Popular Astronomy *(1950))*

Now in his glory, with the Pleiades that he had struggled to capture from Yerkes still accessible and the whole Milky Way Galaxy to explore, Barnard wasted no time. He rushed to set up the Bruce telescope in a small wooden structure on a hillock on the trail between the Monastery and the Observatory workshop. Within this improvised structure he had mounted together on the cement pier the 10-inch Bruce and 6 ¼-inch photographic telescopes, as well as the 3.4-inch doublet, the lantern lens and the 5-inch visual guide scope. The first plates were exposed on the night of January 27, 1905.

At first, Barnard was prone to experience night terrors of the kind he had sometimes experienced years earlier when observing Jupiter with the 5-inch Byrne refractor in Nashville. The Mt. Wilson Solar Observatory was still in the early construction stage, and sometimes, while working in the small Bruce Observatory, he was the only person on the mountain. Moreover, since the small observatory was separated by some distance from the Monastery, which was hidden from view by heavy foliaged spruce trees, even when others were about he found himself, "essentially isolated from the rest of the mountain," as he later recalled:

> I must confess that at times, especially in the winter months, the loneliness of the night became oppressive, and the dead silence, broken only by the ghastly cry of some stray owl winging its way over the canon, produced an uncanny terror in me, and I could not avoid the dread feeling that I might be prey any moment to a roving mountain lion. The sides of the [Bruce] observatory were about five feet high, so that it would have been an easy thing for a hungry mountain lion to jump over it and feed upon the astronomer. So lonely was I at first that when I entered the Bruce house and shoved the roof back I locked the door and did not open it again until I was forced to go out.[18]

Fortunately, with the coming of spring (and the increasing accessibility of the southern Milky Way), the loneliness and oppression that he had experienced during the winter began to lift. A good part of this had to do with the reawakening of insect life, which "began its notes in the chaparral":

> The dread of the night soon passed away and the door was left open and it became a pleasure to sit and listen to the songs of nature while guiding the telescopes in long exposures, heedless of all beasts of prey. No one knows what a soothing effect these "noises of the night" have on one's nerves in a lonely position like that on Mount Wilson.[19]

Generally, Barnard found the skies at Mt. Wilson much better than at Yerkes—after the drenching rains early in the season, the dust which had been a matter of concern completely settled out, and Barnard began to obtain magnificently deep plates of the southern Milky Way. In all, he would expose some 500 plates—154 with the 10-inch Brashear doublet, 151 with the 6 ¼-inch Voigtländer doublet, 110 negatives with the 3.4-inch Clark and another 90 with the lantern lens. As usual he worked like a fiend. For many years he had been used to getting by on a few hours of sleep a night, but on Mt. Wilson he often gave up sleep altogether. Adams recalled:

> Barnard's hours of work would have horrified any medical man. Sleep he considered a sheer waste of time, and for long intervals would forget it altogether. After observing until midnight, he would drink a large quantity of coffee, work the remainder of the night, develop his photographs, and then join the solar observers at breakfast. The morning he would spend in washing his plates, which was done by successive changes of water, since running water was not yet available. On rare occasions he would take a nap in the afternoon, but usually

he would spend the time around his telescope. He liked to sing, although far from gifted in the art, but reserved his singing for times when he was feeling particularly cheerful. Accordingly, when we at the Monastery heard various doleful sounds coming down the slope from the direction of the Bruce telescope, we knew that everything was going well and that the seeing was good.[20]

Barnard at the Bruce 10-inch photographic telescope on Mount Wilson. (Credit: University of Chicago Photographic Archive [apf6-04469], Hanna Holborn Gray Special Collections Research Center, University of Chicago Library)

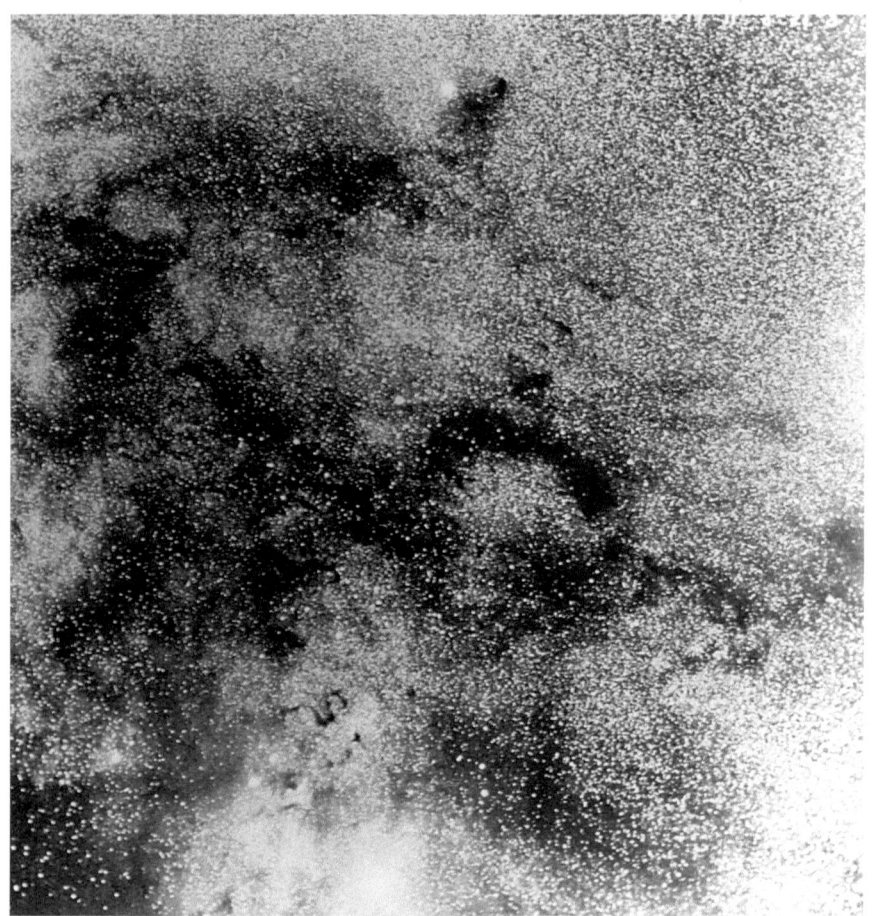

Region North of Theta Ophiuchi, photgraphed by Barnard at Mt. Wilson on May 8, 1905. He wrote of this area of the sky that "it is so extremely puzzling that one attempts a description of it with hesitation. That most of these dark markings which, in a word, ornament this portion of the sky are real dark bodies and not open space can scarcely be question.... What their true nature is does not seem clear." Plate 19 of A Photographic Atlas of the Milky Way

Region of Theta Ophiuchi and eastward, imaged by Barnard on Mt. Wilson on June 30, 1905. He wrote of the markings on this plate: "It is clearly shown that there is some kind of matter in the great vacancy, which is feebly luminous over its entire area...It is evident that we are not looking out into space through an opening in the Milky Way." Plate 21 of A Photographic Atlas of the Milky Way

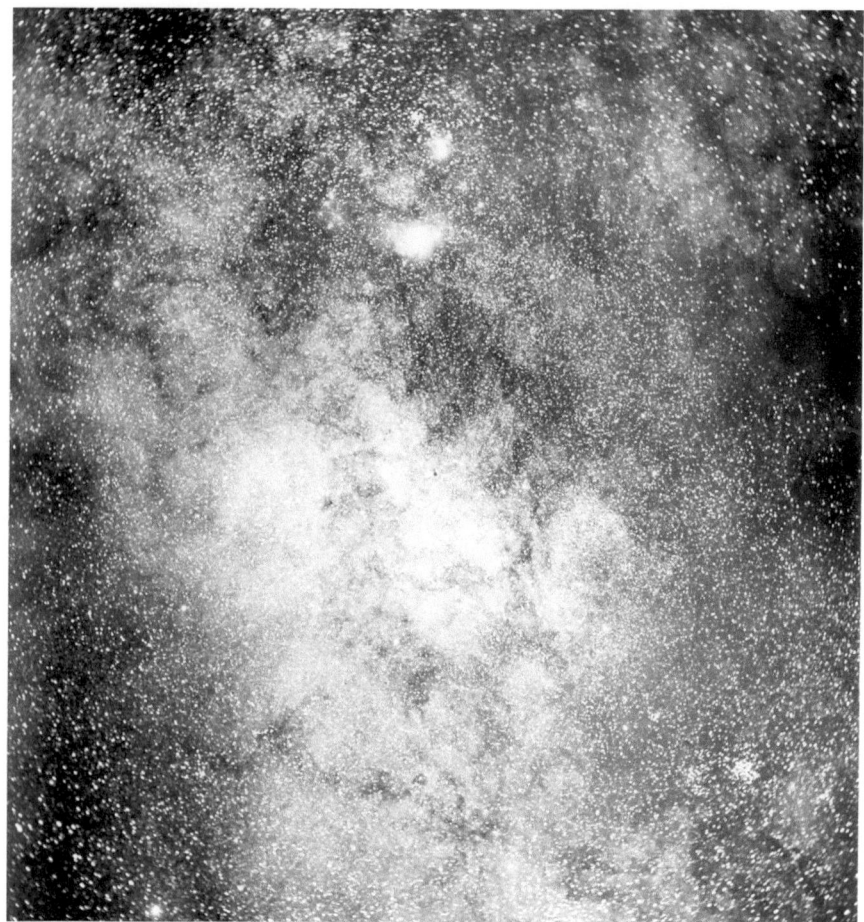

The Great Star Clouds in Sagittarius, imaged by Barnard on Mt. Wilson on July 6, 1905. Plate 26 of A Photographic Atlas of the Milky Way

From Yerkes, Frost, on hearing from Barnard of the excellent conditions he was enjoying, wrote, "It is a pleasure to know that you are having such fine weather for work, but I hope you will not overdo, and you will give up some clear nights when you need sleep."[21] Barnard could always be counted upon to ignore such advice. He knew, moreover, that he might never again have such a grand opportunity to explore the star clouds, diffuse nebulae and strange dark regions which he had first recorded with the Willard lens and regarded as vacant spaces but would eventually conclude, after years of study, were clouds of obscuring matter (and, as we know now, interstellar dust).

Despite his initial trepidation in being alone on the mountain at night, Barnard "fell in love with the mountain and everything connected with it," Adams recalled.

He was fascinated by the views, studied the birds, measured the growth of yucca stalks, and treasured the sight of a deer. I remember his excitement on winter morning when he came in to breakfast and announced that he had just seen a wildcat walking through the snow outside his bedroom window… Barnard's devotion to the mountain may judged by the fact that during four months of his stay he made but one trip to the valley. This was to Sierra Madre to see a notary and to have his hair cut, after which he turned around and started back up the trail. His health was excellent at the time, and to those who knew him in later years it will be a surprise to learn that he once clambered down the steep walls of the ridge below the Monastery, crossed the deep canyon, and climbed the side of Mount Harvard to the toll road, perhaps as difficult a trip as any around the mountain top.[22]

Among the hardships of living and working on the summit at this time was the unavailability of water. Any needed, including that Barnard used for developing his plates, had to be packed up the mountain by burro from Strain's Camp, named after the first pioneer who had settled on the mountain. It was located on the north side of the mountain, where there were springs. The old burro which dutifully performed this laborious task was named Pinto, and Barnard became especially fond of him when he discovered that his hair was much finer than a human's—thus exceptionally well suited for making cross wires for a guiding telescope.

Doubtless Barnard would have loved to stay on Mt. Wilson with Hale, Adams, Ellerman and Ritchey, and escape the unhealthy winters of southern Wisconsin. But as Barnard was the Yerkes Observatory's most celebrated astronomer, Frost would never have agreed to give him up. Moreover, except in photography where he broke new ground, he was above all else a skilled practitioner of the tried and true classical methods of astronomy, such as work with a filar micrometer. Though hardly the most romantic branch of astronomy (and certainly far less so than discovering new satellites or planets), Barnard's contributions were significant. At Lick he had published what long remained the most authoritative measures of the diameters of planets and the larger satellites of the Solar System, and he was the first to succeed in making direct measures of the diameters of the four largest asteroids.[23] All credible, but he had nothing to do with astrophysics which was, of course, the overriding mission of the new Mount Wilson Observatory. So in mid-September 1905, after a stay of 8 months, Barnard packed up the Bruce telescope and his precious plates and, escorted down the mountain early one morning by Adams, retraced the journey by train back to Wisconsin. He would spend the rest of his life analyzing, learning from, and preparing the plates for publication. One can only imagine that for once a sense of contentment might have settled, if only briefly, on this hard-driving, insecure, neurotic, brilliant man. So often appearing almost childlike in the ways of the world, he possessed the insight of genius when he surveyed the starry sky, and had by now established himself as the greatest astrophotographer of his time. The Milky Way photographs were his greatest achievement, a supreme monument that even today a century after his death inspires awe.

Meanwhile, during the time he spent on Mt. Wilson photographing the Milky Way, the planet Mars, coming to opposition on May 8 when it shone like a glowing coal in the constellation Libra (just beyond the reach of the Scorpion's claws), asserted a commanding presence in the night sky. Lying outside of the scope of his Milky Way researches which he pursued with single-minded intensity, he probably gave it little heed, but at least at times he must have regretted not having the telescopic means of bringing it closer. Doubtless he must occasionally have thought wistfully of the magnificent views of the planet he had obtained with the great Lick refractor in the late summer of 1894, and wondered if he would ever see such things again. However, a new challenge had surfaced that same year, 1905. Percival Lowell's assistant C.O. Lampland had obtained photographs of the planet which, according to his employer, showed fine details, including even the canals. Though seeing conditions in southern Wisconsin left much to be desired for visual work, the great refractor, with its long focal length, offered a great advantage for planetary photography. The next opposition was not until July 1907, which gave Barnard time to look thoroughly into the technical issues of photographing planets. Barnard would see to it that Lowell and Lampland would not have the field to themselves forever.

Notes

1. E.E. Barnard to Sir William Huggins, Dec. 25, 1898; Royal Astronomical Society Library.
2. E.E. Barnard to George Davidson, May 14, 1899. Davidson. George. Papers, Bancroft Library, University of California, Berkeley.
3. E.E. Barnard, observing log book. Barnard. Edward Emerson. Papers, University of Chicago Library, Hanna Holborn Gray Special Collections Research Center.
4. R.H. Tucker to his mother, February 17, 1901. Tucker. Robert Hawley. Papers. Mary Lea Shane Archives of the Lick Observatory, University of California, Santa Cruz.
5. E.E. Barnard to George Davidson, March 26, 1900. Davidson. George. Papers, Bancroft Library, University of California, Berkeley.

6. E.E. Barnard, unpublished lecture notes. Barnard. Edward Emerson. Papers, Special Collections and University Archives, Jean and Alexander Heard Library, Vanderbilt University.

7. E.E. Barnard to William Rainey Harper, July 26, 1897. Barnard. Edward Emerson. Papers, Special Collections and University Archives, Jean and Alexander Heard Library, Vanderbilt University.

8. E.E. Barnard to John A. Brashear [1899]. Barnard. Edward Emerson. Papers, Special Collections and University Archives, Jean and Alexander Heard Library, Vanderbilt University.

9. Hale was an astronomical entrepreneur extraordinaire. He was interested in science in his pre-adolescent years, and his father William Hale, who had made a fortune building the elevators that made the towers of the Chicago skyline possible during the years after the Great Fire, encouraged his precocity by building a private brick laboratory for chemical research and later an observatory equipped with a 12-inch refractor, at the family mansion at 4545 Drexel Boulevard in Kenwood, on Chicago's south side. He went to MIT, where he volunteered to work in photography with Edward C. Pickering at the Harvard College Observatory, and for his senior thesis built the first successful spectroheliograph for studying the Sun in the light of specific elements. When the University of Chicago began to take shape in Hyde Park, under "boy president" William Rainey Harper, himself "a bundle of energy" and only 35, Harper recruited the obviously gifted Hale, at 24, to become an associate professor of the University of Chicago, and with funds provided by streetcar magnate Charles Tyson Yerkes established the Observatory at Williams Bay around the 40-inch refractor. Hale's chief research interest was the Sun (and, by extension, the stars), but the interest was not just scientific; he was quite subject to seasonal mood swings, and though content to remain in Chicago while his father was still alive, the latter's death in 1898 removed the most important obstacle to leaving the gloomy Midwestern winters. He managed to get himself appointed to a committee to advise how the Carnegie Institution might use a portion of a $10,000,000 gift from Andrew Carnegie (equivalent to some $360 million today) to establish a high-altitude solar observatory and large reflecting telescope in a southerly location which would, in Hale's words, "give extraordinary results, and greatly advance our knowledge of the Universe." Quoted in Helen Wright [7]. Hale received $10,000 in seed money to investigate suitable sites; the ambitious project itself had not yet been approved. As a first step a former Lick astronomer now at the University of Michigan, W.J. Hussey, was sent west to scout sites, and decided that Mt. Wilson (elevation 5886 feet) near Pasadena was the most promising site.

10. On Hale's probable manic-depressive illness, see: William Sheehan and Donald E. Osterbrock [8].

11. E.E. Barnard to G.E. Hale, December 29, 1903.; Hale. George Ellery. Papers, California Institute of Technology Archives and Special Collections.

12. G.E. Hale to E.E. Barnard, March 9, 1904. Hale. George Ellery. Papers, California Institute of Technology Archives and Special Collections.

13. E.E. Barnard to G.E. Hale, March 23, 1904. Hale. George Ellery. Papers, California Institute of Technology Archives and Special Collections.

14. E.E. Barnard to G.E. Hale, December 29, 1904. Hale. George Ellery. Papers, California Institute of Technology Archives and Special Collections.

15. Wright, *Explorer of the Universe*, p. 188.

16. Ibid., p. 123.

17. E.E. Barnard, observing log book; Barnard. Edward Emerson. Papers, University of Chicago Library, Hanna Holborn Gray Special Collections Research Center.

18. E.E. Barnard, unpublished manuscript; Barnard. Edward Emerson. Papers, University of Chicago Library, Hanna Holborn Gray Special Collections Research Center.

19. Adams, "Early Days," p. 97.

20. Adams, "Early Days," p. 98.

21. E.B. Frost to E.E. Barnard, July 27, 1905. Barnard. Edward Emerson. Papers, Special Collections and University Archives, Jean and Alexander Heard Library, Vanderbilt University.

22. Adams, "Early Days," p. 97.

23. The micrometric measures of the diameters of Saturn's ball and ring system, and of its satellite Titan, and of the Galilean satellites of Jupiter are published in: E.E. Barnard [9, 10]. Compared to spacecraft measures, Barnard's are often good to a few percent. His micrometric measures of the four largest asteroids are found in: E.E. Barnard [11]. His values in miles, compared to modern ones (in parentheses) are: Ceres 485 (588), Pallas 304 (320), Juno 118 (148), Vesta 248 (325). Previous investigators had assumed that the asteroids all had the same albedo, which wasn't in fact true; Barnard was the first to actually measure the diameters directly, and to estimate their albedos. Setting the albedo of Mars equal to 1, he found as follows: Ceres 0.57, Pallas 0.88, Juno, 1.67, and Vesta 2.67. Incidentally, one of the difficulties with these measures (and much discussed by Lowell) involves "irradiation," a tendency for bright objects to seem to bleed over into adjacent dark areas. This is very evident, for instance, in the polar caps of Mars, which appear to project into the surrounding space. To deal with this, Barnard used a micrometer that was modified to give a double image, and by bringing the two disks he was trying to measure into contact, the irradiation embarrassment was made to disappear.

References

1. Wells HG (1898) The war of the worlds. London, William Heinemann, p 3
2. Bacon JM (1901) Wadesborough [sic], North Carolina. In: Maunder EW (ed) The Total solar eclipse 1900: report of the expeditions organized by the British astronomical association to observe the total solar eclipse of 1900, May 28. Knowledge Office, London, pp 16–17
3. Barnard EE (1901) The total eclipse of the sun in Sumatra. Popul Astron 9:528–544
4. Mitchell SA (1928) "With Barnard at Yerkes observatory and at the Sumatra eclipse," Edward Emerson Barnard memorial number. J Tenn Acad Sci 3(1):27
5. Barnard EE (1905) The Bruce photographic telescope of the Yerkes observatory. Astrophys J 21:35–48
6. Adams WS (1950) Early days at Mt. Wilson—II. Publ Astron Soc Pac:97–115
7. Wright H (1966) Explorer of the universe: a biography of George Ellery Hale. E. P. Dutton and Co., New York, p 161
8. Sheehan W, Osterbrock DE (2000) Hale's 'Little Elf': the mental breakdowns of George Ellery Hale. J Hist Astron 31:93–114
9. Barnard EE (1895) Micrometrical measures of the planet Saturn, and measures of the diameter of his satellite Titan. Made with the 36-inch equatorial of the lick observatory. Mon Not R Astron Soc 55(7):367–382
10. Barnard EE (1895) Filar micrometer measures of the satellites of the four bright satellites of Jupiter. Made with the 36-inch equatorial of the lick observatory. Mon Not R Astron Soc 55(7):382–390
11. Barnard EE (1895) Micrometrical determinations of the diameters of the minor planets Ceres (1), Pallas (2), Juno (3), and Vesta (4), Made with the Filar micrometer of the 36-inch equatorial of the lick observatory, and on the albedos of those planets. Mon Not R Astron Soc 56(2):554–563

8

Percival Returns

Contents

My golden spurs now bring to me,
And bring to me my richest mail,
For tomorrow I go over land and sea
In search of the Holy Grail;
Shall never a bed for me be spread,
Nor shall a pillow be under my head,
Till I begin my vow to keep;
Here on the rushes will I sleep
And perchance there may come a vision true
Ere day create the world anew.
—James Russell Lowell, "The Vision of Sir Launfal"

New Directions

The fact that the criticisms of Lowell's Venus work so nearly coincided with the onset of his breakdown is undoubtedly significant, and shows that at this point his work had reached a crisis over the issue of replicability—the ability of other astronomers to reproduce his results.

When founded in 1894, the Lowell Observatory was concerned broadly with the study of the properties and evolution of the bodies of the Solar System, with special emphasis on whether any of them, but especially Mars, might have developed life. As original as was this program of research, Lowell and his assistants pursued it at first mostly with traditional instruments and with techniques of peering through the eyepiece and sketching planetary surface detail revealed fleetingly in the best moments.

Historians of science Steven Shapin and Simon Schaffer, in their influential study of Robert Boyle, describe how Boyle employed three different "technologies" or "knowledge-producing tools" in his efforts to win converts to the experimental philosophy being espoused by the newly founded Royal Society [1]. According to Shapin and Schaffer, Boyle's first technology, a *material* one, entailed arguments for the physical integrity of the air pump and allowed others, by being able to construct working air pumps themselves, to replicate Boyle's experimental claims. Second, as a supplement to this material technology, Boyle adopted a *literary* technology whereby he aimed to make his readers into vicarious witnesses of his experimental trials. Finally, he employed a *social* technology, which involved the construction of open laboratory spaces where witnesses could observe experiments and equipment at first hand.

Clearly, Lowell's pursuit of data about the planets combined similar elements. His material technology included, most notably, his telescope, the superiority of the atmospheric conditions he claimed for Flagstaff, and a number of techniques to enhance the seeing including the use of various diaphragms to stop down the objective lens of the 24-inch Clark to optimize the match of aperture to the atmospheric conditions. (Equally important, though not much emphasized, was the reduction of chromatic aberration, which both blurred the detail and contributed to the vivid colors that diversified the Martian disk.) His literary technology included vivid passages in which he emphasized his relation to pioneers and explorers and made his readers virtual eyewitnesses to the process of scientific discovery (as in the passage quoted in Chap. 4, in which he observed the flashes from the polar cap of Mars). His social technology included parading the testimony of a number of observers possessed of a supposedly "innocent eye," such as Leonard, Morse, George Russell Agassiz, Barrett Wendell and others. Above all, there was the striking testimony of his highly detailed maps and globes.

The 24-inch Clark refractor, being used to observe Mars in December 2021. The relatively short focal ratio of 16.3, owing to Lowell's desire to save money by fitting it into the dome of the 18-inch Brashear used in 1894, means that the instrument suffers badly from chromatic aberration and is often stopped down to get the best views. (Photograph by William Sheehan)

Though Barnard, who had observed the planets with large telescopes under excellent conditions, could wish, as he told George Ellery Hale, that the linear markings on Mars, Venus, Mercury, and Jupiter's satellites could be "wiped out from the literature of astronomy,"[1] the point would have been lost on non-specialist readers. Never having looked through a large telescope themselves, they had to trust the eyewitness reports of those who had. They generally would not have suspected that much of this detail, represented so boldly on drawings and maps, might be illusory, and seldom realized that the maps of Mars represented composites of many sessions of observing in which only a few features were present at a given time. Nor did most of them have more than a passing acquaintance, if any, with the similar features reported on other bodies than Mars.

As he gradually eased back into astronomical work in 1900, Lowell ordered a spectrograph from the Pittsburgh optician John Brashear. On his return to Flagstaff, he made it his first priority to press the spectrograph into service in buttressing his work on Venus with a new, presumably more objective and unchallengeable, form of material technology.[2] Indeed, as David Strauss concludes, "the skepticism about the Venus markings clearly changed the direction of research at Lowell."[3] Lowell, of course, had no intention of mastering the spectrograph himself. Instead, he planned to assign the task to the technically adept and resourceful Douglass.

Douglass, however, by this time was secretly embracing heresy. During Lowell's four-year absence from the Observatory, his assistant had not only come to doubt the superiority of Flagstaff seeing but also to suspect that many of the linear markings recorded on Mars and other planetary bodies might be illusory. He wrote to Dr. Joseph Jastrow, a psychologist at the University of Wisconsin at the beginning of 1901:

> I would have written you long before but for Mr. Lowell's indifference to taking up the psychological question involved in astronomical work..... I have made some experiments myself bearing on these questions by means of artificial planets which I have placed at a distance of nearly a mile from the telescope and observed as if they were really planets. I found at once that some well-known planetary appearances could, in part at least, be regarded as very doubtful....[4]

At the beginning of 1901, Lowell was in England, where he observed Nova Persei with Sir Robert Ball, Director of the Cambridge University Observatory. Though a mathematician not an observer, Ball was a compelling lecturer who would begin a highly lucrative lecture tour of the United States in October in which he expressed strong support for Lowell's views about Mars. Lowell wrote to Douglass, "Isn't this new star a marvel for magnitude. There has been nothing like it since Tycho Brahe!"[5] In the same letter he announced that he was now completely recovered from his illness and would be arriving in Flagstaff to resume direction of the Observatory in time for the February 22, 1901, aphelic opposition of Mars. Douglass can hardly have welcomed the thought of once more coming under the thumb of his strong-willed employer, and shared with his erstwhile mentor William H. Pickering, now back at Harvard, his misgivings about Lowell's "strong personality, consisting chiefly of immensely strong convictions," and of his questionable methods of research. Similar observations had previously been made by Basil Hall Chamberlain, cited earlier, regarding Lowell's deductive approach to the Far East. Lowell, he told Lafadio Hearn, would argue "from some general notion, such as the supreme virtue of 'modernity,' the 'impersonality of Orientals,' etc.—he has only 3 or 4—and then bend the facts to suit the preconceived idea."[6] Douglass hit exactly the same nail on the head, venturing the opinion that "It appears to me that Mr. Lowell has a strong literary instinct and no scientific instinct. I had supposed that he was anxious to acquire the latter and nothing could have given him a stronger incentive than this Venus business."[7] Moreover, Douglass saw no evidence that Lowell had learned anything from his mistakes. Pickering expressed sympathy and assured him that Lowell's work was generally held in low regard by the astronomical profession but thought it best that Douglass stay put, "unless the loneliness is too oppressive."[8]

Neither Chamberlain's comments nor Douglass's—so far—circulated back to Lowell. However, the same day Douglass wrote to Pickering he also did so to Lowell's brother-in-law, William Lowell Putnam, II. Douglass did so "in confidence," though he must have known that it would be hard for Putnam to maintain that indefinitely. "I am deeply attached to Mr. Lowell and would like to see his name have the highest renown," Douglass told the Observatory's trustee. However, he worried about Lowell's lack of appreciation of the scientific method, also his tendencies to "carelessness … looseness of generalization verging toward misstatement… [and] neglect of alternative hypotheses." He instanced specifically Lowell's refusal to take seriously any explanation of the Martian canals apart from the "habitation hypothesis." In conclusion, Douglass wrote:

> His work is not credited among astronomers because he devotes his energy to hunting up a few facts in support of some speculation instead of perseveringly hunting innumerable facts and then limiting himself to publishing the unavoidable conclusions, as all scientists of good standing do, in whatever line of work they may be engaged…. I fear it will not be possible to turn him into a scientific man.[9]

On his return to Flagstaff, Lowell—not yet aware of his assistant's apostasy—luxuriated in the moment, writing to his sister Elizabeth, "I am so much at home here ... and yet no one I know knows it."[10] For a time things seem to have gone on as before. However, though Douglass had asked that Putnam keep his letter "between ourselves," in the end, the whole matter could not be kept private. In July 1901, Putnam showed Lowell the letter, and Lowell reacted with his usual decisiveness. Douglass was dismissed immediately and without explanation. In desperation he appealed for reinstatement. He reminded Lowell of his promise to allow him to work with the new Brashear spectrograph Lowell had ordered, but Lowell was unmoved. In October he wrote a chill response: "I have at present all the assistants I want at Flagstaff."[11] Douglass had no choice but to begin casting around for positions elsewhere, but they were few and far between and it didn't help that he had been cast as "Lowell's man." He attempted, through Pickering, to rejoin the Harvard Observatory; tried to get hired at the U.S. Naval Observatory, from which See had come and gone; he even approached W.W. Campbell at the Lick Observatory asking whether there were any positions in California he might fill, declaring that "Mr. Lowell and I have had the break-up that I have thought possible since the summer of '96."[12] Unfortunately, nothing was available.[13]

Lowell Rebuilds His Staff

After Douglass departed, Lowell was, at least temporarily, left in the lurch. However, neither then nor later was he ever alone. In addition to servants, always necessary to the patrician lifestyle Lowell had enjoyed from infancy, he had with him the one assistant he could not afford to be without—Wrexie Louise Leonard. Though his personal relations with her have always been the subject of (ultimately unsatisfying) speculations, and though she and Lowell seem to have been involved romantically as well as professionally, that is rather beside the point, for there can be no question of her utter indispensability to his work life. From the time he hired her in 1893, her role had become more and more complex, and by the first years of the twentieth century she would personally handle the observatory's accounts, keep the bank records in order, and dispatch checks on Lowell's orders. She was also in charge of purchasing, which meant buying such things as observatory stationery in Boston and furniture in Chicago. In addition, during his frequent absences in Boston or Europe, she could be relied on to keep up a steady (and often rather gossipy) line of communications between him and his staff. Not without literary skill, she wrote speeches for him, as well as going over publisher's proofs for Lowell and other staff members. As noted earlier, she also observed the planets with Lowell during the early years, though for whatever reason, after 1901 she seems to have been relegated to a more secondary role—perhaps for no better reason than that Lowell feared he might be ridiculed for relying on his secretary for such data, and a woman to boot.[14] In more relaxed mode, she regularly took part in picnics and adventurous forays into the wonderlands of northern Arizona Territory, with Mormon Lake, Walnut Canyon and its cliff dwellings built by the ancient Sinagua people, and Oak Creek Canyon being favored destinations. In general, life in Percival's eyrie was pleasant indeed.

Percival Lowell, Wrexie Leonard and Godfrey Sykes, at Mormon Lake, early 1900s. (Credit: Lowell Observatory Archives)

Above, Wrexie Leonard and a group at Oak Creek Canyon. (Credit: Lowell Observatory Archives)

Above, Wrexie Leonard among rocks, possibly at Walnut Canyon. (Credit: Lowell Observatory Archives)

Percival and an unidentified man hiking in snow, possibly in San Francisco Peaks. (Credit: Lowell Observatory Archives)

For all her many roles, she was not the one to operate the spectrograph for him. However, Lowell had the good fortune that, at almost the exact moment Douglass had been dismissed, he was approached by his former assistant Wilbur A. Cogshall, a casualty of the See affair and now at Indiana University, who reminded Lowell of a commitment he had made early that spring to hire Vesto Melvin Slipher, usually referred to simply as "V.M.," a recent Indiana University graduate, as an assistant. At first Lowell was reluctant to honor his promise to Cogshall. "As regards Mr. Slipher," he wrote, "I shall be happy to have him come when he is ready. I have decided, however, that I shall not want another permanent assistant and take him only because I promised to do so, and for the term suggested. What it was escapes my memory. If, owing to this decision, he prefers not to come, let him please himself."[15] Slipher did indeed come, arriving a month later, and by that fall was working to attach the spectrograph to the 24-inch refractor—with rather limited success. In frustration, in December 1901 Slipher asked Lowell for permission to go to Lick Observatory to learn from an expert in spectrography such as W.W. Campbell, but Lowell turned him down:

> I think it would be inadvisable for you to go to the Lick at present. When you shall have learnt all about the spectroscope and can give as much as you take it will be another matter and at that time I shall be very glad to have you go.[16]

Lowell's rejection of Slipher's request was probably mainly a reflection of his fierce independence, but it also probably owed something to fear that his new assistant—at the moment freshly graduated, completely inexperienced, and likely to bend to his will—did not want to risk his coming into contact, as Douglass had, with astronomers elsewhere who might lead him to doubt the Observatory and its work. In the end, Slipher, who came from a farm in Indiana and was used to tinkering with machinery, managed without outside assistance.

On first arriving on Mars Hill, Slipher moved into in the original four-room house that had been erected in 1894, while Lowell continued to live in town. However, Lowell had now decided to establish a permanent residence on Mars Hill, and employing Godfrey Sykes as contractor expanded the original cottage for his own and Miss Leonard's use while building an astronomers' residence in 1902 to take care of Slipher and other employees. The "Baronial Mansion," as it was called, would expand to at least 18 rooms by the time of Lowell's death.[17]

The Baronial Mansion, probably about 1907, with the Clark dome behind it and to the right. Lowell's office was in the small addition to the left, which faced into the morning sunlight. (Credit: Lowell Observatory Archives)

The Search for Credibility

It was largely through V.M. Slipher's efforts—and those of another assistant, Carl Otto Lampland, also an Indiana University graduate who joined the staff in October 1902—that Lowell would gradually bend toward a a somewhat more rigorous cast to his own work, though he would never entirely succeed in winning the esteem of his "peers" in the astronomical establishment. He did, however, make some effort to do so. Thus, in the month Lampland came on board, Lowell received appointment as non-resident professor of astronomy at the Massachusetts institute of Technology (MIT). He gave six lectures on the Solar System at MIT in December, which would fill a slender volume published the following spring [6]. Being equal parts textbook and popular exposition, the book tends to fall between the stools, and does not appear to have attracted much notice—though it did glean a favorable, if rather perfunctory, review in the *New York Times*, as well as in *The Dial, Engineering Literature*, and *The Week's Progress*.[18] He included a good deal of mathematics, most of which could have been easily worked up from basic treatises on celestial mechanics and the theory of tides, while attempts to work down from the exactness of mathematical deductions to often questionable observations seem almost whimsical. As usual, Lowell displays literary skill in glossing over the gaps in his arguments, and makes a first foray into what would become a major focus of later work, a scheme of planetary evolution which Lowell and Morse called "planetology." It is formulated rather crudely in quaintly anthropomorphic terms, as in the following passage which follows his chapter on Mercury and begins that on Mars:

> Mercury presents us one phase of planetary development; Mars another, quite different. The two represent stages in world-life as distinct as those of gray hair and brown in human life.
>
> Whatever the absolute ages of the several planets, their relative ages, as measured intrinsically, decrease pretty steadily with their distance form the Sun. Mercury is old; Mars, middle aged; Jupiter young.
>
> World-life has its earmarks of time as human life has, and betrays them quite as patently.
>
> Lack of atmosphere, colorlessness, changeless attitude toward the Sun, are the signs of old age in a planet. Mercury shows all these tokens of senility. Mars presents a very different picture.
>
> Color is a telltale trait; for it is a sign that surface development still goes on. Lack of atmosphere alone prevents vegetation, and this, coupled with unalterableness of face presented to the Sun, weathers the surface to a neutral gray. Such a body shows but the bleached bones of a once living world...[19]

The book is also notable for including a first mention, based on the aphelia of comet orbits catalogued by Flammarion and others, of a possible trans-Neptunian planet—a subject which would of course occupy him obsessively in later years.

If E.E. Barnard ever read *The Solar System*—and there is no reason to believe he would have—he would likely have found it in equal measure amusing and perplexing. At the same time Barnard was writing of "however much the general surface features of Mars may be misrepresented, and however much the canals, single and double, may be illusory,"[20] he would have encountered head scratchers such as the following:

> Having got the hint, for it was scarcely more than that, during his first season, the opposition of 1877, [Schiaparelli] then showed that element of genius without which very little is ever accomplished, the persistence to follow up a clue. As Mars came round again he attacked the planet in the light of what he had already learnt, and first confirmed and then extended his discovery. This he continued to do at each succeeding opposition. The more he studied, the stranger grew the phenomena he detected. And it is to his everlasting credit that he did this in the face of the skepticism and denial of practically the whole astronomic world. He won. The voices that ridiculed him are all silent now. To-day the canals of Mars are well-recognized astronomic facts, and constitute one of the most epoch-making astronomic discoveries of the nineteenth century.[21]

Even more doubtful would have seemed the following passage in which the non-resident professor of astronomy at MIT recalls the 1880s-era dogma about the superiority of small telescopes to large ones for planetary observations, and emphasizes that despite their failure to be seen in large telescopes the canals could be made out easily enough in even the most modest of glasses:

> Contrary to what the layman thinks, the size of the instrument is the least important factor in the process. As in most things, the man is the essential machine; and next in desirability to the presence of man is the absence of the atmosphere. In good air, with fair attention, the canals are not very difficult objects. Indeed, the surprise is that they were not detected long ago. Under suitable conditions a four-inch glass will show them perfectly. Steady air is one essential; steady study another.
>
> In appearance they are unlike any other phenomena presented in the heavens. Pale pencil lines, deepening on occasion to India ink, seem to cobweb the continents. Their tone depends on the seeing, in the first place, and on the season, in the second. Their width is invariable throughout, and their directness something striking. Measurable width they have not; it is only by depth of tint that their importance is inferred. But their most amazing attribute is their geometric character. They seem to be generally arcs of great circles drawn from certain salient points on the planet's surface to certain other equally salient ones.
>
> Their number appears to be legion. Schiaparelli discovered 104. But the better the planet is seen the more of them come out. About 350 have now been mapped at Flagstaff, and the number is only limited by our penetration. Like the asteroids, the larger ones have already been detected. Each opposition now brings out smaller and smaller specimens.[22]

Though evidently completely ignored by professional astronomers, one can imagine how it would have been received had it been noticed, for the same year *The Solar System* came out, William H. Pickering published *The Moon: A Summary of the Existing Knowledge, of our Satellite, with a Complete Photographic Atlas*. Pickering's findings even included "canals" on the Moon. Barnard's review in the *Astrophysical Journal* began unsparingly, and would have applied, no doubt, with equal force, to the then-current ideas about vegetation and canals on Mars:

> The Moon, … it would seem, might offer a rich field for careful and original investigation with sufficiently powerful telescopic means.
>
> Professor William H. Pickering has taken up this subject in recent years, and his results are so startling—bordering, as they do, on the sensational—that one hesitates to accept them, or rather to accept his conclusions.

He claims to have found evidences, not only of present volcanic life, but of of snow and ice, clouds and vegetation.

Perhaps if these discoveries had come slowly and one at a time, with long intervals between, they might have been received with better grace; but they have been turned out by wholesale, and almost any place on the Moon would seem to be conspicuously productive of one or more of the above phenomena; even the fateful canals of Mars are found to be denizens of the Moon also…

… Professor Pickering gives comparisons, side by side, of drawings and photographs of certain craters to prove the existence of the lunar canals. The photographs used in this comparison are so excessively enlarged that the details are mere blotches and it is hard to say what they represent. In these pictures, plates E and F, the resemblance between the drawings and the photographs are certainly very vague, and an interested mind could possibly trace out the canals or any other desired feature, but to the unimpassioned mind the various canals shown in the drawings have to be imagined on the photographs.[23]

Venus Again

Easily the most remarkable thing about Lowell's lectures and book, at least to the attentive reader, is not what is included but what is omitted. There are, as expected, chapters on Mercury, Mars, Jupiter and Saturn, and even asteroids and comets. Venus is notable only by its absence.

The reason for the omission, evidently, is that at the time he was preparing these lectures Lowell was still obsessing about the harsh criticisms of his Venus work. He had observed Venus again in 1901, but apparently did not find the results conclusive, and so—after agonizing over the matter, and despite once telling Douglass never to admit a mistake—at last, in July 1902, he did the hitherto unthinkable: he admitted a mistake. He wrote a brief note to the German journal *Astronomische Nachrichten* in which he announced: "Continued observations have convinced me that the spoke-like markings are probably not upon the surface of the planet but are optical effects of a curious and—astronomically speaking—of a hitherto unobserved kind."[24] A further irony: Lowell admitted that artificial planet experiments had played a crucial role in bringing about this retraction. They would also help debunk his Martian researches, though he never conceded the latter. Instead, he proceeded to launch his own line of attack consisting of an extensive series of experiments to determine the fineness of lines just visible to the eye, based on measures of telegraph wires seen from a long distance. He did establish that the eye is able to make out very fine wires indeed. However, his argument was actually rather circular; what he had affirmed was that the eye could indeed make out extremely fine lines on the surfaces of planets, provided that there really were extremely fine lines (and not something else) on the surfaces of planets.

Perched at a rather dizzying height on this small platform reachable only by ladders, Lowell sketched artificial planet disks through the 24-inch refractor in an attempt to reassure himself of the reality of the markings seen at Flagstaff. As a result of these experiments, Lowell briefly doubted his Venus markings, and published a retraction, which in turn was soon retracted. (Credit: Lowell Observatory Archives)

Lowell seated beside the 24-inch Clark refractor with the Brashear spectrograph attached. (Credit: William Sheehan Collection)

He also worried about the apparent spectrographic demonstration, in April 1900, by the respected Russian astronomer Aristarch A. Belopolsky at the Imperial Observatory, Pulkovo that Venus had a "short" day, in contrast to the synchronous rotation that Lowell had deduced from his observations of the spoke system. Obviously, if this held up, his Venus work would have to be discarded wholesale. However, by the following March 1903, Slipher had at last proved himself "clever enough to have gotten over the snags" of using the spectrograph, and in refutation of Belopolsky's result seemed to find evidence of a very slow rotation for the planet (though not necessarily, as Lowell claimed, a synchronous rotation; i.e., equal to the period of revolution, 225 days). Lowell didn't care for nuance; as far as he was concerned, the case was closed, and Slipher's result so conclusive as to make it "unnecessary" ever to repeat them.[25]

Between January and July 1903, Lowell was once more scrutinizing Venus with the Clark refractor, severely diaphragmed as usual. Examination of his observing log book shows that though a few of his sketches resemble the grotesque caricatures of earlier years, what he usually shows are often a pair of prominent bright cusp caps and one or two fragmentary lines sweeping in from the terminator. They are underwhelming, to say the least. Despite any obvious continuity in the series, Lowell proclaimed that the lines that were now appearing were identical with the earlier spoke system (and even used one of the names from his 1896 map, "Hero," for one of them). In addition, he threw down the gauntlet to the skeptics. He had, he insisted, taken every precaution, and declared, "These markings came out at times with a definition to convince the beholder of an objectiveness beyond the possibility of illusion" [2, p. 186]. He would later—in an undated note entitled "Hints on 'Seeing'"—declare his fealty to the truth of what was revealed in what he sometimes referred to as "revelation peeps": "Fine detail on the planets only comes by glimpses. Experience enables one to tell the stamp of the true from the imaginary and experience further shows that of the few doubtful cases almost all stand for a reality not the reverse." That was Percival in a nutshell. Often wrong, but never in doubt.

It is the east and Percival is the sun. Percival, dressed for the day, commands a prospect which includes the San Francisco Peaks from a porch of the Baronial Mansion. The Clark dome can just be made out behind the pine tree on the right. Photo taken about 1903. (Credit: Lowell Observatory Archives)

And Mars…

There was something circular in this argument, to be sure; it was something like that which he had used in vouching for the superiority of Flagstaff seeing, based on the fine planetary details visible there. They were seen, therefore the seeing was good; the seeing was good, because they were seen. The conviction of the reality of the markings was in fact to be deployed against one of the strongest challenges yet against the Martian canal theory—by E. Walter Maunder, the astronomer at the Royal Observatory, Greenwich, assigned to photographing and measuring the areas and positions of sunspots on the solar disk on every clear day, who had written in 1894: "We have nor right to assume, and yet we do habitually assume, that our telescopes reveal to us the ultimate structure of [a] planet" [8, p. 251]. Maunder had offered in evidence that the canals might not represent the ultimate structure of the Martian surface a observation in 1892 by a Mr. Gale, of Paddington, New South Wales, in which one canal had broken up into a chain of "lakes" on a night of superb definition, and wondered whether others might similar represent small features too small or too close together to be visualized separately and so blended together into apparent lines.

By 1903, Maunder's interesting thesis had received a considerable amount of support. The British army captain Percy Braybrooke Molesworth, a very skilful observer and leading member of the British Astronomical Association, used a 12 ½-inch reflector that year from Trincomalee, Ceylon (now Sri Lanka), within only a few degrees of the equator, found the amount of detail on Mars "bewildering," and suspected that "the broad effects one draws are simply the combined results of myriads of small details, too minute to be appreciated separately…. I cannot help being certain that our present instruments are quite incapable of dealing with the details of Mars, and that even the best and most careful drawings give an utterly wrong idea of the configuration of its surface. The eye interprets as well as it can, but the task is beyond its power."[26] Similar views were being developed by Vincenzo Cerulli, like Lowell a man of means who in 1890 had constructed a private observatory near Teramo, Italy, which he named Collurania (Urania Hill). With a fine 15.5-inch refractor, the largest in Italy after the 19-inch refractor at Brera that Schiaparelli had used in later years, he began as one of the most prolific recorders of the canals; he did not, at first, suspect that they might not be real, but the moment of truth for him came on the night of January 4, 1897, when there were "some moments of perfect definition [in which] Mars appeared perfectly free of undulation." Under these conditions, Cerulli watched with astonishment as the prominent canal Lethes "lost its form of a line and altered itself into a complex and indecipherable system of patches."[27] From that point onward, he became a leader of the anti-canal school that maintained that the canals were an illusion that would disappear as soon as the planet was examined with sufficient optical power under ideal conditions.

In 1903, Maunder himself reappeared on the scene, with a dramatic experiment conducted using naïve subjects (possessed, one might say, of the "innocent eye") recruited from cadets, aged twelve to fourteen, at the Royal Hospital School in Greenwich [10]. Known as the "cradle of the Royal Navy," and located in what is now the National Maritime Museum, it was dedicated to educating the children and grandchildren of seafarers in the Royal and Merchant Navies, and providing training in seasmanship and navigation to allow them to follow future maritime careers. So the subjects were a respectable lot. In a series of experiments carried out between July 1902 and May 1903, Maunder and his associate, the school's headmaster J.E. Evans, showed students a series of prepared disks on which the large dark areas shown on maps of Mars were left intact but the usual canals were replaced with more naturalistic shapes—winding river-like marks, minute dots, faint speckles. The altered maps were hung in front of the classroom and the students, without prompts as to what they were supposed to see, were instructed to sketch it as faithfully as they could, "putting in all that they could see and nothing of which they were not sure." They were not allowed to leave their desks in an attempt to get a closer view. The tests were carried out thirteen times, and many of the students—especially those at intermediate distances—drew illusory straight lines in places none existed; those farther back left these areas blank, while those closest to the front often managed to make out something of the irregular details. The experiment was instructive in illustrating the ladder of perception that Lowell had described—first the main dark areas such as the Syrtis Major were made out in the Christiaan Huygens era, then more details in the Beer and Mädler era, and finally canals in the Schiaparelli era. The implications of the Evans/Maunder experiment—as well as of the observations by the likes of Cerulli, Molesworth, and others—was that Lowell had interrupted his ladder of perception too early, and that the ultimate structure of the surface was yet to be revealed in details beyond the level of the canals. When Maunder presented his paper to the Royal Astronomical Society and the British Astronomical Association, the response was quite favorable—only a single member at the BAA meeting, Mark Wicks (who later wrote a work of science fiction book *To Mars via the Moon*, dedicated to Lowell), protested in Lowell's behalf. Meanwhile, the growing influence of Maunder's reaction led E.M. Antoniadi, no longer Flammarion's assistant at Juvisy but still serving as Director the Mars Section of the BAA, to include two maps in his BAA report on the 1901 opposition: one showed a few canals, and these as broad, diffuse streaks, the other the usual mass of canals. He labeled the latter: "Chart of Mars. Of all Authenticated, but not necessarily Objective, Details."[28]

Lowell seems to have missed the entire point of what Evans and Maunder were attempting to show: not the limiting visibility of actual fine lines, but the extent to which other kinds of detail, at or just below the threshold of visibility, might *appea*r to be lines. Offended by the experiment's *lèse majesté*, he responded to the "illusion" Or "Small Boy Theory" with ridicule:

…Because some boys from the Greenwich (Reform or) Charity School, set to copy a canal-expurgated picture of the planet, themselves supplied the lines which had preceptorily been left out, the Martian canals have been denied existence; which is like saying that because a man may see stars without scanning the heavens, therefore those in the sky do not exist. As to the instructions the boys received we are left in the dark. It looks as if some leading questions had unconsciously been put to them. At all events, English charity boys would seem to be particularly pliant to such imagination, for when Flammarion retried the experiment with French schoolboys, and even inserted spaced dots for the canals in the copy, not a boy of them drew an illusory line.

The fact is, this is one of those deceptive half-truths which is so much more deleterious than an unmitigated mistake. Under certain circumstances it is quite possible to perceive illusory lines, due either to shadings otherwise unmarked and thus synthesized or to immediately precedent retinal impressions transferred to places where they do not belong by rapid motion of the eye, as I had myself discovered before the English experiment had been tried. But, as I have also found out, these effects are produced only at the limit of vision, and in that limbo of uncertainty the whole art of the observer consists in learning to distinguish t he true from the false. Strength of impression, renewed effect in situ, and a peculiar sense of reality or the reverse enable him to adjudge the two… But, furthermore, and fatally to the theory here in question, the Martian canals when well seen are not at the limit of vision as its framers supposed, but well within that boundary of doubt; so that the premise upon which the whole theory gives way. Under good atmospheric conditions the canals are comparable for conspicuousness to many of the well-recognized Fraunhofer lines and are just as certainly there.[29]

Lowell advanced much the same line of argument in vigorous defense of the markings on Venus:

Nothing was set down without a caveat until I had assured myself of the certainty of [a marking's] non-subjective existence. Two points I examined specifically: one the objective assurance of the psychical perception; the other the space-prolongation of an impression by movement of the eye. With regard to the first point, experiments on the visibility of a wire show that it is possible by direct consciousness to part the true from the spurious. Although it is possible to see illusory lines, it is also possible to become cognizant of the fact. If one pays attention, an hallucination of the sort may be found to differ from a presentation of fact by the absence of the sense of reality… With regard to the second kind of illusion, the transference of a perception from one point to another by motion of the eye, experiments have led me to believe in the possibility of its production at times near the limit of vision … due to continuity of

impression, for the eye retains an impression for the twentieth of a second…
That it may be produced, and that it may also be precluded by holding the eye
still is sufficient. This plan I adopted. It is not so easy as one might imagine, for
bent on detection, the eye has a roving drift hard to hold in check.[30]

In fact, Lowell deceived himself in thinking he could somehow hold the eye's
"roving" in check. Nystagmic movements of the eye occur unconsciously and
continuously, and if they were, in fact, held in check—if the image was
arrested from movement as has been done for instance by attaching small mir-
rors to the eyeball—the image would disappear altogether.

A Martian Mania

The year 1903 is significant for seeing the most sustained effort Lowell had
made thus far in monitoring the phenomena of the Red Planet. He was at the
Observatory for the better part of a year, rather than only for brief periods
interrupted by regular returns to Boston as before, and despite some attention
paid to Venus, as noted, Mars was his all-consuming preoccupation. At the
opposition of March 29, 1903, the planet's distance from the Earth was a not
very favorable 95.7 million kilometers (59.5 million miles). Nevertheless,
Lowell scrutinized it from January until July, with an almost monomaniacal
intensity. He thus belied remarks he would make a Memorial Day speech at
the Courthouse in Flagstaff honoring the town's Civil War veterans. "In any
active brain," he said, "ideas are constantly originating. They crowd and jostle
on another for recognition. Now it is a curious and instructive fact of psychology … that any and each of these ideas is a force which, unless inhibited by
another idea, will instantly and inevitably produce its own particular effect….
We see with monomaniacs where one idea acquires such momentum as to
bear down all others before it. For by constant repetition any idea gets stronger and stronger."[31] Unhealthy in politics, where one faction gains too much
control, or in the mind, Lowell himself nevertheless gave in to a growing
monomania for Mars, and as with the repetition of planetary detail *in situ* that
in the end convinced him of its reality, he in the end succumbed to the very
tendency he warned against.

In large part, Lowell's giving in to this obsession reflects the sense of confidence and security he found in his Mars Hill redoubt, where insulated from
the slings and arrows of critics and surrounded by loyal and admiring allies he
was master of all he surveyed. There, as he feared for politics, one faction—the
Lowell faction of believers in literal canals on Mars—gained complete

control; he felt empowered like a monarch, enjoyed absolute sway, and from his Observatory on high could look down with disdain on any and all who dared challenge his inspired dictates. This tendency to rather frightful arrogance had already been noted by his associates in Japan. Basil Chamberlain wrote during his last summer in Tokyo: "He has become more certain of everything. I cannot stand ... the way he has of jumping at some general idea or theory, enunciated as a theorem," after which he "selects and bends facts to underprop that generalization."[32] This, of course, was exactly the same thing that Douglass had picked up on. But Douglass was now banished, replaced by young astronomers whose fealty was assured. His arrogance only grew worse.[33]

During the 1903 opposition of Mars, Lowell made 372 separate drawings of the planet, following the features as they rotated into, out of, and back into view (the entire cycle, because of Mars's 40 min longer rotation period than the Earth's, which made its features appear to fall backward from night to night, taking 40 days). His sketches all are stylistically very much the same, indicating the dark areas in rather cursory schematic form; they are entirely perfunctory and serve merely to establish the orientation of the moment to the planet's turning globe. His focus was on the spidery lines, now single, now double, all diverging from and converging on, as the case might be, the smaller dark patches or spots he had taken to describing as "oases."

Because the planet was kept under surveillance as continuously as possible meant that the sketches, rather than reflecting only the static view at any one time, captured over the intervals in which they came into view how "these strange things show themselves to be subject to change. That is, they take on a kinematic character."[34] This kinematic character was first revealed in the behavior of a few specific canals, before being applied to the system as a whole.

The first was the Brontes, a canal that cut a huge swath from south to north from its origin at Sinus Titanum to Semnon Lucus, the southernmost of the Propontis congeries of spots. (We use Lowell's names for these features which really do evoke not just points on a map but places on the planet; note that he uses "lucus," for forest, rather than the earlier Schiaparelli approved term "lacus," for lake, reflecting Lowell's particular way of interpreting the nature of these markings.) It was conspicuously continuous throughout its whole length at the end of February. But then—as if "actuated by a spirit of contrariety," Lowell quipped—despite the decreasing distance of the planet, it grew enfeebled in its northern part, just when it ought to have grown more prominent. Then, when it next came around for scrutiny at the end of March and the beginning of April, it reasserted itself again, to become as conspicuous in mid-July as it had been in February.

Others were the Amenthes and the Nepenthes-Thoth, which occupied the area around the Libyan continent wedged under the pointed western side of Syrtis Major. Here Schiaparelli had in the 1880s noted significant changes which, according to the prevailing view of the time, he had interpreted as produced by flooding from the seas. The Thoth, which followed a curving track from the dark protuberance Lucus Moeris on the western side of Syrtis Major to another dark spot Aquae Calidae at the southernmost apex of the wedge of Casius, had been conspicuous, and often geminated, throughout the 1880s; but Lowell could find no trace of it in 1894, and it continued to remain, as Lowell put it, "in hibernation" during the two subsequent oppositions in 1896–97 and 1901 (1899 being missed because of his illness). At the beginning of 1903 there was still no sign of it; instead another canal, the Amenthes, extended from the point of Syrtis Minor to Aquae Calidae, reigning solitary in this part of the disk. However, surprises were in store: by the end of March the Amenthes had begun to fade until it had quite vanished by the end of April; while in the meantime the Thoth, the Nepenthes, and a third canal, the Triton, had all strengthened, and by the end of May had all geminated; while in July, the Amenthes reappeared, showing alongside the Thoth-Nepenthes, and become indeed the stronger of the two, suggesting, as Lowell put it, that "the lines were in process of relapsing into the *status quo ante*."[35]

Amenthes alone in February.

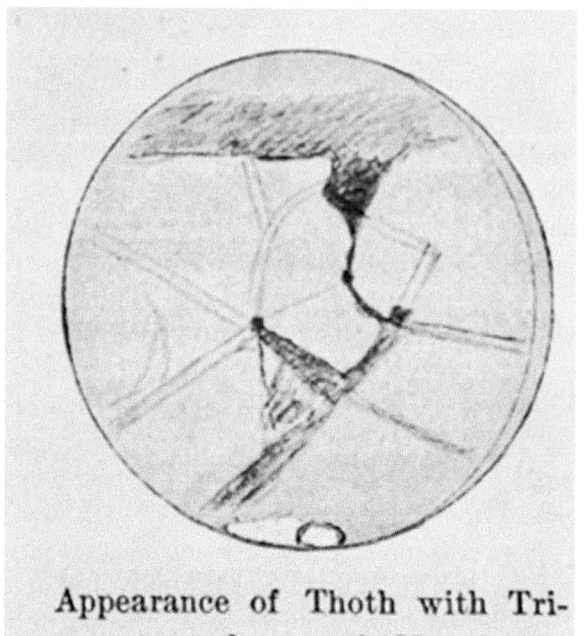

Appearance of Thoth with Triton and curved Nepenthes.
Amenthes vanished. April 20.

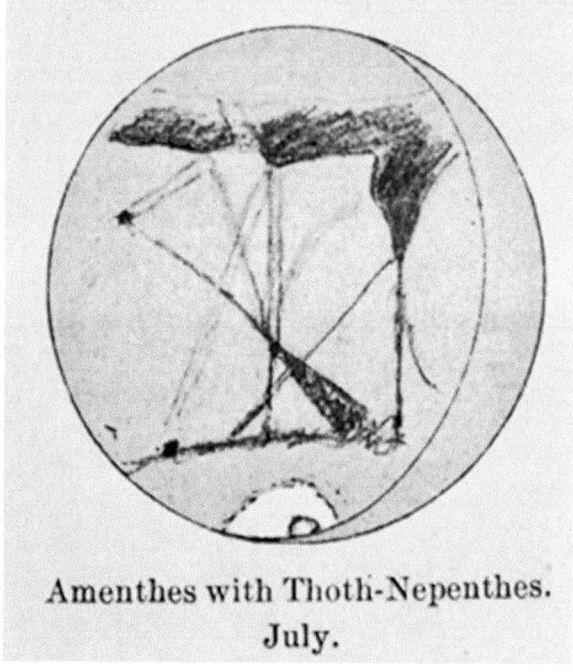

Amenthes with Thoth-Nepenthes.
July.

Lowell sketches of Mars, 1903, as reproduced from Mars and Its Canals *(Boston, 1906),*
purporting to show changes in the Amenthes and Thoth-Nepenthes canals as described
in the text

These striking examples of the changes the canals—at least some of them—exhibited suggested to Lowell the idea that the canals, like the main blue-green patches (Mare Erythraeum, for instance), might also be seasonal in nature. To investigate this notion, he proceeded to plot the percentages of visibility of each of 109 canals recognized during the several months of observation—an investigation involving some 10,900 measures from the sketches in all—percentages that were then plotted upon coordinate paper in which the horizontal direction represented time and the vertical the amount of the percentage. "The curve," he explained, … "would be [the canal's] history graphically represented … [and] could be considered the canal's cartouche,—after the manner of the ideographs of the Egyptian kings,—symbolizing its achievements and distinguishing it at once from others" [13].

To Lowell's mind, the cartouches produced a breakthrough of the first order in the understanding of the Martian phenomena, and almost—so he thought—a Q.E.D. in regard to his theory of the canals as irrigation channels. The cartouches seemed to put beyond doubt that the canals participated in the general "wave of darkening" sweeping alternately from the planet's poles to the equator earlier established (or at least strongly suspected) from the behavior of the dark areas. The development clearly indicated the progression of the canals' development, and even allowed a remarkably precise calculation of the rate of advance—51 miles per day, or 2.1 miles per hour. He effused, "[T]he mental ear detects the sound of water percolating down the latitudes." When he told his staff of the discovery, he exulted, "I think I have found the law governing the development of the canal system." Further, the cartouches revealed that the wave of darkening that proceeded from the north pole to the equator extended beyond into the south; the identical process occurred in reverse a half Martian year later. There was a delay between the time the cap melted and the canals began to appear—first those near the cap, then the others, in order of distance from it, marking a "stately march down over the face of the disk."[36] He wrote,

Thus we reach the deduction that water liberated from the polar cap and thence carried down the disk in regular progression is the cause of the latitudinal quickening of the canals. A certain delay in the action … seems to negative the supposition that what we see is the water itself.

On the other hand, vegetation would respond only after a lapse of time necessary for it to sprout,—a period of, say, two weeks,—and such tarrying would account for the observed delay.

Vegetation, then explains the behaviour of the canals…

… [I]t is a vernal quickening peculiar to Mars which knows no counterpart on earth.[37]

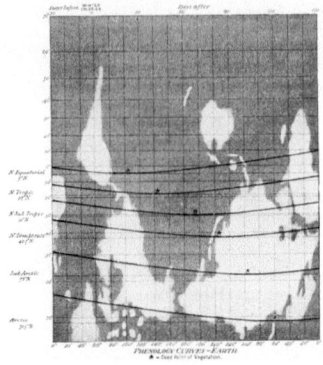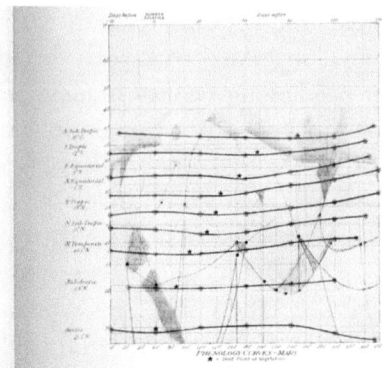

A map of the Earth (left), with south printed at the top to correspond with the telescopic appearance of the surface of Mars on which cartouches showing the wave of canal-development have been drawn. The stars mark the epoch of the dead-point of vegetation at successive latitude; time increases toward the right of each diagram. On Earth the vegetation develops directly in response to the return of the Sun. On Mars, the development is in the opposite direction, showing—said Lowell—that vegetation there responds instead to the effect of water received from the melting polar cap. These diagrams were first published by Lowell in the Proceedings of the American Philosophical Society

It had no counterpart on Earth because the transport of the water on Mars seemed to proceed in both directions—from north to south and some distance across the equator, then back from south to north again; a state of affairs that could not be explained by the action of gravity, since an irrigation ditch could not possibly transport water downhill in both directions. But Lowell had an answer for this as well. The water had to be pumped. If pumped, Mars had technology, and where there was technology, there had to be intelligence. In other words, Mars was the abode not only of life—in the form of vegetation—but of intelligent life.

In a way, Lowell's cartouches represented an intellectual *tour-de-force*. But they also showed the workings of a powerful intellect running far beyond anything justified by the rather marginal perceptual data available to it. That data was presented to shore up the "habitation hypothesis"; no alternative was considered. His approach was consistent with the impression of his friends of growing intellectual arrogance—"He has become more certain of anything"—or, in modern neuropsychological terms, what has been referred to as the "false tagging theory," which posits a dysfunction of the prefrontal cortex, specifically the ventromedial prefrontal cortex, which leads in some subjects to what has been called a "doubt deficit" [14].

The doubt deficit and inability to even see evidence that suggested alternative hypotheses is illustrated by Lowell's reaction to an observation made on the evening of May 25, 1903. V.M. Slipher was observing Mars with the Clark telescope, and noticed a conspicuous projection at the terminator of Mars. Similar projections had been observed from time to time since 1890, especially at Lick Observatory, and invariably attracted a strong reaction from the press. Douglass had observed something similar in 1900. On this occasion, Slipher immediately called Lowell to the telescope to examine it, and the latter found it "not whitish but ochre-orange, closely assimilated in tint to the subjacent parts of the disk, the region to the north and west of the western end of Deuteronilus." From its color Lowell deduced that it was "not a cloud of water-vapor," but a cloud of dust."[38] The dust cloud was evidently short-lived; by May 28 all trace of it had vanished. In general, Lowell regarded such events as rare as hen's teeth, and despite correctly recognizing that this was a dust cloud, failed to incorporate the notion of pervasive windblown dust in his interpretation of the Martian seasonal phenomena. He would never accept that dust could hide the usual markings for weeks at a time, though this had been the case in October/November 1894 and would be again in July/August 1909. Thus, he failed to grasp the truth that the "wave of darkening" and changes in the visibility of features such as the Brontes and the Amenthes/Nepenthes-Thoth might be due to their variable covering and uncovering by the transport by variable seasonal winds of thin coatings of dust.

Miss Irva Struthers

In the latter summer of 1903, as Mars finally retraced its steps and retreated to inaccessible remoteness from the Earth, Lowell returned back East and found himself relaxing at a resort in York, Maine. Now approaching fifty, and long since given up by friends and family as a confirmed bachelor, he embarked on what seems from the outside looking in to have been a rather sudden and determined change of direction for someone who had so long insisted on not being tied down. A woman of social-register standing from Philadelphia, first spied at tea but who at first pretended not to notice him, Miss Irva Struthers became a serious object of Lowell's romantic interest. He courted her seriously, wrote two dozen love letters to her (a typical phrase: "When will her Ladyship be at home … to a Martian") and proposed. He submitted to his fate with the calm of one to whom suspense was more intolerable than despair, giving her only 2 weeks to decide. She turned him down—mainly, it seems, on religious grounds; she was, Lowell wasn't. And that was that.

"A man of many minds and many moods," as Wrexie Louise Leonard called him. Above, in a formal photograph taken on June 3,1904, Lowell strikes a relaxed view. Credit: Lowell Observatory Archives. Below, a more intellectual pose: Lowell photographed on June 4, 1904 by Boston photographer James Edward Purdy. (Credits: Lowell Observatory Archives and Library of Congress's Prints and Photographs division)

Lowell, showing a more arrogant face, in a photograph on June 29, 1904. At the end of the following month, he began his affair with Edith Petit. (Credit: Lowell Observatory Archives)

Within 8 months of the end of this rather feverish but failed effort, on his latest trans-Atlantic voyage at the end of July 1904, be became attracted to another woman full of witty banter and with a strong interest in the arts, Edith Petit. She now became his regular companion as together they made the rounds of fashionable cosmopolitan resorts such as St. Moritz and Paris. Together they joined a group of mildly decadent American expatriates led by Ralph Curtis and including Bernard Berenson, who had just published his best-known work, *The Drawings of the Florentine Painters,* and was busily grabbing up great works of Renaissance art for Isabella Stewart Gardner (another relative, through Percival's cousin John Lowell Gardner) to decorate her mansion in Boston. They belonged to the social set so fashionably captured in the portraits of John Singer Sargent which, after falling out of favor for a few decades, have become fashionable once again. Lowell enjoyed flirting and shopping with Miss Petit, and taking in some of the grand views in the Swiss Alps, and later admitted that he might have fallen hard for her did it not emerge that she had lost her virginity 4 years earlier. (She would, he said, have been "a rare full flower but for her awakening.")[39] Nevertheless, they made an attempt at sexual intimacy; the best Lowell could do was, he admitted, "a half performance."[40] He also took up with a Miss Erna Stevenson, a 30-year-old woman who seems to have been physically the aggressor and with whom Lowell again made a brave attempt to engage sexually (twice, in a hut off a trail in the Swiss Alps and in a room in St. Germain, near Paris). However, Lowell failed to find any pleasure in these attempts, and seemingly preferred voyeuristic pleasures to the real thing. "In place of sexual pleasure as evidence of his success in overcoming inhibitions, Lowell came away with a trophy in the form of photographs of his nude companion."[41]

Though astronomical themes were cast into the background that summer, they did not disappear altogether. He and Miss Petit travelled to Milan in August, in the hope of meeting Schiaparelli at Brera as in 1896. In retirement now, Schiaparelli was not in Milan but at Monticello, where he had a country villa; so Lowell, not without difficulty, made his way there. For some time he had been trying to win the "Master Martian" over to his own views about Mars, the two men corresponding in French (neither Schiaparelli nor Flammarion for that matter spoke English; but Lowell had been fluent in French since childhood). Lowell had sent the first numbers of the Lowell Observatory *Bulletins* to him; one announced Slipher's work on the rotation of Venus, another described Lowell's own work on the cartouches of Mars. Schiaparelli praised both, and in particular found the latter "a remarkable attempt." He added, no doubt to Lowell's great satisfaction, "Your theory regarding vegetation is becoming more and more likely."[42] In addition, also in August, Lowell wrote to Flammarion from the Grand Hotel Nation in Lucerne to introduce Miss Leonard, who was about to visit Paris herself. "It will give her great pleasure to see you," Lowell wrote, "and since she herself has made interesting observations on Mars in Flagstaff and in Mexico, you will find her interesting and interested—she knows you both from reading and from a literary point of view."[43] He met up with Camille and Sylvie Flammarion at the end of September, before returning to the United States. Before his departure he cast his eye back over the preceding year—with its failed marriage proposal to Miss Struthers, and its pleasant but inconclusive flirtations with Miss Petit and Miss Stevenson—and lamented in his "Journal—1904" on September 5: "I miss the only solace of life, work." Two days later, he wrote: "And so I go back to my only solace. Alone! Yes, always alone, though I had hoped for the opposite. Perhaps that [work] I may make a success; my own life is such a dismal failure."[44]

Lowell on his pedestal. A photograph taken by his companion Miss Erna Stevenson at a park at St. Cloud, near Paris. From: P. Lowell "Journal—1904." (Credit: Percial Lowell Papers [MS AM 3166, III], Houghton Library, Harvard University)

My Only Solace

By the beginning of 1905, Lowell was back in harness, his dalliance with Venus over and determined once more to take up Mars. His redoubt on Mars Hill consisted, as Ralph Curtis would say, of "discomfort and cave-dweller's luxuries."[45] The "cave-dweller's luxuries" of the Baronial Mansion were certainly a step down from the amenities featured in Curtis's palazzo on the Grand Canal in Venice or the luxury hotels Percival stayed in when abroad, but it was luxurious enough, at least by standards of the Wild West. The Lowell archives contain a host of invoices which show his soft-spot for French cuisine and luxurious items including salted almonds, piccalilli and chow-chow, macaroons in brandy, vanilla syrup, Roquefort cheese, French sardines, and Chianti and, especially, case after case of La Cresento Margaux wine.[46] In addition, there were always cigars on hand—apparently, after Mars, cigars were a particular passion of his. Lowell on one occasion noted that "the cigar … was an indispensable item that provided 'the only excuse for a dinner'" and that "everything … seems the better through the pleasing mist of the cigar."[47] On the other hand, even with several fireplaces, the BM must have been challenging to heat, and it hardly ran itself—in order to keep the place running smoothly required the attentions of a full-time cook and a retinue of servants, about whose inefficiencies and insolence Lowell was apt to grumble. Lowell also laid out a garden (horticulture being one of the interests he shared with his late father). As Lampland later recalled, he enjoyed

> great success with many flowers and … especially fine displays of hollyhocks, zinnias, and a considerable variety of bulbs. Gourds, squashes and pumpkins were also great favorites. You will remember one year the especially fine collection of gourds and that bumper crop of huge pumpkins, many prize specimens being sugar fed. At times Dr. Lowell could be seen in the short intervals he took for outdoor recreation, busy with his little camel's hair brush pollenizing some of the flowers…. Then the frequent, almost daily, walks on the mesa. Certainly he knew all the surrounding country better than anyone here. He would refer to the different places such as wolf Canyon, Amphitheatre Canyon, Indian Paint Brush Ridge, Holly Ravine, Muellin Patch, etc…. Trees were an endless source of interest to him…. Cedars or junipers seemed to be favorite subjects for study, though other varieties or kinds were not overlooked….[48]

But mostly it was astronomy, and the 24-inch Clark from which he could enjoy his escape into the realms of the "wonder-world of stars." His work on Mars in 1903 had given him an enhanced position in the astronomical world, which included being awarded the Prix de Janssen (the highest award of the Société Astronomique de France, for his work on Mars in 1904), and at least from an outside perspective he seemed to return to the scene as one who

... trod on silk, as if the winds
Blew his own praises in his eyes,
And stood aloof from other minds
In impotence of fancied power.[49]

Another formal portrait from 1904: Lowell standing by a chair. Perhaps one can guess from this something of his imposing demeanor on the lecture stage. (Credit: Lowell Observatory Archives)

Lowell and his staff, 1905. From left: Harry Hussey (Lowell's porter), W.L. Leonard, V.M. Slipher, Lowell, C.O. Lampland, and J.C. Duncan, the first Lawrence fellow at LowelLowell who was, assigned to the early Planet X search. (Credit: Lowell Observatory Archives)

The red planet would come to opposition on May 8, 1905, with an apparent diameter of 17.3 secs of arc. Though far from the maximum possible, Mars was nevertheless closer than it had been since 1896. Moreover, mere distance was not the determinative factor in the visibility of the canals. Lowell had written in *The Solar System*:

> They are not always equally visible. Sometimes they are conspicuous, sometimes scarcely discernible even to a practiced eye. And this is not mere matter of distance. The best time for seeing the planet is not the best time for detecting the canals....
>
> ... [T]he distant encounters ... are not so unfavorable as they are thought. For another factor besides nearness affects the reckoning. The planet's axis is tilted to the plane of its orbit at an angle of 25°, and is so faced that the southern hemisphere is presented to us at the time of closest approach. Now the canals lie chiefly in the northern hemisphere. In the next place, it is then the northern winter, and careful comparison reveals the fact that the conspicuousness of a canal is a function of the Martian time of year, becoming pronounced in summer and fading out in winter.[50]

His visual surveillance of the planet was to be even more extensive than 2 years earlier. Observations were carried out continuously from mid-January, when the disk was only 6.4 secs. of arc across, through their peak in early May, until mid-August when it had shrunk to only 10.0 secs. of arc. Though his early observations showed little, as expected given the small size of the disk, that little was enough for him to telegraph to Morse, "Canals already seen and in agreement with theory." He invited Morse to come to the Observatory for an extended stay so that he could witness for himself the seasonal changes in the main blue-green areas and the canals, and advised that he ought to time his arrival with the advent of the doubles, which "do not come on till the early part of June and to miss them would be to see the play minus Hamlet." Morse would arrive in Flagstaff in May, and stay for a month.

In addition to Mars, another interest—soon to rival it—became evident for the first time. In *The Solar System*, he had briefly mentioned some work by Flammarion on families of comets belonging to the giant planets, including several comets associated with Neptune, and intimated that the latter might serve as a "fingerpost" to a still more distant planet or planets unknown. From a casual interest, by February 1905 he had begun actively to pursue the possibility. His first step was to request up-to-date charts and photographic surveys of the sky. Further, he charged John C. Duncan, the first holder of the short-lived Lawrence Fellowship (named for Percival's mother), with searching for with a 5-inch wide Voigtländer lens expressly obtained for the purpose the following September. (By then, Lowell was once more overseas, in England). In addition, his literary output that year was remarkable, even for him. His first order of business was to finish off the backlog of summaries of observations at previous oppositions, those of 1898–99 and 1900–01 oppositions observed by Douglass and that of the 1903 opposition he had observed, for the third (and final) volume of the observatory *Annals*, and to publish articles in *Popular Astronomy* on the 1903 work on Venus and Mars.[51] He also unveiled a new line of defense of the Flagstaff planetary observations which he had been testing out privately on sympathetic colleagues since 1903.[52] It was clearly designed to rebut the influential negative testimony of Barnard who, despite having been generally credited with remarkable eyesight for faint stars, comets, and nebulae, had failed to see either the markings on Venus or the canals of Mars. In an article, "On the Kind of Eye needed for the detection of planetary detail," again published in *Popular Astronomy*, Lowell suggested that observers like Barnard simply did not have the right kind of eye for the detection of planetary details:

> … The eye is the portal to perception; through it is determined what shall enter the brain. But there are two ways in which eyes may admit information; by being sensitive to light or by being acute to form. The two qualities are quite distinct… [17, p. 82]

… Now in the study of planetary detail perceptivity to form is more essential than receptivity to light. The eye needed is therefore the acute not the sensitive one. I have seen this well exemplified in the case of two observers working practically side by side—the one could detect the Martian canals the easier but chiefly as streaks, the other when he caught them, saw them well defined. It thus behooves a man to test his own eyes before he presumes to deliver an opinion upon what can or cannot be seen by another. Indeed the asserting of a negative is not only an inconclusive but a most dangerous proceeding….[53]

In fact, his claim that the acute and the sensitive eye did not go together does not seem to have been tested empirically at the time.[54] In his own case, the exceptional acuity of his eye was attested by the noted Boston ophthalmologist, Dr. Hasket Derby; however, at the same time Lowell claimed to have an eye prodigiously sensitive to faint stars, with the ability to count sixteen in the Pleiades.[55] Apparently the two kinds of sight went together in Lowell but not in ordinary mortals like Barnard.

The whole case is somewhat reminiscent of that of N-rays, whose existence was reported in 1903 by the highly respected French physicist René Prosper Blondlot and became, with the canals of Mars, "the best and most widely analyzed example of tribal delusion in the history of science" [20]. Defenders of the disputed phenomenon adopted a methodology similar to that of Lowell, with one of them admitting after the N-rays had been debunked: "… [Y]ou always see the expected effect when you are persuaded that the rays exist, and you have a priori the idea that such an effect can be produced. If you add another observer to control the observation, he sees it equally (provided he is convinced): and if an observer (who is not convinced) sees nothing, you conclude that he does not have sensitive eyes."[56]

To return to Lowell's campaign of Mars observations. Each set of sketches—there would be several hundred disk drawings in all—has the usual Lowellian form, showing the main details rapidly filled in and the canals registered one moment only to vanish the next—evanescent things, seen in glimpses, recognized only with foreknowledge of what was there and revealed for an instant, and in a sudden flash. The canals were elusive until, once caught, they were encountered at every turn; rather like the gods of Japan, whose acquaintance came only little by little to Westerners as Lowell had described in *Occult Japan*:

It is to be remembered that what no one is interested to reveal may stay a long while hid. For, with quite Anglican etiquette, the Japanese never thought to introduce their divine guests and their foreign ones to each other. Once introduced, the two must have met at every turn. Indeed, the visitants from the spirit-world remind one of those ghost-like forms of clever cartoonists, latent in the cultures of more familiar shapes, till, by some chance divined, they start to view, to remain ever after the most conspicuous things in the picture.[57]

A difficult object of scrutiny. This photograph, taken by E.C. Slipher only a few days before opposition on December 4, 1911, when the apparent diameter of the disc was 18".3, shows that even at a better-than-average opposition—it is here larger than it ever got at the opposition of May 1905, for instance—the planet appears smaller than a medium-to-small crater on the Moon. Slipher notes (in Mars: the photographic story, *p. 76), "This serves to emphasize the meticulous care necessary to discover the hundreds of details that have been recorded on Mars..." It also vividly illustrates just how challenging it was to record fine Martian details photographically, as they lay close to the size of the silver grains in the emulsions of the plates used at the time. (Credit: Lowell Observatory Archives)*

Lowell's 1905 map of Mars (Mercator projection), for use in identifying markings named in the text. It is useful to keep in mind that the detail shown on this map was supposed to have been discerned on the tiny disk shown in the preceding image of Mars next to the Moon. (Credit: Lowell Observatory Archives)

Lowell's observations from 1905, inscribed on a six-inch wooden globe. One of these globes—as much an objet-d'art as a scientific artefact—at each opposition he successfully observed, beginning in 1894. (Credit: Lowell Observatory Archives)

Activity around the date of the 1905 opposition became frenetic, and Lowell's spirits received a boost when his old friend Morse arrived on Mars Hill for the month-long stay previously discussed, in which, without being prejudiced by studying drawings and maps by other observers or being told in advance what Lowell and other astronomers had seen, Morse would put the "innocent eye" of an intelligent but naïve observer to see what Mars had to offer in the reputedly superlative Flagstaff seeing.

Adventures in geology: During Morse's visit to Flagstaff in May–June 1905, they made at least one memorable excursion, to the Grand Canyon. Here, Lowell, left, and Morse are make their way along the famous Bright Angel Trail; O'Neill Butte is the prominent formation in the left middle of the picture. (Credit: Lowell Observatory Archives)

"A good cigar is a smoke." Percival Lowell poses with a book in one hand and a cigar in the other, while gazing out the window of his study, in an east-facing room of the Baronial Mansion. This photo was taken in May–June 1905. (Credit: Lowell Observatory Archives)

Morse and Lowell enjoy a smoke in Lowell's study in the east room of the Baronial mansion during Morse's visit in May–June 1905. (Credit: Lowell Observatory Archives)

Lowell in 1905, holding one of the Mars globes he drew up at each opposition of Mars.

Morse's entries in his observing log book were later published in his charming and unabashedly pro-Lowell book *Mars and Its Mystery*. They provide vivid testimony at the manner in which some of the canals were revealed to him, as shown by the following excerpts:

May 14. Midnight. Saw the planet for the first time. A beautiful luminous disk with shades of tone dimly visible. Southern pole cap white and seen.

May 15. Certain details sufficiently distinct to make out dark areas, and at times a line or two.

May 16. Occasional flashes of a few lines... With no better seeing conditions than last night, more details came out, and for the first time I am encouraged to believe that each day an improvement will take place. I saw enough to make my first drawing....

May 31. Saw a little more than I saw last night but did not see a trace of things that Mr. Lowell and his assistants apparently saw without effort. I realize that it requires a special training to observe the flickering evanescent markings on Mars...

June 5. I find a slow advance in my ability to see the markings though it is exasperating that the janitor of the Observatory talks about plainly seeing certain details which he indicates to me by a sketch, and looking at the region I can see no trace of a canal or anything else.

June 7. Seeing very good and in my observations tonight added another canal. It is a most difficult matter to catch the fleeting lines as they appear with startling distinctness to instantly vanish again...

June 13. In my observations to-night added one new canal and completed another, and was able to detect one that Mr. Lowell had not seen during the evening—a well-known one he says. It simply shows that one must continually observe as the lines flash out for a single instant.[58]

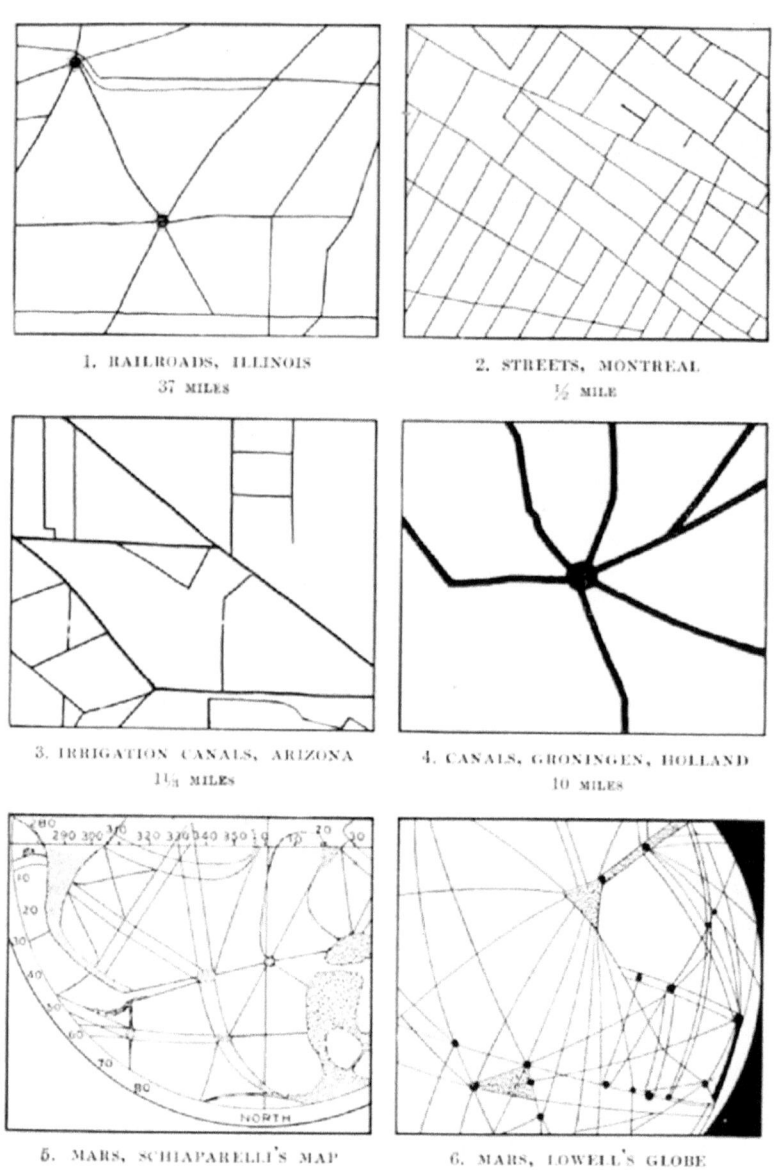

1. RAILROADS, ILLINOIS
37 MILES

2. STREETS, MONTREAL
½ MILE

3. IRRIGATION CANALS, ARIZONA
11⅓ MILES

4. CANALS, GRONINGEN, HOLLAND
10 MILES

5. MARS, SCHIAPARELLI'S MAP

6. MARS, LOWELL'S GLOBE

A plate from Morse's Mars and Its Mystery, *showing comparisons between various structures on Earth and the features recorded by Schiaparelli and Lowell on Mars as evidence that the latter were artificial works constructed by the Martians. (From: Edward S. Morse,* Mars and Its Mystery *(New York: Macmillan, 1906))*

Morse's experience in seeing the canals in glimpses, "revelation peeps," as Lowell called the moment-to-moment flashes in which the canals appeared a few at a time, underscored one of the most enduring aspects of the phenomenon. The fact that the canals—single or double—were seen only in glimpses, and never as a complete system but only a few at a time (despite the impression given by the maps), was in fact the critical point about them; and yet the point was not fully grasped by either the Lowellians or the anti-Lowellians. The latter had recognized that the actual markings on Mars might be *spatially* discontinuous (as demonstrated in the Evans-Maunder experiments) but had not appreciated the implications of their being *temporally* discontinuous, seen staccato in moments (usually lasting a fraction of a second) of exceptional seeing.

What this meant is that the the canal-filled maps by Lowell and others were not comparable to "still life" paintings, the genre that perhaps was assumed by most readers. They did not show what was seen "steadily and whole." Rather the maps were a cumulative impression of what passed too quickly for the eye to grasp, like the images of a galloping horse in the pre-Eadward Muybridge stop-action era. They were a kind of palimpsest, in which one momentary impression was superimposed on the next as if each were recorded on translucent vellum paper and the sheets stacked one on top of the other. Again, never were more than few "canals" seen at a given time; from moment to moment and sketch to sketch they differed in presentation (and location). This is evident in every page of Lowell's and others' observing log books. It also undercut Lowell's seeming ability to locate each detail of the Martian geography—including the canals, of which the Ganges, the Nektar, Nilokeras I and II, Phison, Gihon, Dijoun, Ganges, Laestrygon, Lethes, Jamuna, Amenthes, Thoth, Hiddekel and a host of others all appeared in map after map and list after list as if they were well-known places on the surface of the Earth; their positions seemingly accurately vouchsafed from one observation and indeed from one opposition to the next. And yet, the precision was, like many of the features themselves, at least partly illusory; if one overlays all of Lowell's maps (as done in the figure below), the canals form a tessellated pattern of fine gossamer filaments that practically covers every square degree of the surface, a testimonial to the subliminal nature of details seen in glimpses.

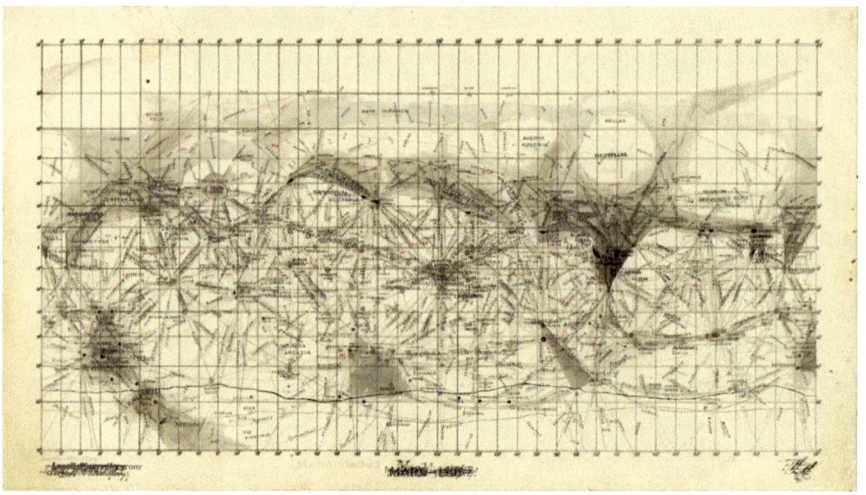

A Lowellian palimpsest of canals: Lowell's Mars maps from 1896 to 1911 superimposed. (Credit: Joel Hagen)

Notes

1. E.E. Barnard to G.E. Hale, October 26, 1905. Hale. George Ellery. Papers, Mount Wilson and Palomar Observatory Library.
2. Briefly, in 1902, even Lowell had come to doubt his markings, and announced in a note published that year in *Astronomische Nachricthen*, no. 3823, pp. 129–132: "Continued observations have convinced me that the spoke-like markings are probably not upon the surface of the planet but are optical effects of curious and—astronomically speaking—of a hitherto unobserved kind." He cited the artificial planet observations as having played a role in changing his mind. However, he soon retracted his retraction, announcing, after a series of further observations of the planet in 1903, in which he took "special care … against self-deception," that "the lines making in from the terminator which constitute the spokes … appeared again in the same places they had occupied in 1897 and 1901. This alone is very strong evidence of their reality. In the next place, these markings came out at times with a definition to convince the beholder of an objectiveness beyond the possibility of illusion." See P. Lowell [2, p. 186]. This was to remain his firm position for the rest of his life—though in later years he expended little effort in publicly defending them.
3. Strauss, *Percival Lowell*, pp. 194–195.
4. A.E. Douglass to J. Jastrow, January 9, 1901; Lowell Observatory Archives. So far as is known, Jastrow never responded. Though he later wrote a book about human error, it contains nothing about the controversies about Mars. See: J. Jastrow [3].

5. P. Lowell to A.E. Douglass, February 27, 1901; Lowell Observatory Archives.

6. Basil Hall Chamberlain to Lafcadio Hearn, January 10, 1893; Koizumi, *Letters*, p. 2.

7. A.E. Douglass to W.H. Pickering, March 8, 1901; Lowell Observatory Archives.

8. W.H. Pickering to A.E. Douglass, March 27, 1901; Lowell Observatory Archives.

9. A.E. Douglass to W.L. Putnam, II, March 12, 1901; Douglass. Andrew Ellicott. Papers, University of Arizona Library.

10. P. Lowell to Elizabeth Lowell, April 12, 1901; Lowell Observatory Archives.

11. P. Lowell to A.E. Douglass, October 8, 1901; Lowell Observatory Archives.

12. A.E. Douglass to W.W. Campbell, August 23, 1901; Campbell. William Wallace. Papers, Mary Lea Shane Archives of the Lick Observatory, University of California, Santa Cruz.

13. Eventually Douglass—after a stint in local mining ventures, running an assay office in Flagstaff, and standing as a Republican candidate for probate judge—he was elected, despite the fact that he lacked legal training and experience, for his "good judgment and sterling honesty"—Douglass finally recovered from the ruins of his career. In the year of his dismissal from the Lowell Observatory, he had taken up, as a hobby, an interest in using tree rings tree rings to establish historical climatic variations and the chronologies of ancient archaeological sites, thereby laying the basis of the science of dendrochronology, and eventually was appointed the first director of the University of Arizona's Steward Observatory in Tucson. There, he continued his tree ring research, and also discovered that seeing conditions there were better for astronomical purposes than in Flagstaff—establishing a trend that has continued to the present day of establishing major astronomical facilities in southern rather than northern Arizona. On Douglass's later career, See Webb, *Tree Rings and Telescopes*. Also: W. Sheehan and G.W. Lockwood [4].

14. However, see: W. L. Leonard [5].

15. P. Lowell to W.A. Cogshall, July 7, 1901; Lowell Observatory Archives.

16. P. Lowell to V.M. Slipher, December 18, 1901; Lowell Observatory Archives.

17. It was known as the "BM" for short. The original entrance room of the cottage became the dining room and parlor, and the dark room became Lowell's wine cellar. The first expansion of the house was to the west (the side away from the Clark dome); it included Lowell's study and next to it a bathroom with a sunken bathtub in which it was possible, even for a six-footer like Lowell, to lie at full length. A kitchen, with a music box contraption that played metal disks, a breakfast room, bedrooms and den were then added, and connected to the original rooms by a narrow passageway that served also as the butler's pantry. The bedroom next to the den was Lowell's bedroom, with a brick fireplace in the corner, and there was another fireplace, faced with native malapai, in the den. Three new bedrooms were added, including Leonard's, which also had a brick fireplace. This first expansion also included an under-the-floor passageway from

the door just in front of the study and sunken-tub bathroom to the north porch just in front of the entrance to the three new bedrooms, and provided a way to get from Lowell's study to the bed rooms without going through the kitchen or middle bedroom. (It was long rumored on Mars Hill that this passageway allowed Lowell to visit Leonard's bedroom without detection). In addition to the small personal study of Lowell's between the front living room and bath room, an all-purpose office and study was added at the east end of the building, and which was used as the main study until 1910 when the "old library" was built just down the hill, east and a little north of the "BM." It was never luxurious, and walls of unpeeled logs furnished breeding grounds for insects, while packrats could often be heard scurrying about in the attic. These and more details are found in Henry Giclas, "Henry's reminiscences." Manuscript dated November 15, 1991; Lowell Observatory Archives.

18. "Prof. Lowell's New Book," New York *Times*, July 18, 1903; "Our planetary companions," *The Dial*, July 16, 1903; "Lectures on the Solar System," July 16, 1903, *Engineering Literature*, supplement to *Engineering News*; "Our stellar neighbors," *This Week's Progress*, August 1, 1903. Thanks to David Baron for pointing these out to me.

19. Lowell, *The Solar System*, op. cit., pp. 47–48.

20. Barnard, "The South Polar Cap of Mars," p. 249.

21. Lowell, *The Solar System*, p. 54.

22. Ibid., pp. 56–58.

23. E.E. Barnard [7, pp. 359–360]. Barnard notes (p. 360) that Pickering's chapter on vegetation and the lunar canals indirectly proves that the canals on Mars cannot be the work of intelligent life, quoting a passage where Pickering says: "Sixthly, since no water in liquid form can exist on the Moon, this fact will enable us to rule out many seductive but erroneous hypotheses. Seventh, and lastly, we know that so little air and water vapor exist there that we can confidently also rule out all aid in the construction of these formations [the "canals"] from intelligent or intellectual life .... They merely weaken the strongest argument hitherto found for the existence of highly intelligent life on Mars."

24. P. Lowell (1902) "The Markings of Venus," *Astronomische Nachrichten* no. 3823, columns 129–132.

25. Hoyt, *Lowell and Mars*, p. 135.

26. Quoted in William Sheehan [9].

27. Ibid., p. 125.

28. E.M. Antoniadi [11]. By this time, Antoniadi had come to believe that at least some of the canals represented contrasts between differently shaded regions or the edges of indefinite half-tones.

29. Lowell, *Mars and Its Canals*, pp. 202–203. Again, somewhat off point, Lowell carried out a series of experiments on the visibility of telegraph lines at the Observatory, in collaboration with Slipher and Lampland. See: P. Lowell (1903) "On the Visibility of Fine Lines." *Lowell Observatory Bulletin* no. 2.

For Lowell, whose visual acuity was apparently somewhat better than that of his younger assistants, the minimum width of a telegraph line that could be detected by the naked-eye was 0.″69 (below 0.″59 the object produced no effect). From these results he estimated that even at a poor opposition (apparent diameter 14″.0), using the 24-inch refractor with his usual power of 310 x, it ought to be possible to make out a canal some ¾ mile across; at a very close opposition and using a magnification of 450 X, a canal 3/16-mile across might be visible. Of course, given the reality of loss of light and definition in the telescope compared to the naked-eye, this probably exaggerated somewhat what was possible; still, Lowell thought that ½ mile was a reasonable estimate of the narrowest line visible on Mars. This gives some idea of what he thought he was seeing on Mars. He also noticed that below this threshold "there are stimuli perceptible so faint and so fleeting as to …. leave only an indefinite subconsciousness of their presence which the brain is unable to part from its own internal reverberations." He found that this "narrow limbo, this twilight of doubt," in the region between 0.″59 and 0″.69, but anything above that and the brain "was cognizant of objectivity as such." Thus, as he claimed with the markings on Venus, there was a threshold above which "the observer was almost always able to distinguish the real wire from the illusory ones, as long as he could recognize anything at all." Of course, all this depended on the existence of actual lines being present, and did not really address the premises behind the Evans/Maunder experiment. His argument seems to come down to this: If there are actual fine lines on Mars, one will see them, and since once sees them, there must be actual fine lines on Mars. As with his confidence in the superior seeing conditions in Flagstaff which allowed fine planetary detail to be seen, from which it followed that, since such detail was seen seeing conditions in Flagstaff must be superior, there is a circularity to his arguments.

30. Lowell, "Venus 1903," pp. 184–185.
31. P. Lowell (1903) "Commemoration Day Address," May 30; manuscript, Lowell Observatory Archives.
32. Basil Hall Chamberlain to Lafcadio Hearn, August 5, 1893; Koizumi, *Letters*, p. 34. I am reminded of the comment of Lord Melbourne regarding the historian Thomas Babington Macauley: "I wish I could be as cocksure of anything as Tom Macaulay is of everything." In fact, Lowell and Macaulay were much alike. According to historian Hugh Trevor-Roper, introduction to Lord Macaulay [12]: "Macaulay owed his success, in part, to the absolute clarity of his views and his style. This clarity proceeded from his own conviction. From the age of twenty-two, as far as we know, his convictions were fixed: he never changed his mind. And because his convictions were strong and clear, he tended, when faced with the evidence on any subject, instantly to range it in accordance with them…. He was no doctrinaire but an exceptionally well-read man of the world. Nevertheless, … however elevated or sophisticated his expression of it, his judgment was always influenced by his convictions…. They caused him to make

dogmatic judgments in fields where he was unqualified to speak and would never stoop to learn. To the modern reader, the dogmatism of his comments, in such fields, can be even more offensive than their ignorance or insensitivity."

At p. 30, "With his clear mind ... [and] his absolute certainty, Macaulay never wavered in his judgement. ... Maucaulay would survey all the facts, take past convictions into consideration, allow extenuating circumstances, and then, in his own favorite word, 'pronounce.' But judgment, once decided, was firm. There were no qualifications."

Much of this of course can be said about Lowell as well. In Maucalay's case, these characteristics have been ascribed, at least in part, to his prodigious memory—"he knew the whole of *Paradise Lost* and the whole of *Pilgrim's Progress* by heart, and could take in at a glance, and retain in his mind, a whole printed page, and discharge it again in his conversation" This tended to make him confident in what he remembered, but also he "seldom, if ever, detached a fact from the pattern in which, and through which, his memory had preserved it." Lowell also obviously was possessed of quite an extraordinary memory, and something similar may have been involved in his tendency to subordinate details to the same rigid and recurrent patterns of interpretation.

33. Apart from Lowell's own work, the most significant results achieved at the Lowell Observatory in 1903 was by V.M. Slipher, who discovered a series of absorption bands in the spectra of Uranus and Neptune which neither he nor anyone else could identify at the time. See: V.M. Slipher (1904) "On the Spectrum of Uranus and Neptune," *Lowell Observatory Bulletin* no. 13. The identity of the bands, as was finally revealed in 1931 by the German physicist Rupert Wildt, proved to be methane and ammonia gas.

34. Lowell, *Mars and Its Canals*, p. 281.

35. Ibid., p. 322.

36. Ibid., p. 176

37. Ibid., pp. 177–178.

38. P. Lowell (1903) "A Dust Storm on Mars." *Lowell Observatory Bulletin* no. 1. This publication was later reprinted as Note 13 in Lowell, *Mars as the Abode of Life*, pp. 256–265. Lowell's failure to appreciate the extent to which the atmosphere of Mars is affected by suspended dust led him to underestimate the contribution of dust to the albedo of the atmosphere of Mars and to overestimate its density, in the calculation given in *Mars as the Abode of Life*, pp. 238–240.

39. Ibid., p. 39.

40. Ibid., p. 40.

41. Ibid., p. 41.

42. G.V. Schiaparelli to P. Lowell, December 4, 1904. Translated from French by Jennifer Putnam. Lowell Observatory Archives.

43. P. Lowell to C. Flammarion, August 20, 1904. Juvisy Observatory Archives.

44. P, Lowell, "Journal-1904"; Houghton Library of Harvard University

45. Ralph Curtis to Barrett Wendell, March 15, 1915; Wendell. Barrett. Papers, Houghton Library, Harvard University.

46. Margaux is still one of the most prestigious appellations of the Bordeaux wine region. A 750 ml bottle of classified first-growth Chateau Margaux from 2009 was priced in 2023 at $1400.

47. Strauss, *Percival Lowell*, p. 37.

48. Constance Lowell to C.O. Lampland in A.L. Lowell, *Biography of Percival Lowell*, pp. 153–154.

49. From: Alfred Tennyson, "A Character" (1830), lines 21–24. Tennyson wrote this about a rather Lowell-like character, Sunderland, who was known as the ablest debater in the Cambridge Union, and whom Tennyson regarded (with obvious disdain) as the type of the intellectual dilettante or self-pleased aesthete.

50. Lowell, *The Solar System*, pp. 58–59.

51. P. Lowell (1904) "Venuus 1903"; P. Lowell [15, 16].

52. P. Lowell to W.A. Cogshall, September 23, 1903; Lowell Observatory Archives.

53. Ibid., p. 94. Lowell worked this into a longish article, "Means, Methods and Mistakes in the Study of Planetary Evolution," the text of which is dated April 13, 1905. In addition to the acute vs. sensitive eye argument, he insisted (as he had since 1894) that "large telescopes are not always to be preferred for planetary work. On the contrary, small ones are not only sometimes but nine times out of ten more efficient, more powerful in planetary visual research … than large ones," and that at Lowell Observatory diaphragms of 18, 12, 9, and 6-inch diameter had greatly improved the seeing." These arguments were to be the cornerstone of his later arguments against observers equipped with large instruments who failed to see the canals as he saw them—notably, E.M. Antoniadi in 1909.

54. It would have been quite easy to have done so. The author's first published paper was on this subject. See: W. Sheehan [18]. The results were opposed to Lowell's conclusion.

55. According to A. Lawerence Lowell, *Biography of Percival Lowell*, p. 61, Derby claimed that Percival's eyesight was the keenest he had ever examined. The claim about seeing stars in the Pleiades is found in E.P. Martz, Jr. [19].

56. Jean Becquerel, quoted in Gratzer, *Undergrowth*, p. 23.

57. Lowell, *Occult Japan*, p. 98.

58. Morse, *Mars and Its Mystery*, pp. 160–165. Morse's—and indeed Lowell's—experience of seeing the canals during flashes of exquisite seeing (lasting a fraction of a second) comported with that of Morse's friend the Swampscott, Massachusetts electrical engineer Elihu Thomson, who had built a 10-inch refractor and used it to observe Mars before its July 1907 opposition. Thomson wrote to Morse, "You will be interested to know that for an hour last night between eleven and twelve the air was so steady here that my ten-inch telescope put on Mars, not only showed a great wealth of detail but also some of the canals, and at times, only at times, a network appearance of them. I have been

at it for 3 weeks on every fair night, but not until last night was there definite result, though at times the effect was as if there existed markings too evanescent to be made out. I despair of getting another such night in this climate. Perhaps the smoke, and general stirring up by bonfires and fireworks [on the Fourth of July], had temporarily worked the atmosphere to uniformity. At any rate, it did not last more than an hour.

"The proof that there is no illusion consists in the fact that only at the moments of very great steadiness could the detail be seen:—canals and all. If the effect were the result of optical illusion, it should have appeared on other nights and on last night it should have been present even when the disc was only fairly steady.

The reality is exactly the contrary. To see this finer detail demands the maximum of steadiness, and therefore the appearance when seen is that of the true markings of Mars." Elihu Thomson to E.S. Morse, July 5, 1907; Lowell Observatory Archives.

The fact that what the eye grasps in only brief moments no matter how chiseled and engraving-like the image during those moments may not represent the complete reality seems not to have occurred to Thomson.

References

1. Shapin S, Schaffer S (1985) Leviathan and the air-pump: Hobbes, Boyle, and the experimental life. Princeton University Press, Princeton, pp 25–26
2. Lowell P (1904) Venus 1903. Popul Astron 12:184–190
3. Jastrow J (1936) The study of human error: false leads in the stages of science. D. Appleton-Century, New York
4. Sheehan W, Lockwood GW (2020) Seeking an inconstant constant: the quest to discover the variability of the sun from William Herschel to Andrew Ellicott Douglass. J Astron Hist Heritage 23(1):63–88
5. Leonard WL (1907) Drawings of Mars. Popul Astron 15:387–388
6. Lowell P (1903) The solar system. Houghton, Mifflin and Co, Boston
7. Barnard EE (1904) Review of W.H. Pickering's the moon. Astrophys J 20:359–364
8. Maunder EW (1894) The canals of Mars. Knowledge (November 1):249–252
9. Sheehan W (1996) The planet Mars: a history of observation and discovery, Tucson, University of Arizona Press, p 125
10. Evans JE, Walter Maunder E (1903) Experiments as to the actuality of the 'canals' observed on Mars. Mon Not R Astron Soc 58:488–499
11. Antoniadi EM (1903) Report of the Mars section, 1900–1901. Memoirs Br Astron Assoc 11:85–142

12. Macaulay L (1968) The history of England. Washington Square Press, London/New York, pp 26–27
13. Lowell P (1908) Mars as the abode of life. New York, Macmillan, p 171
14. Asp E et al (2012) A neuropsychological test of belief and doubt: damage to ventromedial prefrontal cortex increases credulity for misleading advertising. Front Neurosci 6:100
15. Lowell P (1905) Brontes, a study in Martian 'canal' development. Popul Astron 13:1–8
16. Lowell P (1905) The Thoth and the Amenthes. Popul Astron 13:251–260
17. Lowell P (1905) On the kind of eye needed for the detection of planetary detail. Popul Astron 13:92–94
18. Sheehan W (1980) The eye and the astronomical observer. J Assoc Lunar Planetary Observers 28:150–154
19. Martz EP Jr (1938) William Henry Pickering 1858–1938—an appreciation. Popul Astron 46:299–310
20. Gratzer W (2000) The undergrowth of science: delusion, self-deception and human frailty. Oxford, Oxford University Press, p 1

9

Canali! Canali!

Contents

> *Bravo! I telegraphed you and bravo! I repeat. Your despatches cause our hair to stand on end and our voices to stick in our throats…*
> *—Percival Lowell to David Peck Todd, July 1907*

Canals Photographed?

Lowell and Morse had a splendid time, smoking cigars in the Baronial Mansion office, enjoying fine dining and wine, and hiking the Grand Canyon's Bright Angel Trail, exploring the ancient cliff dwellings at Walnut Canyon, and pondering from afar the Painted Desert. The latter's "lambent saffron" always reminded Lowell of the telescopic tints of the Martian globe and suggested to him that in both cases they attested to presence of "pitiless" deserts. Moreover, near the middle of Morse's stay, came an announcement that would startle the world. As reported in a small item on the front page of The New York *Times*:

© The Author(s), under exclusive license to Springer Nature Switzerland AG 2024
W. Sheehan, *Parallel Lives of Astronomers*, Springer Biographies,
https://doi.org/10.1007/978-3-031-68800-3_9

CAMBRIDGE, Mass., May 27.—A telegram was received at the Harvard Observatory to-night from Prof. Percival Lowell, Director of the Lowell Observatory at Flagstaff, Ariz., stating that the canals of Mars have been photographed there for the first time.

The photographer was Lowell's hard-working assistant C.O. Lampland, who had for some time been diligently and unobtrusively working to develop a new form of material technology (following Shapin and Schaeffer's scheme), the planetary camera, in an effort to obtain improved photographic images of Mars. At the time, the results of photography on the planet had been extremely meager. The first photograph to capture Mars's image, taken by Benjamin Apthorp Gould at the Cordoba Observatory in Argentina in 1879, shows only a featureless blur of light. The first images of any use were obtained by W. H. Pickering at Arequipa in 1890 and 1892, and show the polar caps and limb clouds. However, the detail of such images was far less than could be grasped by a visual observer, and were far below the threshold of resolution needed to capture fine details such as the canals.

The reason for this was, as Lowell had pointed out in his first published article about Mars in 1894, that when it comes to faint objects such as stars and nebulae the photographic plate excels and shows far more than the eye, but for planetary detail the eye is superior. "For the difficulty of photographing such detail," he wrote, "is not simply, as the inquirers suppose, a question of a driving clock timed first to the Earth's rotation and then to the planet's pace, which alone would require more perfect apparatus and a more complicated one than any yet devised. The deep difficulty lies in our own atmosphere, which is never steady enough; what is disclosed one minute being swamped by light waves the next. The attentive eye registers each glimpse, the photographic plate only the aggregate, and in the composite picture thus obtained the bad obliterates the good. With faint stars there is no such loss and every gain. For light is all that is wanted, however it be got. Now the eye cannot add impressions; the camera can. If therefore the camera be exposed long enough it will reproduce what the eye could never detect. But, in the case of detail, the longer the plate is exposed the more certainly will the detail be lost. Until, therefore, we rise above our atmosphere or find an absolutely faultless spot, we shall never be able for planetary work, to match the eye by any film, however sensitive or accurately driven."[1] Maunder himself had admitted that Lowell's arguments based on his (and others') drawings were bound to be to some extent subjective and open to interpretation. The only final arbiter of truth—objective renderings of the surface, as might be attained by photography—was unattainable; simply because, "Mars, unfortunately, does not lend itself to photography."

When Maunder said this, in 1903, no photograph of Mars ever taken showed more than a small smudged disk—the smudging due to the constant motion of the image as it rode on a turbulent ocean of air. In 1903, Lowell would have agreed that photographing details such as the canals was impossible. Nevertheless, he was encouraged by Lampland's experiments, which included the use of color filters and the insertion of a secondary lens tube to increase the focal length of the 24-inch Clark from 32 feet to 143 feet, with a corresponding enlargement of the image. Most important of all was the introduction of a sliding plate-holder. This would allow the observer to obtain a series of twenty or more tiny images in rapid succession on a single plate some 3 ½ inches square, simply by turning a ratchet and squeezing a shutter bulb. The expectation was that by scattering Mars images like bullets from a machine gun across the plate, Mars could be commanded to be "motionless, I beg you!," and at least a few of these exposures would occur during moments of the steadiest seeing and fine details revealed. The combination of very great ratio of aperture to focus (1:71.5) and insensitivity of the plates made for a very slow imaging system, and it took 2–4 s to register Mars at all. Thus, it was hardly surprising that, though the 1903 images showed the larger albedo features, they were, as Lowell put it, "incommunicable of canals."[2]

The 1903 results were encouraging enough for Lowell to believe success might be just around the corner. After the opposition he wrote to Lampland from Boston, "How gets on the photographing … of Mars for next opposition? We *must* secure canals to confound the skeptics."[3] Only thus could the doubters be finally and forever silenced. And in 1905 Lampland did—*mirabile dictu*—apparently achieve the impossible. Images of Mars on a Lampland plate taken on May 11, 1905, just 3 days after opposition, seemed to show a few of the broader canals. (Interestingly, Lowell had claimed in 1902 that "none of the canals had any measurable width," but since then had silently and without fanfare shifted his position and now alleged that "the canals … are by no means of a uniform width. Indeed, they are of all sizes, from lines it would seem impossible to miss to others it taxes attention to descry.")[4] Of these canals some thirty-eight in all would eventually be recognized on photographs, including even a broad double, Nilokeras, as Lowell later announced in his book *Mars and Its Canals*:

> Many pictures were taken on each plate one after the other, both to vary the exposure and to catch such good moments of seeing as might chance. Seven hundred images were thus got in all; the days of best definition alone being utilized. The eagerness with which the first plate was scanned as it emerged from its last bath may be imagined, and the joy when on it some of the canals could

certainly be seen. There were the old configurations of patches, the light areas and the dark, just as they looked through the telescope … and there more marvelous yet were the grosser of those lines that had so piqued curiosity, the canals of Mars… By Mr. Lampland's thought, assiduity and skill, the seemingly impossible had been done.[5]

The Clark dome in winter. This photograph, taken before Lowell arrived for the 1905 Mars campaign, was taken in what are typical winter conditions in 7000-foot elevation Flagstaff. (Credit: Lowell Observatory Archives)

Lowell telegraphed the result to Harvard, and by the end of summer 1905 the sensational news had been picked up and carried by newspapers, magazines, and journals in both the United States and Europe. Typical of the press reaction was a story by the science reporter for the Hearst newspapers, Garrett P. Serviss, who remarked the "surprising news" from Flagstaff and declaimed,

"The photographic demonstration—if it is one—that the canals exist is in itself a great step in advance sufficient to place Mars once more in the forefront of interest."[6]

Percival Lowell, studying photographic negatives on the porch of the Baronial Mansion, August 15, 1905. (Credit: Lowell Observatory Archives)

Percival Lowell, relaxing with a cigar on the porch of the Baronial Mansion, August 15, 1905. (Credit: Lowell Observatory Archives)

The photographic demonstration of the canals' existence would win Lampland the prestigious medal of the Royal Photographic Society in 1907. At the end of 1905, Lowell—worrying that Maunder's theories might be sapping the confidence of his English colleagues in the British Astronomical

Association and Royal Astronomical Society—departed on a brief visit to England. There he deposited a set of direct photographic prints made from Lampland's original negatives which would be studied by experts such as William H. Wesley, assistant secretary of the Royal Astronomical Society and a respected authority on astronomical photographs. The latter noted that the tiny images—no more than 1/8-inch across—on which appeared markings of the most delicate shades, "could not be satisfactorily reproduced … by any photomechanical process and I therefore give a drawing which I have made from them" [1] Wesley's drawing showed only details about which there could be no doubt, thus the "seas" and several of the broader "canals" which only with some poetic license could be describes as lines at all. As for the finer canals which Lowell had claimed to be "continuous lines and not a synthesis of other markings," Wesley wrote:

> Doubtless a photograph has no imagination, but imperfect definition applies equally to photographic and visual observations. That which is seen or photographed imperfectly as a smooth continuous line may be full of small irregularities, and may not be strictly continuous. It may, indeed, be a row of dots or discontinuous marks. We can only say that in the *main* these run in straight lines. Major Molesworth, an assiduous and careful observer of Mars, says that at moments of best definition the "canals appear to him more like streaks "made on very rough paper with a round-pointed crayon or stump, rather than an ink-line drawn with a pen."[7]

Barnard Again!

While Lowell was campaigning in England, the Observatory welcomed an unexpected visitor, who, as its greatest living practitioner, was even more an authority on astronomical photography than Wesley or anyone else. On the way back to Chicago from Mt. Wilson, Barnard stopped in Flagstaff, which lay along the rail route back East. Though Lowell was gone, V. M. Slipher and C. O. Lampland were on hand to show him around, and one or the other or both showed him the negatives on which the canals were supposed to have been recorded. What Barnard said to them is unknown, but we know his verdict from a letter he wrote to Hale: "I am perfectly sure you would have agreed that they did not show the canals as claimed."[8] Hale replied, "I would rather have your opinion … than [that] of anyone else," and added that he regretted that "the facts are not generally known, since everyone will now be convinced of the reality of the canals on the supposition that the photographs cannot lie."[9] As always reticent to engage in public controversy, Barnard did not publish anything about the famous Lowell Observatory photographs, but if he had he likely would have offered similar criticisms to those he had published in his review of Pickering's atlas of lunar photographs in 1904:

> After experimenting with the instrument, it was found best [by Pickering] to reduce the aperture to 6 inches... [The] extremely great ratio of aperture to focus—1:270—was necessarily very slow, besides interfering seriously with the separating power of the instrument. Even with the bright Moon the exposures in some cases were of two minutes' duration....
>
> The definition is seldom good in these photographs. Doubtless this is due to unsteadiness during the long exposures made necessary by the relatively small aperture. Though this seriously detracts from a close study of details, yet it does not materially affect their study as a chart and for the examination of the general features. ...
>
> In reference to the ice and snow [of which Pickering claimed to have found evidence on the Moon] ... he gives two reproductions from the same full Moon negative in which one is printed out until only the very brightest regions remain. By this method Professor Pickering endeavors to separate the "snow-covered regions from the ordinary bright, so-called volcanic regions. By this heavy printing there are left regions of extra brightness, while the generally bright regions have nearly all disappeared, leaving the snow areas alone visible. Is not this a trick of photography itself in which the very brightest regions are differentiated from the rest by excessive over-printing? Might not this as readily be carried farther and still brighter regions alone be left, and, if so, what would they be called? How these residual bright regions thus revealed can be called "snow" does not yet seem quite clear. It would appear to be simply a survival of the brightest.[10]

On returning to Yerkes, Barnard remounted the Bruce telescope in its small dome between the Main Building and his house on Lake Geneva. After his Mt. Wilson idyll, he had returned just in time for another miserable Midwestern winter, of which he wrote in American vernacular to Wesley in April 1906. "This has been an awful bad winter for observing, no clear weather at all hardly."[11] The oppression of that winter was lightened somewhat, however, by the arrival of his niece, Mary Rhoda Calvert (Ebenezer's oldest daughter), who came from Nashville to join the Barnard household. A self-effacing, highly intelligent woman, she would devote herself entirely to helping Barnard and Rhoda with their household affairs, and later became Barnard's personal assistant, helping him in the office with correspondence and computations. She would remain at Yerkes as chief computer and photographic technician for decades after Barnard's death. Her greatest accomplishment would be to see through to completion the needed revisions of Barnard's monumental *Atlas of the Selected Regions of the Milky Way*, left unfinished by the great perfectionist at his death in 1923. (It appeared only in 1927.)

As spring finally arrived in southern Wisconsin, Barnard returned to his routine work with the 40-inch refractor, obtaining a series of visual observations of Phoebe, the ninth satellite of Saturn. Discovered photographically by W.H. Pickering in 1898, the ninth satellite was extremely faint, and for Barnard to be able to place the wire of the micrometer on it at all was a real tour-de-force. Barnard also continued his long series of micrometric measures of the fifth satellite of Jupiter and of the sixth satellite, discovered photographically by Charles Dillon Perrine at Lick in 1904. Most importantly, he began to sort out his Milky Way plates, selecting the best ones for future publication, and scrutinizing them closely for hints as to the true interpretation of the mysterious dark markings, especially those in Ophiuchus and Scorpio, that he regarded as belonging to "the most puzzling region that I know of in the sky." Here was found a nearly starless chasm beginning near Theta Ophiuchi, which turned west in shattered form, then strengthened into a definite lane extending for another 15 degrees and finally connected with the "remarkable vacancy in the dense sheeting of small stars" near Rho Ophiuchi. These features were more complicated than the apparent "holes" he thought he could discern so clearly elsewhere in the Milky Way; they contained "vacancies within vacancies." He wrote in an article published in *Popular Astronomy* in 1906:

> The blending of this great nebula into the surrounding regions, where it seems to mingle with the material of the vacancies, makes it hard to tell where the nebula leaves off…. There is a slight suspicion that certain outlying whirls of nebulosity have become dark and that they are the cause of the obliteration of the small stars near [2, p. 581].

Was it possible for a bright nebula to die out and become dark and obscure? Barnard seems to have been drawn for a moment to the possibility, only—as usual exercising his extreme caution in not overinterpreting his observations—to pull back from the prospect. "I think this is fanciful however," he wrote,

> For the irregular vacancy in which it lies connects readily with the vacant lane running east to the region of [Theta] Ophiuchi. No one would suspect for a moment that this lane is anything but an actual vacancy among the stars.[12]

At this point, Barnard began to worry that the vacant lanes of the Milky Way might be like the Martian canals, entirely subjective, due to scarcity of stars alone rather than "channels in a bed-work of nebulous substratum." If all the stars were removed, he asked, would the lanes still exist? The only way was to investigate further with the Bruce. During the frigid month of January 1907, he trained the telescope on the sky north and east of the Pleiades, where earlier photographs with the Willard lens had caught dim suggestions of dark lanes extending far to the east. He exposed plates to this region for five and a half hours, and when the plates emerged from the developing tray, the lanes showed up unmistakably. They were not only devoid of stars but darker than the surrounding sky, and satisfied Barnard that they would still be there even if the stars were all removed. In addition to the dark lanes, the plates showed a large nebula, apparently in a hole almost devoid of stars, from which one of the lanes straggled away to the southeast for several degrees. He described these remarkable features in a paper completed on March 9, 1907, "On a Nebulous Groundwork in the Constellation Taurus," one of a series of probing papers on the dark markings in the Milky Way appearing in the prestigious *Astrophysical Journal*. The final line of the passage quoted here appears most un-Lowellian:

> The pictures seem to show that the brighter part of this nebula is only a small portion of it, and that the nebula is feebly luminous over most of the vacancy…. The feebler portions of the nebula would almost suggest the idea that a large nebula exists here, but the major portion of it is dead or non-luminous, and that it actually causes the apparent vacancy by cutting out the light from the stars, while the few stars visible are perhaps on this side of the nebula. I give this simply as what the picture would suggest to one, and not as what may really be the truth. [3, p. 220]

When he wrote these words, Mars already was brightening in the night sky, and would soon be under closer scrutiny by the "Lowell Expedition to the Andes" than at any time before or since. Lowell's views were carrying everything before them. Meanwhile, it had been years since Barnard pronounced on the subject. His was, however, comparatively, a "still small voice." No one had ever seen the planet as well as he had with the Lick 36-inch refractor in the late summer of 1894, and yet…

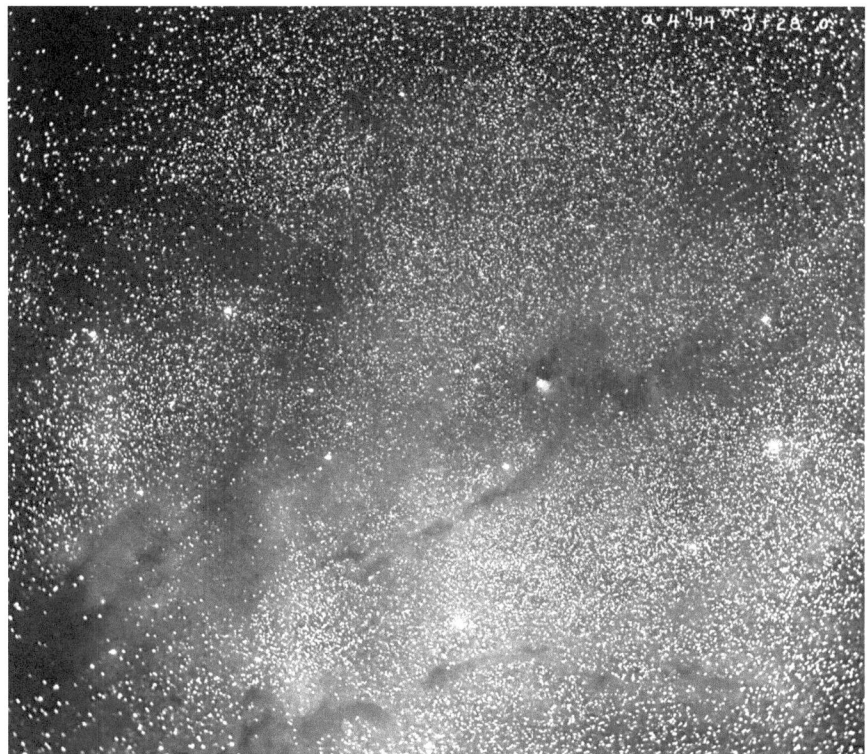

One of the most important plates Barnard ever took of the Milky Way, and one of only a few included in the Atlas of the Selected Regions of the Milky Way *(Plate 5) to have been taken at Yerkes Observatory. This plate, with an exposure of 5 h 29 min, was taken on January 9, 1907, and is centered on R.A. 4 h 16 min and declination 27° 57'*

Canali, Canali!

In June 1906, Lowell finished work on his *magnum opus* on the red planet, *Mars and Its Canals*, of which most of what was new had been obtained during the 1903 campaign, though supplemented somewhat by additional work along similar lines from 1905. As the book worked its way through the press toward publication in December 1906, Lowell—evidently regaining the erstwhile manic level of productivity he had known in the days before his breakdown—gave eight Lowell Institute lectures at MIT in October and November. The first was more thronged than any lecture in the long history of the Institute had ever been. According to Leonard, "Standing room was nil, and demands for admission were so numerous and insistent that repetitions were arranged for the evenings. At these repeated lectures the streets near by were filled with motors and carriages as if it were grand opera night."[13] These lectures later appeared in book form as *Mars as the Abode of Life*.

He also lectured at Sayles Hall at Brown University in the first week of January 1907. There, astronomer Winslow Upton introduced the distinguished speaker to an aspiring young writer, Howard Phillips Lovecraft. Lovecraft would never forget it. He stood in awe of this figure he regarded as "an astronomical giant," and yet, as he long afterward admitted to his correspondent Rheinhart Kleiner:

> Now here is the amusing part—I never had, have not, and never will have the slightest belief in Lowell's speculations; and when I met him I had just been attacking his theories in my astronomical articles with my characteristically merciless language. With the egotism of my 17 years, I feared that Lowell had read what I had written! I tried to be as noncommittal as possible in speaking, and fortunately discovered that the eminent observer was more disposed to ask me about my telescope, studies, etc., than to discuss Mars. Prof. Upton soon led him away to the platform, and I congratulated myself that a disaster had been averted![14]

At the time he met Lovecraft, Lowell was already meditating plans for the coming opposition of July 1907—the first really favorable one of the twentieth century, when the planet would approach within only 38 million miles of the Earth (62 million kilometers), or closer than it had been since 1892. At first he was planning only for Lampland to push forward the photography of Mars from Flagstaff, which he had thus far so ably advanced in 1905. Thus, in September 1906 he announced that he was planning to forego observations with the 24-inch Clark for several months to have the lens refigured, because "for the very best photographic work such as Mr. Lampland and I have to do next year on Mars, a refiguring of the glass is advisable."[15] Lowell and Lampland did indeed take the lens to the Clarks' Cambridgeport, Massachusetts, workshop. While there, they attended a reception for Commodore Robert E. Peary, a somewhat Lowell-like figure who had once written to his mother, "I *must* have fame," and who had just returned from a much publicized but unsuccessful attempt to reach the North Pole.[16] In addition to having the Clark refigured, Lowell was exploring the possibility of building a large reflector that could compete with the 60-inch reflector George Willis Ritchey was completing for the Mount Wilson Observatory. In discussions with Ritchey himself, Lowell explored the possibility of spending $55,000 (worth almost $2 million today) to set up his own 84-inch reflector at Flagstaff, claiming that with such an instrument "in the air at Flagstaff we ought to get photographic conditions which for many years will be unapproached."[17] Nothing came of this particular idea, though Lowell continued to reflect on the reflector for several more years. In the meantime, a different proposal soon came to the fore in his fertile and active mind.

The 1907 opposition would occur with Mars some 28 degrees south of the Celestial Equator. This would make conditions difficult from Northern Hemisphere observing stations (including, of course, Yerkes) as the planet would never attain very great altitude in the sky and so would have to be viewed along a considerable path-length of the Earth's atmosphere. As with Pickering and Douglass's expedition to Peru in 1892, the best conditions would be found in the Southern Hemisphere. Realizing this, Lowell began to give serious consideration to a bold proposal from his old crony David Peck Todd, now professor of astronomy at Amherst College, who had just acquired an 18-inch refractor for the college observatory. Todd agreed to disassemble it, crate it up, and ship it to South America overland via Panama (whose canal would not open until 1914) for a campaign of Mars observations, provided Lowell agreed to cover expenses. Lowell was immediately captivated, and so the "Lowell Expedition to the Andes" was born. Such was the celebrity of the undertaking that no more address than "Lowell Expedition to the Andes" was needed to assure mail delivery to the astronomers en route.[18]

In addition to Todd and wife Mabel Loomis Todd, the team included recent Lawrence fellow and latest staff member Earl Carl Slipher, Vesto's 24-year-old brother. Equipped with a duplicate of Lampland's planetary camera and carefully coached by Lampland in its use, E.C. would serve as the expedition's photographer (and launch a more than five-decade long career as one of the world's foremost planetary photographers). Full of buoyant hopes, Todd gave an interview in the Boston *Evening Transcript* on the eve of the expedition's departure in which he declared: "We are going to South America in hope of getting the facts, and to obtain information on the question of the Martian canals…. Is Mars inhabited? This is a question my wife has often asked me, but do you know I've never been able to answer it? Maybe when I come back…."[19] Meanwhile, on May 6, 1907, Lowell telegraphed a last message to Todd, "Don't forget tooth brush and telescope may be useful," as well as a few lines of doggerel:

> All Hail!
> To you about to sail—
> May you avail
> In our Martian grail
> And then to me the matter mail.[20]

The seven-ton telescope was shipped from New York via Panama to the Chilean port city of Iquique; from thence it traveled by train 65 km further inland to the nitrate-mining center of Alianza, located 1200 m above sea level, where it was set up in open-aired, domeless splendor under burnished, rainless skies. Todd evoked the otherworldly bleakness of the scene:

> The region is an utter desert; the moon itself could not reveal greater barrenness—not a tree or a flower or a blade of grass for miles, not even moss or lichens…. The clustered dwellings of the workers in the nitrate fields … gave an almost populous effect to the barren landscape … Around the whole settlement stretched the solemn, brown, impressive pampa, undulating to the great mountain border, the Andes, its peaks here and there snow-capped, lofty, and magnificent. Here … [is] one of the best astronomical stations ever occupied.[21]

Rain at the site was (and is) virtually unknown, and distant thunder heard several times a day proved not to be thunder at all but the rumble of earthquakes in the mountains. Despite occasional fog and haze, most of the time the air was sublimely still and clear. Under these conditions, with the planet high overhead, the image in the eyepiece was ravishing. According to Todd:

> Every observer, whether professional or [amateur], was amazed at the wealth of detailed markings that the great reddish disk exhibited. Its clear-cut lines and areas were positively startling in their certainty: the splendor of the first visual glimpse in steady air can never be forgotten…. Nearly everybody who went to the eyepiece saw canals; and once I fancied I heard even the bats, as they winged their flight down the pampa, crying, "Canali, canali, canali!"[22]

The expedition members included David and Mabel Todd; Slipher; A.G. Ilse, an engineer from Alvan Clark and Sons; and Todd's student R.G. Eaglesfield. Though all the observers tried their hand at sketching the planet, Slipher's sketches were especially skilful and full of Lowellian-type details. But it was the photographs that would make the greatest impression. Though Todd was the nominal leader of the expedition, and he and Mabel did their best to take as much of the credit as they could in the self-promoting reports they submitted for publication in the press, the greatest contribution was made by young Slipher.

The "Lowell Expedition to the Andes," shown at their observing site near Alianza, Chile. From left, a servant; Mabel Todd; A.G. Ilse, an engineer from the Alvan Clark and Sons firm; Todd; R.G. Eaglesfield; and E.C. Slipher. (Credit: David Peck Todd Papers [MS 496B], Yale University Library)

Indeed, from 1907, thanks mainly to his results, historian Maria Lane has said, "photography supplanted cartography … as the proper standard of proof for Mars representations. The build-up of expectation regarding the Lowell photographs focused on their purely objective quality and their ability to resolve long-standing disputes among astronomers over the existence of the canals."[23] Slipher obtained 13,000 tiny images of the planet on his plates, and from the first, saw—or thought that he saw—canals on the best of the prints. Todd did not see all this detail but was nevertheless willing to accept Slipher's word that it was there, and excitedly reported to Lowell that Slipher had even managed to record a number of the more prominent doubles. Lowell replied: "The world, to judge from the English and American papers, is on the *qui vive* about the expedition, as well as about Mars. They send me cables at their own extravagant expense and mention vague but huge (or they won't get 'em) sums for exclusive magazine publication of the photographs."[24] As yet, Lowell's hat had been thrown over the fence well ahead of his having a chance to see any of Slipher's prints. As soon as the choicest of them gradually began to trickle back, he found them "beyond expectation fine," and enthused, "The canal stock … has already risen in consequence."[25] Lowell possibly hoped that they would deliver a crushing blow to the canal critics, but if so, his hopes were unfulfilled. Slipher's 1907 images fell short just as—and for the same reasons—Lampland's of 1905 had done. The original images were of course tiny, only slightly larger than the 1905 images at 3/16 inch across, and inevitably, those who examined them (the eyewitnesses, as it were) differed as to what was shown in them, as before. Those who believed in canals saw canals; those who did not saw randomly aligned grains of silver in the photographic emulsion (which are on the order of 0.05 microns across). It is also pertinent to note here the strange fact—not fully realized at the time—that both silver halide photography and the human eye tend to record uncannily similar spurious features.[26]

However, the rumor of them proved more potent than the reality, and triggered a bidding war for publication rights won by the *Century*—not least because they agreed to publish, in addition to Lowell's photographs, his series of eight lectures on "planetology" given before the Lowell Institute the previous autumn (which would become the book *Mars as the Abode of Life*); *McClure's* and *Harper's* had declined. Lowell admitted that "to reproduce them *tel quel* would be beyond present processes to accomplish short of great loss of detail," and proposed to his friend George Russell Agassiz that the latter might retouch them for the purpose—the latter was "the best man to do it, if you will. Will you?"[27] However, the associate editor at the *Century* vetoed the idea on the grounds "it would certainly spoil the autographic value of the photographs themselves. There would always be somebody to say that the results were from the brain of the retoucher."[28] Which was, of course, perfectly true. Another suggestion—to have direct prints from Slipher's plates pasted into each copy of the magazine's December issue—was briefly considered; however, in the end, this was rejected as impracticable.

The use of an enlarging camera (or examination of the images with a microscope) only served to magnify the separate grains of silver scattered here and there over the surface, and if pushed too far, caused the picture itself to be lost altogether (as the woefully near sighted Schiaparelli, to whom Lowell sent some of the prints, discovered first hand when he attempted to examine Slipher's images closely with a small microscope).[29] A more definitive solution was to use a larger telescope than had been used in 1905 at Flagstaff and in 1907 in Chile (for instance, the Yerkes refractor), so that larger negatives could be obtained.

Barnard made some effort to study Mars during the 1907 apparition. However, because of the far southerly declination, he was unable to get satisfactory results, either visually or photographically. His observing books include entries such as: "The outlines of the 'seas' are strong and well defined. The air is like running water in front of the planet." And: "Have [tried a diaphragm]. Have taken [it] off… Full aperture now—though the image is not so well defined, I can once in a while really see it better than with 15 inches…. The planet is very low and the air is moving across it in waves."[30]

His efforts with photography were better, and marked a new era for an astronomer whose visual observations of the planets had been almost legendary but who was in the process of largely relegating visual work to the dustbin in favour of photography. At present, the visual observer still had the upper hand, for as Barnard wrote, "Better conditions are required for successful work in this direction than for visual observations. One can do much visually under conditions where the best definition is only momentary; but for these … photographs any break in the definition for even a single second during the exposures means injury or total ruin to the image" [6, p. 471].

Barnard's experiments in 1907 involved taking advantage of the long focal length (65 feet) of the 40-inch refractor, combined with using a Brashear enlarging lens at the focus to magnify the image about two-and-a-half times (making an equivalent focal length of 160 feet). The exposures, of three or four seconds, were made with a yellow colored filter made by Yerkes staff photographer Robert James Wallace.[31] The plates used were 3 × 1 ¾-inch Cramer Instantaneous *Iso* plates (the fastest available at the time). Despite the fact that the exposures were of only a few seconds' duration, Barnard found it necessary to guide the telescope to keep the image stationary, which he did by bisecting the polar cap by spider-three cross wires in the focus of the long guiding finder (61½-foot focus) of the 40-inch. The main dark areas were clearly registered, but the images—despite being larger and clearer than the ones obtained by Lampland in 1905 and Slipher in 1907—showed no canals.

The results were never published, yet they demonstrated what was possible. Thus the stage was set for better results at the September 1909 opposition when the canal debate would finally come to a head.

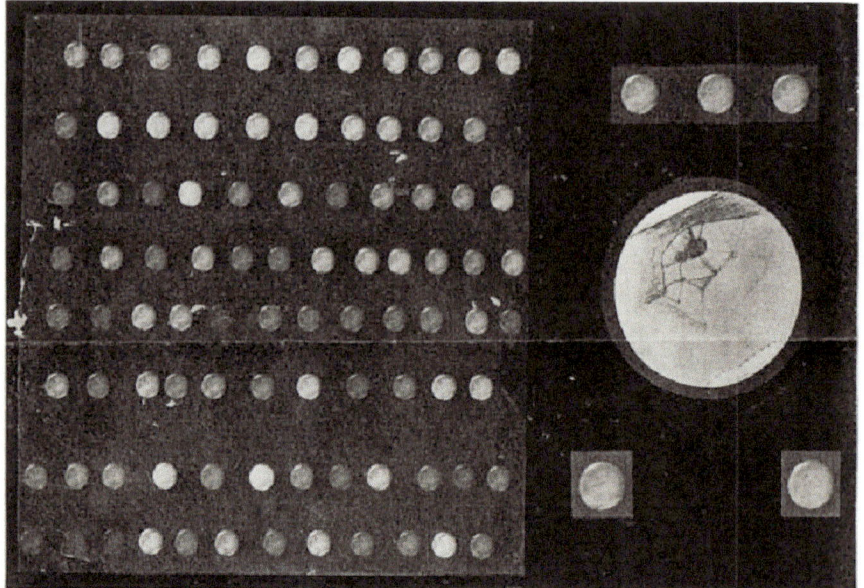

No. 4. SHOWING VARIETY OF INTEN-
SITY OF IMAGE. THE DIFFERENCE IS
DUE TO VARIATION IN TIME OF EX-
POSURE

REGION OF THE SOLIS LACUS. LONGI-
TUDE OF THE CENTER OF THE PHOTO-
GRAPH, 90°. ENLARGEMENTS, AND PRO-
FESSOR LOWELL'S DRAWING

Photographs of Mars taken by E.C. Slipher in Chile, reproduced at the original scale and published to illustrate Percival Lowell's December 1907 article, "New Photographs of Mars: taken by the Astronomical Expedition to the Andes and now first published." Century Magazine vol. 75, pp. 303–311:p. 308. Though Lowell published enlargements as well as one of his drawings at right, he warned that the process of enlargement was of minimal use, as it also enlarged the grains of the photograph and "must not be overdone"

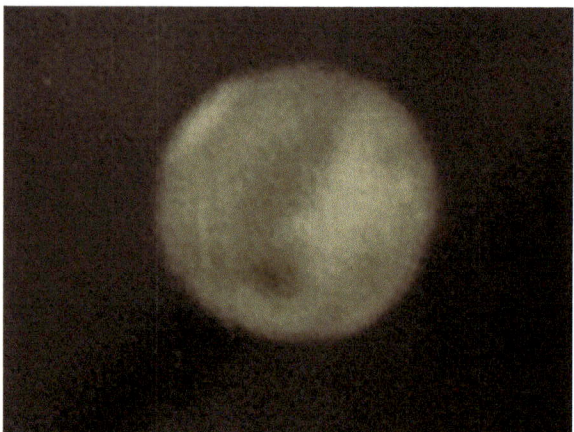

One of Barnard's images of Mars with the 40-inch refractor, July 1907. This image has been reproduced at the original scale, very small at 5/16ths of an inch but still larger than the 3/16-inch images obtained by the Lowell Expedition to the Andes. (Credit: William Sheehan)

Notes

1. P. Lowell (1894a), "Mars," pp. 549–550.
2. P. Lowell, *Mars and Its Canals*, p. 275.
3. P. Lowell to C.O. Lampland, May 16, 1904; Lowell Observatory Archives.
4. The first quote is from Lowell, *The Solar System*, p. 56; the second from Lowell, *Mars as the Abode of Life*, p. 149.
5. Lowell, *Mars and Its Canals*, p. 277.
6. Hoyt, *Lowell and Mars*, p. 182.
7. Ibid.
8. E.E. Barnard to G.E. Hale, October 26, 1905. Hale. George Ellery. Papers, Mount Wilson and Palomar Observatories Library.
9. G.E. Hale to E.E. Barnard, November 9, 1905. Hale. George Ellery. Papers, Mount Wilson and Palomar Observatories Library.
10. E.E. Barnard, "Review of W.H. Pickering's *The Moon*," pp. 359–360. Barnard notes (p. 360) that Pickering's chapter on vegetation and the lunar canals indirectly proves that the canals on Mars cannot be the work of intelligent life, quoting a passage where Pickering says: "Sixthly, since no water in liquid form can exist on the Moon, this fact will enable us to rule out many seductive but erroneous hypotheses. Seventh, and lastly, we know that so little air and water vapor exist there that we can confidently also rule out all aid in the construction of these formations [the "canals"] from intelligent or intellectual life .... They merely weaken the strongest argument hitherto found for the existence of highly intelligent life on Mars."
11. E.E. Barnard to W.H. Wesley, April 4, 1906; Royal Astronomical Society Library.
12. Barnard, "Vacant Regions," p. 581.
13. Leonard, *Percival Lowell: An Afterglow*, p. 25.
14. H.P. Lovecraft to Rheinhart Kleiner, February 19, 1916; in: H.P. Lovecraft (1965), *Selected Letters I: 1911–1924*, (ed.) August Derleth and Donald Wandrei (Sauk City, Wisconsin: Arkham House).
15. P. Lowell to V.M. Slipher, September 11, 1906; Lowell Observatory Archives.
16. This was Peary's 1905–06 expedition, in which he claimed to have discovered a far northern island he called "Crocker Land" (which does not exist) and to have set a Farthest North world record at 87° 06′ on April 21, 1906 (doubtful; the typescript stopped the day before when Peary was no farther than 86° 30′, and the original of the April 1906 record is the only missing diary of Peary's exploration career). He and another explorer, Frederick A. Cook, both claimed to have reached the North Pole in 1909. Peary had friends in high places and his claim was credited by a committee of the National Geographic Society and the Naval Affairs Subcommittee of the U.S. House of Representatives—though a reassessment of Peary's notebook in 1988 by polar explorer Wally Herbert found it

"lacking in essential data." Herbert concluded that Peary did not reach the pole, though he may have come within 60 miles of it. See: Wally Herbert [4]. Cook's claims have also been refuted. "He possessed an outsize imagination and story-teller's gift," writes Allegra Rosenberg. "When he had to turn back short of the pole, he didn't succumb to the cold and go down in history as a noble failure like so many others. Instead, he spent the winter in an ice cave with his Inuit travel-ing companions and spun a story about his own success so inspiring that eventu-ally he seemed to have believed it himself. (Cook insisted he had reached the pole until his death in 1940.)" "The Polar Explorer, and Scammer, Who Should be an American Hero." New York *Times*, December 25, 2023. There are parallels in the lives of both these men—and their gigantic if ultimately unsubstantiated claims—to Lowell's.

17. P. Lowell to G.W. Ritchey, undated; ca. March 1906. The French astronomer Audouin Dollfus once told me that at about this same time Lowell was thinking of acquiring Pic du Midi, which later became legendary as a site for planetary observations. Audouin Dollfus to William Sheehan; personal communication, August 1992.

18. E.C. Slipher wrote to his brother V.M. that "Lowell expedition to Andes" was a sufficient address for mail. "We advertise so thoroughly that this … would find us at any time I think." E.C. Slipher to V.M. Slipher, May 22, 1907; Lowell Observatory Archives.

19. See William Sheehan and Anthony Misch (November 2007) "The Great Mars Chase of 1907," *Sky and Telescope*, pp. 20–24.

20. P. Lowell to D.P. Todd, May 6, 1907; Lowell Observatory Archives.

21. Sheehan and Misch, "Mars Chase," p. 22. Alianza has long since been aban-doned. According to David Baron, who visited the Atacama Desert during an eclipse expedition to Chile in 2019 and located what was left of the tennis court next to the mine managers' building, where the astronomers had set up their telescope "virtually nothing is left of Alianza—anything of value (bricks, lum-ber, metal) was removed when the mining ceased decades ago. The place is ugly, desolate, hot, and desperately arid. The mining was the only reason anyone would think to live there. As you'll recall, the miners were after nitrates, which were found in a layer of caliche some several feet underground. To get at it, the men set dynamite and blasted the earth over many square miles. What is left is a landscape that looks like it had been the scene of trench warfare." David Baron to William Sheehan, personal communication, April 26, 2024.

22. Sheehan and Misch, "Mars Chase," p. 22.

23. Lane, *Geographies of Mars*, p. 56.

24. P. Lowell to D.P. Todd, July 26, 1907; Lowell Observatory Archives.

25. P. Lowell to D.P. Todd, September 6, 1907. Lowell Observatory Archives.

26. See: Thomas A. Dobbins (May 2015) "Saturn's Elusive D Ring" *Sky & Telescope*, pp. 214–215.

27. P. Lowell to G.R. Agassiz, September 16, 1907; Lowell Observatory Archives. There was a precedent, as in 1905, some images retouched by Wesley had been published by the Royal Astronomical Society.

28. R.U. Johnson to P. Lowell, October 8, 1907; Lowell Observatory Archives. As a footnote to this story, Lowell became involved in a dispute with D.P. Todd about the latter's plans to publish an account of the expedition, along with photographs, in the *Cosmopolitan*, which Lowell agreed to only provided Todd's article not appear until after his in the *Century* had done so, and provided that Todd make whatever corrections insisted upon by Lowell. For a time Todd failed to do so, and Lowell went so far as to threaten legal action end. In the end, the corrections were made, and Todd's article appeared, with some of the as usual tiny images reproduced from Slipher's plates. See: David Todd [5].

29. G.V. Schiaparelli to P. Lowell, September 2, 1909. Lowell Observatory Archives.

30. E.E. Barnard, observing log book; Barnard. Edward Emerson. Papers, University of Chicago Library, Hanna Holborn Gray Special Collections Research Center.

31. When Lowell applied to Wallace to make a filter for use in the Martian canal photography, Wallace wrote to Hale: "'for a consideration' I could put them [the 'canals'] in the screen." R.J. Wallace to G.E. Hale, Jan. 30, 1904. Hale. George Ellery. Papers, Mount Wilson and Palomar Observatory Library.

References

1. Wesley WH (1905) Photographs of mars. Observatory 28:314
2. Barnard EE (1906) On the vacant regions of the sky. Pop Astron 14:579–583
3. Barnard EE (1907) On a nebulous groundwork in the constellation Taurus. Astrophys J 25:218–225
4. Herbert W (1989) The noose of laurels: Robert E. Peary and the race to the north pole. Athenaeum, New York
5. Todd D (1908) Professor Todd's own story of the mars expedition. Cosmopolitan 44(4):343–351
6. Barnard EE (1911) Photographs of the planet Mars, etc. Mon Not R Astron Soc 71(5):471–472

10

Wisps, Tores, and Good and Bad Reviews

Contents

[P]robably most readers of this book will take it up with a strong disbelief in the sensational theory of Mars, which has now for some years past been associated with Mr. Lowell's name, [b]ut a perusal of its pages … should at least convince them that Mr. Lowell is no mere faddist, but a scientific observer whose researches are entitled to the greatest respect…
—Review of "*Mars and Its Canals,*" *Journal of the British Astronomical Association*

© The Author(s), under exclusive license to Springer Nature Switzerland AG 2024   **449**
W. Sheehan, *Parallel Lives of Astronomers*, Springer Biographies,
https://doi.org/10.1007/978-3-031-68800-3_10

Illustration by William R. Leigh. From H.G. Wells, "The Things that Live on Mars," The Cosmopolitan magazine, *March 1908. (Credit: Public Domain)*

News from Jupiter and Saturn

Mars was the big story in astronomy in 1907. In rounding up the significant events of the year, an anonymous writer in the *Wall Street Journal* asked, "What has been in your opinion the most extraordinary event of the twelve months? Not the financial panic which is occupying our minds to the exclusion of most other thoughts" but "the proof afforded by astronomical observations … that conscious, intelligent human life exists upon the planet Mars." The writer continued:

> The proof is indeed circumstantial … of the same kind and of the same strength as might convict a criminal accused of murder in cases where there had been no actual witnesses to the deed.… There could be no more wonderful achievement than this, to establish the fact of life upon another planet… That a more complete knowledge of the planet Mars may possibly have a profound effect upon life on our own globe goes without question.[1]

Hardly noticed was another contest that overlapped with the controversies over the photography of canals, which in the end would pit Lowell and Barnard against one another. As part of his developing scheme of "planetology"—an attempt to give Laplacian/Spencerian cosmology a more quantitative shape—Lowell began to pay at least some attention to the giant planets, which he supposed were worlds in the molten stage of planetary evolution and still glowing by their own internal heat. In March 1907, he was contacted by a British amateur astronomer, Scriven Bolton, of Hyde Park, Leeds, who had recorded a series of "wisps or lacings across the bright equatorial belt" of Jupiter, and asked Lowell to try to confirm them.[2] Lowell gladly took up Bolton's request, and within a month had done as he could always be counted upon to do and had confirmed Bolton's class of spidery markings. They were as delicate, in their way, as the canals of Mars—but at least in this case, they were completely non-controversial, since such markings had been seen not only by Bolton but by other observers before and since. Of their character, Lowell confidently affirmed, "The wisps are not wisps of cloud, since they are dark, not light, but gaps strung out in the clouds themselves" [1]

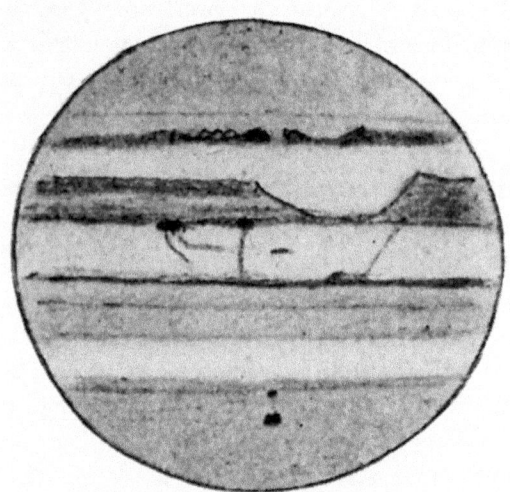

DRAWING OF JUPITER BY DR.
LOWELL.   APRIL 12, 1907.

Above, drawing by the Leeds amateur astronomer and illustrator Scriven Bolton,
February 21, 1907, using a 10-inch reflector, showing what he called "wisps" in the
bright equatorial zone of the planet. (Note that the drawing shows the shadow of
satellite II partly covered by Io, whose own shadow follows it.) (Credit: John A. Rogers,
British Astronomical Association. Below: One of Lowell's sketches, confirming the exis-
tence of the wisps. From: P. Lowell, Evolution of Worlds *(1909))*

It may have been Bolton's impetus that led Lowell to take a peek into what was going on at Saturn as well, given that in 1907–08, the Earth and the Sun were due to pass through the ring-plane for the first time since October 1891. The significant dates were:

1907
April 17. Disappearance Earth in plane of rings
July 26. Reappearance Sun in plane of rings
October 4. Disappearance Earth in plane of rings
1908
January 7. Reappearance Earth in plane of rings

The April passage through the ring-plane occurred with Saturn too close to the Sun to be studied. From then until July, the "dark" side of the rings was on view from the Earth, and the shadow of the rings appeared as a dark band along the planet's equator. During this interval, Lowell made only a single observation, on June 19, 1907, but it was enough to reveal what he called "a peculiar feature … for the first time." The shadow of the rings appeared "dusky rather than dark, [and] presented, when first looked at, a triple appearance. On more careful scrutiny this proved to be due to a narrow black line that threaded it medially throughout its length" [2, p. 186].

Rather surprisingly, nothing was done to follow up on the strange phenomenon, but "other work [occupied] the observers," as Lowell explained. The "Lowell Expedition to the Andes" was underway, and in addition, Lowell and his staff were kept busy following up asteroids being discovered *en masse* during the photographic search for "Planet X," which had been proceeding quite secretly since the summer of 1905 with a photographic survey of the sky along the ecliptic carried out under Indiana University Lawrence fellows John C. Duncan, E.C. Slipher (before he was assigned to the Andes expedition) and Kenneth P. Williams, who on discontinuation of the program in September 1907 returned to Indiana where he taught celestial mechanics.[3] Of the Lawrence fellows, only one, Slipher, was hired for the permanent Lowell staff. Also at the end of June, Lowell made a quick trip back East to receive his first honorary LL.D. degree, from Amherst College (henceforth he would have the dignity of being "Dr. Lowell").[4]

Percival Lowell in academic regalia, on the occasion of receiving the LL.D. degree from Amherst College in June 1907. (Credit: Popular Astronomy)

While Lowell was rather madly gadding about and focused on supervising and publicizing the attempts to photograph the canals from South America, Barnard kept a steady eye on Saturn. The Midwestern winter just past was unusually bad, and in addition, conditions remained poor during the spring and early summer. Barnard was not finally able to observe Saturn until July 2, when he first caught a view of the "dark" side of the rings with the 40-inch. He jotted in his observing log book: "[T]he entire surface of the ring was easily seen, though the Sun was not then shining on its visible surface. Where it was projected against the sky, the ring appeared as a greyish hazy or nebulous strip." In addition, he noticed two nebulous pale-grey "condensations" on each side of the planet [3, pp. 346–347]. Barnard immediately deduced that what he was seeing was not the actual sunlit edge of the ring but the oblique surface shining by sunlight "percolating" through the particles making it up. However, the condensations remained perplexing.

Meanwhile, Lowell had worked out something quite extraordinary and unexpected about the rings. On the one occasion he looked in June, he noticed no irregularities on the thread-like ring (though admittedly, he said, "it was not as critically scanned perhaps as it might have been"). However, in November, he noticed "agglomerations," symmetrically placed along the ring on either side of the globe. They were clearly the same as the "condensations" Barnard saw in July.

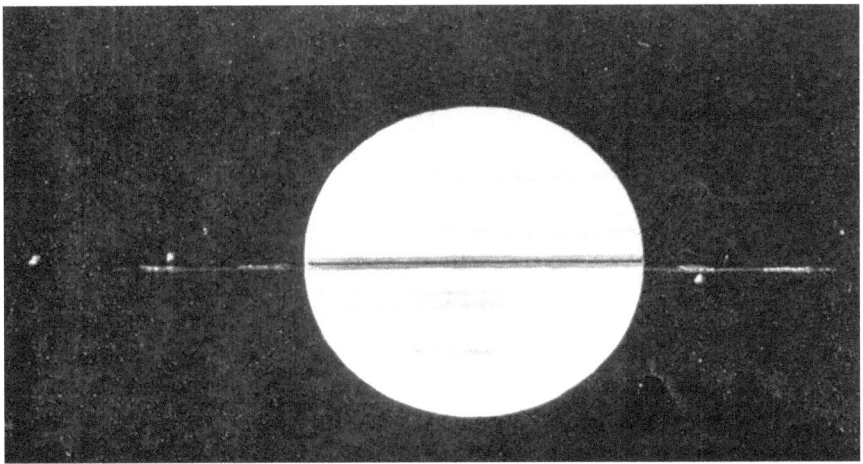

Lowell's finished drawing, based on his observations of November 1907, showing the dark core in the median of the shadow of the ring on the ball, seen only at Flagstaff, and the brightenings along the thread of the edgewise ring which he interpreted as "tores." (From: Lowell Observatory Archives)

On the basis of this sparse observational evidence, Lowell proposed a rather grandiose theory: Rings rings B and C were not flat rings at all but instead contained "tores," using a term derived from botany for the raised receptacle of the whorls of a flower. The black core of the ring's shadow was the shadow of the flat ring A bordered by particles of B and C rising as "agglomerations" above and below its plane. As usual, instead of publishing only in scientific journals, he announced his results to the *New York Times*. The agglomerations, he said, consisted of debris formed from collisions among the billions of orbiting particles making up the rings. From this it could be inferred that, because such collisions must involve a loss of orbital momentum among the colliding particles, "the ring system is in the process of falling in upon the planet."[5] The headline that followed read, "Rings of Saturn are Falling In," and was pounced upon as quite preposterous by astronomers including Simon Newcomb, who pointed out that the rings had been shown by a number of experts, including the great Scottish physicist James Clerk-Maxwell, to be stable, or at least not discernibly "falling in." Lowell responded (in a letter published in the *Times* on November 21) that the criticisms had been "founded on a misapprehension of what had been observed at Flagstaff," and that he had never meant to suggest that the process involved was anything other than "a very slow [one], … going on since the ring system formed and [destined to] go on for a long time to come" [4].

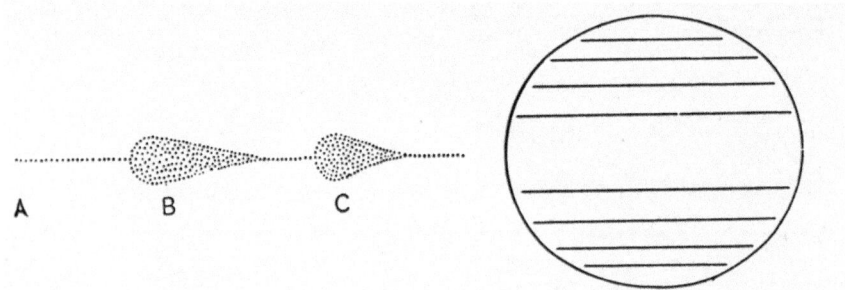

THE TORES OF SATURN.   Not drawn to scale.

Lowell's interpretation of the tores. (From: The Evolution of Worlds, *1909)*

All of this was more smoke than light, and except at Lowell Observatory, the tores were soon forgotten. However, meanwhile, Barnard remained on the case. From early July, the rings remained quite unchanged, but on Christmas night 1907, he noted in the 40-inch refractor that the thread-like ring appeared much thinner than it had looked 2 weeks earlier. By then southern Wisconsin was in the throes of its usual harsh winter weather. Barnard, despite having been unwell for some time, continued to observe, and proceeded to critical observations as the Earth made its final passage through the ring-plane. On January 2, 1908, he got up from his sick bed long enough to note that the rings were still visible in the 40-inch, though "very thin" and with a satellite at each end. Without using an occulter to block the glare of the bright globe of the planet and render faint objects visible, he found it nearly impossible to see any trace of the rings on the sky. However, the condensations were feeble but present as slightly brighter parts of the ring. On January 5, the rings were still fainter, and no trace of them could be seen without the occulter, while on January 6, though in poor seeing, they were completely invisible. Then, on January 7, they were easily visible again; however, without any trace of the condensations. On January 8, in hazy conditions and through breaks in cloud, the rings could be seen faintly on both sides of the planet. Barnard concluded that the Earth's passage through the ring-plane had likely occurred at about the time of his observation on January 6.

In Barnard's view, the observations when the Sun and Earth had been on opposite sides of the rings (those made prior to the passage through the ring-plane) put beyond any doubt just how thin they must be. He wrote, "At such times, though the very oblique surface of the ring[s] was visible, nothing could be seen of the edge … which should have been seen as a thin rim of light. This sunlit edge … was not visible at any time, though it was always looked for carefully when the seeing was best [5].

In contrast to Lowell, who as always had rushed into publication on the basis of a few observations, Barnard struggled to get everything right, and reworked a first draft of his November observations by including an account of the additional observations up to and including those involving the final ring-plane passage and reappearance. He dated the draft January 10, 1908, and the following day sent it off to Thomas Lewis of the Royal Observatory, Greenwich, with the comment: "I have been sick in bed … and am up for a bit to day [sic] to get this off but shall have to go back to bed. I managed to get the observations of Saturn by taking big risks and wrapping up good to go into [the] big dome."[6] Though Barnard may have gone back to bed for a while, he refused to put his paper to bed. His papers in the archives of the Royal Astronomical Society show that he submitted constant revisions and corrections, which must have come close to driving the poor secretary, W.H. Wesley, mad. However, in the end he managed to solve the mystery of the "condensations." He always had misgivings about the use of the term (as well as Lowell's "agglomerations" and Lick Observatory astronomer Robert G Aitken's "knots," since these terms implied local accumulations of material). Availing himself of the scientific method, he considered several possibilities to explain them (rather than leaping at the first notion to come to mind), and by a process of elimination arrived at the one that seemed most likely. They were not true condensations on the ring, but only bright places appearing brighter because of contrast or "irradiation" (the apparent bleeding of a brighter object into its darker background). That they could not be material condensations was proved by their complete disappearance when the rings were edgewise. Were they true masses in the rings, such as "whorls" or "tores," they would have been most prominent at such times.[7]

Acting on a hunch, Barnard lined up the positions of the condensations with the carefully measured dimensions of the components of the ring system. He discovered—*Eureka!*—that the condensations fell precisely in the positions of the boundaries of the crepe ring and the Cassini division. From this he deduced that the Cassini Division was not as entirely devoid of particles as it looked, but rather:

> If …the Cassini division were filled with particles as closely clustered as they are in the crape ring a satisfactory explanation of the condensations would be that they were simply due to the sunlight shining through and illuminating the particles in the [crepe] ring for the inner condensations, and a similar effect of the Sun shining through the Cassini division and illuminating the particles in it would produce the outer condensations. The fact that the inner and outer condensations were essentially of the same intensity would require that the particles should be as closely clustered in the Cassini as in the [crepe] ring.[8]

This was, of course, the complete solution to the mystery, and is completely confirmed by spacecraft observations—including, most spectacularly, those made by the Cassini orbiter. The spacecraft captured an image taken from behind the planet with the Sun eclipsed by the globe of Saturn and the rings resplendently illuminated by forward-scattered sunlight exactly in the manner Barnard had recognized. Did Lowell congratulate Barnard on his triumph? Unfortunately, no; instead he vapored on about the black medial core and the tores, which he continued to maintain—against Barnard—were true "agglomerations," and explained as products of the gravitational perturbations of the inner satellites, especially Mimas, on the ring particles, a topic to which he was to return again and again in later years and expend a great deal of ingenuity to account for.[9] In fact, however, the core and tores were merely illusions due to poor seeing. Even Lawrence Lowell, in writing his biography of his late brother, passed over them quickly. "It is," he wrote, "needless to dwell more upon them."[10]

Barnard's views on the whole sorry subject were expressed privately (with a rather strained attempt at humor) to George Ellery Hale:

> I see that Lowell is still on a "tore" or should I say a "tare." His "invisaging" is forthrightly doing up things in the shape of Mars and Saturn—poor Saturn! One has no sympathy for Mars who has always been guilty of lending himself passively to all the Brenners, *etc.*, that have held out their hand to him. But Saturn![11]

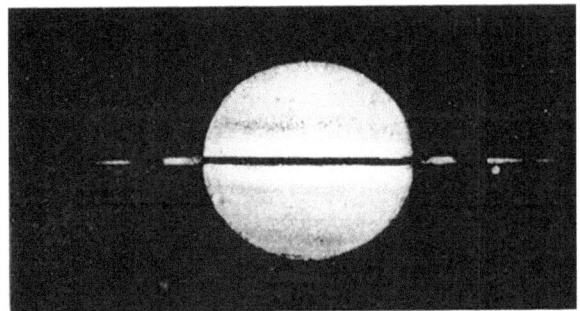

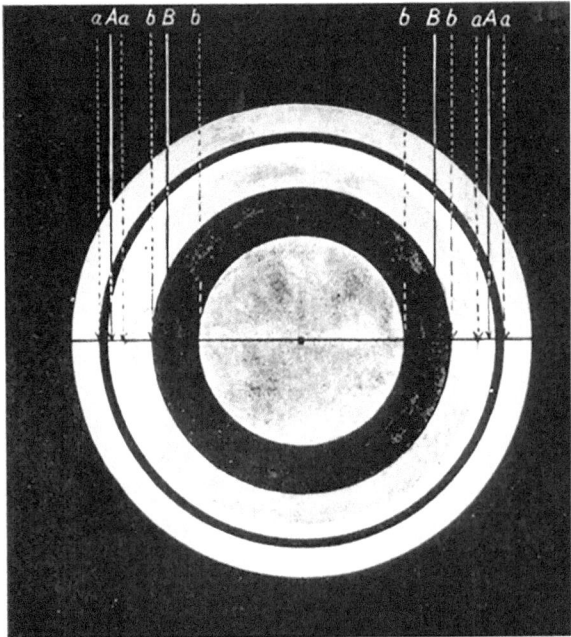

Above, Barnard's drawing of Saturn, November 25, 1907, with the 40-inch refractor, showing the so-called condensations on the ring's unilluminated face (i.e., the Sun and Earth are on opposite sides of the ring surface). Below, his diagram showing that the positions of the condensations agree with those of the crepe ring (Bb) and Cassini division (Aa). (From Monthly Notices of the Royal Astronomical Society, *vol. 68, plate 10)*

A composite image taken by the Cassini spacecraft on July 19, 2013, showing corre-spondence between bright ring features visible by forward scattered sunlight and the same in Barnard's drawings (above). In the original image the Earth is visible as a tiny blue dot but cannot be seen at this image scale. (Credit: NASA)

From Peak of "Mars Furor" to the Eve of the Downfall

Despite Lowell's dalliance with Jupiter and Saturn, Mars, as usual, dominated the headlines in 1907 and into 1908. Indeed, to judge solely from the volume of correspondence about the planet in the Lowell Observatory Archives, which begins to increase in 1905, the peak was clearly in 1907. Thereafter, it continued—though at a declining rate—in the ensuing years until Lowell's death.[12] Though the great sensation was the supposed photography of the canals from South America, Lowell's *magnum opus* on the red planet, *Mars and Its Canals*, published in 1906, received considerable attention. The reviews generally were favorable. Lowell had written a dedication to the great Italian astronomer (who gave his permission), whose influence was visible on every page, and whose approval and support meant more to him than anything else:

To
G.V. Schiaparelli
The Columbus of a New Planetary World
This Investigation Upon It
Is Appreciatively Inscribed.[13]

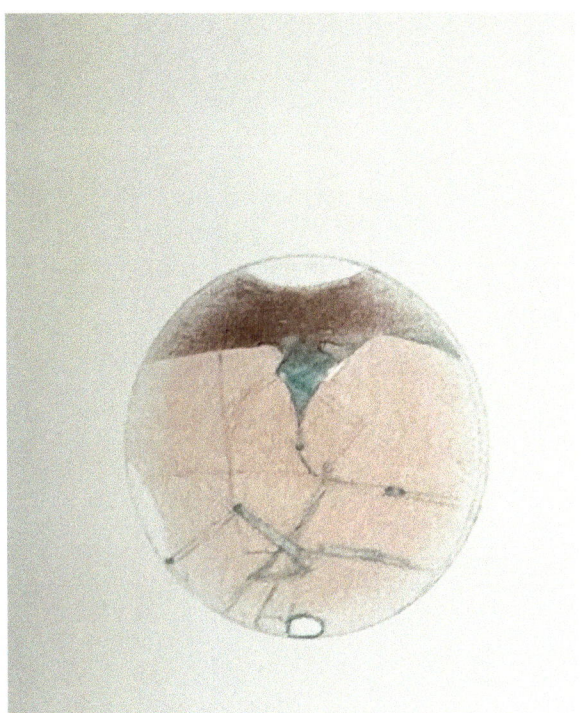

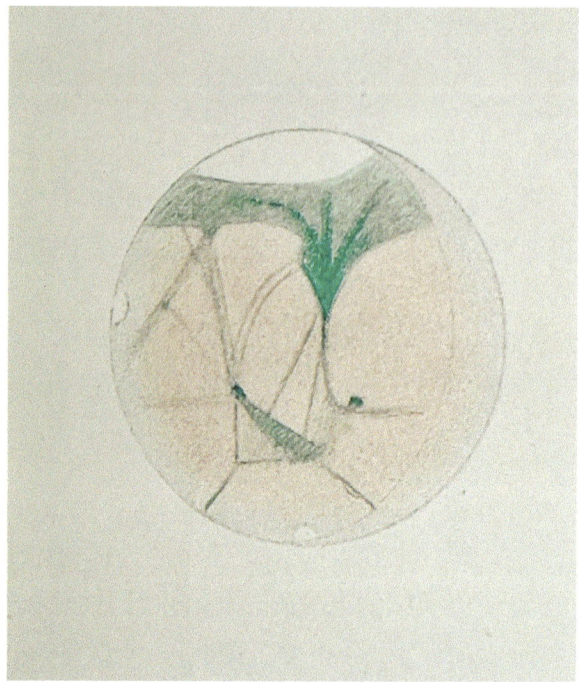

Lowell's drawings of Mare Erythraeum, purporting to show marked changes in color in 1903. The drawings above were made on May 7, 1903 (Martian seasonal date January 16), and June 30, 1903 (Martian seasonal date February 22). In the first, the area appears chocolate-tinted, in the second, blue-green. Lowell noted that the chocolate-brown "precisely mimicked the complexion of fallow ground," the blue-green that of vegetation. Thus, he concluded that a fallow field in winter had become lush with vegetation with the arrival of early spring. (From: Mars and Its Canals*)*

The British writer W.E. Garrett Fisher, author of several science articles for the famed 11th edition of the *Encyclopædia Britannica* but no expert on astronomy, claimed that the reality of the canals had "finally been determined by the successful photography of some [of them]. It is no longer possible to assert that these curious markings are due to some optical defect of the eye or instrument of those who have seen them."[14] The same author argued in a review of *Mars and Its Canals* in the London *Tribune* that Lowell's "conclusions … are still, of course, only a working hypothesis, but those who read Mr. Lowell's book without prejudice will admit that up to the present no other explanation has been offered which can for the moment compare with it."[15] According to the London *Times*, "Mr. Lowell has certainly produced a mass of evidence of a most striking and interesting kind." The Daily *Mail* noted, "Its extraordinary interest and fascination hold the reader spellbound.

Beyond all dispute it is the most striking astronomical work published for twenty years."[16]

By the time *Mars and Its Canals* appeared, Lowell had delivered the eight Lowell Institute lectures on "Mars as the Abode of Life" (which appeared in book form in December 1908). A review of the lecture series in the Boston Evening *Transcript*, whose writers included Lowell's friends Edward H. Clement and John Ritchie Jr., was nothing less than fulsome:

> It seems impossible that Mr. Lowell can raise another girder more grandly impressive and expressive of the whole fabric, or take another step in his scientific syllogism that will hold us any tighter in his logic. He has practically reached already his "Q.E.D." The thing is done, apparently, except for filling in the detail. But with his racy, epigrammatic brilliancy of style, his delicate, quiet humor, his daring scientific imagination—all held in check by instinctive modesty of good breeding, gayly throwing to the winds all professional airs and mere rhetorical bounce—his course will be no doubt as charming to the end as it has been steadily illuminating even for the illuminati.[17]

Lowell also welcomed a long-sought spectrographic result—V.M. Slipher, from the Flagstaff field station in the San Francisco Peaks, taking advantage of unusually dry winter conditions in January 1908, had confirmed, apparently, the existence of water vapor in the Martian atmosphere. He did so by using new red-sensitive photographic plates to record the intensity of a strong water vapor band (called the *a* band) in the far red part of the spectrum of Mars, and comparing it to the same band in the spectrum of the airless and moistureless Moon (the latter serving as a control for the purely sun-reflected spectrum). Many early spectroscopic observers, including Pierre Jules César Janssen of France, William Huggins and Edward Maunder of England, Pierre Angelo Secchi of Italy, and Heinrich Kayser of Germany, as far back as the 1860s, had thought that they could detect water bands in the atmosphere of Mars. However, bearing in mind that the same bands appear in the Earth's atmosphere and that Mars and the Moon do not appear exactly together in the sky, it was difficult for these observers—working visually, and having to maintain their estimate of the strength of the bands in memory—to avoid a considerable amount of subjectivity in their results. Indeed, Lick astronomer W.W. Campbell had carefully repeated this investigation in 1894 from 4200 foot Mt. Hamilton, with negative results. But Slipher had the advantage of the new red-sensitive plates, and saw—or thought he saw—a strengthening of the *a* band in his plates, proving the presence of Martian water vapor—a great result, if it held. In fact, it would not hold; it would be overthrown by Campbell the following year.[18]

Meanwhile, Lowell was moving forward with a scaled back version of the 84-inch reflector project he had discussed with Ritchey. Instead he proposed to acquire a 40-inch reflector, to be built by the Clark firm at a cost of $10,800 (about $345,000 in today's purchasing power). The goal was to have it available for use at the next Mars opposition, on September 24, 1909, when the planet would attain an apparent diameter of 23.8″ and stand much farther north of the Celestial Equator than in 1905 and 1907. Lowell was clearly hoping to refute the testimony of observers with larger telescopes who could not see the canals, at what would be the last really favorable opposition of his lifetime. Unfortunately, the 40-inch telescope was not ready until after Mars's opposition was past—and then Lowell came up with the decidedly bad idea to mount it underground, thinking a subterranean situation would provide the most stable seeing conditions. This, however, a ludicrous idea; in fact, the exact opposite is true—as most turbulent airflow is located near ground level, it would have been far better to mount the telescope on a tower high in the open air.[19]

Nevertheless, arguably the most surprising—and in the long run, most consequential—event of the year 1908 was not, in fact, a strictly astronomical one. It was Lowell's decision, made as usual with great precipitation—to marry his Boston neighbor Constance Savage Keith. Though well known for his "many interests and many moods," he was also a man of many errors. Though from this point in time, given the lack of documentation, we can no longer discern just what exactly motivated him to enter into matrimony at the age of 53, Constance would come to be regarded in retrospect as one of the errors, given that after Lowell's death she would attempt to break his will and in the process tied up his estate for many years.[20] At the time, no one could anticipate that, and all that is known of the circumstances of the wedding appears in the following special dispatch to Boston *Herald*, datelined New York, June 10, where one reads:

> *Percival Lowell, astronomer and orientalist, married Miss Constance Savage Keith, daughter of Beyrer Richmond Keith of Boston, this afternoon in St. Bartholomew's Church, Madison avenue and Forty-fourth street....*
> BUT   FEW   INTIMATES   IN   BOSTON   KNEW   OF INTENTION TO MARRY...

Word was somewhat slower in reaching Flagstaff. There was no special dispatch. However, the Coconino *Sun* reported on June 19 that the newlyweds

… left immediately on a honeymoon tour … and are expected to arrive here some time next September. Information concerning the wedding is meager, and it can only be conjectured that the study of Mars will be superseded by a study [of] Venus.[21]

If a study of Venus commenced at this time, there is no record of it in the observing log books. (Presumably the Observatory's cow, "Venus," does not count.)

We do know that, instead of visiting Canada and the Northwest as originally planned, Mr. and Mrs. Lowell summered in England and Europe—one highlight being a balloon ascension undertaken in a spirit of emulation of Camille and Sylvie Flammarion, who had made a romantic night flight *en ballon* for their honeymoon in 1874. Constance joined Lowell, his cousin the meteorologist A, Lawrence Rotch, C. S. Rolls, and Captain S.A. Cloman for a balloon ride over Hyde Park, London, taking advantage of their lofty position to photograph the paths in Hyde Park below them, as their system, Lowell claimed, resembled that of the canals of Mars. In October, on returning from Europe, they set out for Flagstaff on what was to be Mrs. Lowell's first visit to the couple's sometime home on Mars Hill. It was then that Mrs. Lowell learned first-hand how rigid her husband could be:

I was to meet him on the train for Flagstaff leaving the South Station [Boston] at 2 p.m.; anxious to impress him with my reputation for being punctual, I boarded the train about ten minutes before two. Percival came into the car, holding his watch in his hand, just about two minutes before two. He turned to me: "What time were you here?" I answered triumphantly: "Oh, I got here about ten minutes ago." His reply was: "I consider that just as unpunctual as to be late. Think how much could have been accomplished in ten minutes!" I have never forgotten that remark. Percival never wasted minutes.[22]

It seems clear that Lowell had entered the state of matrimony on condition that his lifelong independence and devotion to his work was not to be interfered with. Indeed, the marriage, as far as can be judged, made him something of a "married bachelor"—one of the things that theologians have said that even God Himself cannot create (as a contradiction in terms) but which Lowell seems to have eased into quite naturally. Lowell was the dominant party, Constance the subordinate one. But this had, of course, been the case in Lowell marriages forever, including that of his parents Augustus and Katharine. Given the ages of the couple (Lowell was 53, Constance 42), offspring were not to be expected. For that matter, it is impossible to imagine Percival Lowell as a doting father.

115

Above: Dr. and Mrs. Lowell on their honeymoon. This portrait was taken in Berlin. Below: Lowell and Camille Flammarion, at Juvisy, riding in Lowell's new Stevens-Duryea. The chauffeur is not shown. Both 1908. (Credits: Lowell Observatory Archives)

The Resistance Forms

Lowell seemed to be riding higher than ever during the year 1908. However, as his biographer David Strauss notes, "Lowell's apparent success in photographing the canals, providing evidence of water vapor on Mars, and attracting the support of prestigious intellectuals … galvanized the American astronomical establishment to launch a counterattack to refute Lowell's claims for the canals. Leading mainstream astronomers used their positions as power brokers within the discipline to put to rest what they regarded as a serious threat to the public's faith in science."[23] The mainstream astronomers would eventually include Campbell; Hale; the Smithsonian Astrophysical Observatory's Charles Greeley Abbot; the Director of Yerkes Observatory, Edwin Brant Frost; the Director of Harvard College Observatory, Edward C. Pickering; and Simon Newcomb, the now retired Director of the U.S. Nautical Almanac Office; and—though often operating behind the scenes—Barnard.

The effort of what Strauss defines as "The Establishment" did not, at least at first, see the canals being "wiped out from the literature of astronomy," as

Barnard once had hoped. A number of astronomers offered versions of the illusion theory that had so far been most prominently advocated by Maunder. Vincenzo Cerulli, at the height of the "Mars furor" in the mid-1890s, had been a prolific recorder of canals from his private observatory at Teramo, Italy. However, having in moments of exquisite seeing witnessed canals breaking up into fine details, he became convinced that the lines were illusory. The canals only represented an intermediate stage of perception of the actual Martian markings.[24] More influential were the criticisms of Newcomb, who wrote a provocative paper on "The Optical and Psychological Principles involved in the interpretation of the so-called canals of Mars."[25] The paper, published in *The Astrophysical Journal*, was quite technical, and analyzed the way that various aberrations in telescope optics might blur an image to produce illusory effects such as "canals." Newcomb got deep into the weeds in this paper, and so did Lowell in an inevitable rejoinder, also published in The *Astrophysical Journal*—one of the few occasions since the journal's founding by Hale and Keeler in which he was allowed "in" by the referees—which included something of an adventitious argument *ad hominin* against Barnard's authority as an observer. Newcomb, recalling Barnard's observations of Mars in 1894, had put him forward as the world's best observer of planetary detail, but Lowell discounted information "received at second hand" (i.e., from Barnard):

> [Newcomb] states that the background upon which the "canals" are seen is not uniform in the case of Mars and that therefore lines on paper are not a true criticism. This is an error due probably to his reading that the "seas" were a jumble of markings impossible to decipher. This jumble is the very canal and oasis system imperfectly seen, as I can state from having seen it resolved. It, therefore, cannot be used as an argument against its own detection after the fact—especially in the light regions where uniformity of tint is the rule. He was arguing from the observations of a sensitive and not an acute eye; a very pregnant source of mistake. For experience shows that an eye good for faint star and satellite work is constitutionally defective for planetary detail and vice versa; a fact dependent apparently upon the size of the retinal rods and cones. [11, p. 131]

Newcomb responded to Lowell's paper, and Lowell responded to Newcomb's. The matter was finally dropped, but only after Newcomb had transported the "Small Boy Theory" from the Greenwich Hospital School to Yerkes. There Barnard and Philip Fox took turns drawing, and confirmed the small boy side of the argument.[26] In truth, the real problem was the canals themselves, which defied optical analysis because they belonged not to the realm of normal science but to what the American chemist Irving Langmuir would famously call "pathological science"—the "science of things that aren't so." Langmuir defined pathological science by the following criteria:

1. The maximum effect that is observed is produced by a causative agent of barely detectable intensity, and the magnitude of the effect is substantially independent of the intensity of the cause.
2. The effect is of a magnitude that remains close to the limit of detectability.
3. There are claims of great accuracy.
4. Fantastic theories contrary to experience are suggested.
5. Criticisms are met by *ad hoc* excuses thought up on the spur of the moment.
6. The ratio of supporters to critics rises up to somewhere near 50% and then falls gradually to oblivion [14].

Wallace Weighs In

The ratio of supporters to critics of Lowell's theory decreased further when 84-year-old Alfred Russel Wallace, co-discoverer (with Darwin) of the theory of evolution by natural selection and since 1902 living in semi-retirement, turned what had started as a review of *Mars and Its Canals* into a short book.

Wallace had long been interested in the question of extraterrestrial life. (His book *Man's Place in the Universe*, the first serious attempt of a leading biologist to consider the probability that life existed on other planets, had appeared in 1903, and had argued—from a Christian perspective—that Mars was not inhabitable and that the only life in the Solar System was on Earth.)[27] In contrast to Darwin (and Lowell), he did not have the good fortune of being born into inherited wealth, and supplemented his annual governmental pension with occasional work for hire, such as reviews.[28] He might have made short shrift of *Mars and Its Canals* but for the added stimulus of an article Lowell published at the end of 1906 in the *Philosophical Magazine* [18]. Here, Lowell claimed by means of a highly erudite mathematical argument to demonstrate that, notwithstanding its much greater distance form the Sun and its excessively thin atmosphere, Mars apparently possessed a climate "on the average equal to that of the south of England."[29] Wallace was not without a few late-Victorian weaknesses of his own. He shared with Sir Arthur Conan Doyle, Camille Flammarion, Sir William Crookes and Sir Oliver Lodge a belief in spiritualism. Also, he suffered a famous failure of nerve when it came to his own theory of evolution by natural selection in ascribing the origin of all species except one—man—to its action. In his view, man and man alone was God's special creation and possessed of a soul. Nevertheless, Wallace was extremely skeptical of Lowell's theories. Apart from his achievements in geography, anthropology and biology, Wallace had early in his career received a thorough training as a civil engineer, so he was in a good position to tease apart the weaknesses of Lowell's arguments about the Martian canal system.

So far as is known, Wallace never looked at Mars through a good telescope, but it was not Lowell's basic observational claims that Wallace disputed. He conceded Lowell was in possession of a fine large telescope and enjoyed seeing conditions that made him "perhaps more favorably situated than any astronomer in the northern hemisphere,"[30] and admitted the strangeness and artificiality of the markings he had discerned on the Martian surface—only to question, outright, Lowell's theory as to their meaning:

> The one great feature of Mars which led Mr. Lowell to adopt the view of its being inhabited by a race of highly intelligent beings … is that of the so-called "canals"—their straightness, their enormous length, their great abundance, and their extension over the planet's whole surface from one polar snow-cap to the other. The very immensity of this system, and its constant growth and extension during fifteen years of persistent observations, have so completely taken possession of his mind, that, after a very hasty glance at analogous fact and possibilities, he has declared them to be "non-natural,"—therefore to be works of art—therefore to necessitate the presence of highly intelligent beings who have designed and construct them. This idea has colored or governed all his writings on the subject. The innumerable difficulties which it raises have been either ignored, or brushed aside on the flimsiest evidence. As examples, he never even discussed the totally inadequate water-supply for such world-wide irrigation, or the extreme irrationality of constructing so vast a canal-system the waste from which, by evaporation, when exposed to such desert conditions as he himself descries would use up ten times the probable supply…. The mere attempt to use open canals for such a purpose shows complete ignorance and stupidity in these alleged very superior beings; while it is certain that, long before half of them [the canals] were completed their failure to be of any use would have led any rational beings to cease constructing them.[31]

Alfred Russell Wallace, as he looked at about the time he wrote Is Mars Inhabited? *(Credit: Natural History Museum, London)*

Instead of Lowell's theory to explain the canals, Wallace offered a somewhat Pickeringesque pet theory of his own, according to which the canals are cracks produced by the contraction of a heated outward crust upon a cold non-contracting interior. No one remembers that. What they do remember is the awesome erudition he marshalled to counter Lowell's suggestions about the temperature, climate, and atmosphere of Mars.

In the *Philosophical Magazine* paper, Lowell had published a calculation based on the laws of heat and various measurements or estimates of the various terms used in these calculations to derive the mean surface temperature of Mars. Reading it, even today, one is apt to feel one is being made the mark of a game of three-card monte, and one rather suspects Wallace must have done so. He lays out some of the variables involved in the whole intricate calculation:

> Radiation, that is loss of heat, is going on concurrently with gain, and the rate of loss varies with the temperature according to a law recently discovered

[Stefan's law], the loss being much greater at high temperatures in proportion to the 4th power of the absolute temperature. Then, again, the whole heat intercepted by a planet does not reach its surface unless it has no atmosphere. When it has one, much is reflected or absorbed according to complex laws dependent on the density and composition of the atmosphere. Then, again, the heat that reaches the actual surface is partly reflected and partly absorbed, according to the nature of that surface—land or water, desert or forest or snow-clad—that part which is absorbed being the chief agent in raising the temperature of the surface and of the air in contact with it. Very important too is the loss of heat by radiation from these various heated surfaces at different rates; while the atmosphere itself sends back to the surface an ever varying portion of both this radiant and reflected heat according to distinct laws. Further difficulties arise from the fact that much of the sun's heat consists of dark or invisible rays [i.e., infrared and ultraviolet], and it cannot therefore be measured by the quantity of light only.

From this rough statement it will be seen that the problem is an exceedingly complex one, not to be decided off-hand, or by any simple method....[32]

Divested of mathematics, the key assumption of Lowell's argument turns out to be that the comparative amounts of heat received by Mars and the Earth depends on their very different amounts of atmosphere, and this estimate depends almost wholly on the comparative albedos (reflectivity) of the two planets—that of Mars being given as 0.27, that of the Earth (based on an entirely different method) as 0.75, from which Lowell concludes that nearly three-fourths of the solar heat Mars receives reaches the surface and determines the temperature, while the Earth gets only one-fourth of its total amount. Invoking a few additional and, in Wallace's view, questionable assumptions, Lowell manages to bring the mean temperature of Mars up to 48 °F. This was almost exactly the same as that of the southern half of England. It also fit perfectly with Lowell's oft-stated conclusion that the presence of liquid water and water vapor were proved by the behavior of the snow caps and especially of the bluish band around the edge which could be due to nothing but soil moistened by meltwater. Thus, even in the polar regions, the temperature had to be above the freezing point of water during a portion of each year. Further, but through a different line of argument, Lowell derives the probable density of the atmosphere of Mars as equivalent to 2½ inches of mercury (85 millibars), one-twelfth that of the Earth.

Wallace's analysis of Lowell's paper ends with a devastating critique of Lowell's whole methodology. "It is evident," says he, "that mathematical calculations founded upon such uncertain data cannot yield trustworthy results...."

Everywhere [in this paper] we meet with figures of somewhat doubtful accuracy. Here we have somebody's "estimate" quoted, there another person's "observation," and these are adopted without further remark and used in the various calculations leading to the result … quoted. It requires a practised mathematician, and one fully acquainted with the extensive literature of this subject, to examine these various data, and track them through the maze of formulae and figures so as to determine to what extent they affect the final result.[33]

Wallace finds a crucial error in Lowell's calculations. In claiming that the thin atmosphere of Mars allows more solar radiation to reach the surface, he had completely overlooked the enormously increased loss of heat by direct radiation in such an atmosphere. Thus Wallace reaches the conclusion that from the scantiness of the atmosphere alone Mars cannot possibly have a temperature as high as the freezing point of water, and that lacking liquid water—the first essential of organic life—"Mars, therefore, is not only uninhabited by intelligent beings such as Mr. Lowell postulates, but is absolutely UNINHABITABLE."[34] (Regarding the ice caps, he rejected altogether Lowell's view that they consisted of water-ice, instead favoring that of the Irish physicist Johnstone Stoney that they consisted of frozen carbon dioxide.)[35]

Though recent writers including Carl Sagan and Stephen Jay Gould have come to regard *Is Mars Habitable?* as a minor masterpiece,[36] in its time it was one of Wallace's least appreciated works. Indeed, it received hardly any attention at all. Not least, in contrast to Lowell's "racy, epigrammatic style," his was (as the above excerpts suggest) ponderous and highly technical. It was decidedly not apt to make any impression on the general public. Moreover, the delay in its publication meant that it missed striking while the iron was hot. Lowell was completely dismissive of it, and may not even have read it carefully. To the English amateur astronomer (and Lowell admirer) Mark Wicks, Lowell called the book "such a mass of misstatements and inadequate knowledge on the whole subject that it hardly seems worth replying to—nice man as Dr. Wallace is."[37] Later, in a letter Lowell sent to the editor of *Nature*, he wrote, "misstatements cannot be too carefully avoided in science, especially when a man, however eminent in one branch, is wandering into another not his own."[38] He had seemingly forgotten that he himself had once come rather stormily into astronomy from Oriental studies, and in any case, would be reminded of his own interloper status when, a year after Wallace's review, Lowell's Lowell Institute lectures of October–November 1906, *Mars as the Abode of Life*, appeared, and came in for reviews that make Wallace's seem mild in comparison.

Blackwelder Attacks

Mars As the Abode of Life was puffed by the publishers as "science that reads like romance." Ahead of publication, the Boston *Herald* announced that "nothing more audacious or captivating to the poetic and moral imagination is now underway among men" than Lowell's astronomical investigations.[39] More significant was the reaction of the aged Schiaparelli. Despite an early tendency to hold Lowell at arm's length, he had been charmed and eventually won over. In a letter of March 1909, the man whom Lowell called his "Martian Master" and who had only a little more than a year to live, thanked his American colleague effusively for sending him a copy of his latest book:

> I should before everything thank you for the new volume that you have just sent me on our planet. My attention as above all captured by the second part, that which contains the notes on various questions; I see there that you leave no corner of Areography without bringing the light of your studies....[40]

In his latest book, Mars is still front and center, but Lowell has begun to pivot by placing his discoveries about that small planet against broader themes of planetary evolution, framed in the now familiar Laplacian/Darwinian/Spencerian terms but packaged under a new name, planetology. Planetology is presented as a new field of science in which the development of worlds is determined, ultimately, by their initial mass. It is mass that defines the amount of internal heat a world initially possesses, and also its subsequent rate of cooling. The Moon at first seems to mark an exception to the scheme, since from its small mass, it ought to have cooled quickly and soon become completely frozen over. However, the features of its surface bear witness to an eruptive stage, meaning it must once have possessed significantly more heat than it does now. He suggests that the Moon did not, however, emerge as a separate body; instead, as sketched in an old theory of George Howard Darwin (Charles's son) based on the latter's study of tides, it formed a part of the Earth's mass, with the two rotating together as a single pear-shaped body in about 5 h, before the Moon separated and carried with it some of heat of the parent body that led to the volcanic stage. Thus, he says, the Moon "had its birth in a rib of Earth." The apparent exception actually turns out to support the scheme.[41]

Of the other bodies of the Solar System, Jupiter and Saturn are at the youthful molten stage of their development, still hot enough internally to radiate their own heat; Mars, smaller than the Earth, has cooled more rapidly and lost its seas and most of its moisture; while Mercury and the Moon have run their evolutionary course and are now dead worlds. Mars, Lowell claimed, is now a desert planet, as the Earth would one day be, "rolling a parched orb through space." His argument was based on a process Lowell called "desertism," which seems to have been rather loosely based on the writings of the late Yale University geologist James Dwight Dana, who had claimed that over geologic time the Earth had become steadily drier with the lands increasing in area. Thus, according to Lowell:

> Study of Mars proves that planet to occupy Earthwise in some sort the post of prophet…. It enables us to no mean extent to foresee what eventually will overtake the earth in process of time; inasmuch as from a scrutiny of Mars coming events cast not their shadows, but their light, before.
>
> It is the planet's size that fits it thus for the rôle of seer. Its smaller bulk has caused it to age quicker than our earth, and in consequence it has long since passed through that stage of its planetary career which the earth at present is experiencing, and has advanced to a further one, to which in time the earth itself must come, if it be not overwhelmed beforehand by other catastrophe….[42]

The future fate of the Earth was evident in the "present great extent of the Martian deserts" which dominated its aspect in the telescope. Here Lowell, inspired no doubt partly by distant vistas of the Painted Desert as seen from the San Francisco Peaks, pulled out all the stops and produced a literary set-piece:

> Beautiful as the opaline tints of the planet look, down the far vista of the telescope-tube, they represent a really terrible reality. To the bodily eye, the aspect of the disk is lovely beyond compare; but to the mind's eye, its important is horrible. That rose-ochre enchantment is but a mind mirage. A vast expanse of arid ground, world-wide in its extent, girdling the planet completely in circumference, and stretching in places almost from pole to pole, is what those opaline glamors signify. All deserts, seen from a safe distance, have something of this charm of tint. Their bare rock gives them color, from yellow marl through ruddy sandstone to blue slate. …. But this very color, unchanging in its hue, means the extinction of life. Pitilessly persistent, the opal here bears out its attributed sinister intent.[43]

Finally, he came to his observations of the canals and the inferences derived from them (one can imagine the hair on the necks rising of those in the Huntington Hall auditorium as he spoke these words):

> … Not only do the observations we have scanned lead us to the conclusion that Mars at this moment is inhabited, but they land us at the further one that these denizens are of an order whose acquaintance was worth the making…. Their presence certainly ousts us from any unique or self-centred position in the solar

system, but so with the world did the Copernican system the Ptolemaic, and the world survived this deposing change. So may man. To all who have a cosmo-planetary breadth of view it cannot but be pregnant to contemplate extra-mundane life and to realize that we have warrant for believing that such life now inhabits the planet Mars.

A sadder interest attaches to such existence: that it is, cosmically speaking, soon to pass away. To our eventual descendants life on Mars will no longer be something to scan and interpret. It will have lapsed beyond the hope of study or recall. Thus to us it takes on an added glamour from the fact that it has not long to last. For the process that brought it to its present pass must go on to the bitter end, until the last spark of Martian life goes out. The drying up of the planet is certain to proceed until its surface can support no life at all. Slowly but surely time will snuff it out. When the last ember is thus extinguished, the planet will roll a dead world through space, its evolutionary career forever ended.[44]

At the very outset of this literary farrago, Lowell introduced a speculation that closely resembled the Chamberlin-Moulton planetesimal hypothesis of the beginning of the Solar System. It had been hinted at as early as 1903 then developed in quantitative detail in 1905 by Thomas Chrowder Chamberlin, chairman of the geology department at the University of Chicago, and Forest Ray Moulton, a noted celestial mechanician at the same institution. The basic idea—that the planets formed from material wrested from the Sun during an encounter with a nearby star—was intuitively appealing, though ultimately untenable. The gaseous material rent from the star would not possess suffi-cient mass to condense into planets, but would instead simply dissipate and be lost to space. Lowell seized on it in part because of its dramatic possibilities. In his version:

> So far as thought may peer into the past, the epic of our solar system began with a great catastrophe. Two suns met. What had been, ceased; what was to be arose. Fatal to both progenitors, the event dated a stupendous cosmic birth.
>
> It is more than likely that one or both of the colliding masses were dark bod-ies, dead suns, such as now circle unseen in space amid the bright ones we call the stars. Probable this is, for the same reason that the men who have been far outnumber the men who are. It is not to be supposed that the two rovers actu-ally struck, the chances being against so head-on an encounter; but the effect was as disastrous. Tides raised in each by the approach tore both to fragments, the ruptured visitant passing on and leaving a dismembered body behind in lieu of what had been the other. That the stranger continued on its way is shown by the present moment of momentum of our system. For it is very small, and the fact can be proved to mean that after the encounter its matter still lay massed for the mot part in a single centre. Thus, what had been a sun was left alone, with its wreckage strewn about it....[45]

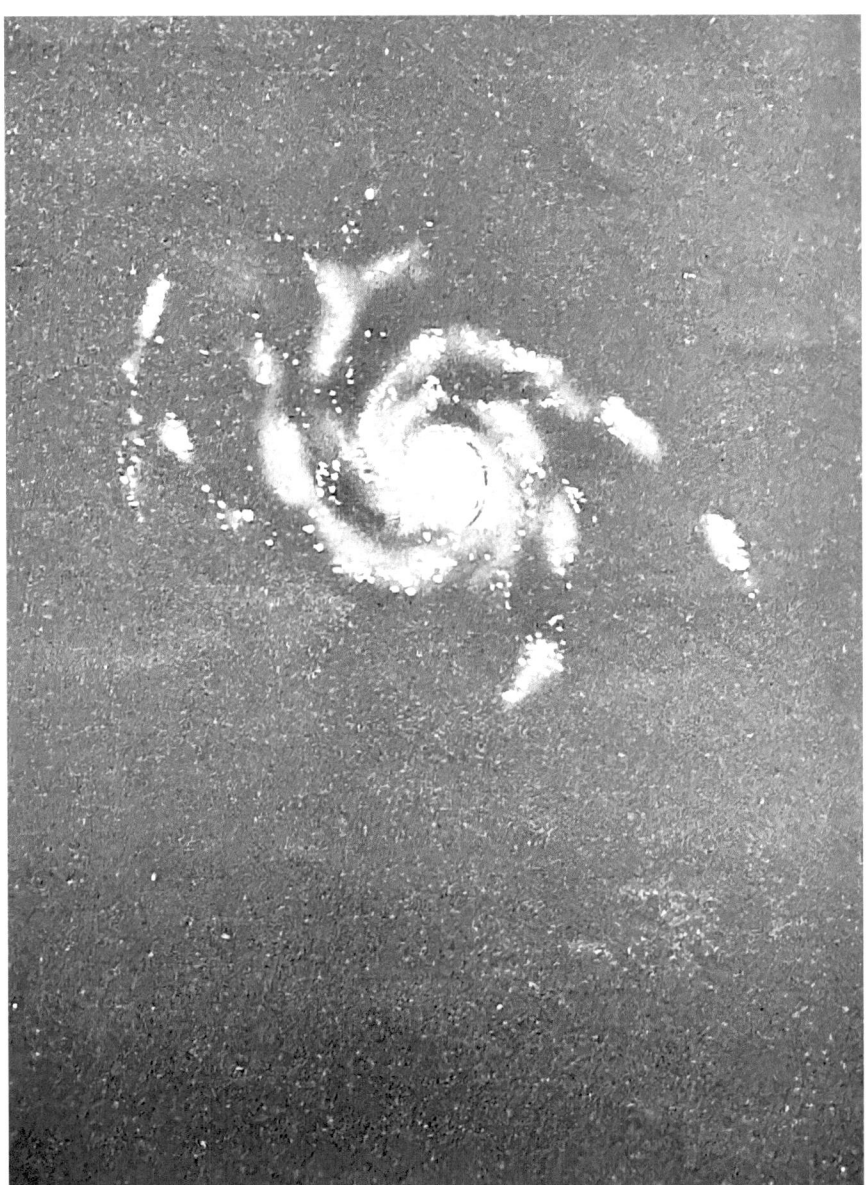

A figure, based on George Willis Ritchey's photograph of the spiral nebula M101 in Ursa Major, used by the British astronomer A.C.D. Crommelin to illustrate the Chamberlin-Moulton hypothesis. The caption reads: "The Planetesimal Hypothesis ... suggested by Moulton and Chamberlin. They suppose another sun to have approached ours ages ago, raising great tidal waves, which left the Sun in the form of two spiral arms. Condensations forming in these, and collecting more matter from the dust-streams around them, became the planets." (From: T.E.R. Phillips and W.H. Steavenson, Splendour of the Heavens *(London, 1923))*

Despite the resemblance of his own view to the Chamberlin-Moulton theory (though Lowell did not actually make any claim for his priority), he rather scurrilously attacked the theory at several points—for instance, he rejected the Chamberlin-Moulton explanation of the production of direct rotation by particle impacts and instead relied on tidal forces to reverse the initially retrograde rotations [22]. In addition, Lowell pushed the boat out even further in a course of lectures given at the Massachusetts Institute of Technology in February and March 1909, to be published as *The Evolution of Worlds* in December 1909. Here again he presented the inherently sensational idea of the catastrophic origin of the Solar System in the near-approach of another star to the Sun (popularized in the press as "Professor Lowell's Dark Star"), illustrating his ideas with images of spiral nebulae by the British amateur Isaac Roberts and erstwhile Yerkes astronomer George Willis Ritchey. He suggested that these nebulae—turning up in the hundreds of thousands and eventually millions on deep plates—might be relics of other, similar encounters of dark stars with suns in other parts of the Galaxy to those that had given rise to the Solar System. Thus, he declared:

> Suppose, now, a stranger to approach a body in space near enough; it will inevitably raise tides in the other's mass, and if the approach be very close, the tides will be so great as to tear the body in pieces along the line due to their action; that is, parts of the body will be separated from the main mass in two antipodal directions. This is precisely what we see in the spiral nebula. [23]

The most prominent "white nebula" in the sky: The Great Spiral in Andromeda, M31, photographed by George Willis Ritchey with the 60-inch reflector at Mount Wilson Observatory in 1908. Lowell had assumed that this and other white nebulae were solar systems in formation, but V.M. Slipher's spectrographs beginning in late 1912 proved otherwise. (Credit: Public Domain, Mount Wilson Observatory photograph)

In an article for *The Atlantic Monthly*, "The Revelation of Evolution," which closely follows the reasoning in *The Evolution of Worlds*, Lowell emphasizes the importance of meteorites to understanding the origin of the Solar System:

Without attempting here a picture of what probably took place, let me sketch a line or two of its reconstruction as they have taken shape at midnight to one watcher of the stars. Strange to say, perhaps the latest news about our family has come from the smallest and most seemingly insignificant of its members... meteorites... Their speed and their great numbers show that they are cosmic bodies like the planets themselves, the unswept-up remnants, in fact, of what once strewed space, and out of which the planets were formed. They are thus parts of the primal nebula...[24, p. 178]

As for the primal nebula, he assigns it to the class of white nebulae, defined in contrast to the green. He explains the reason for the difference and then reveals the basis of the intricate spiral structure:

Their color, white, arises from their showing a continuous spectrum, and indicates that they are composed in large part at least of solid particles, whereas the *green* tint of the others comes solely from glowing gas. Now, this spectrum is just what they should show were they flocks of meteorites—and such they undoubtedly are.... [F]rom their peculiar structure, we can infer what the process was that scattered the constituents of the once compact ball whose existence the meteorites attest...[46]

As usual, one notices the emphatic nature of his statements. (For instance, of the spiral nebulae, he says "such they undoubtedly are"; this is the style of speaking referred to as "power speech" and shown to be highly persuasive in jury trials.) The irony is that, despite his dogmatic self-confidence, every statement he makes here is at least partly in error, though the errors would sometimes, as here, prove to be singularly productive ones. Lowell would assign his assistant V.M. Slipher the technically difficult task (because of the faintness and dispersed nature of the white nebulae's light) of providing proof positive of his notion that the spirals were rotating nebulae by putting it to test with the spectrograph. Slipher would demonstrate both that the nebulae were rotating[47] and also that they were moving in the line of sight (first demonstrated as a blue shift in the case of the Andromeda Nebula, then later as mostly red shifts, suggesting an expanding universe).[48] The white nebulae were not flocks of meteorites at all. They were—if one wishes to keep with the term—flocks of stars. They were not Solar Systems in formation but other Milky Ways.

In contrast to *Mars as the Abode of Life*, in which Mars is still the principal actor on the stage, in *The Evolution of Worlds* and "The Revelation of Evolution," the red planet has retreated to hardly even a cameo role—in all, only one paragraph is devoted to the red planet. In fact, this was no accident, and attests to a marked shift of Lowell's interests after the Mars furor peaked in 1907 and his marriage the following year. As William Graves Hoyt pointed out:

> This shift of emphasis away from Mars and toward more general planetary studies is reflected by the fact that of the seventy-five observatory Bulletins published while Lowell lived, only twenty-eight do not concern his Martian work in some way, and twenty-three of these appeared after 1907…. Notably, too, the two volumes of his observatory's *Memoirs*, both published in 1915, concern non-Martian matters, the first a trans-Neptunian planet and the second the rings of Saturn. His only other memoir for this post-1907 period was entitled "Origin of the Planets" and was published by the American Academy of Arts and Science[s] in April, 1913.[49]

With this turning to grander themes, Lowell also (not coincidentally) began to show a noticeably more inflated view of himself, in which he presented himself as a persecuted pioneer like Darwin. (He somehow omitted mention of Wallace!) He writes in "The Revelation of Evolution," ostensibly penned for the centennial of Darwin's birth but revealing as much about himself as about Darwin:

> A master-thought lives always—it speaks forever in the echoes it evokes…. In it its thinker immortally lives. For its birth we still remember long after the man himself is dead, and it is to celebrate one such centenary that so many meetings have this year been held. Yet, amid the resounding plaudits, few perhaps were conscious of what it was they cheered, nor among mankind's conceptions how Darwin's stood related to the rest. For the origin of species is as much a mystery as ever; it was for bringing animate existence within universal law that Darwin's memory is kept. It was as part of cosmogony's master-key that the thought that led to its unlocking was great. For in truth the rise of the organic is but the latest chapter in a serial of the sky…
>
> Evolution is nothing more nor less than the mainspring of the universe. Grand in its simplicity, it is the one fundamental fact on which all we know depends. From its influence nothing can escape; for it has fashioned everything, from nebula to man….[50]

Lowell runs through the list. He races through the history of astronomy (in which he praises the pre-Socratics but dismisses Ptolemy as "that arch-mediocrity" who "imposed upon the world, to dominate it for more than a millennium, a false earth-centred astronomy by means of an imposing book"). He describes the development of the nebular hypothesis from Kant and Laplace to recent developments by Edouard Roche and Sir George Howard

Darwin about the influence of tides in the formation of the Moon, mentions his erstwhile assistant T.J.J. See's work supposedly showing how tides satisfactorily explain the genesis of double stars, and goes on to articulate again the idea of the formation of the Solar System by the encounter of the Sun with a passing "tramp." Then comes the section on meteorites and the white nebulae. Next he speculates, on the basis of some recent work involving radium done at the Cavendish Laboratory in Cambridge, about the way that life might have sprung from inorganic matter.[51] He turns, without resorting to the first-person pronoun, to the significance of his own work on Mars:

> Just as researches on this Earth all point to the bringing forth of life by a planet as the necessary outcome of its own career, provided its physical condition be right, so has investigation in the sky. From our island home in space we may peer across to other islands voyaging through the void, and by telescopic help mark what there is going on. In very different stages we find them, of their own evolutionary career. Such diversity of itself attests a general development.... Now, within the last few years, research has brought to light testimony that our nearest of solar kin has had its organic history too. Upon the planet most likely to support such existence at the present moment, other than ourselves, study has disclosed features which cannot be explained except as evidence of trans-planetary life. Pregnant with thought this is, for it brings corroboration of the whole evolutionary process from beyond the confines of our native earth. That the inorganic should develop into the organic on a single planet might perhaps be accidental, but not on two. From Mars comes the cosmic assurance that it is Nature's law.[52]

Unfortunately, Lowell's bubble was about to burst. On or about April 23, 1909, an issue of *Science* under that date fell through the mail slot at the Lowell Observatory, which included a devastating review of *Mars as the Abode of Life*. It was written not by a fellow astronomer but by a young geologist, Eliot Blackwelder, then at the University of Wisconsin and later at Stanford. Blackwelder's case against Lowell was, ironically, the same Lowell himself had made to the editor of *Nature* regarding Alfred Russel Wallace's review of *Mars and Its Canals*: "misstatements cannot be too carefully avoided in science, especially when a man, however eminent in one branch, is wandering into another not his own."[53] Lowell was about to be hoist by his own petard, for as William Graves Hoyt points out, "by going outside astronomy, and especially into geology, to establish his doctrines of planetology he had merely "laid himself open to attack on his flanks and ... swelled the ranks of his attackers."[54] With planetology, he had simply overextended his intellectual baggage train.

Blackwelder the geologist noted, for one thing, that Lowell's geology was now decades out of date. That, however, was perhaps only to be expected of an interloper without the necessary training, knowledge and expertise:

It is not surprising that Mr. Lowell, an astronomer, should have only a layman's knowledge of geology, but that he should attempt to discuss critically the more difficult problems of that science, without, as his words show, any understanding of the great recent progress in geology, is astonishing and disastrous. One can not but recall the old adage that "fools rush in where angels fear to tread." [25, p. 659]

Blackwelder finds much fault in Lowell's sweeping generalizations from geology and paleontology. However, he finds a more serious problem in Lowell's seeming ignorance of such fundamentals as the nature of metamorphic rocks, of the mass of evidence opposed to the postulate of a warm, damp, clouded, seasonless climate for Earth during the Carboniferous Period, and his profound ignorance of contemporary botany. He saves, however, his strongest criticisms for Lowell's theory of "desertism," noting that James Dwight Dana's ideas on which they were based were long out of date, and that, says Blackwelder, were Dana still alive, he would no doubt himself reject these views. The current thinking of geologists was that "there have been fluctuations of land and sea throughout recorded geologic history, and … no general tendency" of the sort claimed by Lowell. Now Blackwelder has got up a head of steam and continues apace:

Having assured his readers that the Earth is drying up and that it will sooner or later "roll a parched orb through space," he cited as proof the alleged fact that deserts are increasing in size. This is the beginning of the dreadful end which "is as fatalistically sure as that to-morrows sun will rise, unless some other catastrophe anticipate the end." Here again the proverb applies, "a little knowledge is a dangerous thing."… Had he inquired into the recorded facts of geologic history, he would have learned that deserts have existed in many parts of the world ever since the earliest times, wherever topographic and atmospheric conditions were favorable. It is not probable that our present deserts are more extensive than those of the Permian period, during which the saltiest of salt lakes partially covered the site of Germany.[55]

The examples given are enough to "show what kind of pseudo-science is here being foisted upon a trusting public," says Blackwelder, by "a highly educated man of distinguished connections and some personal fame" who "writes in a vivid, convincing style, with the air of authority in the premises." In fact, however, he is little more than a charlatan:

> The average reader naturally believes him, since he can not, without special knowledge of geology and kindred sciences, discern the fallacies. He has a right to think that things asserted as established facts are true, and that things other than facts will be stated with appropriate reservation.... The misbranding of intellectual products is just as immoral as the misbranding of the products of manufacture. Mr. Lowell can not be censured for advancing avowed theories, however fanciful they are, for it is the privilege of the scientist nor for making unintentional mistakes, for that is eminently human. But I feel sure that the majority of scientific men will feel just indignation toward one who stamps his theories as facts; says they are proven, when they have almost no supporting data; and declares that certain things are well known, which are not even admitted to consideration by those best qualified to judge. Censure can hardly be too severe upon a man who so unscrupulously deceives the educated public merely in order to gain a certain notoriety and a brief, but undeserved, credence for his pet theories.[56]

There is no way of knowing whether Barnard read this review, though it seems likely. If so, he might well have recalled what he had written (but never published) in his youthful *Mars: His Moons and His Heavens*, which anticipates the care he took in making and interpreting his observations and the judicious manner. Writing in 1880, he had said, "Man is too quick at forming conclusions."[57]

Reeling from Blackwelder's review, Lowell adds a searing footnote to "The Revelation of Evolution," in which he refers to Blackwelder as a "geologist out West" (as opposed to those belonging to the Eastern establishment, such as Agassiz at Harvard and Dana at Yale), and defends himself as having, like Darwin, "[dared] to synthesize the facts." Darwin's case shows that "the chief function of genius is to change the world's point of view." It was the general resistance to this invaluable service (resistance that Blackwelder embodies in his own case) that made the loneliness and neglect of geniuses so sad. From the phase of "myopic mediocrity" that is generally the condition of the species, he writes,

> ... rises now and then a man [who is a] mutation, the biologist would call him, a sport. And the world's sport he usually is, or worse. The fact that he differs from his fellows is cause for condemnation at once.... Only intrinsic excellence enables such mutations to survive; the fact that they can stand alone in self-sufficiency.

Now the chief distinction between this man and his mates is ... simply that he is intrinsically distinct. Not so much that his intellect is keener, nor that his energy is great, though both these are ancillary to the result, but that he sees things untrammeled by the prejudices of his time. He rises superior to the crystallized conceptions of the race... Bowing neither to custom nor authority, he perceives facts undistorted by the glasses society has self-imposed. ...

Each century starts with the ideas it was bequeathed. To alter them is no easy task. For whether right or wrong their own inertia carries them on... Especially is this the case where the new idea affects man's vanity by lessening his self-esteem. Darwin's ideas did both, and many of us can remember the storm of opposition they evoked. ... Hardly a man of his own generation accepted his ideas; not a man of the next but subscribed.

But there is, unfortunately, a sad side ... which we shall do well to let sink into our hearts. A genius speaks to one world, only to be heard by another. The world of his own day turns a deaf ear to him, the world of the generation that comes after acclaims...[58]

Here he goes from arrogance, often in evidence even during the Far Eastern period, to delusions of grandeur. He had anticipated this development in the chapter "On Exploration" with which he had opened *Mars and Its Canals*:

To be a pioneer in thought is to stand ... alone with nature, not for a few minutes, but for life. The isolateness of the few great minds of each generation of men is utterly undreamed of, for want of understanding, by those about them...

As if this loneliness by nature were not enough, it must needs be accentuated by man. For he rises in such cases in chorus to condemn. Consider Darwin, in patient study, testing the working out of natural selection and adding fact to fact, only to have the whole denounced as ridiculously absurd. Think you the denunciations of the master while living are wholly compensated by the plaudits after he is dead? The loneliness of greatness is the price men make the genius pay for posthumous renown.[59]

Lowell's rejoinder to Blackwater appeared only in August, and in the meantime appeared a train of letters to *Science* extending from the spring to the fall, with mingled voices being heard until the whole came to resemble one of the sestets of one of Mozart's buffo operas. First the egomaniacal and disgraced Thomas Jefferson Jackson See, now in exile at Mare Island, wrote to defend Lowell "fervently if somewhat irrelevantly," says Hoyt. In addition, and rather adventitiously, See criticized Chamberlin and Moulton on the grounds they had simply produced an "inconsistent and destructive" criticism of Laplace's nebular hypothesis (while ignoring the complete mish-mash of unsystematic and disorganized "cosmogonic" ideas he, See, had himself published over the previous 25 years).[60] Next Yale University geologist Joseph Barrell offered qualified praise for Lowell (not of his geology but his other contributions) while singling See out for the distinction of having penned a "personal, befogging and dogmatic rejoinder," whereupon Moulton joined in and without mentioning Lowell at all pointed to See's actually rather thin record of accomplishments (as well as adding to certain unspecified personal "aberrations").[61] At last, Lowell wrote a letter to *Science* in which he rather tediously and *ad nauseam* repeated the points made in "Revelation," insisting his geology was taken from "recognized" if mostly out-of-data sources, "Dana, [Sir Archibald] Geikie, Dr. Lapparent and recent research," and that "only the weaving together is new" [26, p. 338] He then attacked Moulton at length. The Chicago celestial mechanician would have the final say at the end of November claiming that Lowell had in fact largely borrowed (unacknowledged) key aspects of the very planetesimal hypothesis in which he had claimed to have found fatal errors, and wondered "whether in his forthcoming book on 'The Evolution of Worlds' he will not give additional proof of his affection for the planetesimal theory, though perhaps under some other name, or in some nameless form more congenial to that mysterious 'watcher of the stars' whose scientific theories, like Poe's visions of the raven, have taken shape at midnight" [27].

The "mysterious 'watcher of the skies'" at this point retreated from the pages of *Science*, but continued to persevere, before more friendly audiences, on the theme of the persecuted pioneer. Shortly after Moulton's letter appeared, he lectured in Prescott, Arizona Territory:

> Every new departure from a beaten path has its youth against it, and the more so in proportion as it is destined to revolutionize man's thought.... Only the accustomed and the commonplace do men take kindly at once. The strange terrifies them. It is with ideas in men as with unfamiliar sights in beasts. Both shy at first at what they have never seen before. Scientist and layman alike are afraid to commit themselves to that upon which they have not been brought up.[62]

Lowell, who was in many respects, such as politics and investments, as deep-dyed a conservative as could be, would without intended irony rail against conservatism as "troglodytic," and declaim against it as "a self-confessed euphemism for dull." He would add: "For conservatism proceeds from slowness of apprehension. It may be necessary for certain minds to be in the rear of the procession, but it is of doubtful glory to find distinction in the fact.[63] And yet at that very moment he was in fact falling back toward the rear of the procession. He had hardly begun his own campaign of Mars observations in 1909 when the erstwhile *bête noire* of the spokes of Venus, Eugène Michel Antoniadi, had already issued the third of a series of "interim reports" on Mars in the *Journal of the British Astronomical Association* based what he had been able to see with the Grande Lunette at Meudon Observatory near Paris. What he had seen included not only what was "new" and "revolutionary" and "strange," but demonstrated quickness, not slowness of apprehension, and from Lowell's point of view, "unpardonable improper" ideas. If, to use an analogy, observation is the infantry and theory the cavalry of science, Antoniadi's infantry was breaking through Lowell's center, while the cavalry (which included the camera and spectroscope) stood inactive and watched uselessly from a far-off hill.

Percival Lowell would be confronted with views of Mars that were strange and unfamiliar, views to which it was impossible to commit himself. They would be utterly different from those in which he had held to with the intensity of delusion. He could not embrace them; he could only shy away from them, and deny, to the bitter end, what went against the satisfaction of his own greatest needs.

Notes

1. Quoted in Hoyt, *Lowell and Mars*, p. 13.
2. P. Lowell to Scriven Bolton, March 4, 1907;. Lowell Observatory Archives.
3. Williams had an interesting subsequent career. On his return from Lowell, he became an instructor of mathematics in the Department of Mathematics at Indiana University, left in 1911 to attend Princeton University, then returned to Indiana in 1914 to resume his teaching career in the Department of Mathematics (including a stint as Chair from 1938–1944) until his retirement in 1957. Simultaneously, he served as a first lieutenant with the Indiana National Guard, and in 1916, under Brigadier General John J. Pershing, saw action in the Mexican Border Expedition against the paramilitary forces of Francisco "Pancho" Villa, establishing, on his return to Bloomington, Indiana, a Student Army Training Corps with himself in command. This was soon to be renamed the Reserve Officers Training Corps (R.O.T.C.). In later years, he became an amateur historian of the Civil War, and published *Lincoln Finds a General*, five volumes (of an intended seven) about the Union general Ulysses S. Grant.
4. A photograph of him in academic robes and mortar board taken on this occasion was later used by the Italian-American painter Ercole Cartotto to produce a full-length painting that would long grace the walls of Lowell House at Harvard University.

5. "Saturn's Rings are Falling in," New York *Times*, November 9, 1907.
6. E.E. Barnard to T. Lewis, January 11, 1908; Royal Astronomical Society Library.
7. Barnard, "Observations of Saturn's Ring," pp. 355–356.
8. Barnard, "Additional Observations," pp. 365–366.
9. P. Lowell [6]. This was Lowell's first article on the tores written after Barnard published his papers in the *Monthly Notices*, and shows that he must have given it only cursory attention, for he writes (pp. 139–149), apparently just before submitting his own paper, and completely missing Barnard's whole point: "Since this article was written Barnard has published his observations with his explanations. His explanations, however—for he gives two,—one that the eye sees through the underside of the rings and that such light is greatest where the rings are densest, the other the exact opposite, that the light is most where the ring is least crowded—are self-condemning on several counts, one for instance that the inner condensation does not fall by his own showing on the ansal position of any part of ring A, but wholly on the crape ring. Each explanation might account for one agglomeration alone but for that very reason fails for both. The presence of the gaps is another fatal objection to them." Though Lowell said little more about the "tores," he seems to have continued to have had complete faith in them, and much of his later work was concerned with his attempts to show in detail how the inner satellites had sculpted various fleetingly glimpsed divisions in the rings. This effort culminated in P. Lowell (1915) "Memoir on Saturn's Rings," *Memoirs of the Lowell Observatory*, vol. 1, no. 2, in which the theoretically derived positions of gaps in the rings agree rather precisely with those observed at Flagstaff.
10. A.L. Lowell, *Biography of Percival Lowell*, p. 134. At the kind invitation of Lick Observatory Director Joseph Miller and former Director and historian of astronomy Donald E. Osterbrock, the author had a chance to observe the August 10, 1995 ring-plane passage with the 36-inch refractor at Lick Observatory. At first, in bad seeing, a soft-edge but prominent ring shadow with a dark core appeared that corresponded to what Lowell had seen at Flagstaff in 1907. The impression of a cloud of particles spreading out from the ring-plane was certainly convincing, but later, in moments of exquisite seeing, the effect vanished and the shadow narrowed to engraved sharpness. It was clear that the Lowell effect is caused by minor atmospheric turbulence widening the shadow and diffusing the edges. See: W. Sheehan and S.J. O'Meara (January 1996) "Crossing Saturn's Rings," *Sky & Telescope*, pp. 106–110: p. 107.
11. E.E. Barnard to G.E. Hale, undated (but received by Hale on June 1, 1908). Hale. George Ellery. Papers, Mount Wilson and Palomar Observatory Library.
12. As noted in Hoyt, *Lowell and Mars*, p. 209.
13. Lowell had for years been carefully cultivating a friendship with Schiaparelli, sending him the Observatory publications and, in recent years, photographs, which the Italian astronomer, now suffering from extreme near-sightedness, had to examine with a microscope. The latter responded on June 30, 1905 to the

news that the canals—or at any rate many dark lines—had been photographed: "I would never have believed that it would be possible, and I doubt that we could arrive at something similar in our climate with such an agitated atmosphere. The climate of the Flagstaff Observatory must be very exceptional and the observers of very extraordinary skill." Regarding the dedication to Schiaparelli, Lowell had asked permission in a letter of (about) June 27, 1906: "Dear colleague and friend—I have just put my hand to the conclusion of a new book about the planet Mars and I would like to show my respect by dedicating it to you to whom it owes the magnificent success of this research." Schiaparelli of course accepted "the mark of esteem with which you propose to honor me," but admitted later (February 1, 1907) to being somewhat embarrassed by the comparison with Columbus, given the latter's "terrible ambition, which was the principal cause of the misfortunes in which the great Admiral was overwhelmed." He did endorse the book as "a small masterpiece," and on the subject of areography alone—the mapping of the planet,—he wrote, "You have walked straight and steady … where many times I have stayed in uncertainty, embarrassed by the unexpected and by the strangeness of what I have seen. Your book contains a true program of research for observers to come… I am now more persuaded by you that the hypothesis of oceans and continents cannot represent the facts, and the hypothesis of vegetation is, for the moment, the most probable of many. In any case, even the most skeptical should recognize in it an excellent working hypothesis." This from a man who had been unable, owing to deteriorating eyesight and the increasingly poor seeing conditions in Milan, to observe the planet to any real purpose since 1890. See: Jennifer Putnam and William Sheehan [7].

14. Quoted in Hoyt, *Lowell and Mars*, p. 201.
15. Quoted in Ibid., p. 207.
16. Excerpts of reviews, back matter in A.R. Wallace, *Is Mars Habitable?* and P. Lowell, *Mars as the Abode of Life*.
17. Excerpt of review, back matter of P. Lowell, *Mars as the Abode of Life*.
18. Three months after obtaining the spectrograms that putatively showed water vapor in the atmosphere of Mars, Slipher, in April 1908, wrote to Campbell. Mostly he asked the Lick Director for advice about reflecting telescopes, but he also mentioning his spectrographic results *in passim*. Campbell congratulated him for his apparent success, but expressed an interest in examining Slipher's spectrograms for himself. A few days later, Slipher sent copies of the spectrograms. Immediately Campbell saw that the extended red sensitivity of Slipher's plates decreased rapidly with wavelength right at the location of the *a* band, which made it difficult to see the band, and in addition, Mars and the Moon at the time the spectrograms were taken were far apart in the sky, so that their light had travelled along dissimilar path lengths in the Earth's atmosphere. Campbell doubted, therefore, that the *a* band on Slipher's spectrograms showed any real strengthening at all.

At this time, April 1908, Campbell was already beginning to plan (though he did not tell Slipher this) an expedition to Mount Whitney to obtain spectrograms of Mars under the most favourable conditions. It took over a year to plan, but finally, after much preparation, on September 1, 1909, Campbell and a recent Lick Ph.D., Sebastian Albrecht, obtained a good series of exposures on Mars and the Moon with the two bodies at nearly the same altitude. When the results were analyzed, the *a* band was found to be weak in all the spectra, and certainly no stronger in Mars than in the Moon. This meant both that there was very little water vapor above Mount Whitney and that there was almost none in the atmosphere of Mars. Campbell's negative result would bear the test of time: only in 1963 were tiny amounts of Martian water vapor finally detected further in the infrared by extremely sensitive spectroscopy on 100-inch reflector at Mount Wilson. See: Donald E. Osterbrock [8].

19. Ritchey, then starting to plan the 100-inch reflector for Mt. Wilson, proposed to Barnard setting up the telescope in a shelter on wheels, which could be rolled over for protection during the daytime and in inclement weather, then pulled away at night. G.W. Ritchey to E.E. Barnard, September 7, 1909; Barnard. Edward Emerson. Papers, Special Collections and University Archives, Jean and Alexander Heard Library, Vanderbilt University. Barnard had long suffered greatly from "dome seeing" at Yerkes must have thought it a capital idea, and once told E.M. Antoniadi he "hoped the great Yerkes dome could be mounted on a rail, so that he could push it away and work in the open air...." Quoted in Sheehan, *Immortal Fire*, p. 396. Antoniadi cited Ritchey's opinion that the fact that the Meudon refractor he used with such great success in 1909 is located at a considerable height (24 m) above ground was a great advantage. E.M. Antoniadi (1975), *The Planet Mars*, trans. Patrick Moore (Shaldon Devon, UK: Keith Reid), p. 17.

20. Putnam, *Explorers of Mars Hill*, p. 80.

21. Ibid., pp. 74–75.

22. Quoted in A.L. Lawrence, *Biography of Percival Lowell*, p. 135.

23. Strauss, *Percival Lowell*, p. 220.

24. See William Sheehan [9, p. 125 and p. 130]. Though Schiaparelli often seems to have straddled the fence, he was not unsympathetic to Cerulli's views, and wrote to him a letter in July 1907 in which he compared the various stages of seeing the Martian surface details to those of seeing a printed page from various distances. In a first stage, A, the vision is confused and the page appears as a grey square; at a next stage, B, this view is replaced with a vision of geometrical lines; at a third stage, C, one suspects breaks and irregularities; finally, at stage D, one is able to read the individual letters. Thus, Schiaparelli concludes: "[T]he first observers of Mars, to 1860, lived in stage A. Since this epoch, Secchi, Kaiser, and Dawes came near to stage B, finding some lines.... In the years after 1877 the view produced in me and others was stage B—a vision apparently complete and accurate of single and double lines on the planet. Now, thanks to you [Cerulli],

we are entering stage C; the naïve faith in the regularity of the lines is shaken, and we have the prospect of yet another stage, D, in which the appearance of lines will resolve into forms of a different order—closer to the true structure of the Martian surface. But will this, then, be the final truth? No'; for of course as optics continue to improve, the process will proceed to other stages of vision, or illusion. My thanks to you for the progress you have realized along this stairway." (Translation: W. Sheehan.) The further stage, D, had been realized by Barnard in 1894 and would be by E.M. Antoniadi in 1909.

25. S. Newcomb [10, p. 2]. For this paper, Newcomb transported the "Small Boy Theory" to Yerkes, where both Barnard and Phillip Fox took turns viewing through an opera glass at a distance of 100 feet eight or nine scale drawings of Mars on which markings had been overlaid on a mottled surface. In the result, broken lines looked continuous, while straight black lines appeared blurred.

26. S. Newcomb [12], where he says, "One word to correct a possible misapprehension of the bearing of my argument. So far as it goes, the canals of Mars might be fine lines of inky blackness. It only seeks to show that there is an indefinite number of other features which an observer may train himself into interpreting as fine dark lines, and that the actualities on Mars may therefore differ widely from the observer's optical inference." Lowell's response to Newcomb's response is at P. Lowell [13]. The whole controversy produced lasting ill-feeling between Newcomb and Lowell, so that, as Newcomb wrote Campbell on traveling through Flagstaff in June 1907: "Six months ago I could not have entertained the idea of going through Flagstaff without stopping. But it was the middle of the night, and Lowell did not seem overanxious to have me look at Mars through his telescope." S. Newcomb to W.W. Campbell, June 30, 1907; Campbell. William Wallace. Papers, Mary Lea Shane Archives of the Lick Observatory, University of California, Santa Cruz.

27. Wallace rejected the possibility of intelligent life anywhere else in the Universe, on the grounds that human beings were "the unique and supreme product of this vast universe" and that "the universe was actually brought into existence" for the purpose, specifically, of supporting human life. See: A. Russel Wallace [15].

28. A. R. Wallace [16]. For background, see: Robert W. Smith [17].

29. Wallace, *Is Mars Habitable?* pp. v–vi.

30. Ibid., p. 7.

31. Ibid., pp. 103–105.

32. Ibid., pp. 44–46.

33. Ibid., p. 51.

34. Ibid., p. 109. Lowell's estimate of the albedo of the Martian atmosphere was the most critical parameter in his calculation, but because he did not recognize that dust storms occur on Mars and that suspended fine dust is always present to a greater or lesser degree, he underestimated the brightness—and hence overestimated the thickness—of Mars's atmosphere. The effects of local dust storm activity on both ground and air temperatures on Mars were studied by InSight

at Elysium Planitia during the Local Dust Storm of 2018. See: D. Viúdez-Moreiras, C. E. Newman, F. Forget et al. [19].

35. Wallace, *Is Mars Inhabitable?* p. 52.

36. See: Carl Sagan [20]; Stephen Jay Gould, "War of the Worldviews," *Natural History* 12/96-1/97, pp. 22–33.

37. P. Lowell to Mark Wicks, February 27, 1908; Lowell Observatory Archives. On Wicks, see: Robert Crossley [21]. From what little is known about him, he seems to have been knowledgeable about pipe organs, and in 1887 published an often-reprinted book, *Organ-Making for Amateurs*, though astronomy was clearly an avocation, and in 1911 he published *To Mars via the Moon: an astronomical story* (London: Seeley and Co.), which was dedicated, rather effusively, to Lowell. He declared in the dedication that "many years' careful study of the various theories which have been evolved has convinced me that the weight of evidence is in favour of Professor Lowell's conceptions, as being not only the most reasonable but the most scientific; and that they fit the observed facts with a completeness attaching to no other theory."

38. P. Lowell to editor, *Nature*, March 19, 1908; Lowell Observatory Archives.

39. Hoyt, *Lowell and Mars*, p. 233.

40. G.V. Schiaparelli to P. Lowell, March 26, 1909. Draft in Archives of the Brera Observatory. (Trans. Jennifer Putnam)

41. Lowell, *Mars as the Abode of Life*, p. 27.

42. Ibid., pp. 111–112.

43. Ibid., p. 134.

44. Ibid., pp. 215–216.

45. Ibid., pp. 3–4.

46. Lowell, "Revelation", p. 178.

47. The first case was the "Sombrero Galaxy" (NGC 4594) in Virgo. V.M. Slipher (1914) "The Detection of Nebular Rotation," *Lowell Observatory Bulletins*, vol. 2, no. 62.

48. V.M. Slipher (1913) "The radial velocity of the Andromeda Nebula," *Lowell Observatory Bulletins*, vol. 2, no. 8, is the first one in which this is noted. At the end of 1914, at the seventeenth American Astronomical Society meeting in Evanston, Illinois (at which Edwin P. Hubble was presented), he presented fifteen radial velocities, and received a standing ovation. Technically, of course, the spiral nebulae were not rotating: the stars within them were revolving in a common plane and direction, giving the appearance of a single object in rotation.

49. Hoyt, *Lowell and Mars*, p. 240.

50. Lowell, "Revelation," p. 174.

51. The Irish physicist J. Butler Burke added radium salts into sterilized beef broth, and produced a profusion of small mutating shapes he dubbed "radiobes." He classified them as existing somewhere on the border of living and non-living. In fact, subsequent experiments at Cavendish showed that they were nothing more than sulfide precipitates, therefore completely inorganic. These experiments had

been completed even before Lowell had set pen to paper, which shows just how rapidly these fields undergoing change at the time.

52. Lowell, "Revelation," p. 179.
53. P. Lowell to editor, *Nature*, March 19, 1908; Lowell Observatory Archives.
54. Hoyt, *Lowell and Mars*, p. 256.
55. Ibid., p. 660.
56. Ibid., p. 661.
57. E.E. Barnard (1880) "Mars; his moons and his heavens." Unpublished manuscript, Vanderbilt University Archives.
58. Lowell, "Revelation," pp. 180–182.
59. Ibid., pp. 182–183.
60. T.J.J. See (1909) "Fair Play and Toleration in Science," *Science* (May 28), pp. 358–360. Hoyt, *Lowell and Mars*, p. 258.
61. F.R. Moulton (1909) "Remarks on Recent Contributions to Cosmogony." *Science* (July 23), pp. 113–117.
62. P. Lowell (1909) "The Lowell Observatory and Its Work." Text of lecture delivered in Prescott, A.T., November 24; Lowell Observatory Archives.
63. Lowell, *Evolution of Worlds*, pp. 179–180.

References

1. Lowell P (1909) The evolution of worlds. New York, Macmillan, p 106
2. Lowell P (1907) Tores of saturn. Lowell Obs Bull 1:186–190
3. Barnard EE (1908) Observations of Saturn's Ring at the time of its disappearance in 1907, made with the 40-in. refractor of the Yerkes observatory. Mon Not R Astron Soc 68:346–360
4. Lowell P (1907) Saturn's tores. Sci Am 97(24):441–442
5. Barnard EE (1908) Additional observations of the disappearances and reappearances of the rings of saturn in 1907–08, made with the 40-inch refractor of the Yerkes observatory. Mon Not R Astron Soc 6(360–366):364
6. Lowell P (1908) The tores of saturn. Popular Astron 46:133–146
7. Jennifer Putnam and William Sheehan (2021) A complicated relationship: an introduction to the correspondence between Percival Lowell and Giovanni Virginio Schiaparelli. J Astron Hist Herit 24(1):179–227
8. Osterbrock DE (1989) To climb the highest mountain: W.W. Campbell's 1909 Mars expedition to Mount Whitney. J Hist Astron 20:77–90
9. Sheehan W (1996) The planet mars: a history of observation and discovery. University of Arizona, Tucson
10. Newcomb S (1907) The optical and psychological principles involved in the interpretation of the so-called canals of mars. Astrophys J 26(1):1–17
11. Lowell P (1907) The canals of mars, optically and psychologically considered: a reply to professor Newcomb. Astrophys J 26(3):131–140

12. Newcomb S (1907) Note on the preceding paper. Astrophys J 26:141
13. Lowell P (1907) Reply to professor Newcomb's note. Astrophys J 26:142
14. Langmuir I (1989) Pathological science. Res Technol Manag 32(5):11–17
15. Russel Wallace A (1903) Man's place in the universe. McClure, Phillips & Co., New York, p 315
16. Wallace AR (1907) Is mars habitable: a critical examination of Professor Lowell's book "Mars and Its Canals," with an alternative explanation. Macmillan, London
17. Smith RW (2015) Alfred Russel Wallace, extraterrestrial life, mars, and the nature of the universe. Vic Rev 41(2):151–176
18. Lowell P (1907) A general method for evaluating the surface-temperatures of the planets; with special reference to the temperature of Mars. Philos Magazine, Ser 6(14):161–176
19. Viúdez-Moreiras D, Newman CE, Forget F et al (2020) Effects of a large dust storm in the near-surface atmosphere as measured by InSight in Elysium Planitia, Mars. JGR Planets. https://doi.org/10.1029/2020JE006493
20. Sagan C (1980) Cosmos. Random House, New York, pp 108–109
21. Crossley R (2011) Imagining Mars: a literary history. Wesleyan University Press, Middletown, p 91
22. Brush SG (1996) Fruitful encounters: the origin of the solar system and of the moon from Chamberlin to Apollo. Cambridge, Cambridge University Press, p 48
23. Lowell P (1909) The evolution of worlds. New York, Macmillan, p 24
24. Lowell P (1909) The revelation of evolution: a thought and its thinkers. Atl Mon 104:174–183
25. Blackwelder E (1909) Mars as the abode of life. Science 29(747):659–661
26. Lowell P (1909) Mars as the abode of life. Science 30(767):338–340
27. Moulton FR (1909) A reply to Dr. Lowell. Science 30(775):631–641

Contents

There are celestial sights more dazzling, spectacles that inspire more awe, but to the thoughtful observer who is privileged to see them well there is nothing in the sky so profoundly impressive as these canals of Mars. Fine lines and little gossamer filaments only, cobwebbing the face of the Martian disk, but threads to draw one's mind after them across the millions of miles of intervening void.
—*Percival Lowell,* Mars as the Abode of Life, *p. 146*

Barnard insisted in a letter to Hale that reports of the canals on Mars, as well as the markings on Venus, Mercury, and the satellites of Jupiter, should be "wiped out from the literature of astronomy."
—*David Strauss,* Percival Lowell: The Culture and Science of a Boston Brahmin, *p. 224*

W. Sheehan, *Parallel Lives of Astronomers*, Springer Biographies,
https://doi.org/10.1007/978-3-031-68800-3_11

A Blurry Mars

Mars, at its September 28, 1909, opposition would approach closer to the Earth than it would again in either Lowell's or Barnard's lifetimes. In addition, it would be located just 4 degrees south of the Celestial Equator (which was far better than the 28 degrees south in 1907), and so much more favorably placed for observation at Northern Hemisphere observatories.

As usual, efforts to observe the planet at Flagstaff began well before opposition—in April, with the disk only 7–8 arcseconds across. E.C. Slipher, veteran of the "Lowell Expedition to the Andes" and the member of the staff most interested in Mars observations, found the albedo markings fairly normal looking until May 25, when he found the Pandorae Fretum invisible and Hellas dull. We now know that in fact a planet-encircling dust storm was getting underway, which developed over the summer and caused fading (to near invisibility) of the main markings including even the Syrtis Major. We do not know this from Lowell Observatory records. Lowell attributed the planet's bland and lemony appearance to poor seeing, and admittedly this may, in part, have been true, since from late June until late September Flagstaff experiences the annual "monsoon" season, when clouds and rain are common. In addition, Lowell had a lot of other things on his mind that summer. But Slipher too seems to have fallen off the pace.

British Astronomical Association Mars Section Director and historical researcher Richard McKim suggests on the basis of his sleuthing the Flagstaff observing records for dust storms:

> From Slipher's 1922 work I discovered that he tended to explain a lack of detail by bad seeing, whereas in fact at that time there was dust activity. Slipher recorded "'seeing awful" for the last four days of 1909 May in the Flagstaff logbook. This may have been true, but on June 3: "'The Solis Lacus is lighter than the dark region above—much so." Some notes on polar rifts follow on June 15 and 18. Then nothing till July 25, when: "The planet is darker than 3 days ago," and finally some more typical work from August 16. This hiatus suggests an unwillingness to admit that the Martian atmosphere [dust storm activity] was responsible for the lack of albedo detail.[1]

In other words, according to McKim, neither Lowell nor Slipher were able to accommodate evidence which went against their long-standing beliefs about the cloudlessness of the Martian skies. The lack of comprehension that the main markings were hidden under a pall of dust parallels that in October/November 1894 when Barnard and others witnessed a large regional dust storm. Lowell's brain (and Slipher's) simply refused to believe that the main Martian markings could disappear in this manner.

Indeed, the cynosure that strange summer was not Mars but Venus, and then in association with a highly anticipated vicennial celebration marking the opening of Clark University (Worcester, Massachusetts), at which Lowell would receive his second LL.D degree. Clark had opened on October 2, 1889, with graduate departments in mathematics, physics, chemistry, biology, and psychology. Originally scheduled for July, the event was pushed back to early September—and thus would butt up against the Mars opposition—in order to accommodate a special overseas guest: Sigmund Freud. Freud had been invited largely because the university president, G. Stanley Hall was a psychologist. He has been described as "an eccentric and an enthusiast," and was, like Lowell himself, "more an indefatigable publicist and advocate of novel ideas than an original researcher" [1]. Also like Lowell, he courted controversy. Reviewing Hall's bulky two-volume treatise *Adolescence*, the educational psychologist Edward L. Thorndike criticized it for being "chock full of errors, masturbation and Jesus." Hall was, Thorndike concluded, "a mad man."[2] Lowell's inclusion in the festivities had been arranged by an old friend, William E. Story, a mathematics professor at Clark University who had spent 2 weeks in Flagstaff with Lowell in 1894 and had been honored with the dedication to Lowell's 1895 book *Mars*.[3] Lowell was, however, far from the main attraction. His honorary degree was but one of twenty-two bestowed. The other recipients included Freud, of course; his Swiss protégé Carl Gustav Jung; and Nobel laureate Ernest Rutherford, who had just completed, with Hans Geiger and Ernest Marsden, his demonstration of extreme deflections of alpha particles shot at a thin gold foil that showed the existence of the atomic nucleus.

Lowell talked about Venus on the evening of September 8, and—still smarting from Blackwelder's criticism of him as outside his lane by venturing too far into geology—began by defensively insisting on the particularity of his own specialization:

> The special object of the observatory which I have the honor to represent is the study of planets of our Solar System, beginning with their present state and passing thence to their evolutionary history. So extended to-day is the astronomic field that to do good work one must specialize his endeavor, restricting himself to one particular branch of it and incidentally refraining, we may add, from discussing that of which he has not expert knowledge. Now research on the planets constitutes one such division, making as it were, an entity in itself. For diverse as the planets are to-day, they are all the result of one particular evolutionary process and knowledge of each member throws light upon the development, past or present, of the others.

It is popularly imagined that our gaze is concentrated on Mars, to the exclusion of much else, and that we are particularly concerned with its habitability. That this is a popular fallacy I shall show you tonight. For we shall contemplate together another planet in the light that study of the past 13 years at the Lowell Observatory casts upon it, and we shall see not only that such a study has indicated it not to be habitable, but that the question of habitability has not in the least affected our research. In short, to us habitation by organic life or non-habitation is merely an incident in the study of a planet's history, which we view with as strict scientific impartiality as we do the presence or absence of water-vapor in its air. We are concerned solely with the facts, a romantic enough revelation in themselves. [2 , pp. 521–522][4]

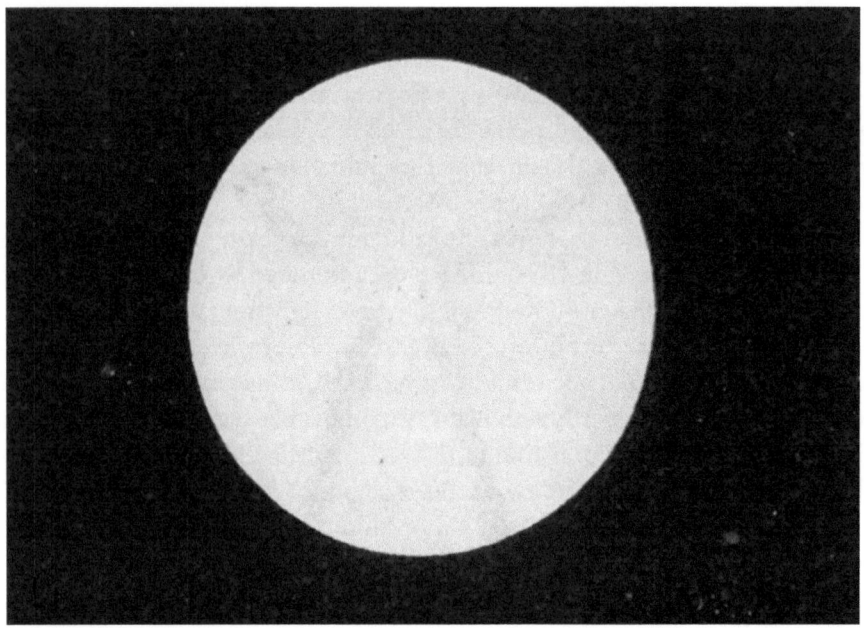

Though including the 1896–97 sketches of the spoke system, **The Evolution of Worlds** *includes only one of more recent vintage: this rather feeble rendering from April 12, 1909. It appears to have been done on a separate paper as it is not found in the Lowell observing log books. (From: The Evolution of Worlds, p. 79)*

Members of the audience expecting something new would likely have slumped back in disappointment, as Lowell simply rehashed old ideas about the planet circulating since 1896–97. Even his drawings of the planet were mostly recycled; a few sketches from April 1909 by himself and the Slipher brothers were presented, but they showed the spoke system of Venusian markings only feebly, perhaps betraying a lack of conviction even on the artist's part. It seems that the audience responded with only polite applause, with even the Boston *Evening Transcript* noting only that the talk "was cordially welcomed."[5] We do know that Robert Hutchings Goddard, the future pioneer of rocketry who was beginning graduate studies at Clark that fall, was in attendance. A native of Worcester, at the age of seventeen he had read in serialized form H. G. Wells's *War of the Worlds*, and been captured for life by a dream of going to Mars [3]. He wrote in his diary, "Professor Lowell's lecture on Venus in the evening. Good."[6]

To an audience forever seeking fresh stimulation, psychoanalysis proved to be more compelling than planetology. Freud delivered five lectures (in German) on consecutive days—the only ones he would ever give in the United States—with unexpectedly large crowds including the doyen of American psychology (and Lowell's sometime fellow neurasthenic) William James. Freud left Worcester feeling that psychoanalysis had reached the point of being "not a delusion any longer; it had become a valuable part of reality."[7] Though this proposition would continue to be debated, planetology was also—despite Lowell's defensiveness about it—something of the same order; a mass of speculations, provocatively set forth, of which some might be true, but of which the whole was hardly disciplined enough to qualify as a proper science. Some of planetology's findings might even be illusions, of the sort Freud would later analyze in his book about the basis of religion, *The Future of an Illusion*. It was a tendency of primitive man (his term) to project what he knows and wishes for into the world outside himself. "It is natural to him," Freud would say, "something innate, … to regard every event which he observes as the manifestation of beings who at bottom are like himself. It is his only method of comprehension. And it is by no means self-evident, on the contrary it is a remarkable coincidence, if by thus indulging his natural disposition he succeeded in satisfying one of his greatest needs."[8] So Lowell, too, in his quest for intelligent beings on Mars, had indulged his natural disposition, and satisfied his greatest needs, despite the disclaimers he had offered in his vicennial Venus lecture.

Mars from Yerkes and Meudon

Though the news about Mars from Flagstaff was sparse, there was a great deal of activity going on elsewhere. At Yerkes, Barnard photographed Mars at a level surpassing all previous efforts. He used the same set-up he had tried in 1907—a yellow color filter and a Brashear enlarging lens at the focus of the great refractor, with exposures of three or 4 s needed to register the image even on the fast Cramer Instantaneous Iso plates. He wrote:

In all the exposures, though of only a few seconds' duration, it was necessary to guide the telescope to keep the image stationary. This was done by bisecting the polar cap by cross-wires (spider threads) in the focus of the long guiding finder (61–1/2 feet focus) of the 40-inch telescope. [5]

Though he began imaging the planet in mid-July, when the planet's disk was some 14 arcseconds, he must have been puzzled by the results. Though the polar cap was clearly visible, the rest of the disk was blank—a result, of course, of the dust swirling around the planet at the time A month later, even with the disk grown to a generous 18 arcseconds, details remained faint, and Barnard decided to suspend his Mars photography until conditions improved.[9]

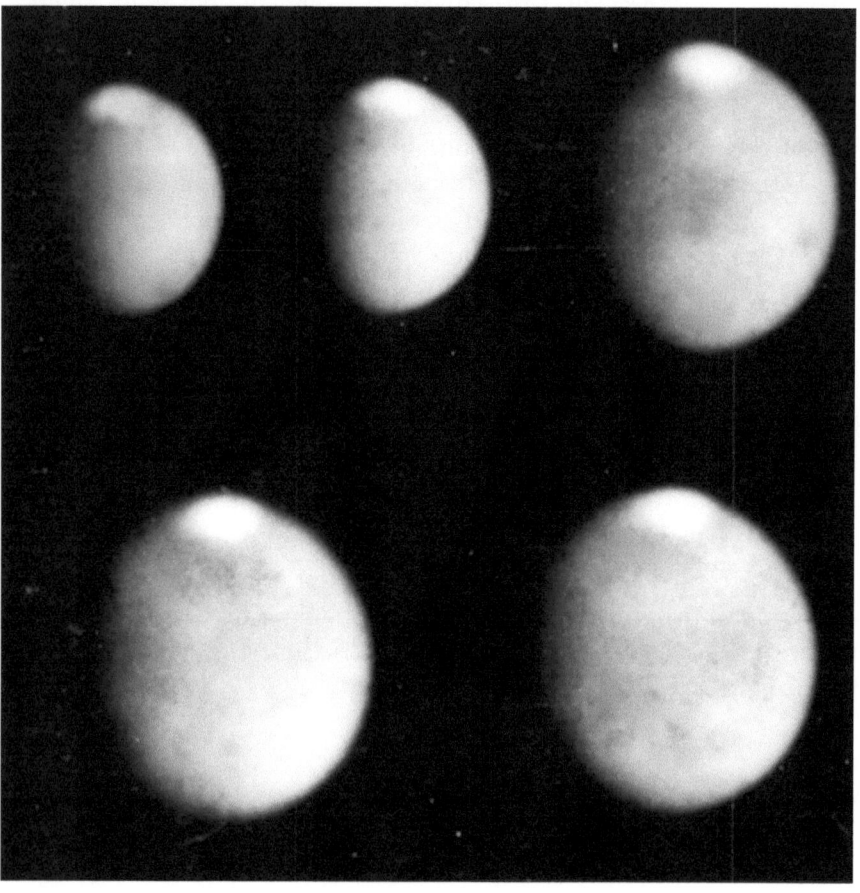

Barnard's images of Mars, showing the fading of the usual markings due to the great dust storm of 1909. Above, left and middle, July 14, 1909; right, August 18, 1909. Below, both August 18, 1909. These were not published during Barnard's lifetime. (From: Kuiper and Calvert (1947). Astrophysical Journal, vol. 105)

Slipher was also imaging Mars, without much more to show for himself. So, for the time being, the task of maintaining continuous records of the phenomena of the red planet was mostly carried forward by enthusiastic amateurs making visual observations. The members of the Mars Section of the British Astronomical Association took the lead. An early report came from Ellison Hawks, a colleague of Scriven Bolton's, who reported on July 23 that with Bolton's 12-inch and 18-inch reflectors, "As soon as work was commenced, it was noticed that the markings of the surface of the planet were extremely faint, and the colouring of the disk in general appeared of a much lighter tint than usual. That ruddy glow, which so characterises the 'fiery planet,' seemed to have entirely disappeared, its place being taken by a pronounced yellow tint. Even to the naked eye this change of colour was apparent to a certain extent, for the planet shone in the heavens with a lustre more akin to that of Jupiter than what we generally expect Mars to present."[10]

The man whose work on Mars during the 1909 opposition was to become legendary was also an amateur, Eugène Michel Antoniadi. He was Camille Flammarion's former assistant and, though still based in Paris, continued to serve (as he had since 1896) as Director of the British Astronomical Association's Mars Section. Until August he was away in Athens for a chess tournament, but upon his return he found an urgent letter awaiting him from another French amateur astronomer, René Jarry-Desloges. Like Lowell a man of independent means, Jarry-Desloges was also a devotee of Solar System studies, and set up several observatories selected for their favorable seeing, such as Mont Revard in Savoie and Le Masseqros in Lozère. His assistant Georges Fournier, with the 37-cm Schaer refractor at Mont Revard, had been among the first to notice the unusual faintness of the Martian albedo features, and Jarry-Desloges had wanted to enlist Antoniadi to the cause of monitoring further developments. Antoniadi's interest was piqued, and set up his personal telescope, an 8 ½-inch reflector by the British instrument-maker George Calvert, on the lawn in front of Flammarion's observatory at Juvisy, where he made his first observations on August 12, 1909.

The extraordinary feebleness of the albedo markings was apparent at a glance. In a first "Interim Report" to the BAA, which he submitted at the beginning of September, he wrote: "Now, it so happened that the great interest attaching to this opposition as further enhanced by Mars itself, the planet having displayed lately phenomena altogether without a parallel in the records of the past; and, while vast changes had transformed here and there the appearance of its dusky areas, a gloomy, yellow veil, obliterating the markings, enshrouded immense tracts of the Martian surface" [8 , p. 427]. Despite the faintness of the details, he had noted 32 canals, with the Indus and the Titan being "seen very easily"; all the rest were, as usual, glimpsed only "in fractions of a second."[11]

E. M. Antoniadi

E.M. Antoniadi, photographed on board the ship Norse King *during an 1896 eclipse expedition. This shows him the year he became Director of the BAA Mars Section. (Credit: Dr. Richard McKim and Council of the British Astronomical Association)*

E.-M. Antoniadi: Savant of the Skies

Here we must digress, briefly, to properly introduce Eugène Michel Antoniadi, whose name has already appeared occasionally in these pages and whose greatest role was as Lowell's nemesis (and Barnard's "bull dog") in the battle over the Martian canals.

Antoniadi was born to Greek parents in Constantinople (now Istanbul) in 1870 and christened Eugenios Mihail Antoniadis, which he later naturalized to the French Eugène Michel Antoniadi.[12] By the time he was in his late teens, he had already shown great aptitude and skill in drawing and architecture, had mastered several languages including not only modern Greek and the classical languages but French and English—his favorite model for English prose was Gibbon's *Decline and Fall of the Roman Empire*—and became a keen amateur astronomer. With a 75 mm refractor set up in the Princes' Islands in the Sea of Marmara, Antoniadi began making excellent drawings of sunspots and of Mars, Jupiter, and Saturn, and sent them for publication in the *Bulletin* of the Société Astronomique de France (SAF). The latter had been founded by Flammarion in 1887, and Antoniadi was a member from 1891. In 1890, hearing of the formation of the British Astronomical Association (BAA), he became one of the original members, having already published papers (perhaps in Gibbonesque style) in the *Journal of the Liverpool Astronomical Society*. In 1892, he upgraded his instrumentation to a 4 ¼-inch (108 mm) refractor by Mailhat, and became active in the BAA Mars Section, with four of his drawings being reproduced in the Section's first Mars Memoir.

The following year, 1893, he achieved a breakthrough. Flammarion invited him to come to Juvisy to serve as an assistant, and Antoniadi eagerly accepted. Presumably he made the long journey from Constantinople to Paris on the Orient Express—perhaps in company with his cousin Basil Zaharoff, the notorious arms dealer, who made the journey so often that he had his own railroad car reserved for the purpose.

Antoniadi arrived soon after Flammarion's publication of *La Planète Mars et ses conditions d'habitabilité*, the book which had so electrified Lowell on his return from Japan at the end of 1893.[13] No doubt Antoniadi found Flammarion inspiring to work for, as they both shared a particular interest in Mars. In addition to observing with Flammarion's 9-inch Bardou refractor, Antoniadi had access to Flammarion's very extensive library, whose extraordinary holdings continue to turn up surprises to the present day. (A copy of the *Nuremberg Chronicle*, worth millions, was found during a recent inventory; it was immediately deposited for safekeeping in the Bibliothèque de France.) In addition to his work for Flammarion, which included making many beautiful water-colors of the planets, Antoniadi in 1896 succeeded B.E. Cammell as Director of the British Astronomical Association Mars Section, in which capacity he published preliminary reports as well as the definitive Memoirs of the Section's work. The latter included maps summarizing what had been seen at each opposition, and during this era—the height of the "Mars furor"—each is tessellated with canals.

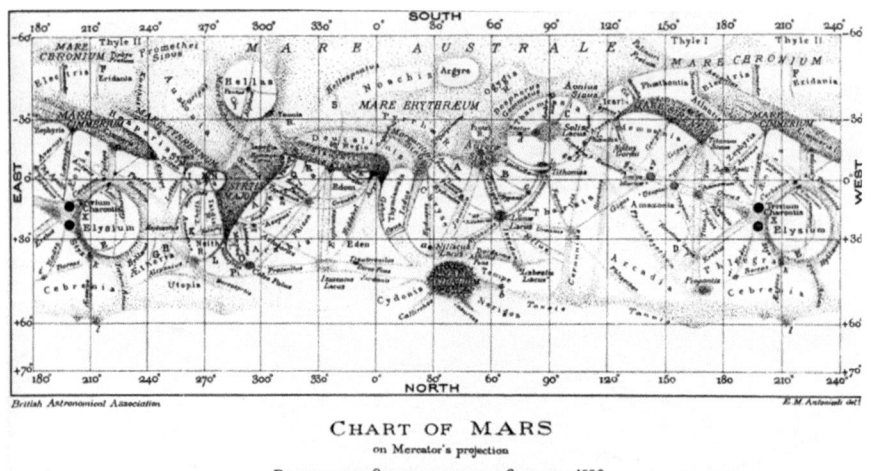

Antoniadi's first Mars map as Director of the BAA Mars Section, 1896, showing features reported both by himself and other members of the BAA Mars Section. (Credit: British Astronomical Association)

In some ways Antoniadi and Flammarion were too alike to get along indefinitely. They were both highly methodical, fastidious men, with strong personalities, and Antoniadi had from the first maintained some degree of independence by living not at the Observatory but in Paris. In addition, Antoniadi came increasingly to resent the contract he had signed, which enabled Flammarion to take credit for his work. Eventually, the stress of his situation began to affect his health.[14] Nervous and high-strung, he considered moving to England, and at one point confided to Annie Maunder (the wife of BAA founder and canal skeptic E. Walter Maunder) that he "could not stand M. Flammarion any longer."[15] He was rescued by marriage in 1902 to one Katherine Sevastopolos, a member of a well-to-do Greek family in Paris who also had Constantinople roots and connections. Following his marriage, Antoniadi no longer needed to work for a living. He resigned his position at

Juvisy, finished writing the BAA Mars Section *Memoir* for 1901 (in which he intimated that the canal network as depicted by Lowell was an illusion, "the result of the eye straining to see myriads of unresolvable details, or giving illusory hard edges to the intersections of half-tones of different intensity"), and he and Katherine moved into a flat at 74, rue Jouffroy, Paris. There he occasionally observed with the 8 ½-inch Calver reflector, but for the most part his interests lay outside of astronomy. He began studying chess seriously, and eventually would achieve "near grandmaster" status. His main preoccupation at the time, however, which would bring him back to Constantinople for 4 ½ months in 1904, was a massive project involving drawing and (for the first time, and with the approval of Ab dul Hamid II himself) photographing the Mosque of Hagia Sophia. In Constantinople, he secured 1008 negatives, and made as many sketches, later redrawn and included, together with many watercolors, in his three-volume *Atlas of the Mosque of St. Sophia*, published in a limited edition in 1907. It was his first book, the only one he ever published that is non-astronomical in its subject matter, and the only one he wrote in Greek. Clearly, he spared neither time nor expense, and paid a Paris printer 12,000 francs (about $75,000 in current dollars) for the plates. What is most interesting about it, for our purposes, was its use of the painstaking and time-consuming method of stippling for illustrations intended for publication, and his artistic philosophy of allowing no liberties with the exact and accurate representation of the subject matter before him. He was a realist, not an impressionist, and though he obviously developed his own style, he was closer to Meissonier than to Monet. The same stipple technique and exacting fidelity to the subject would later characterize his fully mature drawings of Mars.

Hagia Sophia, as painted by Antoniadi in 1904. Note the stippling technique, used in many of his planetary drawings. From: Eugène Michel Antoniadi, Ekphrasis tês Hagias Sophias, 3 vols. (Istanbul: P.D. Sakellarios, 1907–09)

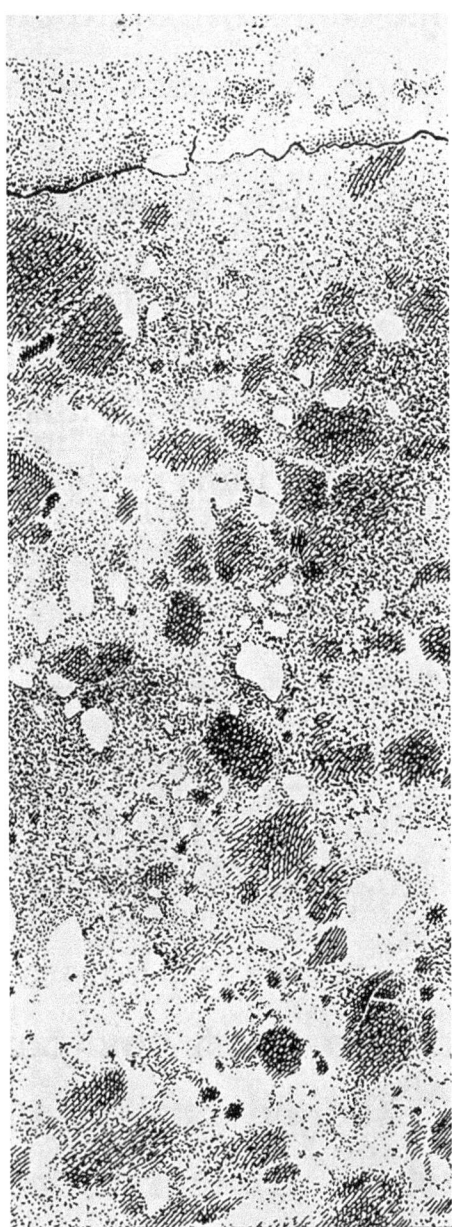

E.M. Antoniadi's stipple rendering of the appearance of the surface of a marble revetment at Hagia Sophia, showing the style he would later employ in drawing Mars. From: Eugène Michel Antoniadi, Ekphrasis tês Hagias Sophias, 3 vols. (Istanbul: P.D. Sakellarios, 1907–09)

"I Draw Well"

It was characteristic of Antoniadi's personality to jump from one intense interest to another. After leaving Flammarion, he largely abandoned astronomy for several years. It may have seemed, even to himself, that he was finished, that he had done all he would ever do. But in 1908, Antoniadi suddenly reappeared in the *Monthly Notices of the Royal Astronomical Society* with a careful—and on the whole very favorable—analysis of one of the Lowell photographs of Mars, that of July 11, 1907, in which he represented the "trustworthy" details in one of his signature stipple drawings. Seventeen canals are shown, as those "more or less discernible on the images [and] corresponding, with one exception, to canals of Schiaparelli's charts" [10 , p. 112]. On coming across Antoniadi's article, Schiaparelli told Lowell, "[T]hese attempts [show] that your photographs can already, for the quantity of details, be placed on the same line as visual observations."[16] He also suggested that, given Antoniadi's obvious skill in drawing, Lowell would do well to put other images into Antoniadi's hands for similar treatment. However, Lowell—perhaps still bearing a grudge because of Antoniadi's harsh criticisms of his Venus observations—rejected the idea; he preferred that Schiaparelli—despite his failing eyesight, and the fact that he now could use only one extremely myopic eye—be the one to examine the Lowell Observatory Mars photographs for fine details. ("You and you alone should create the representation of what might see in the photographs of Mars that I sent you," he wrote. "Because, in spite of what you write to me about your poor eyes, you can still see much better than Mr. Antoniadi or anyone else who has not studied the planet well.")[17] An opportunity was missed.

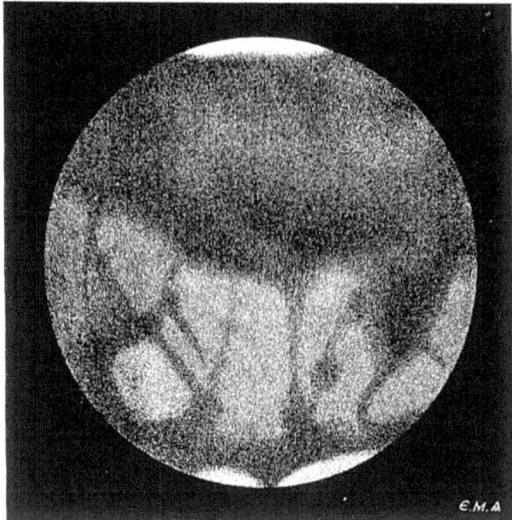

FIG. 1.—Drawing of Mars, showing all trustworthy details visible on photographs of the planet taken by Prof. Lowell on 1907 July 11. ω=250°.

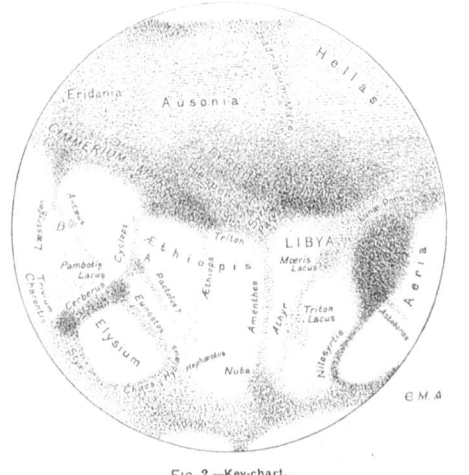

FIG. 2.—Key-chart.

Antoniadi's drawings showing "all trustworthy details" visible on a Lowell Mars image from July 11, 1907. (Credit: Monthly Notices of the Royal Astronomical Society, 1908)

Antoniadi himself produced only two telescopic drawings of Mars in 1905 and three in 1907; none is remarkable. It was only Jarry-Desloges's letter about the strange fading of the albedo features—including even the Syrtis Major—that truly galvanized him, as he realized that Mars was now presenting phenomena such as never had been seen before.

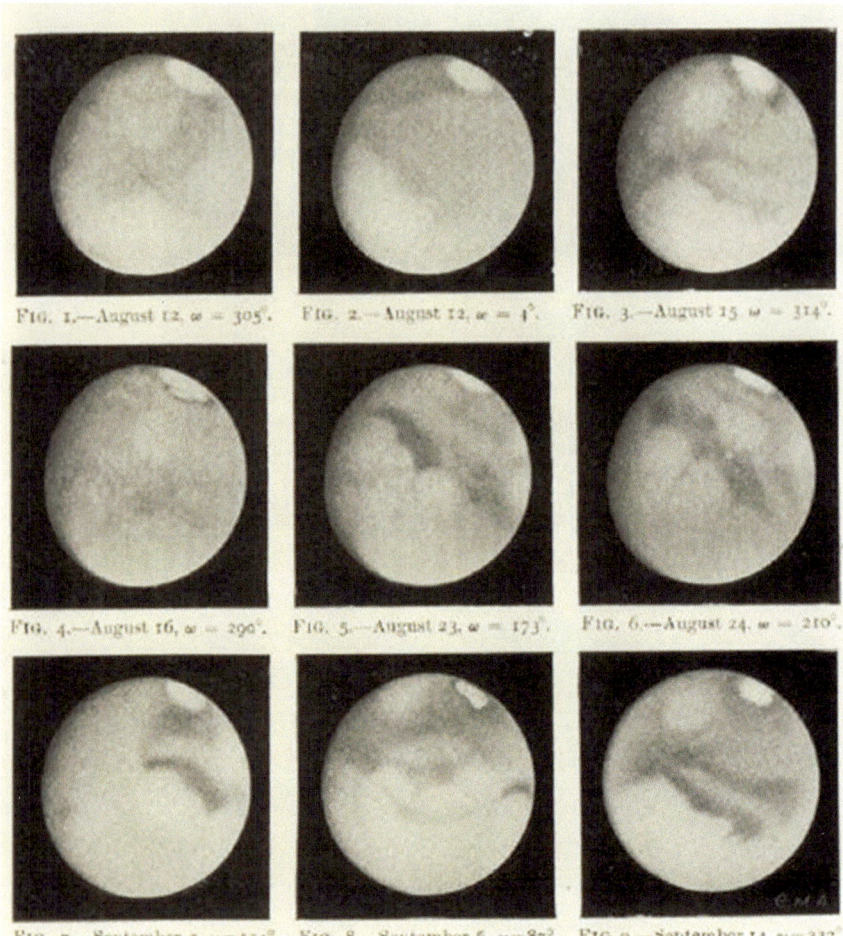

FIG. 1.—August 12, ω = 305°.   FIG. 2.—August 12, ω = 4°.   FIG. 3.—August 15, ω = 314°.

FIG. 4.—August 16, ω = 290°.   FIG. 5.—August 23, ω = 173°.   FIG. 6.—August 24, ω = 210°.

FIG. 7.—September 3, ω = 134°.   FIG. 8.—September 6, ω = 87°.   FIG. 9.—September 14, ω = 332°.

DRAWINGS ILLUSTRATING THE PALLOR OF THE MARIA IN AUGUST 1909, AND
THE GRADUAL RECOVERY OF THEIR INTENSITY, ACCORDING
TO THE DIRECTOR'S OBSERVATIONS WITH THE 8½-IN. REFLECTOR.

Stages of the planet-encircling dust storm as recorded by E.M. Antoniadi using an 8 ½-in. Calver reflector at Juvisy Observatory. The Syrtis Major and Sinus Sabaeus, which are usually among the darkest features of the planet, are faint in the first four drawings (August 12, 15 and 16), but by September 14 they have largely regained their usual strength. Note that these drawings were made using the fine stipple technique Antoniadi had used in his work on Hagia Sophia. (Credit: Richard McKim)

"I Thought I Was Dreaming"

As Antoniadi began a regular series of observations in August and September at Juvisy, a letter arrived in his mail slot that would definitively alter the history of Mars observations. Henri-Alexandre Deslandres, a solar astronomer, Director of the Meudon Observatory and then serving as president of the Société Astronomique de France,[18] invited him to come to Meudon to use the 32.7-inch (83-cm) Henry Brothers' refractor (known as the "Grand Lunette") on Mars. At 39, the same age Lowell had been when he had taken up the study of the planet, Antoniadi's long apprenticeship was finally over; he now found himself summoned to a stage worthy of his great abilities.

The building housing the Grand Lunette was built in 1877 on the ruins of an ancient castle almost completely destroyed by the Prussians during the Siege of Paris in 1871. It rises on the edge of a high terrace that slips unimpeded to the Meudon Park below. The great refractor consists of a twin telescope, one for visual observations, the other for photographic observations, and is still the largest refractor in Europe. The sheer drop on the terrace side, which looks out over the city of Paris, causes the seeing to be generally very good for objects to the east of the meridian, even those that are low and rising; to the west, where there is no such drop, air currents from the ground cause the seeing to deteriorate rapidly as objects approach the meridian.[19]

Antoniadi's first night on the Grand Lunette was September 20, 1909, with Mars rising above the city of Paris. As Antoniadi biographer Richard McKim points out, it would have been rather tricky to use the instrument in 1909. There was no rising elevator floor like those installed for the great refractors at Lick or Yerkes; instead, a small unsteady-looking platform was hoisted up and down the dome wall by a system of giant rails and pulleys. Riding the platform off the floor, Antoniadi would have seen only a long vertical strip of starlight through the dome's narrow slit as the eyepiece end of the Grand Lunette approached, darkness swallowing up everything else around him.[20]

As well, on ascending to his lofty perch and preparing to put eye to the eyepiece, Antoniadi found that his options for observing were rather limited. Someone had borrowed most of the eyepieces, and 320 X was regrettably the highest magnification to hand. Though as with any large refractor, chromatic aberration could not be entirely avoided—it produced a bluish-violet ring fringing the image when the eye-end of the focusing tube was pushed in, a purple-red fringe when it was drawn out—Antoniadi did not find it greatly annoying in the case of Mars.[21] Also, he found that, "notwithstanding the great length of the refractor [16 meters], its focusing varied but little during the observations through the passage in front … of masses of air of differing densities" [12 , pp. 78–79]. Without being interrupted constantly to focus the image, he found the right hand, which he used for drawing, freer with the 32.7-inch than had been the case with the 8 ½-inch Calver on the lawn at Juvisy.

The view of Mars he had on his first night of observations, September 20, would become the stuff of legend and seem like a revelation. It was a *Eureka!* moment in which he would largely grasp, as never before, the character of Mars's hitherto enigmatic surface. It was also a matter of incredible beginner's luck. Not only was the hemisphere of the planet (that in which Syrtis Major stood on the central meridian) now almost completely clear of the dust that still hid the markings of the other hemisphere, but the seeing that night was preternaturally still and calm. Of the nine nights he would observe with the Grand Lunette that autumn, on only two was the seeing found to be satisfactory, on one other good, but on all the rest "boiling." Of all these nights, September 20 was the best. It was so not only of 1909 but of Antoniadi's entire long career as "*astronome volontaire à l'Observatoire de Meudon*," which would include the oppositions of 1911, 1924, 1926, and 1928–29. (The Grand Lunette was out of commission from 1912 until 1924, so Antoniadi then observed only with the 8 ½-inch. Calver reflector in 1914 and 1916; he completely skipped observing Mars in 1918-20-22).

Of this singular night of vision, Antoniadi would later write:

At first glance cast through the 32 ¾-in. … the Director thought he was dreaming and scanning Mars from his outer satellite. The planet revealed a prodigious and bewildering amount of sharp or diffused natural, irregular, detail, all held steadily; and it was at once obvious that the geometrical network of single and double canals discovered by Schiaparelli was a gross illusion.[22]

Antoniadi shown here taking notes at the desk on the observing platform of the Grand Lunette. This photo, from the 1920s, is the only known to show him observing. (Credit: William Sheehan Collection courtesy of Audouin Dollfus)

The sky transparency over Paris that night was not very good; there was a fog over the city, and this might have led an inexperienced observer to feel discouraged. However, in Antoniadi's experience, it was almost always during light mist that the images at Meudon were steadiest, and allowed the full aperture to be used to advantage. (It does not, by the way, appear that the telescope was even equipped at the time with a diaphragm of the sort used constantly by Lowell and his assistants at Flagstaff). By contrast, on nights when the sky was very transparent, with brilliant, strongly-twinkling stars, the images tended to be highly agitated, and the 32.7-inch. Refractor showed the details no better than a 12-inch or smaller instrument would show them. On this night of nights, the seeing remained rock steady for a period of several hours. Experiencing something akin to an astronomical miracle, Antoniadi could hardly believe his eyes:

> The first glance … was a revelation.… I did not believe that our present means could ever yield us such images of Mars. The planet appeared covered with a vast and incredible amount of detail held steadily, all natural and logical, irregular and chequered, from which geometry was conspicuous by its complete absence. *Syrtis Major* was approaching the central meridian, and seemed expanded into a huge cornucopia, twice severed by dusky bridges.… A gigantic triple bay (*Deltoton Sinus*) extended N. of *Hammonis Cornu*. *Lacus Mœris* … [enclosed] a "peninsula" and an "island," constantly visible. Libya and Hesperia appeared shaded, Mare Tyrrhenum like a leopard skin! … A maze of complex markings covered the S. part of *Syrtis Major*, and, although these were held quite steadily, no trace whatever of "canals" in the dark regions could be detected.[23]

Through his many years of experience observing with small telescopes, Antoniadi had developed a basic approach to observing, which was now habitual. His method was to concentrate on a particular small region on the disk and await the most favorable instants to catch the fine details. He then registered these details in his memory. He then did the same with each adjacent region until, having stored all the shapes in their relative positions, he withdrew from the eyepiece and sat in front of a table with a lit control panel on the observation deck where he quickly set down on paper the memorized impressions before they began to fade.[24] His long experience as a draftsman, including during the years he had worked on Hagia Sophia, allowed him to exactly and effortlessly render all the forms and to build up the image from the component shapes. Many an observer, over a long career, has had the experience of nearly perfect planetary images, only to sigh, "I wish I could draw like Antoniadi." There was, alas, only one Antoniadi. As a draftsman of planetary detail, he has never been surpassed.

With perfect seeing lasting for some 7 h, Antoniadi did not need to build up the image from the component shapes revealed only in favorable moments. Instead, it was as if he were sketching a still life, flowers in a vase. Since in the usual conditions of seeing fine planetary details were visible only in moments of steady seeing and by "glimpses," Antoniadi was enjoying a rare chance to see Mars as it really was.

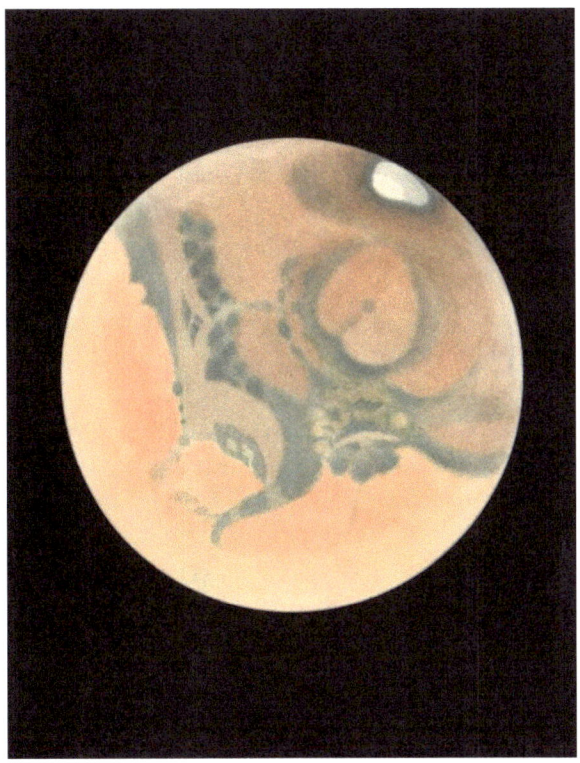

One of several versions of Mars by Antoniadi, recording the appearance of the planet in perfect seeing conditions on September 20, 1909. (Credit: William Sheehan Collection courtesy of Patrick Fuentes, Juvisy Observatory)

As noted, September 20 was not the only good night Antoniadi experienced. On October 6 he described the seeing as "glorious," and he made out a delicate unevenness of tone in Mare Sirenum and resolved the faint pear-shaped Solis Lacus into at least six irregular patches. The night of October 11 was also good, and allowed confirmation of much of what was seen on October 6 (though with some differences that he attributed not to the seeing but to actual changes in the Martian features owing to the ongoing effects of clouds). On the six other nights he observed at Meudon, the images were more or less agitated, though even under these conditions he found that there were moments when good views were to be had. They were preceded "by a period of slight rippling of the disk, after which the undulations would cease *suddenly*, and the becalmed image of Mars would reveal a host of bewildering irregularities."[25] Even on these bad nights the full aperture of the great refractor outperformed that of smaller instruments. He found nothing in his experience to support the usual Flagstaff practice of diaphragming the lens. Antoniadi's first published notice of his success at Meudon was sent on September 21 to a Greek journal published in Athens, and published on September 28: "[N]arrow black canals … were not visible at all, although details much smaller … were quite plain in that telescope."[26]

By then, others were achieving significant results. On 5 days at the end of August and the beginning of September, Mars and the Moon would be close together in the sky. W.W. Campbell, director of the Lick Observatory since 1901, and Charles Greeley Abbot of the Smithsonian Institution went to the top of Mount Whitney, the highest mountain in the Continental United States, in order to obtain spectrograms of the two bodies, which could be compared for possible intensifications in the bands due to water vapor in the atmosphere of Mars. As discussed earlier, V.M. Slipher had claimed to have detected water vapor in the atmosphere of Mars in January 1908, when the air over Flagstaff was very dry but, as Campbell noted, Mars and the Moon had then been far apart in the sky. Though the matter was mainly scientific, it also likely had a personal side. As Lick Observatory historian Donald E. Osterbrock has written:

> Campbell probably detested Lowell even more than did most other contemporary astronomers. As a poor boy who had grown up on a Midwestern farm, and had achieved his education and position only by tremendously hard work and effort, Campbell undoubtedly started off with a good deal of antipathy for the wealthy New England aristocrat. Lick Observatory, where Campbell had begun his research career, had been strongly anti-Lowell from the beginning. Most of the educated people Campbell knew who were not astronomers found Lowell's ideas on a dying race on Mars far more interesting than Lick Observatory's quantitative researches on stellar motions and spectroscopic binaries. Most importantly, all of the rigid Campbell's training had taught him to believe that scientists collected data, framed hypotheses, made predictions, and tested them quantitatively. They published their results in factual form in scientific journals where other scientists could read them, analyze them and further test them, until the final scientific "truth" emerged as the consensus of the knowledgeable

researchers in the field. Scientists did not leap to conclusions and build observatories to prove them, or publish smoothly written articles that gave only one side of the story in glossy magazines written for nonspecialists. Campbell believed that the famous Boston author, whom he called "a private astronomer by the name of Percival Lowell," needed to be taught a lesson, and that he was the man to do it…. [17 , p. 80]

Campbell's result was essentially negative, though he framed his conclusion very carefully. What he had found did not mean that Mars has no water vapor, only that the quantity present, if any, must be very slight. Thus, he maintained:

> Any water vapor in the Martian atmosphere must have been much less extensive than was contained in the rarefied and remarkably dry air strata above Mount Whitney. These observations do not prove that life does not or can not exist on Mars. The question of life under these conditions is the biologist's problem rather than the astronomer's. [18]

The Mount Whitney results were obtained using a low-dispersion spectrograph, but in January 1910, Campbell confirmed them using a spectrograph built around a high-dispersion diffraction grating, which allowed each individual water-vapor line in the spectrum to be resolved and for the ones in Mars's atmosphere to be separated from those in the Earth's by their relative radial velocity to one another. As George Russell Agassiz pointed out in print, Lowell had actually proposed this method in a Lowell Observatory *Bulletin* in 1905, and it had even been tried by V.M. Slipher in that year; but Campbell thought so little of Lowell's work that he had never looked at any of these papers. Campbell's result, again negative, would stand the test of time, and only in 1963 would minute amounts of Martian water vapor be detected by extremely sensitive spectroscopic methods further out in the infrared.

In addition to spectrographs, cameras were being trained on Mars as never before, and realizing better results than in July and August when Barnard's images had revealed only dust. With the Martian atmosphere finally clearing, Barnard himself, on September 24 and 28, obtained images on which—as Antoniadi later wrote—the delicate details seen at Meudon appeared, "beyond the range of Schiaparelli's or Lowell's refractors."[27] Also in September, a French aristocrat, bachelor, and wealthy scientist, Aymar Eugène de La Baume-Pluvinel, came on the scene. La Baume-Pluvinel had his own research laboratory in his Castle at Marcoussis, and had traveled on several eclipse expeditions between 1888 and 1905. With his sometime collaborator Fernand Baldet, he ascended on foot to the top of Pic du Midi, elevation 2877 m (9440 feet) in the Haute Pyrenées, for the purpose of photographing Mars near opposition. The excellent conditions on the Pic for astronomy had first been noted by Benjamin Baillaud, the director of the observatory at Toulouse, who on his visits to the Pic had found the images of planets, stars and nebulae "always good, most often very good, and frequently of extraordinary beauty" [19]. Filled with enthusiasm, he managed to secure funds to set up a 10-m dome (afterwards known as the "Baillaud Dome"), housing a 50 cm (20-in.) Newtonian reflector and a 23-cm refractor (the latter 6 m long) on a single mounting. Later experience would show that Baillaud's faith in the site was entirely justified. After a long period of bad weather, as noted by frequent observer Audouin Dollfus, "the sky will suddenly clear, and remain for a period of several weeks. The telescopic seeing then reaches perfection [as the air flow becomes laminar], and in the second part of each night the images are rock steady" [20 , p. 240]. Indeed, Pic du Midi was the latest observing station—after Lick, Mt. Wilson, and Meudon on the best nights—found to enjoy consistently better conditions than Flagstaff.

La Baume-Pluvinel and Baldet brought good weather with them, and inspired a hope that might compete with the vaunted results of the Lowell expedition to Chile in 1907. However, their methods were somewhat different. Instead of doing as E.C. Slipher had done (and would do again in 1909), recording long sequences of photographs continuously all night long, La Baume-Pluvinel and Baldet did as Barnard was doing—they watched the seeing at the eyepiece, and only made exposures in the steadiest moments. After making about twenty exposures on each plate, which took about 15 min, they processed the plate in the darkroom and quickly perused it. Then the same operation was repeated an hour later, by which time Mars had rotated some 15° in longitude. After 6 h, the planet had rotated through 90°. After a few weeks the entire surface of the planet had been photographed. The observations ended on October 20, with a total of 79 plates having been secured in this way. As the photographs took place at the same time Antoniadi was making visual observations at Meudon, there was considerable overlap, and as in the case of Barnard's photographs, no canals showed up on any of the Pic du Midi images. Nor did they show up on images obtained by George Ellery Hale with the newly installed 60-inch reflector at Mt. Wilson. Almost overnight, one of Lowell's strongest arguments in favor of the canals—that they had been photographed—seemingly went the way of the Dodo.

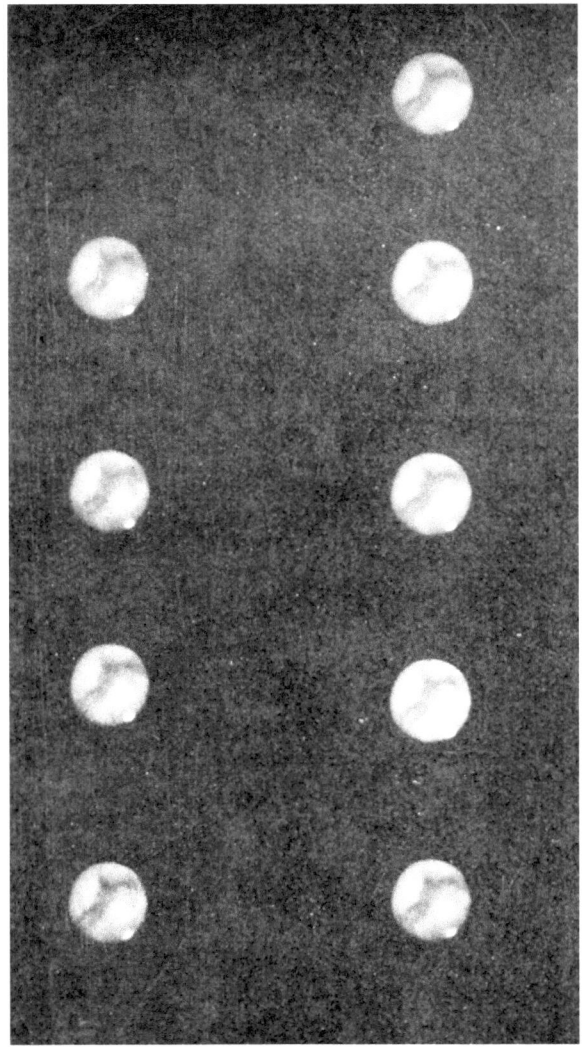

Barnard's best images of Mars, September 28, 1909, at 16 h 46 min G.M.T. (From: Monthly Notices of the Royal Astronomical Society, vol. 71, plate 10)

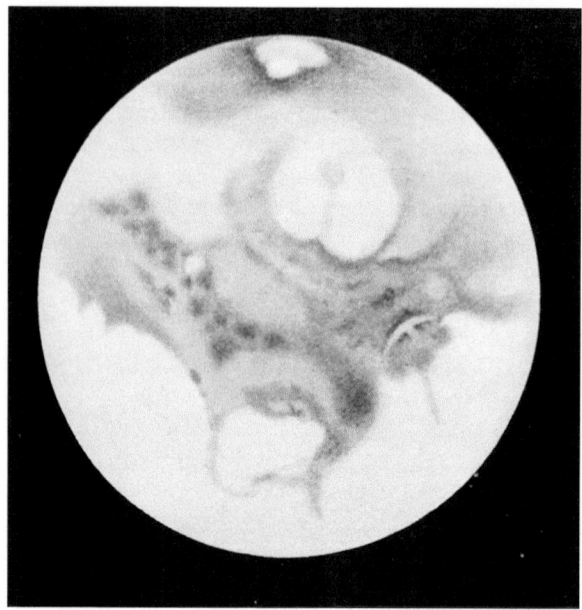

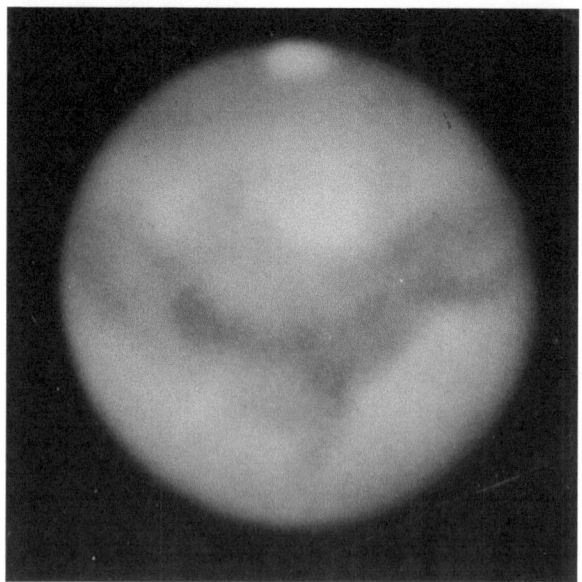

Antoniadi's drawing of September 20, 1909, compared with a Barnard photograph of September 29, 1909, showing remarkable agreement. (Credit: William Sheehan collection)

A Clash of Titans

The 1909 opposition of Mars will be forever remembered as the climactic contest between Antoniadi and Lowell—the former to become the most persistent critic of the canals, the latter their determined champion. For some years the two men had not been on speaking terms, chiefly because of Antoniadi's harsh criticisms of Lowell's Venus observations in 1897 (criticisms that, as we have noted, arguably contributed to Lowell's breakdown). However, on September 9—at almost at the exact moment Lowell was lecturing about Venus in Worcester, and just days before Antoniadi first looked through the Grand Lunette—the latter wrote, in his elegant and unblotted hand, apologizing for the manner (if not the matter) of his earlier comments, which he admitted had been "somewhat sharp and displaced." He appealed to Lowell to send his publications and suggested that in return he would do something for Lowell: "I draw well, and if I can be of any use to you in that line, I hope you shall not hesitate to apply [to] me in need."[28] Presumably he had in mind applying himself to Lowell's more recent photographs in the way he had the one from 1907 (and as he would later do with Barnard's), but Lowell again turned down the offer, while seeing a chance to get his erstwhile keenest critic onside. As soon as he returned from Worcester to Flagstaff, he immediately mailed off to Paris some of the *Lowell Observatory Bulletins* and maps, and mentioned to his correspondent that *Mars and Its Canals* had just been translated into French. (This would hardly have been of any use to Antoniadi, as his English was impeccable). When he wrote to Lowell, Antoniadi had mentioned his imminent opportunity to observe Mars with the Grand Lunette, and Lowell, who clearly regarded Antoniadi as very much an amateur (and even neophyte) offered unsolicited advice. "I am glad that you are to use the Meudon refractor," he wrote, "but remember that you will have to diaphragmed [sic] it down to get the finest possible details. Even here we find 12 or 18 inches the best sizes."[29] (This letter was dated September 26; it is inconceivable that it would have had any effect on Antoniadi, who was a man of strong independent views, but in any case, by then Antoniadi had already begun to see for himself.)

On October 9 Antoniadi responded to Lowell's letter of September 26, including four beautifully executed original pencil sketches showing Mars under various conditions of observation. He told the American astronomer:

> …Of course, the Flagstaff skies must give you an ideal definition. Here, in N. France, we seldom have really good images.

It might interest you to hear that my Meudon work has surpassed all my expectations. I hasten, therefore, to send you 4 of my best drawings, obtained with the full aperture (33 inches) of that splendid refractor. These drawings are by far the best I have ever made, and I do not doubt that you have seen the same things along the whole line. With the exception of the linear "canals" (Anubis, Phison (double, very probably), Poros (double?), Euphrates, and your Labotas of 1894 (this "canal" was very dark and marked on Sept. 20), all seen by flashes of 1/3 of a second), all the other markings I show were held steadily, and the tremendous difficulty was not to *see* the detail, but accurately to *represent* it. Here, my experience in drawing proved of immense assistance, as, after my excitement, at the bewildering amount of detail visible, was over, I sat down and drew correctly, both with regard to form and intensity, all the markings. On September 20 … however, one third of the minute features I could not draw; the task being above my means….

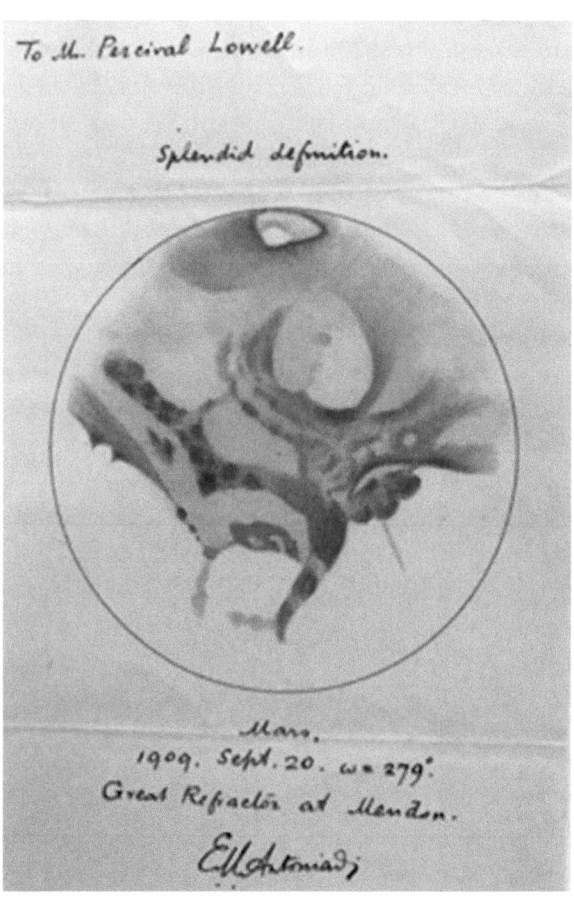

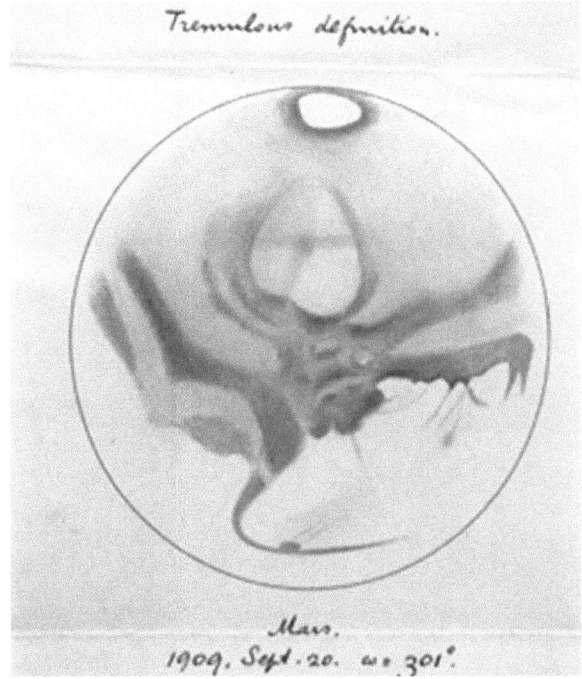

Tremulous definition.

Mars.
1909, Sept. 20. ω = 301°.

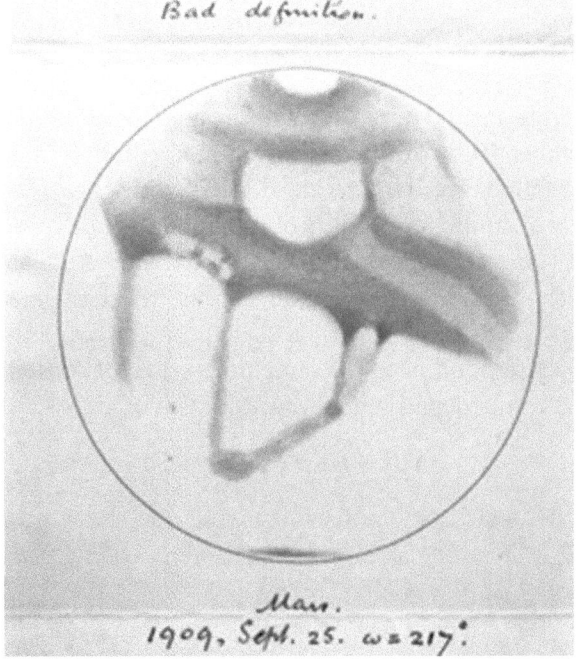

Bad definition.

Mars.
1909, Sept. 25. ω = 217°.

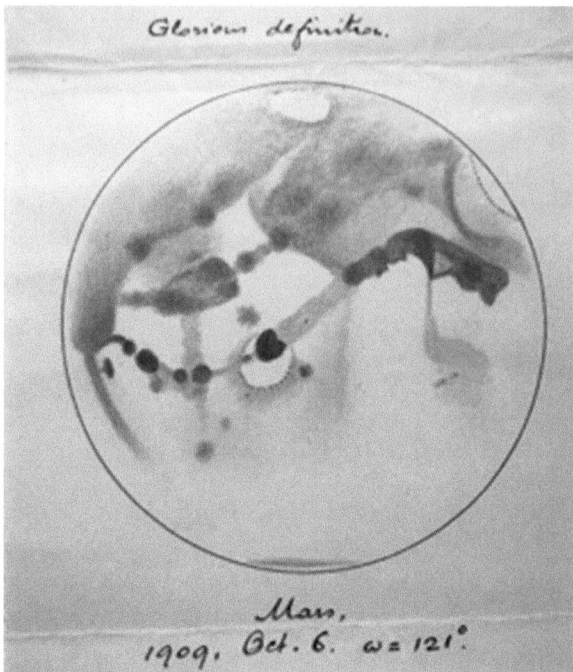

Four sketches by E.M. Antoniadi, showing the planet in "splendid," "bad," "tremulous," and "glorious" definition; sent to Percival Lowell with his letter of October 9, 1909. Lowell would later write, "The one you marked tremulous definition strikes me as the best. It is capital." (Credit: Lowell Observatory Archives)

Antoniadi wrote again 2 days later, adding an account of his further observations on October 10, with a pencil sketch made in only "moderate" seeing but showing an abundance of interesting detail. For instance, "Lacus Phoenicis, black, and really of annoying visibility on [October] the 6th, was almost invisible last night! Can we doubt then of the obliterating effect of the clouds of the planet? Tithonius [Lacus] was also confuse[d] through such obliteration."[30] Antoniadi sent Lowell a pencil sketch, but to Barnard—with whom Antoniadi was corresponding at the same time—a color version, which is not only accurate as to the details but a beautiful work of art.

A color drawing by Antoniadi inscribed "to my dear Barnard" showing Mars on October 11, 1909, in "moderate" seeing. Comparing this to Antoniadi's October 6 drawing, among those sent to Lowell, it is evident that cloud has obliterated Lacus Phoenicis and Tithonius Lacus. (Credit: Richard McKim and the BAA Mars Section Archives)

Canali Novae

For some reason, the mails seem to have been unusually slow that fall, and this letter did not reach Flagstaff until late October. Lowell was gone for a few days at the end of the month (see below) and did not finally get around to answering until November 2. By then—completely oblivious to the revelations at Meudon—he had published one of the strangest articles ever written about Mars:

On September 30, 1909, when the region of the Syrtis Major came round again into view after its periodic hiding of six weeks, due to the unequal rotation periods of the Earth and Mars, two striking canals were at once evident to the east of the Syrtis in places where no canals had ever previously been seen. Not only was their appearance unprecedented, but the canals themselves were the most conspicuous ones on that part of the disk....

The record books were then examined, when it appeared that not a trace of them was to be found in the drawings of August, July, June, or May.... That they had not been observed in previous years was then conclusively ascertained by examination of the records of those years. The record of canals seen here is registered after each opposition in a fresh map of the planet's surface. This has been done since the beginning of the critical study of Mars at this observatory in 1894. Now when these maps came to be scrutinized for [these] canals, each of them showed blank of any such features. Nor had any observer previous to 1894 recorded them, as the observatory library of the subject bore witness. Schiaparelli had never seen them nor had his predecessors or or successors. This determined definitely that no human eye had ever looked upon them before. But stirring as it is to know that one is the first to see a new geographic feature on another planet,—akin to the thrill of finding unknown land in our Antarctic regions—a much deeper scientific interest attaches to the question whether a phenomenon previously undiscovered was also previously non-existent; for in that case one has seen something come into being with all that such origination implies.[31]

Not only had the canals not been seen before, he implied that they had not even have existed before. They were not to be found among the 117 canals Schiaparelli had mapped, nor were they, until September 30, 1909, found numbered among the 690 recorded at the Lowell Observatory. Two needles in a haystack had been found, and they were true "canali novae"—canals new to the planet "in fact and function, and as such ... the most important contribution to our knowledge of the planet of recent years."[32] Though he did not quite say so explicitly, this implication was clear. The Martians had been busy, and in these works *extra ordinem naturae* (as he put it) they had been engaged in new construction on a superhuman scale. "Measurement of their dimensions," Lowell affirmed, "shows each of them to be a thousand miles long and some twenty miles wide. The cañon of the Colorado would be a secondary affair in comparison. This fact of size along precludes their being of cataclysmic origin, for no such chasm could suddenly be opened on the earth, where the internal forces are far greater than can possibly the case with Mars."[33]

Though Lowell did his best to hype this "most important contribution to our knowledge of the planet in recent years," the public seems to have ignored him. Says Hoyt, "… his announcement stirred little interest in the popular press, and where it was noted in scientific circles it was met with hard overt scepticism."[34] Most astronomers by now tended to regard such "discoveries" as delusional. Meanwhile, Lowell and his old Flagstaff crony, the recently appointed Associate Justice of the Arizona Territorial Supreme Court, Edward M. Doe, set out by train and stage coach to recently discovered areas of the Petrified Forest, Sigillaria grove and north Sigillaria forest, nine miles north of the now-defunct town by Adamana, where they enjoyed meditations on vistas of geologic time in preference to continued telescopic surveillance of the distant opaline deserts of a distant planet.[35]

Interim Reports of the BAA Mars Section

Antoniadi, like a man possessed, was now churning out a succession of interim reports for the BAA Mars Section. The first, mentioned above, described the fading of the markings due to dust; the second ("On the Meteorology of the Planet"), added further details about the nature of dust storms, and the third dealt with "the nature of the so-called 'Canals' of Mars." Each was a minor masterpiece, and has become a classic in the history of Mars observations.

Antoniadi was the first to suggest that dust storms might be a regular recurring seasonal phenomenon of the planet, since "in spite of the rarefaction of the planet's atmosphere, the feeble power of gravity at its surface, and the probability that five-eights of the visible markings are sandy deserts, [conditions] would be rather favourable to the formation of clouds of dust, especially around the time of perihelion."[36] This went to the heart of the just-observed fading of the dark markings. The 1909 dust storm was the largest since 1877 when the French astronomer-artist Étienne Léopold Trouvelot had recorded a planet-encircling dust storm. (That in October/November 1894, observed by Barnard, had only been a large regional storm.) Trouvelot alone had seen it, since he observed the planet far ahead of the opposition when the event began and other observers, including Nathaniel Green and Schiaparelli, came on the scene only after it had ended. Trouvelot, however, had no idea what he was seeing. Now Antoniadi had acquired the key to unlock many of the secrets of the Martian world.

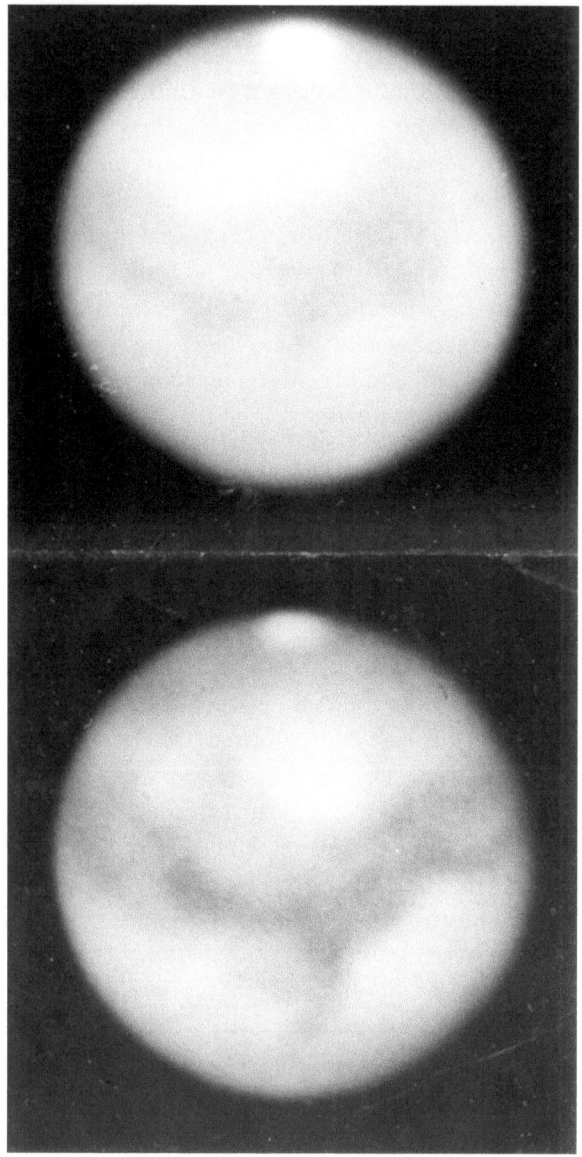

The great planet-encircling dust storm of 1909 subsides. The upper image was taken by Barnard with the 40-inch on September 25, 1909 at 7:10 U.T.; the hemisphere of Mars centered on Margaritifer Sinus still appears partially obscured by dust. The lower image, taken by Barnard on September 29 at 3:25 U.T., shows the Syrtis Major region; the atmosphere is now completely clear. (Credit: William Sheehan collection)

Antoniadi pointed out that the canals were close to the limit of detectability, no matter in what instrument they were observed. Thus, they were no more evident in larger telescopes than in smaller ones, or when closer to the Earth rather than farther away. Antoniadi wrote:

That clouds and haze obliterated the surface of Mars [during the period of observations] there can be no doubt from what has been already stated [in the first and second interim reports of the Mars Section for 1909]; yet this obliteration was not general, so that, if the "canal" network were real, it could not fail to appear through the breaks of the atmospheric veil. But it did not do so … [T]he fugitive impressions of straight lines were … manifestly illusive, since they had the same breadth and intensity in the 32.7-inch O[bject] G[lass] as in the 8 ½-in. mirror [of the Calver reflector]. In fact, the first glance thrown on Mars by the Director [of the BAA Mars Section] in the Meudon refractor on September 20 last showed him that the visibility of the average straight "canal" did not increase *pari passu* with that of the real markings in the great telescope; and that while the latter showed wonderful details which could not be dreamt of in the small reflector—details held most steadily—the appearances of straight lines were as faint, fugitive, and cowardly in the great instrument as in the small one; and, of course, such is the deportment of illusions only. Besides, if the "canal" network were an objective reality at the limit of visibility, we ought to see it by glimpse *as a network*, and not glimpse *severally* its various components.

We are thus led to consider the dangers of glimpsing. A glimpsed object is not as certain as an object held steadily, and, however self-evident or trite such a remark may be, yet it is a very important one to make here. A great many real objects are doubtless glimpsed with inadequate means; but a great many subjective appearances are also glimpsed, and, unfortunately, not recognized as such; and, were due regard to have been paid to the treacherous character of glimpsing, the existence of many celestial marvels would never have been foisted on the scientific world. [21 , p. 28]

The fact that the canals appeared during glimpses whereas the actual Martian details were seen steadily provided the greatest insight into the true nature of the canal phenomenon; only illusions behave that way. The reality of the Martian surface consisted in complex details—"small seas, diffuse bands, amorphous, knotted shadings, edges of half-tones, narrow black lines."[37] He found that many of Schiaparelli's canals had at least some basis in reality, but almost all of Lowell's were simply optical illusions.[38]

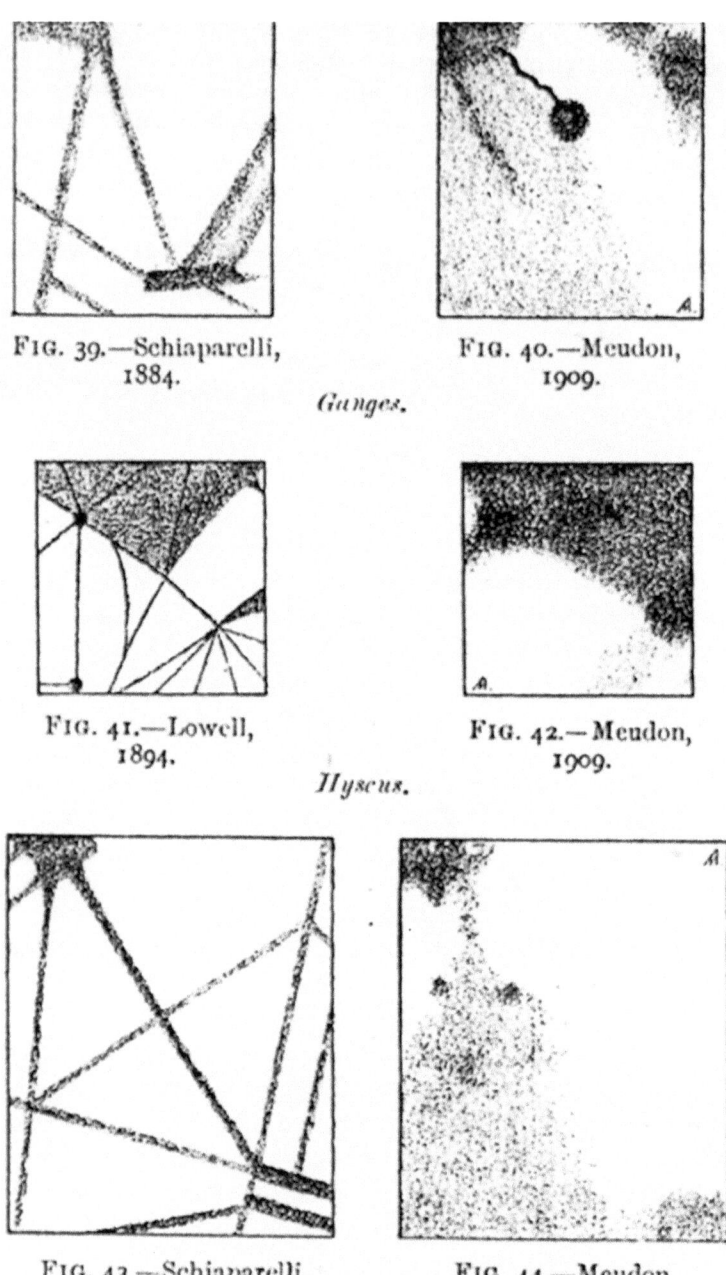

FIG. 39.—Schiaparelli,
1884.

FIG. 40.—Meudon,
1909.

Ganges.

FIG. 41.—Lowell,
1894.

FIG. 42.—Meudon,
1909.

Hyscus.

FIG. 43.—Schiaparelli,
1884.

FIG. 44.—Meudon,
1909.

Tartarus.

Antoniadi side by side comparisons of "canals" shown by Schiaparelli (and in one case Lowell) and the complex and irregular markings seen at Meudon. (From: E.M. Antoniadi (1909) "Impossibility of the Linear 'Canal' Network of Schiaparelli as an Objective Reality on the Planet," Report of the Mars Section, 1909, in: Memoirs of the British Astronomical Association, *Vol. 20, Part II (1915). pp. 39–43)*

Lowell Responds

Meanwhile, even before Lowell got around to answering Antoniadi's letters, the latter submitted interim reports—one on the meteorology of the planet, another on the nature of the canals—to the BAA. They were in press or soon to be when, on November 2, Lowell finally sat down to contemplate Antoniadi's drawings. "The one you marked tremulous definition [and showing canals] strikes me as best. It is capital," he wrote. However, he continued, the planet was "not so well defined" in the others, made in what Antoniadi had labeled "moderate," "splendid," and "glorious" definition.[39] In other words, Antoniadi had drawn the details best when the seeing had seemed worst! Lowell had an explanation for this. "This is the great danger with a large aperture," he wrote, "a seeming superbness of image when in fact there is a fine imperceptible blurring which transforms the detail really continuous into apparent patches. On the other hand, a bodily movement often coincides with the revelation of fine detail. This subject we have carefully investigated here and all of our observers recognize it."[40] To prove his point, Lowell sent his drawing of the two supposed "*canali novae*," where in the place of the new canals Antoniadi had shown only the oblong darkish patch Lacus Moeris (recorded in one of E.C. Slipher's drawing but not in Lowell's) and a broad bland dusky region along the west side of Syrtis Major.

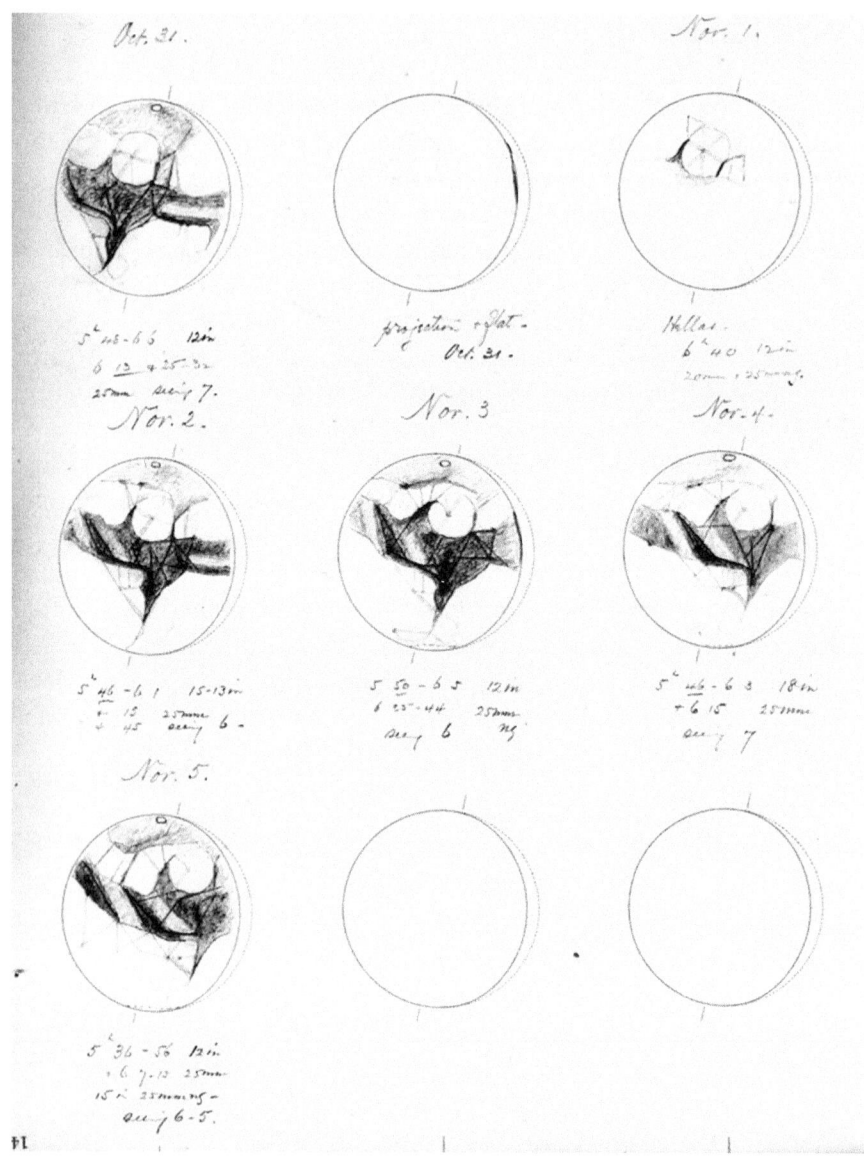

Lowell's sketches of Mars, October 31 and November 1-2-3-4-5, 1909, showing the "canali novae" extending from Syrtis Minor northward across Libya, and best seen in the drawing of November 5. Lowell and Earl C. Slipher had first recorded these on September 30, and seemed to believe that they were actually new canals constructed by the Martians. (Courtesy: Lowell Observatory Archives)

Antoniadi could hardly believe Lowell's letter, which arrived in Paris on November 15, 1909. This was the stuff of delusion. Antoniadi tried to explain to Lowell that there could simply be no possibility that the detail he had seen so definitely and steadily with the Grand Lunette was an illusion due to "blur-ring." On the night of revelation, September 20, the Mare Tyrrhenum had appeared mottled "like a leopard skin." Instead of being laced with canals, the Martian "deserts" showed everywhere "a maze of knotted, irregular, chequered streaks and spots."[41] To be as specific as he could, Antoniadi described the Martian desert Amazonis, where in bad seeing "hideous lines" ("canals"!) were wont to appear in brief glimpses but which on October 6 and again on November 9 had given "an elementary view of the true structure of the Martian deserts."

The image had been slightly tremulous on each date. Suddenly, however, definition became perfect, and a wonderful sight had presented itself for a dozen seconds on both occasions, the "soil of the planet becoming covered with a vast number of dark knots and checquered fields, diversified with the faintest imaginable dusky areas, and marbled with irregular, undulating fila-ments, the representation of which was evidently beyond the powers of any artist."[42] Even held thus steadily—and despite his extraordinary powers as a draftsman—Antoniadi found the details too intricate to draw, but he pro-duced an impressionistic sketch (not published until 1988) in which the details resemble strikingly the patterns of windblown streaks around the cra-tered terrain shown in the *Mariner 9* spacecraft imagery. Such was Antoniadi's acuity as an observer, the retentiveness of his memory for images (eidetic memory), and the skill honed through years of sketching and painting under the vast vault of Hagia Sophia that no other Mars observer would have been remotely capable of capturing even the impression of this intricate detail. But it seems to have made no impression on Lowell whatever. Instead, near the end of his life, he told science writer Waldemar Kaempffert that Antoniadi was "… a man without knowledge of how to observe."[43]

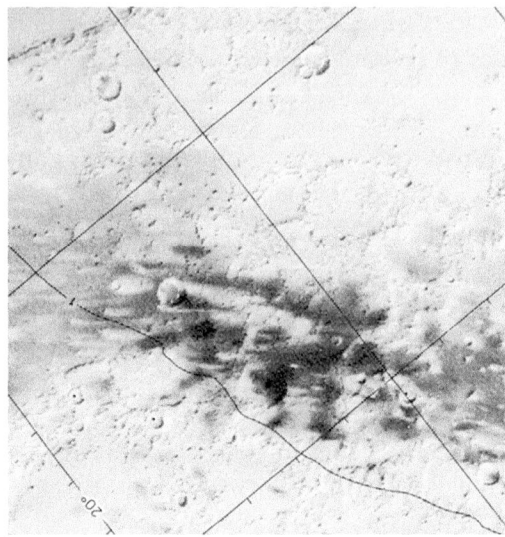

Above: Detail from sketch in letter, E.M. Antoniadi to Percival Lowell, November 15, 1909, showing an impressionistic rendering of markings in the Amazonis desert seen with the 32.7-in. Meudon refractor. Lowell Observatory Archives. Below: Part of a modern airbrush map of Mars based on Mariner 9 imagery, showing the structure of small markings at longitudes 185–190°, latitudes 10–20°N, and so orientated as to make a suggestive comparison with the Antoniadi sketch above. (From: R.M. Batson, P.M. Bridges, J.L. Inge, Atlas of Mars *(Washington, D.C., 1979), MC-15 Elysium quadrangle, shaded relief version)*

Barnard's Views Confirmed

While Lowell returned to his Flagstaff eyrie, Antoniadi continued his barrage of interim reports—a fourth was dated November 4, 1909 (with an addendum on the reality of the geminations of some canals reported by Lowell and E.C. Slipher dated November 26), and a fifth on December 23. By then the opposition was long past, but reports from observers elsewhere were trickling in. From Mt. Wilson, George Ellery Hale, using the 60-inch reflector diaphragmed to 44 inches and a magnifying power of 800 X, wrote, "I was able to see a vast amount of intricate detail—much more than has been shown on any drawings with which I am acquainted. In spite of the very fine seeing on certain occasions … no trace of narrow straight lines, or geometrical structure, was observed. A few of the larger 'canals' of Schiaparelli were seen, but these were neither narrow nor straight… I am thus inclined to [think] … that the so-called 'canals' of Schiaparelli are made up of small irregular dark regions."[44] But the most important endorsement came from Barnard himself, who—despite suffering, as he always did, from his annual attack of ill health occasioned by the ordeal of trying to observe in winter[45]—commented on the same four drawings Antoniadi had sent Lowell and would publish in his next, "Fourth Interim Report": "I notice in these drawings that in general the canals become broader and more diffuse and more irregular than they are generally shown. This harmonizes better with my own views of the planet with large telescopes."[46] Antoniadi was thrilled to receive this communication from the great observer; "no letter," he said, "ever gave me such pleasure as yours."[47]

Barnard was the genius, said Antoniadi, who in 1894 had delivered "a rebuke from which the spider's webs have never recovered."[48] Having the chance to view the planet with a large telescope for the first time, he completely disavowed his previous work with small apertures, and saw with perfect clarity the way the seemingly inextricable problems of the canals of Mars were to be unravelled:

> … We all felt here that the very favourable apparition of Mars now over would greatly increase our knowledge of the planet; and the eyes of the astronomical world were turned to its foremost observer, using the most powerful refractor in the world. In fact, all our hopes hung on you; we were awaiting with feverish anxiety your results; and your great success in Martian photography this year was gratefully heard by everybody interested in our noble science. We are looking forward to the publication of your photographs, which will doubtless open a new era in the history of the planet.
>
> It is an honour to me to see you expressing a favourable opinion on my Meudon drawings, and to find that we are in perfect agreement regarding the appearance of the so-called "canals" of Mars. You must have seen that my published results described these "canals" as diffuse and irregular, and that I consider the geometrical "canal" network as illusive.
>
> I am sorry to see Dr. Lowell persisting in his inextricable "canal" deadlock. His "canals" are too sharp for the blurring enveloping all the real markings of the planet. He stops down a good Alvan Clark objective to so reduced an aperture that its separating power cannot transcend that of a 12-inch refractor; while on the Flagstaff drawings reason is insulted by perspective-defying canals—the most ingenious mode of Martian signalling….[49]

This was written on December 11, 1909. On December 16, Antoniadi, having in the meantime heard from Schiaparelli (now in the last year of his life and apparently fully won over to Lowell's views), wrote again to Barnard:

... The double appearances of Ganges and Jamuna on October 14, and those of Phison and Euphrates, were all very fugitive, lasting almost invariably 1/3 of a second. I could never hold a double "canal" steadily in my life.

The "canals" seen on November 5 and 9 were all held steadily; and, like all real markings of the planet, were irregular. The Titan swelled out into a shading 1000 miles in breadth—a big "canal" at any rate.

... Your photographs will settle many points. I hope that you had many good nights, and that you could photograph sharply all parts of the Martian surface.

I just received a letter from Prof. Schiaparelli, in which he says: —"The polygonations and geminations for which you show so much horror (and, with you, so many others) are a proved fact, against which it is useless to protest." From this I dissent, basing, as I do, all my knowledge of the planet on my Meudon work. I am sure I saw Mars with the 32.7-inch better than Schiaparelli with his 8 and 18 inches, and I consider my evidence weightier than his. Like you, I think that large object glasses have a great superiority over small ones. Indeed, this is a point that needs no discussion. The complete absence of a geometrical network in the 32.7-inch, coupled with the wonderfully detailed appearance of the "seas," is decisive. Lowell saw the "seas" in 1909 as if he were using a 6-inch!...[50]

Other reports slowly trickled in, supporting the idea that the canals represented only an intermediate and largely illusory stage of seeing the planet. Preeminent among all these was that from George Ellery Hale, who wrote to Antoniadi on January 3, 1910. Hale had been observing and photographing Mars with the 60-inch reflector at Mt. Wilson, and described not only his personal results but those of Charles Greeley Abbot, Walter Sydney Adams, Harold Babcock, Andrew Ellicott Douglass (!), Ferdinand Ellerman, Edward A. Fath, Frederick Seares and Charles E. St. John, all of whom agreed with him as to the true character of the Martian features. Immediately on seeing them, Maunder recognized that they "showed as great an advance as compared with Mr. Lampland['s] as his had been upon the earlier attempts [and] [t]hough the markings came out so clearly ... [t]here was no indication of the spidery network covering Mars which Mr. Lowell had acquainted them with."[51] As for the visual observations, Hale wrote:

... A few of the larger "canals" of Schiaparelli were seen, but these were neither narrow nor straight. On one occasion I could see two "canals" which reach out from the extremities of Sabaeus Sinus resolved into minute curved and twisted filaments. I am thus inclined to agree with you in your opinion (which coincides with that of Newcomb) that the so-called "canals" of Schiaparelli are made up of small irregular dark regions.

If the 60-in. reflector had not shown in many parts of Mars details smaller than those recorded by Lowell, it might perhaps be urged that its failure to reveal the narrow "canals" of geometrical pattern might be attributed to the use of too large an aperture or to some defect of the instrument. As a matter of fact, however, I have always found that under the best atmospheric conditions an aperture of 44 inches, or even more, is highly advantageous for the study of planetary details. I have also convinced myself that from an optical standpoint the 60-in. reflector is an essentially perfect instrument both for visual and photographic work. The complexity of the details which it shows on Mars is so great as to defy all attempts to draw them. Under such circumstances, the perfectly "natural" appearance of the planet, and the total absence of straight lines or configurations such as Lowell draws, seem to me significant. … You will notice that my opinion with respect to the advantages of large apertures and the appearance of Mars under good conditions is in excellent agreement with the views expressed by Prof. Barnard….[52]

Barnard himself, comparing Hale's photographs with Antoniadi's drawings, wrote to the latter: "The very remarkable agreement between your drawing of the Syrtis Major region and the magnificent photograph by Mr. Hale ought to interest every student of Mars. It proves to me that your drawings are precise."[53]

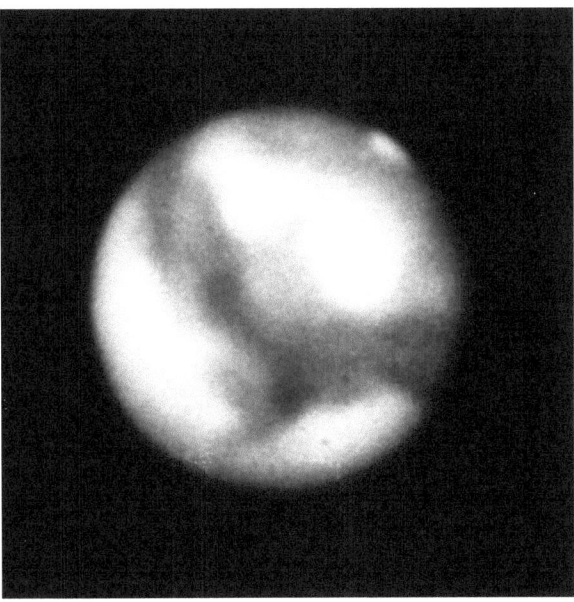

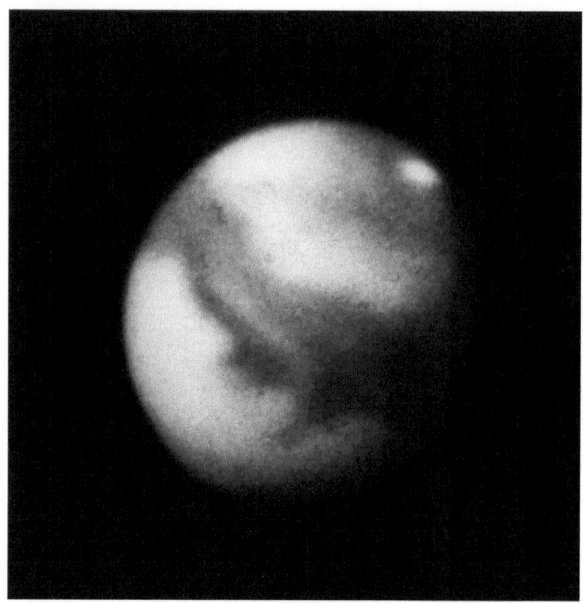

*Mars, photographed by G.E. Hale at the 100-foot Cassegrain focus of the 60-inch reflec-
tor at Mt. Wilson. Above, October 5, 1909; Below, November 3, 1909. These photographs were presented by E. Walter Maunder at the BAA meeting on December 29,
1909, 1910. (Credit: Mt. Wilson Observatory)*

Antoniadi's triumph was apparently complete. However, Lowell refused to
admit defeat—and indeed, he never would. At the end of the latest Flagstaff
campaign, he and E.C. Slipher busied themselves with annexing yet another
world to the empire of lines, adding to Mercury, Venus, Mars, Jupiter and the
satellites of Jupiter yet another vassal—Saturn and its rings. On the night of
September 19, 1909—ironically, the eve of Antoniadi's first magical night of
Mars observations—Lowell, in the company of his old friend George Russell
Agassiz,

> noticed what seemed to be faint lacings traversing diagonally the planet's equa-
> torial belt. Not only was the phenomenon unprecedented, but it was so faint
> and illusive that I was unable at first to assure myself of its objective reality. On
> mentioning my impressions to my assistant, Mr. E.C. Slipher, I found to my
> surprise that he had had suspicion of the same thing on September 9, and had
> even thought to detect a trace of it on the photographs of Saturn taken by him
> afterwards....[54]

Sometime after Agassiz left, the long-awaited 40-inch Clark reflector arrived
on Mars Hill, and was mounted below ground level in the rather perverse

expectation that this would improve the seeing. In fact, it did the opposite. Lowell and E.C. Slipher were putting the new telescope through its paces when another of Lowell's old friends, Robert Wheeler Willson, a professor of astronomy at Harvard, came for a brief stay. The night before he arrived, Lowell and Slipher—despite Lowell's stern admonitions to Antoniadi about the need to stop down the Meudon refractor—were looking at Mars with the full aperture of the new telescope. They did not look in vain. Lowell wrote to Leonard:

> … You will remember our letters and telegrams out here—seeing the Canals with the full aperture of the 40 in. The latter was difficult but—there they were by diligent looking…. This was the night before Professor Willson arrived, since then no seeing to speak of and a great snow storm.—Nevertheless he has seen many Canals ill and one Canal well. He feels he is getting on. He always believed but is now seeing for himself. The blue belt about the shrinking cap was one thing he particularly wished to see and that he has seen. He has been exorcising the imps who make bad seeing and is happy and calm.[55]

Before returning to the East, Willson was granted even greater measures of success. Lowell wrote that Willson:

> … has been having the time of his life thanks to his own good expert eyesight. His strides in detection have beaten all records—blue water round the melting cap, canals seen better and better each night until finally doubles and straight narrow lines. After giving him a long wait nature relented, showing him first a projection [on the limb of Mars] a rare event and then last night seeing 8–9 [on a scale of 1–10], in which E.C. Slipher discovered a new division in Saturn's ring B near the Crape ring which he, Prof. Willson, saw perfectly and finally belts and the oblateness of Neptune… Truly his visit came to an end in a blaze of glory which he greatly appreciated. He burst out in the den [of the Baronial Mansion] with "This is the happiest moment of my life."[56]

On December 8, the wisps and ring details were visible in the 24-inch refractor, and Lowell worked up an E.C. Slipher photograph and sent it to Willson as a souvenir of their night of revelations. Then he was off to England again, bringing reinforcements to his position against the anti-canalists. As usual, these included a cache of the latest photographs, including some of Saturn as well as of Mars, to be presented to the Royal Institution and the Royal Astronomical Society. However, on finding an unexpected gap in his schedule, he decided on very short notice to present them as well to the BAA's meeting at Sion College on March 30, 1910.

This retouched photograph of Saturn shows the wisps and ring divisions as recorded by Lowell on December 8, 1909. It was given by him to Robert Wheeler Willson, professor of astronomy at Harvard, in commemoration of Willson's visit to Flagstaff, in which both he and E.C. Slipher recorded the same features. (Credit: Collection of Historical Scientific Instruments, Harvard University)

Among others, E. Walter Maunder, the Acting Secretary of the BAA and Lowell's old nemesis, was present. With his usual dogmatism, the Flagstaff astronomer set forth the argument he had been honing ever since Antoniadi had published his drawings at Meudon. Tremors in the atmosphere, he said, could cause an image full of really linear details—be they the diffraction rings of a star image or canals on the surface of Mars—into a "mosaic of points."[57] He insisted that, unaware of the fact, Antoniadi had been deceived by what he had observed. To get the real details, he ought to have stopped down the 33-inch aperture. "Even in our air, which is selected for the purpose, and is much better than that at Meudon," he continued, "… the maximum efficiency lies anywhere from 18 to 12 inches." He cited the recent experience of Robert Lundin, optician of the firm of Alvan Clark & Sons, who had come to Flagstaff to install the new 40-inch reflector, with various apertures, who remarked, "Before I came here I did not believe what you said about diaphragming down and getting better images, but I see it is so."[58] Lowell concluded:

There are a great many things which you will find are so, which may not have seemed to be so before you have made prolonged study. There are a great many pitfalls in the path, and the study of Mars is not a study which can be taken up and looked at for one, or two, or twenty nights. It has been said that no one can begin to know the planet after a whole year's work, and I can confirm that. I thought I knew something about Mars when I had been observing it for a whole opposition, but now that I have been observing it for 15 years I can see that what I saw then was there, but I can now see amazingly more.

I am always skeptical myself, because I am not going to be found wrong; and have been skeptical of things I thought I had seen, and then discovered that my assistant had put down those same things in his drawings. In other words, I have always striven to be too careful rather than the reverse.[59]

With this, Lowell ended his speech. H.P. Hollis, in the chair, expressed his gratitude to Lowell. "I do not remember ever at these Meetings having had such a thrilling half hour as that we have just experienced…. We have all read things about Mars sometimes, I think, a little different from what Prof. Lowell has told us, but what we have just heard about his assistant's drawings and about his own scepticism may make some of us feel inclined to change our views." There followed a period of questions and answers in which deference to the sheer force of Lowell's personality and sympathy with his views predominated. Yet Maunder politely but emphatically demurred to the general enthusiasm. "As Prof. Lowell knew," he said,

[the speaker] did not see eye to eye with him as to the actual rectilinear character of the canals, and he deduced an altogether different conclusion from the experiment … [regarding] the smallest angular diameter at which one could detect a telegraph wire…. [S]ince there was so wide a difference between the limit at which we have real definition of the object and the limit of mere perception, even though it be distinct, it followed that markings well within those limits—and Prof. Lowell had told them that the limit for mere perception was one-hundredth that of full definition—then those markings must tend to assume the simplest geometrical form—the forms, that is to say, of the circular dot and the straight line….[60]

By then the evening was getting late, and there was still another talk to hear (the one regularly scheduled before Lowell's add-on), and so—after a few brief comments from another audience member in which Lowell's photographs were preferred to Hale's—Lowell left the hall and disappeared into the night.

One can only imagine how the meeting would have gone if Lowell's nemesis—Antoniadi—had been present. He was in Paris, having just a few days before put the finishing touches to his latest essay, "Further Objections to Prof. Lowell's Canal System of Mars," in which he ridiculed Lowell's practice of "crippling" his objective and showing Mars "not only destitute of intricate details, but also shorn of some of his most prominent features." This he proceeded to demonstrate by a series of comparisons of one of his own drawings, a drawing based on Hale's photograph, and one of Lowell's drawings showing the "canali novae." He concluded:

> We thus reach a decisive and indestructible position. While the Meudon picture is in satisfactory agreement with the Mount Wilson photographs, the Flagstaff sketch cannot really stand the crucial test of comparison. With the theory of interference rings, Prof. Lowell is at a loss to explain how a sharp "promontory" can be seen where there is practically no hint for it; how an irregular, triple "bay," held steadily, can be begotten out of nothing; how a bright area can appear heavily shaded; how two minute circles and three straight lines can be conjured into a huge, uneven, and irregular "lake," 700 miles across; and how the impersonal evidence of photography can readily countenance all these irregular appearances, were they to rest on a simple geometrical configuration. And when an enormous dusky spot, subtending 4 arcseconds … is totally missed, what faith can we have in those fine black networks which have obscured the face of so many members of our [solar] system?[61]

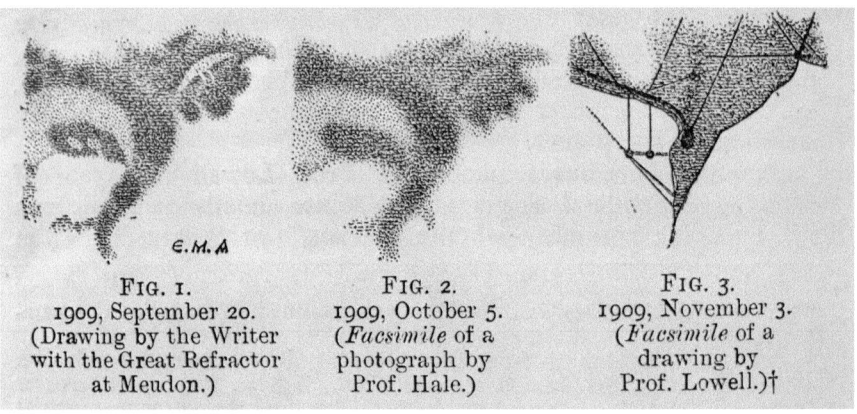

FIG. 1.
1909, September 20.
(Drawing by the Writer
with the Great Refractor
at Meudon.)

FIG. 2.
1909, October 5.
(*Facsimile* of a
photograph by
Prof. Hale.)

FIG. 3.
1909, November 3.
(*Facsimile* of a
drawing by
Prof. Lowell.)†

E.M. Antoniadi drawings showing the Syrtis Major region in 1909. Fig. 1, as observed by Antoniadi at Meudon on September 20, 1909. Fig. 2, as shown in George Ellery Hale's photograph of October 5, 1909. Fig. 3, Lowell's drawing of November 3, 1909. Antoniadi notes: that the form of the Syrtis Major was no longer as Lowell showed it but as W.R. Dawes had recorded it in 1864: "a great 'promontory' invaded it on the left side, beyond which a triple 'bay' expanded as far as Hammonis Cornu to the right; and while the 'continent' of Libya was heavily shaded, Lacus Moeris extended over 18 degrees of longitude, some distance north of the equator.... But what was Prof. Lowell's experience with this part of the planet? That, although straight lines appeared plentiful, none of the foregoing visual and photographic features are recognizable on his drawing." (From: E.M. Antoniadi, "Further Objections to Prof. Lowell's Canal System of Mars" (April 1910))

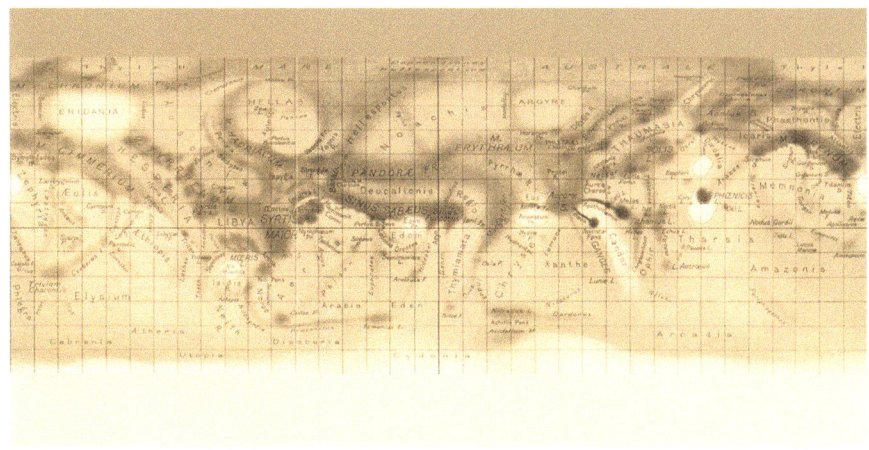

A tinted version of Antoniadi's map of Mars, 1909. The original version appeared in "Report of the Mars Section, 1909," in Memoirs of the British Astronomical Association, *vol. 20, Plate V. This version courtesy Joel Hagen*

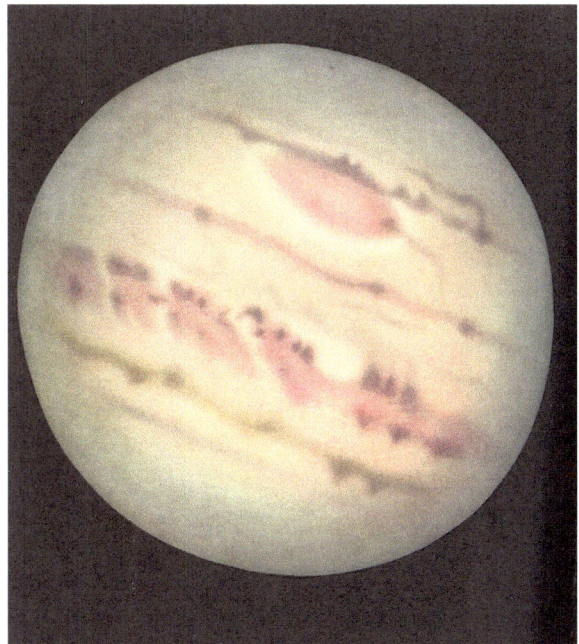

Antoniadi's drawings of Jupiter (above), made on August 5, 1926 with the Grand Lunette, and Saturn, October 7, 1941. (From: The Flammarion Book of Astronomy, *1964)*

Notes

1. Richard McKim, "Telescopic Martian Dust Storms," p. 37.
2. Ibid., pp. 206–207.
3. Over the many years of their correspondence, Lowell had referred to Storey as "Quadric," Storey to Lowell as "Mars."

4. P. Lowell [2 , pp. 521–522]. Lowell's reply to Blackwelder in which he briefly admitted to the accusation about the deficiency of his geology before pivoting against Moulton, appeared in *Science* September 10, so the Venus lecture was clearly part of a wider campaign.

5. Boston *Evening Transcript*, September 9, 1909.

6. Esther C. Goddard and G. Edward Pendray [4]. I am grateful to David Baron for calling my attention to this reference.

7. Gay, *Freud*, p. 207.

8. Sigmund Freud (1961) *The Future of an Illusion*. English translation by James Strachey (New York: W.W. Norton), p. 22.

9. These photographs were finally published in 1947. See: G. Kuiper and M.R. Calvert [6].

10. Quoted in E.M. Antoniadi [7 , p. 362].

11. Ibid., p. 431.

12. Following here, mostly, Richard McKim [9].

13. Seldom has a more fateful collision of a book and a man occurred. Flammarion's book is still in some ways unsurpassed as a source of information about the history of Mars observations up to 1892, and has been translated into English as: Camille Flammarion (2015) *The Planet Mars*. Patrick Moore (trans.), William Sheehan (ed.) (New York: Springer).

14. A.S.D. Maunder to P.B. Molesworth August 31 1900; Molesworth papers, Royal Astronomical Society Library.

15. A.S.D. Maunder to P.B. Molesworth, Dec. 13, 1900; Molesworth papers, Royal Astornomical Society Library.

16. G.V. Schiaparelli to P. Lowell, January 31, 1909. Draft of letter in Archives of the Brera Observatory. (Trans. Jennifer Putnam)

17. P. Lowell to G.V. Schiaparelli, March 16, 1909. Archives of the Brera Observatory. (Trans. Jennifer Putnam)

18. Deslandres is best remembered today as a solar astronomer, who invented, independently of George Ellery Hale, the spectroheliograph.

19. It is worth pointing out that the situation of Lowell Observatory is rather similar: it is located on the northeast (leeward) corner of a plateau, with "Mars Hill," the site of the 61-cm Clark refractor, positioned at the very edge, looking down a 100 m drop to the valley below (which includes the city of Flagstaff). Despite some of the misgivings of Barnard and Douglass about the way the San Francisco Peaks might create turbulence affecting the seeing, and though air rising over the Peaks does tend to ripple the air to the east, Mars Hill is well to the south, and doesn't seem much affected. In fact, the observatory seems to enjoy a high percentage of nights with good to excellent seeing.

Lowell and Antoniadi both used scales to attempt to estimate the quality of seeing. Lowell (and other observers in Flagstaff) used a scale of 1–10, which continues to be used by many American observers of the planets:

1. Very poor images, impossible to see details
2–3. Almost continuous distortion with occasional brief good moments
4–6. More continuous disortions with short intervals of good seeing
7–8. Intervals of perfect seeing with fine scale distortions in between
9–10. Perfect seeing with steady images at high magnification.

Antoniadi, following the experience of using the Grand Lunette in 1909, published the following scale, which has come into almost exclusive use by European (including British) observers and which the author, personally, prefers:

I.   Perfect seeing without a quiver, even with the highest magnifications.
II. Slight undulations, with moments of calm lasting several seconds.
III. Moderate seeing, with larger tremors permitting only medium powers to be used.
IV. Poor seeing, with constant tremulous undulations.
V.   Very bad seeing, scarcely allowing the making of a rough sketch; blurred images even at low power.

20. Richard McKim, "E.-M. Antoniadi, 1870–1944," talk at International Workshop of Communications of Mars Observers conference, Paris, September 18, 2009.
21. E.M. Antoniadi [11]. English translation by Patrick Moore (1975), *The Planet Mars*. (Shaldon, Devon: Keith Reid), p. 29. Some of the colors Antoniadi recorded and regarded as Martian—green areas, and grayish or blue areas turning to brown, lilac-brown and even carmine—were doubtless false color effects produced by the giant lens. This helps to explain why—though he doubted the canals—he continued to believe in the vegetation hypothesis. Lowell, of course, had similarly colorful views of Mars from Flagstaff.
22. Antoniadi (1916) E.M. Antoniadi, "Report of Mars Section 1909," *Memoirs of the BAA*, vol. 20, pp. 25–92:p. 32.
23. Ibid.
24. It is important to point out that memory is not automatic, and what ultimately is remembered depends on selective attention in the moment and to what is encoded in working, or short-term, memory. This involves a different network of brain structures than long-term memory systems. Working memory is fleeting—lasting typically a few seconds at most—and unless an event is encoded in some fashion and meaningfully associated with knowledge that already exists in long-term memory it will not be successfully remembered later. Also, when we remember past events, we reconstruct them with the aid of our memory traces but also with our current belief of what happened. In addition—as is well known regarding eyewitness testimony—retroactive interference changes the memory

of an event through misinformation supplied during the retention interval. This is one of the reasons that defendants like to draw out the legal process, whenever they can, through endless appeals. See: Daniel L. Schacter [13]. Also: Elizabeth F. Loftus [14].

It needs to be emphasized that we remember only what we attend to and what we encode. In some circumstances, only fragments may be encoded—like lines and geometric figures. This type of encoding is fast, and suitable for seeing in glimpses. It is clearly the sort of process that Lowell relied on at the telescope. However, a great deal of information passes unnoticed in this approach. By contrast, Antoniadi seems to have used something like the classical method of *loci*, which relies on placing vivid mental images of what is to be remembered successively in familiar locations (classically, in the rooms of a large house). This more leisurely and systematic process allows a great deal more information about a complex stimulus, like the morphology of planetary surface features, to be grasped. Incidentally, the method of loci also tends to be well developed in chess players, who must plan ahead for several offensive moves while simultaneously anticipating through the use of memory how the other player could counter the planned moves. It should be emphasized here that memory retrieval is ultimately the key process; we cannot know the entirety of what is in our memory, only the portion of it that we are able to retrieve. Also, when we remember past events, our memories are not absolute, like photographs. Rather they are reconstructed with the aid both of memory traces and with our current belief about what really happened. Even in the almost immediate process of retrieving the memory of what was just revealed in the telescope eyepiece, as we prepare to make a drawing, we reconstruct the memory through our belief about what we just saw.

25. Antoniadi, "Fourth Interim Report," p. 79.
26. Noted in E.M. Antoniadi (1909) [15 , p. 140]. This was written 15 days, and published 8 days, before the well-known telegram of Edwin Brant Frost, dated October 6, 1909: "Yerkes telescope too big to show canals." Frost's telegram was in response to one received from Robert Jonckheere, a young supporter of Lowell who had just opened a private observatory with a 14-inch refractor near Lille (a gift from his father on attaining his majority) asking as to the visibility of canals in the great Yerkes refractor. See Michael J. Crowe [16].
27. Antoniadi, *La Planète Mars*, p. 25; Moore, *The Planet Mars*, p. 36.
28. E.M. Antoniadi to P. Lowell, September 9, 1909; Lowell Observatory Archives.
29. P. Lowell to E.M. Antoniadi, September 26, 1909; Lowell Observatory Archives.
30. E.M. Antoniadi to P. Lowell, October 11, 190;. Lowell Observatory Archives.
31. P. Lowell (1911) "The Canali Novae of Mars," In: *Lowell Observatory Bulletins*, vol. 1, nos. 1–50 (1903–1911) (Boston, 1911), pp. 243–245:p. 243.
32. Ibid., p. 244.
33. Ibid., p. 245.
34. Hoyt, *Lowell and Mars*, p. 234. The press seems to have been more taken by reports by British astronomers that a "tremendous south polar cleavage" had

appeared on the planet, followed by "a gloomy yellow veil." The New York *Times* wired Lowell on October 17, 1909, asking whether these phenomena might portend "a catastrophe transcending any ever known on earth" and the "total destruction of any life that may have existed." LOA. Obviously the seasonal rift in the south polar cap was hardly unusual, while the "gloomy yellow veil" was none other than the dust storm that had been observed by Antoniadi and others. Reports about Mars were obviously apt to become sensationalized in paraphrase en route to publication in the newspapers.

35. On his return to Mars Hill, Lowell wrote an account of the visit, "The Newly Discovered Petrified Forest of Arizona"; Lowell Observatory Archives. It was never published. To give some flavor of it, here is the first paragraph: "Drowsy dawn was in the act of becoming open-eyed day when I stepped from my bed-room out upon the veranda of my bungalow on mars Hill. Set on the mesa's edge, the house stands level with the primeval forest to the west, but to the east looks abruptly down three hundred and fifty feet to the valley below and thence out in far prospect over a slow descending pine-topped tableland to the distant desert beyond. The coming glory of the expected Sun would have drawn my gaze eastward even had it not been led there by the thought of the journey I was about to take. For somewhere in the morning of the day to me, as in the morning of the world to it, lay the Petrified Forest of Arizona, a hundred and twenty miles away. To it I was bound and as my eyes ranged the rose-flecked vault to the sharp-cut horizon I was aware on its part of a strange coming to meet me. Above the rim of pale blue silhouette stood out in clear profile against the dawn the two mesas to the north of Holbrook, eighty miles off as the crow flies, raised into sight, and that beyond what I had ever seen them, by the looming due the air. In truth below the horizon, even from my vantage point, they now stood tiptoed by refraction into visibility. Only in the early morning do they ever thus vouchsafe approach, sinking back always as the day wears on; and such matutinal visitation testifies to a steady atmosphere that indicates good seeing. By all tokens, then, this anticipatory showing of themselves should harbinger a later revelation under [propitious] skies....."

36. Antoniadi, "Second Interim Report," p. 24.

37. Antoniadi,"Third Innterim Report," p. 25.

38. Ibid., p. 28.

39. P. Lowell to E.M. Antoniadi, Nov. 2, 1909; Lowell Observatory Archives.

40. Ibid.

41. E.M. Antoniadi to P. Lowell, Oct. 9, 1909; Lowell Observatory Archives.

42. Antoniadi, "Fifth Interim Report," p. 137.

43. P. Lowell to Waldemar Kaempffert, Jan. 31, 1916; Lowell Observatory Archives. Lowell continued to maintain that with the full aperture of the Henry Brothers telescope, Antoniadi had simply been studying an image hopelessly confused and broken up by the seeing. At a meeting of the BAA into which he dropped unannounced and spoke impromptu on March 30, 1910, he argued that when

one looks at the optical image of a star, it consists (owing to the interference of light waves) not of a point but of a circle with concentric rings of light around it, and that this is what appears in good air in a small telescope (2–6 inches aperture). As one increases the aperture, "the rings begin to waver, and finally … break into a mosaic of points." He alleged that the same thing happened with fine linear details such as the canals. As for why the 24-inch refractor (even stopped down to 18 or 12 inches) beat the 33-inch, Lowell said: "The point is not as between 22 and 24 [inches] but between the whole scale all the way down. It has been said that the best Indian of all is the dead Indian, and so the most perfect you can use is no aperture at all. Just as you get down to this nothing, so your image will improve, because the air-waves which disturb are quantities of the first order to observers. The light-waves are quantities of the second order…". He suggested that theoretically it was true that the bigger the aperture, the finer the detail will come out, this all depended on the quality of the air, and insisted that "even in our air, which is selected for the purpose, and is much better than that at Meudon, we can rarely use 24 inches to advantage, and as we diaphragm down we find improvement…."

"Report of the Meeting of the Association, held on March 30, 1910" (1909–1910). *Journal of the British Astronomical Association*, vol. 20, pp. 285–294. The repugnant, but for the time characteristic, saying about the dead Indian is attributed to the Civil War Union cavalry general and pacifier of Indians on the Great Plains, Philip Sheridan. Presumably the British audience present for Lowell's remarks would have realized that by Indian was meant Native Americans rather than subjects of the Indian Subcontinent.

Both Antoniadi and Barnard, who at least had direct experience in the matter, found Lowell's argument completely spurious. Barnard noted, in a letter to Antoniadi on May 27, 1910, "Lowell … forgets that large instruments cannot give definition good enough to show the geometrical canals. He forgets that the best proof of the power of definition of an objective is the separation of very close double stars. In this field, the work by Mr. S.W. Burnham and Professor [Robert G.] Aitken with the 36-inch shows that measurements can be made of double stars which are so close that the Lowell refractor, under its atmospheric conditions, cannot hope to reveal." Quoted in: E.M. Antoniadi (1930) *La Planète Mars* (Paris: Hermann and Cie), p. 27. For his part, Antoniadi cited the example of Cassini's division in front of the ball of Saturn as a line fully comparable to the canals Lowell claimed to exist on Mars, "yet no large glass has ever broken it up into an irregular mosaic, held steadily, and the writer never saw it so sharply defined as with the Meudon refractor…". He concluded, "The hypothesis of broken interference rings, thus being in contradiction with both theory and experience, can no longer seriously be maintained to account for the complex markings on [Mars]." E.M. Antoniadi [22, p. 375]. Lowell, of course, never accepted that he had been worsted in this particular argument.

44. G.E. Hale to E.M. Antoniadi, January 3, 1910; quoted in E.M. Antoniadi [23 , pp. 191–92]

45. His colds and bronchitis were perennial, though the worst case had been in 1907–08 when he had been observing the ring-plane passages of Saturn and following which he had to give up observing altogether for several weeks. At that time Ferdinand Ellerman of Mt. Wilson had urged him to leave Wisconsin altogether during the winters: "I feel awfully sorry for you… What you ought really to do is pack up your material as the autumn approaches and head out into some mild climate and work up your summer observations, and not try to do any observing in winter, but get away from that abominable climate from December to April." F. Ellerman to E.E. Barnard, March 12, 1908. Barnard. Edward Emerson. Papers, Special Collections and University Archives, Jean and Alexander Heard Library, Vanderbilt University. As usual, this sensible plea went unheeded.

46. Quoted in Antoniadi, "Fifth Interim Report," p. 137.

47. E.M. Antoniadi to E.E. Barnard, December 11, 1909. Barnard. Edward Emerson. Papers, Special Collections and University Archives, Jean and Alexander Heard Library, Vanderbilt University.

48. Antoniadi, "Fifth Interim Report," p. 138.

49. E.M. Antoniadi to E.E. Barnard, December 11, 1909. Barnard. Edward Emerson. Papers, Special Collections and University Archives, Jean and Alexander Heard Library, Vanderbilt University.

50. E.M. Antoniadi to E.E. Barnard, December 16, 1909; Barnard. Edward Emerson. Papers, Special Collections and University Archives, Jean and Alexander Heard Library, Vanderbilt University.

51. "Report of the Meeting of the Association, held on Wednesday, December 29, 1909, at Sion College, Victoria Embankment, E.C.," (1909–1910), *Journal of the British Astronomical Association*, vol. 20, pp. 119–130: p. 123.

52. Quoted in Antoniadi, "Sixth Interim Report," pp. 191–192.

53. E.E. Barnard to E.M. Antoniadi, May 16, 1913. Barnard. Edward Emerson. Papers, Special Collections and University Archives, Jean and Alexander Heard Library, Vanderbilt University.

54. P. Lowell [24]. Wisp-like markings were also recorded at the end of the same month by Mentor Maggini, another master of spidery Martian canals, at the Florence Observatory. Lowell could not fail to note the similarity to the wisps of Jupiter recorded in 1907, and concluded hastily (p. 235) "The existence … of lacings or regular persistent rents in these clouds so nearly resembling one another on the two planets, shows that such features are the inevitable outcome of the forces at work under the physical conditions, and give us some clue to the meteorologic processes exemplified there." At least Lowell did not claim to see the exiguous outer ring that had been reported by Georges Fournier at Jarry-Desloges's observatory on Mont Revard and Emile Schaer at the Gebneva Observatory in 1907–08, and which would long remain, in the phrase of

A.F.O'D. Alexander, "a sort of 'Loch Ness monster' of Saturn in which some believe, but of whose reality most astronomers are very sceptical." Alexander [25]. But Barnard decisively put paid to the supposed exterior ring during two evenings of careful observations with the 40-inch refractor on January 12 and 19, 1909. To minimize the effect of glare he occulted the globe by the edge of the telescope field and used a mica occulting device, but was unable to find any sign of it. Thus, he wrote of his results on the evening of January 19: "No evidence of a ring exterior to the bright rings… The shadow of the rings on the ball was sharp and black. Nothing abnormal about the ball or rings." See: E.E. Barnard [26 , pp. 622–623].

55. Leonard, *Afterglow*, p. 111.
56. Ibid., p. 112. Lowell added that he had suspect the belts and oblateness of Neptune the previous January. However, he never seems to have published these results.
57. "Report of the Meeting of the Association, held on March 30, 1910, at Sion College, Victoria Embankment, E.C.," (1909–1910), *Journal of the British Astronomical Association*, vol. 20, pp. 285–294:p. 288. See also endnote 41.
58. Ibid., p. 289.
59. Ibid., p. 289.
60. Ibid., p. 291.
61. E.M. Antoniadi, "Further Objections," pp. 376–377.

References

1. Gay P (1988) Freud: a life for our time. W.W. Norton, New York, p 206
2. Lowell P (1909) The planet Venus. Pop Sci Mon 75(December):521–536
3. Lehman M (1988) Robert H. Goddard: pioneer of space research. Da Capo Press, New York, p 14
4. Goddard EC, Edward Pendray G (eds) (1970) The papers of Robert H. Goddard, volumes I–III. New York, McGraw-Hill, p 102
5. Barnard EE (1911) Photographs of the planet Mars. Mon Not R Astron Soc 71:471–472. (with two plates)
6. Kuiper G, Calvert MR (1947) Note: Barnard's photographs of Mars. Astrophys J 105:215
7. Antoniadi EM (1916) Report of the section, 1909. Mem British Astron Assoc 20:353–420
8. Antoniadi EM (1909) First interim report on the observations of 1909. J Br Astron Assoc 19(10):427–433
9. McKim R (1993) The life and times of E.M. Antoniadi, 1870–1944. J Br Astron Assoc 103:Part I, 164–170; Part II, 219–227

10. Antoniadi EM (1908) Note on some photographic images of Mars taken in 1907 by Professor Lowell. Mon Not R Astron Soc 69(2):110–114
11. Antoniadi EM (1930) La Planète Mars. Hermann et Cie, Paris, p 18
12. Antoniadi EM (1909) Fourth interim report for the apparition of 1909, dealing with the appearance of the planet Mars between September 20 and October 23 in the great refractor of the Meudon observatory. J Br Astron Assoc 20:78–81
13. Schacter DL (1996) Searching for memory: the brain, the mind, and the past. Basic Books, New York, pp 42–43
14. Loftus EF (1996) Eyewitness testimony. Harvard University Press, Cambridge, MA
15. Antoniadi EM (1909) Fifth interim report for 1909, dealing with the fact revealed by observation that prof. Schiaparelli's 'canal' network is the optical product of the irregular minor details diversifying the Martian surface. J Br Astron Assoc 20(3):136–141
16. Crowe MJ (1986) The extraterrestrial life debate: 1750–1900. Cambridge University Press, Cambridge, p 339
17. Osterbrock DE (1989) To climb the highest mountain: W.W. Campbell's 1909 Mars expedition to Mount Whitney. J Hist Astron 20:77–97
18. Campbell WW (1909) Water vapor in the atmosphere of the planet Mars. Science 30(777):474–475
19. Cachon A (1978) The Pic du Midi observatory. Imprimerie Péré, Bagnéres-de-Bigorre, p 23
20. Dollfus A (2010) The first Pic du Midi photographs of Mars. J Br Astron Assoc 120(4):240–242
21. Antoniadi (1909) Third interim report for 1909, dealing with the nature of the so-called 'canals' of Mars. J Br Astron Assoc 20:25–28
22. Antoniadi EM (1910) Further objections to Prof. Lowell's canal system of Mars. J Br Astron Assoc 20:374–377
23. Antoniadi EM (1910) Sixth interim report for 1909. J Br Astron Assoc 20:189–192
24. Lowell P (1910) The wisps of Saturn. Pop Astron 18:232–235
25. Alexander (1980) The planet Saturn: a history of observation, theory and discovery. Dover reprint of 1962 edition, New York, p 319
26. Barnard EE (1909) Recent observations of the rings of Saturn, and their bearing upon some of the phenomena of the disappearance of the rings in 1907. Mon Not R Astron Soc 69(8):621–624

12

Vale Percival

Contents

> *His work on Mars, poor Percy Lowell knew,*
> *Seemed in the process of evaporation;*
> *He knew he needed something new to do*
> *To save his scientific reputation.*
> *Since Uranus's orbit seemed not quite*
> *Explained by Neptune, his observatory*
> *Would see a "Planet X," and he just might*
> *Yet bask in something like Le Verrier's glory....*
> —Robert Bates Graber, Plutonic Sonnets, *CXXXVI*. [1]

© The Author(s), under exclusive license to Springer Nature Switzerland AG 2024
W. Sheehan, *Parallel Lives of Astronomers*, Springer Biographies,
https://doi.org/10.1007/978-3-031-68800-3_12

Mars in Retreat

The 1909–10 opposition of Mars was clearly a defeat for Lowell, though of course, being Lowell, he never would admit it. At the March 30, 1910, meeting of the BAA, Lowell argued that it was completely expected and normal for the Martian albedo features to appear so faint during much of 1909:

> They are seasonal phenomena and that is one reason why such unsatisfactory views were got here in the early part of the season. It was supposed that something was happening on Mars, but the fact is, the periods when Mars approaches closer to the Earth are not favorable ones for observations, because the canal system is not in its flourishing season. The next opposition will be a far better one for the canals.[1]

That meant that there was not much to expect regarding Mars until 1911, when Mars began to retreat through a series of more and more unfavorable oppositions, reaching a low point with the aphelic opposition of 1916. Not until August 1924 would Mars be as close to the Earth as it had been in September 1909. There seemed to be little hope of further progress being made either by those who hoped to prove the existence of canals in Lowell's sense, i.e., irrigation channels, or the skeptics' program of demolishing these "whimsical provocations of truth," as Antoniadi described them to Barnard.[2]

Well aware that under these circumstances the public interest was bound to wane—indeed, it would never in Lowell's lifetime return to the peak of 1907—he to some extent attempted to reinvent himself as he had after his breakdown, when in 1903 he inserted into his researches such more rigorous methodologies as the "cartouches" of Martian canal development, Slipher's spectrographic work, and Lampland's photography. Now his papers become increasingly mathematicised—and more frequently addressed topics other than Mars. A number of papers considered the perturbative action of planets (especially Jupiter) on the semimajor axes of asteroids, and tried to work out the implications for the positions at which the planets themselves had formed.[3] Others, in collaboration with E.C. Slipher, showcased work with the filar micrometer, undertaken in open competition with Barnard, whose work in the field was regarded as authoritative. New values of the dimensions of Saturn's ball and ring system were published, where particular attention was paid to the precise positions of evanescent fine divisions glimpsed on the rings' surface, and a new value for the oblateness of Uranus.[4] Though these measures were straightforward in principle, they needed to be corrected for the effects of chromatic aberration, the precise orientation and finite thickness of the micrometer threads (on the order of $0''.1$) and, in the case of Saturn, factors related to the complicated intersections of light and shade in the ball and rings depending on its axial inclination to the Earth. Then, too, as always, the contribution of the personal equation had to be factored in. The distribution of errors around a mean had to be considered, and the effects of irradiation, where the edges of bright areas in an image appear to bleed over into darker areas.[5] All these corrections *might* lead to more reliable results—or serve merely as so many fudge factors.

Despite the best efforts, the "truth" lay hidden somewhere within a cloud of errors, and even the best measurements with the instruments available at the time never came within less than about 1 or 2 percent of the true (spacecraft-determined) values. To his credit, Lowell's measures showed considerable skill both in manipulating the filar micrometer and elaborately reducing the errors, but as always, he pressed this data into service of conclusions that stood far loftier in the air than the secure moorings of his measurements allowed. This was a problem less of his skill than of what is of even greater importance in an astronomer—judgment. In this work, as in all other work to which Lowell turned his hand, his tendency was to push marginal perceptual data farther than justified in the support of his theories.

During this final period of Lowell's life, he also identified strongly with the mathematical—as opposed to the literary—side of his interest. This, in itself, meant that his later work would not have the public appeal and impact of his Martian theory. The mathematical flavor of his investigations into the structure of Saturn's rings and the possible constitution of its interior, the formation of the planets, and the like, were nowhere more in evidence than in his great search for a trans-neptunian planet—Planet X—to which he devoted more time and energy during the final years of his life than to anything else.

Planet X

"I am an astronomer!" Lowell exclaimed to E.C. Pickering when he and other astronomers stopped in Flagstaff on their way out west to attend the fourth meeting of the Union for International Cooperation for Solar Research, which Hale had organized in August 1910 in Pasadena. By astronomer, Lowell meant it in the old, nineteenth-century sense: someone who actually spent time looking through a telescope, made accurate measurements with a micrometer, and was well-versed in the classical methods of celestial mechanics. He felt increasingly outside the main thrust at time, which emphasized astrophysics. Lowell thought of the latter as nothing more than mindless data-gathering, taking pictures of spectra and making photometric measures of stars to store in huge collections. But, Lowell declared, none of these collectivist efforts would ever lead to great discoveries. For that the highly individualistic and original "genius" was needed, and Lowell left no doubt on what side of the line he fell.

Barnard, with tripod and camera, stops to visit with several astronomers enjoying box lunches on Mt. Wilson, during the Fourth Conference of the International Union in Solar Research held between August 29 and September 3, 1910. (Credit: William Sheehan Collection)

Delegates to the Fourth Conference of the International Union in Solar Research at Mount Wilson Observatory, 1910. Though Percival Lowell pooh-poohed such gatherings, and he wasn't present, it was clearly a going thing, and his assistant V.M. Slipher did attend. Shown in this image are (broadly from left to right): Ellerman, H.C. Wilson, St. John, Larkin, Townley, V.M. Slipher, Fowle, Coblentz, Frost, Idrac, Puiseux, Hartman, Kustner, Slocum, Hamy, Knight, Wolfer, Fath, Rydberg, Hepperger, Fox, Haussmann, Cortie, Turner, Russell, Kayser, Adams, Miller, Ames, Backlund, Konen, Pickering, Fowler, Lampland, Hale, Belopolsky, Deslandres, Schuster, Campbell, Ricco, Mrs. Kapteyn, Bosler, K. Schwarzschild, Mc Adue, Kapteyn, Mrs. Fleming, Watson, Schlesinger, Humphreys, Madrill, J.F. Sanford, Chretien, De La Baume Pluvinel, Fabry, Abbot, Hills, Larmor, Cotton, Dyson, Barnard, King, Newall, Pringsheim, Leuschner, J.S. Plaskett, Gale, Chant, Eversheim, Rotch, W. Mitchell, Stratton, H.D. Babcock, Ritchey, Brackett. Barnard, who was diffident at such gatherings, is barely visible in shadow in the row in the back, underneath the window farthest to the right. (Credit: Public Domain, Mount Wilson and Palomar Observatories, courtesy of AIP Emilio Visual Archives)

Lowell had never forgotten the glow of having been praised by his Harvard professor, Peirce, and took the greatest satisfaction in his mastery of the methods of perturbation theory that he had learned from his admired mentor. It was in connection with this that one must understand his search for Planet X, which eventually grew into an all-out effort in emulation of the famous calculations by the mathematicians Urbain Jean Joseph Le Verrier and John Couch Adams of the position of Neptune, leading to its optical discovery in September 1846.

Lowell's search for Planet X had begun almost casually; already in his MIT lectures on the Solar System in 1902, he had discussed the possibility that a "comet family" whose aphelia lay out beyond Neptune might serve as a "fingerpost" to a planet further on. He assumed that a planet, if it existed, would likely be a giant like Uranus and Neptune, and that it would be an object of the eleventh or twelfth magnitude, possibly revealing itself by its small disk. In that case, it ought not to be too difficult to find it by means of a broad-brush photographic search of the sky along the ecliptic. For this purpose, John C. Duncan, E.C. Slipher, and Kenneth P. Williams, Indiana University graduates and successive holders of the short-lived Lawrence Fellowship, were assigned to photograph the sky with a 5-inch Brashear photographic telescope yielding a field 15 degrees wide, with sharp definition over some 10 degrees. By the time the fellowship was discontinued, in September 1907, no planet had been found. (Clyde Tombaugh would later note that Pluto—the closest thing to such a planet—was then sixteenth magnitude, far fainter than the object Lowell was expecting and at the very limit of the search plates; it was also, because of its highly inclined orbit, far from the ecliptic and outside the range of the photographic coverage of the search.)[6]

The failure of the first search impressed upon Lowell the need for some sort of position indicating where the unknown planet might be hiding among the sky's millions of stars. At first Lowell would attempt to use a "graphical method" of the sort introduced by Sir John Herschel in attempting to understand the way that Neptune had perturbed the motion of Uranus in the years before and after its discovery. This depended upon plotting the residuals—the differences between the theoretical and observed positions of a planet. This is an important concept, and it requires at least a brief parenthesis.

After Uranus was discovered by William Herschel in 1781, a number of pre-discovery observations dating back to one by John Flamsteed at the Royal Observatory, Greenwich, in 1690 were unearthed. These observations allowed for a longer baseline for the computation of the newly discovered planet's orbit. However, within a few years the observed motion of the planet and that calculated for it began to diverge, and despite the best efforts of mathematical astronomers over the next decades to reduce the "residuals" between the two—the difference between observed and calculated positions—by 1845 the difference had increased to a staggering (by astronomical standards) 133 arcseconds. This was an unacceptable state of affairs, and it was this difference that led Adams in England and Le Verrier in France to posit the existence of an unknown planet, to make various assumptions about its orbit, and to undertake a complicated series of calculations in order to find the likely position of the planet.

After Neptune was discovered, it turned out that the orbits Adams and Le Verrier had calculated were not very similar to that actually followed by the planet (though the most important element, the heliocentric longitude—the position of an object on the ecliptic as measured from the center of the Sun—turned out to be not very sensitive to variations in the other parameters, so the calculations had, in that sense, been valid). In order to better understand the situation, Sir John Herschel plotted the residuals for both Uranus and Neptune, and found a peak correlated with the conjunction of these planets in 1822. It seemed from this work that, by extension, the trans-neptunian planet Lowell was seeking might be found by plotting the residuals of the known planets to locate peaks—presumably reflecting conjunctions—and in this way finding, in a quick and dirty fashion, the position of the planet he sought. The whole method depended, obviously, on the residuals—those that remained in the case of Uranus having been reduced, thanks to Neptune, from 133 arcseconds in 1845 to a maximum of 4.5 arcseconds in 1910. (Neptune had not yet been observed long enough to reveal any clues.)

Le Verrier himself had published residuals for Uranus in 1873, which reflected his best efforts to calculate orbits for both Uranus and Neptune based on the observational record going back to 1690. At first, rather than rely on Le Verrier's residuals, Lowell hired a computer at the Nautical Almanac Office, William F. Carrigan, to recalculate them from scratch. By March 1908, Carrigan had finished the first part of the rather protracted and extremely tedious calculations and presented Lowell with recomputed residuals of Uranus for 1780 to 1820. At first Lowell wanted to press on, though possibly—annoyed by how much Carrigan was costing him—he wanted to streamline the work; thus, he added, "I should not take every observation but a few of the best ones only for each [year]."[7] As usual Lowell was stretched thin—negotiating terms for his new 40-inch reflector to vindicate the Martian canals, about to marry, and soon to be off to Europe for his honeymoon. The whole question was rendered moot, though, by the entrance of an unexpected rival into the field—his old collaborator and mentor William H. Pickering. Using a "graphical method" of Herschelian design, Pickering had plotted Le Verrier's residuals for Uranus from 1873, and found an approximate orbit and position for a trans-neptunian planet which he called "O" (because it was the next one out from "N," Neptune).

William H. Pickering, 1914, in a portrait by an unknown photographer. (Credit: Public Domain courtesy of Wikipedia Commons)

On November 11, 1908, Pickering announced his findings at a meeting of the American Academy of Arts and Sciences in Cambridge, Massachusetts, at which Lowell, back from Europe since early October, happened to be present. Though recognizing Pickering's approach to be rather slap-dash and careless, he nevertheless took up the graphical method in earnest, and began graphing Le Verrier's residuals to see what they might reveal. He was at it until the spring of 1909 when other work overwhelmed the effort. (His old friend

Ralph Curtis said of him at the time that he appeared unhappy and over-worked.)[8] Lowell had taken things far enough to recognize that "the problem of a trans-neptunian planet admitted no easy solution" [11]. In addition to beginning to doubt the graphical method, he also had become convinced that Carrigan's calculations were not likely to improve on the residuals published by Le Verrier, and alarmed at how much they were costing him, Lowell dismissed Carrigan.

Lowell's next, momentous step took place in July 1910. At this point he decided to abandon the graphical method and instead to embark on the far more rigorously analytical method based on classical perturbation theory that had been developed by Le Verrier and Adams and had led to the famous discovery of a planet "with the tip of a pen."[9] In taking this step, Lowell—his head "bloodied but unbowed" by the recent beating he had taken from Antoniadi and others over Mars—seems to have hoped to emulate his immortal predecessors and to increase his prestige. Clyde Tombaugh recalled C.O. Lampland telling him, "Lowell wanted desperately to improve his credibility among other astronomers. So, Lowell thought, if he could predict the location of a ninth planet, beyond Neptune, and then find it, it would surely improve his status."[10]

Mars was not forgotten; even in his State Street office, where he and an increasing number of assistants would press ahead with the Planet X calculations, he surrounded himself with maps and globes of the red planet, and perseverated on the now increasingly tired themes of canals and otherworldly irrigation projects. One of those who was forced to listen said, "He leaps from point to point and seems to make no stops at way stations."[11] Except to his oldest and dearest friends—though even perhaps to them when they spoke candidly and privately to one another—these pronouncements seemed to resemble less the prophetic visions of genius than the ravings of a madman.

The mathematical details are abstruse and need not concern us here; the basic procedures adopted by Lowell were those of Le Verrier (which were somewhat more general, if less elegant, than those used by Adams). Lowell's goal was the same as theirs had been: to find the change in the heliocentric longitude, designated by the mathematical symbol, Δv of Uranus due to the perturber, which Lowell referred to as "Planet X." This would lead to a position of the planet along the ecliptic that could ultimately be used to point a telescope. Lowell followed Le Verrier's method, and "set up a series of observation equations (equations of condition) for the residuals of Uranus for various dates (or the means of groups of dates), eliminate Δn, $\Delta \varepsilon$, Δe, $\Delta \varpi$, express them in terms of the perturbations, then combine all the remaining equations, except for certain of the earlier ones, in two groups to be solved for n', ε', e', $\Delta\varpi$'and constants. Then, finally, substituting the solutions in the omitted equations, find for what value of ε' their residuals became a minimum."[12] All well and good (if a tremendous amount of work); but the devil is in the details, and Lowell's 5-year analytical search would produce thousands of pages of calculations that fill twenty large banker's boxes in the Lowell Observatory archives and bear silent witness to a world of pain. The Achilles heel, ultimately, would prove to be the small size (4.5 arcseconds) of the residuals.

This new, more rigorous quest for an unknown planet would challenge—and, he hoped, showcase—Lowell's mathematical abilities. Moreover, in contrast to the Martian theories based on a cobweb of gossamer canals that had so far failed to win over his scientific peers, the discovery of a planet would be a hard, concrete, indisputable achievement. One thinks of Dr. Johnson famously kicking a mile-stone while saying with regard to Bishop Berkeley's theory that there was no matter, "I refute him thus." A planet beyond Neptune, if it existed, was bound to consist of a considerable quantity of matter, and would silence, Lowell thought, once and for all, the critics. By emulating Adams's and Le Verrier's achievements, Lowell could secure for himself an everlasting place among the astronomical immortals. The great Ahab-like quest had begun.

Lowell's favorite portrait, which shows him in his (as usual) sartorial splendor, was taken in 1910 in a studio on Tremont Street, the same avenue where he had been born 55 years earlier. (Credit: Lowell Observatory Archives)

A close-up of Lowell, taken during the same photo shoot as the one above. (Credit: Lowell Observatory Archives)

A Dyed-in-the-Wool Conservative

As the mathematical search for Planet X took shape, it called upon Lowell's expertise in mathematical methods that were entirely worked out in the nineteenth century. He was, in contrast to the astrophysicists whom he regarded as only the latest fashion, a "mathematical astronomer," a member of a dying breed. "It is a popular delusion," he would insist in a 1916 lecture, "that all astronomers must be mathematicians. The fact is they ought to be but are not. The mathematical astronomer is now the exception, due chiefly to the rise of astrophysics."[13] He thus dismissed with contempt distinguished men such as E.C. Pickering, Campbell, Frost, and Hale, who were all partisans of Hale's International Union for Cooperation in Solar Research—an organization perhaps doubly contemptible as it was not only a union (Lowell despised labor unions) but one of astrophysicists, no less.

A Boston Brahmin whose family fortune had been built on Southern cotton and the New England sweat shops of the Lowell and Lawrence textile mills, he was (as were all of his siblings) as solidly a part of the establishment as one could wish. He was every inch a conservative, not only scientifically (much as he tried to say otherwise) but politically. Though he was encouraged by his cronies such as Judge Edward M. Doe to run as the Republican candidate for Senator after Arizona became the 48th state, he had, in fact, always been a dependable supporter of the "Anti-Statehood League," and detested the idea of serving in any kind of political office.[14] Certainly, he could always be counted upon to promote a "Brahmin ideology of elitism and strident nationalism," says his biographer David Strauss. "He lashed out against women's suffrage, labor unions, and such Progressive proposals as initiative and recall, because they would give decision-making power to 'the lowest classes of community—on the pleasing principle that he who possesses least knowledge of, and least interest in, the country is the most eminently equipped to govern it.'"[15] He and the rest of his family were horrified by the 1912 Lawrence Textile Strike (also known as the "Bread and Roses Strike") of immigrant workers in Lawrence, Massachusetts, led by the Industrial Workers of the World. He was strongly anti-immigration and advocated capital punishment on the grounds—as he told the warden of the Arizona State Prison—that it was "both deterrent and prevents criminal propagation."[16] In his view, Prohibition, then being mooted, was quite simply an invasion of individual liberty, and as a hedge against the impending crisis should it pass, he ordered twelve cases of wine along with vast quantities of scotch, beer, and sherry— only to find there was not sufficient room for it in the wine cellar of the Baronial Mansion.[17]

In one of his clumsier attempts at humor in "The Revelation of Evolution," he mocked the "labor-champion who claimed that labor built the railroads because it laid the rails. 'Did you ever think,' said the man he addressed, 'that the end attained depended on where the rails were laid?'"[18] Consistent with this, he envisaged Martian society as organized along the same lines as capital and labor. The Martians had set up not a democracy but an oligarchy of the intellectual elite, as he suggested in an October 1911 lecture, "Two Stars," at Kingman, Arizona Territory, to an audience largely consisting of hard-rock miners. The study of Mars taught us, he declared,

> … not only of a well-ordered community but of how that unity can be fashioned and guaranteed. Think of the intelligence and far-contrivance necessary

to execute and maintain a system of irrigation worldwide in its scope and meticulously dependent upon the seasons for its functional activity. Every drop of water is precious…. The very ablest intelligences that Mars can produce alone are fitted to conceive and direct so universal and vital a matter. Such only can be in command. On the other hand, each member of such a community must equally carry out his part to the utmost of his capability…. From top to bottom each individual has his place and fills it…. To know one's place is the best qualification…. All are given a chance to rise, but only the worthy do. We may be very certain that in the Martian world-economy, the fittest only have survived.[19]

Lowell himself thought the lecture a great success, writing to Leonard afterward, "I made some friends, even among the Socialist miners which was my aim. One of them whose views were quite subversive, now loves me—to my immense surprise…. I had shown how the solidarity of the Martian canal system points to an efficient government in which the best men are at the front and then I went on to show its applicability to us."[20]

At his observatory, too, Lowell saw himself as the "lion," the superintending intelligence who determined where the rails were to be laid. Once, Andrew Ellicott Douglass had failed to recognize this, and had put his head in the mouth of the lion. However, gradually Lowell had mellowed since the early days, and relaxed his grip on at least some members of the staff once they had thoroughly demonstrated their loyalty—especially V. M. Slipher, who was now ably wielding the Brashear spectrograph and using it to make important discoveries. Slipher was henceforth to be relieved of his responsibilities for "Venus," the observatory cow, or urgently filling requests for Shredded Wheat cereal for Lowell's breakfast table, as Lowell had once wired him from Chicago to do.

In a world in which he was subject to severe criticisms, the observatory, both its magnificent setting and the nature of its work, was Lowell's private eyrie (his word), where astronomy was a refuge as well as a passion. Mrs. Lowell later recalled the nature of the routine established when he was in residence, with the reference to the *Traité de Mécanique celeste* attesting to his then utter immersion in the "X" calculations:

As you know, it is not easy for the observing astronomer to lead a strictly regular life in that hours at the telescope often make it necessary to use, for the much needed rest, part of the daily hours usually given to work. His intense occupation with his research problems, however, was broken with great regularity for short intervals before lunch and dinner. These times of recreation were given to walks on the mesa or work in the garden. When night came, if he was not occupied at the telescope, he was generally to be found in his den. It was not always possible for him to lay aside his research problems at this time of day, but he did have some wholesome views on the necessity of recreation and a necessary amount of leisure to prevent a person from falling into the habit of the "grind." To those who came to his den the picture of some difficult technical work near his chair, such as [Felix] Tisserand's [Traité de] Mécanique céleste, will be recalled, though he might at the time be occupied with reading of a lighter character. And occasionally during the evening he might be seen consulting certain difficult parts upon which he was pondering....[21]

The search for Planet X followed a two-track agenda. The first was the massive calculation using Le Verrier's residuals of Uranus from 1873 but carried out with increasing rigor and complexity as time went on. It was (again, in contrast to the case with Neptune, where Adams and Le Verrier had largely worked single-handedly)[22] far too vast for any one person to carry through alone. To support the effort, Lowell organized, at considerable expense, a staff consisting, at the peak of activity in late 1912, of several persons; they worked sometimes under his direct supervision but more often under that of his chief computer Elizabeth Langdon Williams in his State Street office in Boston. They were helped by such state-of-the-art calculating instruments as "Thacher's Cylindrical Slide Rule," billed as the "ultimate cylindrical slide rule," and the "Millionaire Calculator," the first commercially successful motorized mechanical calculator that was advertised as the "only calculating machine on the market... that requires but one turn of the crank for each multiplier or quotient." The "Millionaire" was fast, but big and clunky, and occupied an entire desk. Despite the inconvenience, it was an improvement on doing multiplication and long division by hand.[23]

The calculations led to a series of predictions of where in the sky Planet X might be hidden. The second part of the search was, of course, the photographic search of areas of the sky so indicated. Lampland was the point man on this, and at first, the 40-inch reflector was allocated to the purpose at all times except, as in late 1911, when it was needed for Mars observations.

Lowell's and Williams's calculations progressed slowly at first, with little to report through the fall and winter of 1910–11. On December 10, 1910, Lowell sent a report of progress so far to Lampland. "Miss Williams and I have been pegging away at it, have constructed the curve of perturbation due X if at 47.5 astr. units including all terms of the first power of the eccentricities and examining the most important terms in their squares," he wrote. "We find some interesting things ... but we do not find the planet for except in periods the theoretic curve swears as the residual one. It is of course possible that Leverrier's [sic] theory is not sufficiently exact."[24]

The next development took place on Lowell's fifty-sixth birthday, March 13, 1911, when Lowell telegraphed to Lampland that at long last a position was imminent: "Please begin to photograph ecliptic where south with forty-inch. Hope to wire position in a few days. Calculations tremendously long."[25] Alas, Lowell did not have a position for Lampland "in a few days," and immediately began to worry that the latter might proceed with comparing his plates on his own and stumble across the planet accidentally (as William Herschel had in the case of Uranus). This, of course, would steal the thunder of a mathematical discovery (and his own credit). Even preliminary skirmishes with the problem had suggested that the planet's mass might lie between Neptune's and the Earth's, giving its magnitude as about 12–13 and its disk more than 1″ of arc across. Leonard therefore followed up with a letter in which she told Lampland on March 22, "Dr. Lowell has no objection to your going in for the hunt there as long as he tells you where to look from this end."[26] The point was that, though theory would, of course, necessarily guide the search or else one simply would scan the sky at random, Lampland must not forget that he is merely a tool in Lowell's hands and not an independent discoverer. (Even after Lowell's death, the Observatory was to treat Clyde Tombaugh's discovery of Pluto with something of the same attitude.)

Neither Lowell nor Leonard had cause to worry; no planet revealed itself on Lampland's plates, and in April, Lowell, as Leonard informed V.M. Slipher, continued "working like a slave on Planet X," despite coming down with a severe cold.[27] By the end of the month, Lowell and Williams had calculated that the planet lay at heliocentric 235 degrees. A week later they favored a position four degrees farther to the east, with Lampland instructed to take plates 2 or 3 degrees on either side of the position. These calculations put "X" in Libra moving into Scorpio and apt to be camouflaged among the thickets of stars of the Milky Way. Lampland's plates, exposed several days apart, were sent to Boston. Lowell himself was the one to compare them, laying one atop the other and examining them with a magnifying glass in the hope of finding an object that had moved in the interval.

This was an exceedingly inefficient way of comparing plates containing thousands of star images. Already in 1908, Lampland had urged Lowell to consider the acquisition of a blink comparator, an apparatus specifically designed for comparing star fields that had been invented by physicist Carl Pulfrich at the German firm of Zeiss in 1904. It used an electromagnet to flip a small mirror back and forth so as to redirect the light path successively from a small area of one plate to the corresponding area on a second matched plate. The image of a planet registered in exposures taken several days apart would betray itself immediately by appearing to jump back and forth, making the search far more efficient than the plodding manual method of searching plates with a hand magnifying lens. Indeed, Barnard would use a blink comparator at Yerkes to search plates for faint images of Halley's returning from out beyond Neptune on its forthcoming perihelion passage, and later to discover his famous "runaway" star.

Lowell had examined such a device on his honeymoon visit to Germany that year. At the time he had been cool to the idea, but by 1911 he had changed his mind. Now, on the eve of returning to Flagstaff both for Mars's opposition and in order to assume direct oversight of the observational phase of the search (with Williams placed in charge of the mathematical phase which continued in Boston), he sent an order to Jena for a blink comparator. It would often follow Lowell's peregrinations between Boston and Flagstaff so that at least at first (when discovery seemed possibly imminent) he, personally, would be the one to blink the plates. As the search ground on, he was only too glad to turn the tedious task over to others, as he had in placing Williams entirely in charge of the back-breaking calculations.

Williams was now recalculating residuals of Uranus from the 1903 theory of Le Verrier's erstwhile assistant and now successor, Jean Baptiste Aimable Gaillot, rather than from the 1873 theory of Le Verrier himself.[28] (Apart from the theories of Le Verrier and Gaillot, the only other complete theory of Uranus had been published by Simon Newcomb in 1873; it would have produced a very set of different residuals, but Lowell seems to have refused even to consider it simply because of personal animosity toward Newcomb for ridiculing the canal network.) Williams was clearly a fast worker, for already in July 1911 she had arrived at a preliminary position from Gaillot's residuals—well to the east of the positions Lampland had received only months earlier. Within only a few days, even this latest position had been revised, and put the planet at around heliocentric longitude 210 degrees. Thus, instead of being in Libra or at the border of Libra and Scorpio, the planet's position had slid it over into Virgo. There was a brief surge of excitement when in late July 1911 a "suspicious object retrograding" was caught on Lampland's plates, but the Lowell Observatory Archives give no indication what it might have been. Needless to say, it was not Planet X.

Mars Again: From Flagstaff and Mt. Wilson

The search had to be temporarily set aside, at least by Lowell himself (though Williams and assistants continued their drudge-like toil in Boston) as his attention was diverted to Mars for the November 1911 opposition. His old school chum, now a professor of English at Harvard, Barrett Wendell, visited him at Flagstaff, and Lowell wrote to Leonard, "The canals, you will perceive are much more salient than they were in 1909—as I expected… The Barrett Wendells came on Monday and left on Wednesday, he having seen the canals."[29] In an effort to refute the testimony of Barnard and Antoniadi, who had claimed that the canals disappeared when looked for in large telescopes, Lowell set out with determination to see the canals in the 40-inch reflector that had been delivered late in 1909. Unfortunately, Lowell had handicapped this telescope with one of his brilliant but bad ideas. Deciding it would be a good idea to "sink it in the ground on some hillock to a depth of about 6 feet and then over this to erect a dome,"[30] he hoped to get more stable seeing conditions than with the usual above-ground configuration. It was of course otherwise, and as a result, Lowell made only sporadic observations of Mars with it. Instead he and E.C. Slipher (and Wendell) used the 24-inch, as always diaphragmed, for most of their Mars work.

Wendell may have seen the canals, but his testimony—as that of Judge Doe, Morse, Leonard or other of Lowell's friends or assistants who observed the planet from time to time—was not as important as that of Barnard, who spent the month of November 1911 at Mt. Wilson observing and photographing Mars and Saturn with the 60-inch reflector. (Incidentally, Lowell was on several occasions invited to collaborate with observers elsewhere, either by comparing their views at the telescope or participating in simultaneous observations, but predictably, Lowell always rejected such invitations.)[31]

At Hale's suggestion, Frost gave Barnard a month-long sabbatical around the time of the Mars opposition (November 25) to photograph Mars and Saturn in the 60-inch reflector. Excellent results were obtained, especially with Saturn. Photography was carried out at the Cassegrain focus; visual observations, on the other hand, had to be done at the telescope's 25-foot Newtonian focus, and owing to the time required to change the telescope from the Cassegrain to the Newtonian configuration, it was not possible to use both on a single night. For this reason, only a few visual observations were attempted, but those few were a revelation. For the first time in his life, Barnard enjoyed views not hampered by chromatic aberration like those in a large refractor. The latter had, he said, a "muddy or dirty look." The reflecting telescope's parabolic mirror brought a star or planet to a perfect focus, without any secondary spectrum, so that Mars, in Barnard's words,

> … looks as if cut out of paper and pasted on [the] background of the sky. It is perfectly hard and sharp with no softening of edges. The outline and general definition are much superior to that of a refractor telescope.[32]

Under such conditions the intense blue-greens and chocolate browns reported by observers like Lowell, on which they hung so much supposition, were nowhere to be seen. Instead, the dark markings were "light grey."[33] The aperture was stopped to only 12 inches, because of indifferent seeing; and yet on November 23, Barnard saw the planet better, he thought, than he had ever seen it before. He wrote in his observing log book:

> The Syrtis Major was broken up into a great number of wispy masses. The momentary best seeing gave the impression that the broken masses were still further shattered, so that the whole mass would be a flock of wisps with no continuity of form … whatever. Certainly the true nature of this remarkable region has never been so clearly seen by me before. Even in 1894 with the 36 [inch] of L[ick] O[bservatory] it was not so well seen. The impression I now get of the Syrtis [Major] is … [it is] thoroughly broken up… No trace of any thing resembling a canal either in the dark or bright regions could be seen. I think this is perfectly decisive.[34]

A few nights later, he found the Syrtis Major "… broken up in to cloud like masses with wispy details…. The great mass … preceding the trunk of the Syrtis is almost as conspicuous as the Syrtis [and] seems to be made up of a shredded appearance…. It is not possible to draw the details because they are so complex."[35]

The author observing Mars with the 60-inch reflector at Mt. Wilson in October 2005. (Credit: William Sheehan)

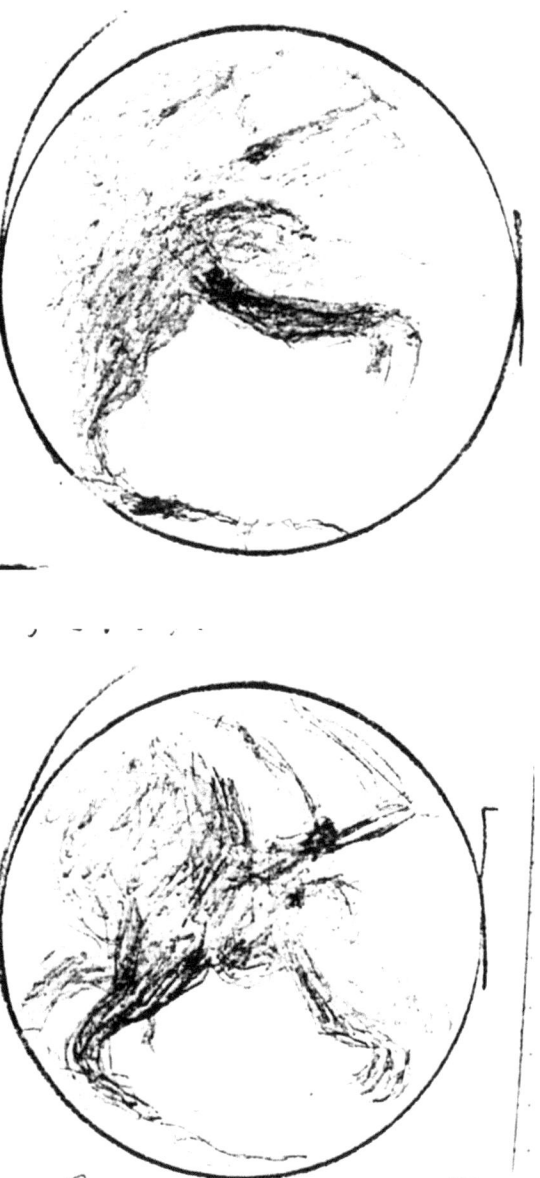

Two rough sketches of Mars by Barnard, made on November 26, 1911, with the Mt. Wilson 60-inch reflector diaphragmed to 14 inches. These are the merest impressions, since, as he noted, "It is not possible to draw the details because they are so complex." (Credit: Barnard observing log book, University of Chicago Library, Hanna Holborn Gray Special Collections Research Center)

Barnard was a good draftsman, especially in comparison to Lowell, Pickering, etc. Of course, he was not in Antoniadi's class (who was?), and in his few rough sketches of Mars as it appeared to him with the 60-inch telescope he admitted to failing utterly to capture all the details visible. However, rough impressions though they were, they helped shore up the epitaph for the Martian canals Antoniadi wrote in 1913:

> Ponderous volumes will still be written to record the discovery of new canals. But the astronomer of the future will sneer at these wonders; and the canal fallacy, after retarding progress a third of a century, is doomed to be relegated into the myths of the past. [19, p. 424]

Henceforth Barnard did not hesitate to pronounce that, with its pure aberration-free images, the 60-inch reflector was not only superior to but vastly superior to either the Lick 36-inch or the Yerkes 40-inch for visual planetary observations. His views of the stars were equally impressive. On scanning the Milky Way with the great telescope, he found that stars looking "like jewels on black velvet," as he told Hale. "The sky was rich and dark, and every star was a glowing, living point of light."[36]

Thus Barnard had a glimpse of what the future held. All great telescopes of the future would be reflectors. The Yerkes refractor would be the last instrument of its kind, and it remains—and is likely always to remain—the largest of its type. After his sabbatical at Mt. Wilson, it must have been bittersweet for Barnard to have to return to Yerkes again, just as southern Wisconsin was descending into the harsh winter conditions. One wonders whether he did not feel a twinge of pain at having been born a generation too early.

The Yerkes Observatory of the University of Chicago,
located at Williams Bay, Wis.

A postcard showing the dome of the 40-inch refractor in winter. The photograph was taken by Barnard, ca. 1916. (Credit: William Sheehan Collection)

A Destiny Narrowly Escaped

Despite the fact that Lowell's theories of Martian life were now clearly on the defensive in scientific circles, they made a deep impression on 36-year-old Edgar Rice Burroughs, a former enlisted solider with the U.S. Cavalry in Fort Grant, Arizona Territory; cowboy; factory worker; gold miner (unsuccessful) and—for six long years—pencil-sharpener wholesaler. Having tried everything else, Burroughs gave up pencils for a pen, and from 1911 tried to make a living as a writer. Between February and July 1912, under the pseudonym Norman Bean, he published "Under the Moons of Mars" in *The All-Story Magazine*, later published between covers as *A Princess of Mars*, introducing his fictitious hero, Captain John Carter of Virginia. Carter would go forth on a series of heroic odysseys to the mysterious red planet ("Barsoom") that would eventually fill eleven novels and encounter the very beings about whom Lowell had primed the public.[37] Burroughs, as well as H.G. Wells before him, was inspired by Lowell, and so would be many science fiction writers after them. One may question Lowell's ability as an observer, or his scientific results. Nevertheless, there is no doubt as to the quality of his literary art, or its ability to stimulate the literary imagination.

The same month "Under the Moons of Mars" debuted, Lowell and his wife left Flagstaff for Europe (taking along with them the Stevens Duryea automobile that accompanied them on all their trips abroad). Immediately on landing in Liverpool, they set out post-haste for the Continent, where their itinerary included a jaunt to Monaco. There, they stayed with Mr. and Mrs. Ralph Curtis at their Villa Sylvia. Curtis, in a letter to Barrett Wendell, expressed his horror on Lowell's "appearing in yellow shoes and a silk hat … in the evening to go with morning overcoat," and lamented that "savants are *hors a loi*."[38] Back in Paris, the Lowells had dinner with the Flammarions, and then—delaying their return to the United States (they had originally booked passage on the maiden voyage of the *Titanic*)—they took in the "beaded eclipse" of April 17, 1912, in the company of La Baume-Pluvinel and others at St. Germain. (Antoniadi and his wife Katharine also observed this eclipse from St. Germain, but—if Lowell and Antoniadi knew of each other's presence there—they were careful to remain well clear of one another.) On returning to Liverpool for their voyage across the Atlantic, the Lowells could only ponder their great good luck in missing the *Titanic* and the fate of such onboard tycoons as John Jacob Astor IV, Benjamin Guggenheim, and Isidor Straus. Writing 2 weeks after the disaster, Lowell admitted that on this occasion he was willing to grant a rare exception to his usual strict insistence on punctuality. "The moral is from such disasters: always delay as long as possible. You will thus prolong your life and may save it! A bad general maxim, however."[39]

Once home again in mid-May, Lowell returned to the Planet X search. He wanted both to pick up the pace of the photographic search in Flagstaff and to up his game on the theoretical side where he and Williams found themselves wading into further complications. Though they had hitherto assumed that X's orbit was nearly co-planar with those of the other planets, the size of the residuals in latitude suggested that the inclination might actually be considerable, and so they had to attempt to compute additional terms they hoped to ignore.

Inevitably, as the calculations became more and more protracted and complicated, the strain began to take a toll. Returning to Boston in September after spending time in Maine, Leonard found Williams "a mere shadow from her perplexing calculations."[40] Nevertheless, she was undaunted, and with a determination equal to Lowell's own, labored on. Acknowledging the mounting difficulty of the calculations, Lowell wrote, "Every new move takes weeks in the doing."[41] At the end of October—at this point activity in Lowell's State Street office was reaching a frenetic peak—Lowell's own health began to give way. Not only did he have to postpone returning to Flagstaff, he was temporarily incapable of making the journey from his West Cedar Street home to his State Street office. His nerves were simply too bad. Leonard reported that he was "weak and run down and must be careful and quiet,"[42] and worried that he might suffer another breakdown such as that which had sidelined him between 1897 and 1900. To her Flagstaff colleagues she confided, "He worries about the work—he wants to be *in it*!" Later, she added, "It is nervous exhaustion, and he is *up and down*! Some days he cannot even telephone. He gets … impatient for things to come from Flagstaff."[43]

Lowell rallied by the end of January 1913 to be "in it" again, at least temporarily, and was once more wiring Lampland positions from Boston, though as usual, none of them stood the test of time. In early February a new set of elements gave a position at heliocentric longitude 58° 4′, placing the planet in Taurus, not far from the Pleiades. No doubt finding the task of keeping up with these ever-changing positions exasperating, Lampland now took his turn breaking down. Lowell advised him to do whatever was needed to recover his health. "I cannot afford to lose you," Lowell wrote.[44]

By the spring of 1913, Lowell "apparently decided to make one final all-out attack on Planet X."[45] He was now extending the time span of his solutions by including observations up through the year 1910. (They had previously stopped at 1903.) He wrote to Gaillot for the latest residuals in latitude and longitude of Uranus; the requested data arrived by June. At once Lowell asked for the same data for Neptune. In addition, increasingly convinced that the eccentricity of the orbit was considerable, he began to consider terms involving e',[2] "unpleasant as the complication was bound to be,"[46] and even introduced some third-order terms. Also, if the inclination was as considerable as he believed it was, this, in combination with the eccentricity, would effectively diminish the mass. Throughout, everything was solved with the rigorous (and time-consuming) method of true least squares.

Always high-strung and now struggling with depression, Lowell left Boston for the summer home of his sister, Katharine Bowlker, at Mount Desert Island, Maine. Nevertheless, Planet X continued to weigh heavily on him. He could not resist monitoring the work in Boston through Williams's telegrams while sending brief reports of the latest positions to Lampland, who had recovered enough to be ready to photograph more star fields. On July 10, 1913, Lowell telegraphed, "Generally speaking what fields have you taken? Is there nothing suspicious?"[47] On August 21 he telegraphed,

> So far best determination for first power e' [eccentricity of Planet X] for present position is two hundred thirty nine degrees and for second power ditto two hundred forty one degrees. Use these. Suspect inclination large probably south. Am personally still on the retired list. Await another excitement proving true. Any news grateful.[48]

He was becoming increasingly irritable and impatient with staff—not so much the other astronomers, but those in more menial positions, such as chauffeurs and servants. He was hard for others to live with because he found it hard to live with himself.[49]

As 1914 approached, Mars in its stern chase with the Earth again became Lowell's priority. He advised V.M. Slipher to "rig up the 40-inch for the best possible visual observations at coming Mars opposition."[50] The opposition occurring in January, and being about as unfavorable as any Mars opposition can be, the results were meager. But Lowell managed in the months before opposition to observe the formation of the north polar cap, which occurred through the deposition of hoar frost in distinct separate stippled areas; saw the blue band of presumed meltwater (known later as the "Lowell band") that formed around the cap as it began to retreat again; and noted the reappearance of the canal Aethiops, a double, after its absence for many years. As it had in 1911, the 40-inch reflector disappointed, and only served to confirm Lowell's by now fixed and irrevocable idea that large telescopes were for ill-suited planetary work. Thus, he insisted, "with its full aperture—as well as when diaphragmed to 30 inches—the [40-inch] was … found to show the canals and oases of Mars as fine geometric lines and dots, thus confirming the results from smaller apertures, and at the same time confirming the statement that the Martian markings are both more easily and better seen with the latter, that is from 12 to 18 inches" [21]. He was satisfied with this conclusion, and spurned a number of suggestions over the years to observe with other telescopes.

In addition to Mars work, various other investigations were underway at the Observatory, as Lowell described in a paper, "Epitome of Results at Lowell Observatory between April 1913 and April 1914," of which the first entry related to Venus: "[C]onfirmation and completion of the detection of the spoke-like markings of the planet by observation of it when west of the Sun up to within 8° of superior conjunction, showing radial markings to exist all round the planet's disk irrespective of azimuth except as determined by local conditions. They are thus a distinguishing feature of the Venusian topography."[51] The longest section relates to the observed seasonal changes in the Martian features (mentioned above) and applauding their agreement with the predictions of the "Lowell theory of the present condition of Mars." Not until number 16 on a list of 31 is the spectrographic discovery by V.M. Slipher that the nebulosity investing the Pleiades star cluster consists of, in what is self-evidently Lowell's not Slipher's prose, "pulverulent matter shining by reflected light of the neighboring stars" (i.e., a reflection nebula); while practically at the rear of the list are the following items:

25. Radial velocity observations of the Andromeda nebula, the first of the spiral nebulae to be so observed, showed it to be approaching the Sun with the extraordinary velocity of 300 kilometers.
26. This discovery was followed by the establishment of the fact that the spiral nebulae as a class have a much higher order of velocity than have the stars. As examples of spirals of high velocity may be mentioned N.G.C. 1068, 4565, 4594, and 5866.[52]

With the 40-inch reflector prioritized for Mars, the Planet X search ground to a halt. Finally, Lowell decided to take advice Lampland had been urging upon him for the previous year. Not only was the large (underground) reflector ill-suited for planetary work, but it had not been a good choice for the planet search because of its small field and the distorted star images produced at the edges of plates exposed with it. Lampland suggested that Lowell acquire a refractor with a doublet lens that would provide a large field (like the Bruce photographic telescope Barnard was using). At last Lowell accepted the advice, and wrote to John A. Miller, director of the Sproul Observatory of Swarthmore College in Pennsylvania, to secure on a temporary basis Sproul's 9-inch Brashear photographic doublet, which had significantly more light-gathering power than the 5-inch Brashear used in earlier phases of the search and a much wider field than the 40-inch reflector. It was delivered to Flagstaff by April 1914 and set up in a temporary dome on Mars Hill.

At almost the exact same moment, the massive calculation was winding up in Boston, and Lowell was able to spare two computers who had been assisting Williams, Thomas B. Gill and Earl A. Edwards, to help Lampland in Flagstaff press forward the photographic search using the Brashear doublet. His two final "best solutions" for "X" were found using mean distances of 43.0 and 44.7 AU, and gave positions (for July 1, 1914) that were 180° apart, at heliocentric longitudes 84°.0 and 262°.8. The first position put the supposed planet in eastern Taurus moving toward Gemini, the second in Scorpio moving toward Sagittarius. Based on the math alone, there was no reason to favor one solution over the other, though since the latter would put the planet in a part of the sky "nearly inaccessible to most observatories," as well as swimming in the swarming star fields of the Milky Way, Lowell decided for practical reasons to attend only to the former. He estimated that his planet had a mass of 1/50,000 that of the Sun, and gave the eccentricity of its orbit as 0.202 for the first solution and 0.195 for the second. Based on analogy with the other members of the Solar System, in which eccentricity and inclination are usually correlated, he guessed that the inclination was likely to be about 10° (a high inclination which, he noted, would make it much more difficult to find). The magnitude might be about 12–13, the apparent diameter of the disk, which might betray its planetary, about 1″ of arc.

By this point Lowell himself seems to have largely lost faith in these endlessly shifting positions. "It must be remembered," he later wrote, "that the actual as against the probable errors of observation might decidedly alter the result."[53] This, in fact, proved to be entirely correct: the actual errors, especially of the pre-1781 observations, were significantly greater than the probable errors Lowell assumed on the basis of the true least-square analysis. Also, terms above the squares in the eccentricities of Uranus and X, though too ponderous to deal with and so "necessarily left out of account," might also alter the result. As we now know (and as seems to have become evident even to Lowell himself), the various positions derived for Planet X were completely illusory; Lowell's and Williams's calculations were based on the weak foundation of measures of Uranus's positions made by eighteenth and nineteenth century observers determining transits with meridian circles. Thus, according to a recent summary:

> As with the marginal perceptual data on which he based his theories about Mars, so with his search for Planet X, the "signal," such as it was, had always been scarcely distinguishable from the noise. In both cases this led to what we would now describe as "Type I errors" (also known as "false positives") [22].

Pluto, found lurking in trans-neptunian space by Clyde Tombaugh at Lowell Observatory in 1930, and at first hailed as a vindication of Lowell's calculations, is far too small to have produced any detectable perturbations of Uranus. It was the entirely serendipitous—if still important—discovery of the largest and brightest object belonging to the "Kuiper Belt." As with Lowell's Mars theories, which were so compelling that they helped inspire a lasting interest of Mars eventually contributing to the Space Age exploration of the planet, so his Planet X obsession led, in the end, to the discovery of a whole vast realm of icy objects on the edge of the Solar System. He made mistakes, but they were fruitful ones.

The Last Trip to Europe

Lowell planned another trip to Europe that spring as he usually did in years that fell between Mars oppositions. His departure was delayed as Constance needed surgery for an ulcer, and then, while she convalesced, he went without her. Even in the weeks before his departure, he obsessed about the X search, pressing Lampland for results: "Don't hesitate to startle me with a telegram 'FOUND," he wrote on May 5.[54] He left the U.S. on May 16.

No telegram "Found" arrived, then or later. In Europe in June, he made the rounds of old friends. Lunching with the Flammarions, he observed that Paris was "hardly itself this year: the tourists are few; the costumes caricatures and generally hideous at that; the [June] weather that of February."[55] One day they attended a Fete du Soleil on the Eiffel Tower, where Eiffel himself was on hand. The next day Lowell showed his latest results on Mars and V.M. Slipher's spectrograms of the spiral nebulae at the Bureau des Longitudes with "*un grand effet.*" Despite all this, Lowell—like Paris—seemed a bit off; Flammarion found him "strangely neurasthenic."

Percival Lowell, dressed in sartorial splendor with top hat, coattails, and cane, during his last visit to London, 1914. As the years went by, he became ever more arrogant, rigid, and fixed in his views. This photo appears to have been retouched to remove some of the lines in Lowell's face, and makes him look far younger than he does in photos from a decade earlier. (From: W. Louise Leonard, Percival Lowell: An Afterglow)

The gathering tensions that would soon lead to war were hard to ignore that summer. Lowell was well aware of the increasingly desperate situation, and booked early passage to the United States from Liverpool by way of Halifax, Nova Scotia. Right up to his departure from England—on Saturday August 1—he was preoccupied with clothes and the fate of his mathematical papers in the event his ship, the *Mauritania*, sank.[56] The following Monday, August 3, the London *Times* blared alarming headlines: "Five Nations at War / Fighting on Three Frontiers / German Declaration to Russia / Invasion of France / German Troops in Luxembourg / British Naval Reserves Mobilized." On Tuesday, August 4: "The Menace of Germany / Campaign Through Belgium / Answer of Great Britain/ Europe Armed /The Russian Frontier / Reported Naval Battle in the Baltic." On Wednesday, August 5: "War Declared / Note Rejected by Germany / British Ambassador to Leave Berlin / Rival Navies in the North Sea / British Army Mobilizing / Government Control of Railways." Percival's sister Amy was in London at the time, and standing with her companion Ada on the balcony of their suite at the Berkeley Hotel overlooking Piccadilly Circus, saw "A great crowd of people with flags … shouting We want war! We want war! They sang the Marseillaise, and it sounded savage, abominable. The blood-lust was coming back, which we had hoped was gone forever."[57] That same evening, the novelist Henry James, writing from Lamb House, Rye, Sussex, characterized the situation in a letter to his old Boston friend Edward Waldo Emerson:

> [W]e sit here in these days & more especially in these nights (for I am writing you very late—have had a longish nap & am not in bed yet,) under the blackness of the most appallingly huge & sudden state of general war. It has all come as by the leap of some awful monster out of his lair—he is *upon* us, he is upon *all* of us here, before we have had time to turn round. It fills me with anguish & dismay & makes me ask myself if *this* then is what I have grown old for, if this is what all the ostensibly or comparatively serene, all the supposedly *bettering* past, of our century, has meant & led up to. It gives away everything one has believed in & lived for….[58]

At midnight, Great Britain declared war on Germany.

Lowell, meanwhile, was far out in the Atlantic. He reached Halifax on August 6, then boarded the train for Boston. With Europe embarked on a gigantic struggle that would see warplanes, poison gas, trench warfare, flamethrowers right out of H.G. Wells, and mass death, Lowell's first order of business was not to lament the collapse of everything he had believed in and lived for but to lament the lack of progress with the X search. No; it being Mrs.

Lampland's turn to develop an ulcer and undergo surgery, he wrote to V.M. Slipher, "I feel sadly of course that nothing has been reported about X, but I suppose the bad weather and Mrs. Lampland's condition may somewhat explain it."[59] Planet X trumped all.

Nor was there any news of X when Lowell returned to Flagstaff for an abbreviated visit in October. The search had, seemingly, stalled somewhere in the outer Solar System, and settled into the same protracted stalemate as the armies on the Western Front.

During this visit, Lowell was photographed on October 17 by visiting astronomer Philip Fox, formerly at Yerkes and now at Northwestern University, "observing Venus" by daylight with the 24-inch Clark refractor. It is Lowell's most iconic image. Though no Venus observations were recorded in the observing log book, there is a record of Lowell's and Fox's foray to Adama (retracing Lowell's and Judge Doe's adventure of October 1909) and down into the Carrizo Wash in the Petrified Forest where they searched for Sigillaria petrified wood. Fox later recalled that on Lowell's tramps "one needed to go a swift pace to keep abreast of him, as he swung along with enormous strides," and how, on this particular excursion,

It was a daring ride down into the Carrizo away from civilization, or perhaps I should say barbarism, for this was in the fall of 1914 when the Germans were burning Louvain and pressing towards Anvers and into France. The daily tramps we took were staggering, with Lowell a difficult pace-maker. Yet we found the specimens, and even now, as I look at samples of them, I remember the ordeal of heat and fatigue which the quest cost us. Blistered cheeks and cracked lips and parched throat,—but a part of the price—were no deterrent to him. When we had gone to the West [for the Solar Union meeting in 1910], Professor Lowell had not been in good health and it was a joy … to see his vigor returning, nourished by the zeal of the quest. [24]

Lowell observing Venus in daylight, October 17, 1914. Photograph Philip Fox. (Credit: Lowell Observatory Archives)

By the time Fox left, Lowell had reached a momentous decision: he was determined to publish the hitherto secret and now completed theoretical side of the Planet X search, the now famous "Memoir of a Trans-neptunian Planet," and planned to present to the American Academy of Arts and Sciences in Boston in January 1915. He was working hard on the manuscript in December, when he announced to Lampland: "I am giving my work before the Academy on January 13. It would be thoughtful of you to announce the actual discovery at the same time."[60]

The "Memoir" *is* a *tour-de-force* of celestial mechanics, but in the end, a great failure and wasted effort. It concludes on a note of anti-climax. Lowell acknowledges that the precision of his predictions was always bound to be compromised by irreducible uncertainties in the residuals:

> … [T]he curves of the solutions show that a proper change in the errors of observation would quite alter the minimum point for either the different mean distances or the mean longitudes. A slight increase of the actual errors over the most probable ones, such as it by no means strains human capacity for error to suppose, would suffice entirely to change the most probable distance of the disturber and its longitude at the epoch. Indeed the imposing "probable error" of a set of observations imposes on no one familiar with observation, the actual errors committed, due to systematic causes, always far exceeding it.[61]

He follows this with an even more humbling (and uncharacteristic) admission of human fallibility, of the limitations of human perception and calculation:

> Owing to the inexactitude of our data, then, we cannot regard our results with the complacency of completeness we should like. Just as Lagrange and Laplace believed that they had proved the eternal stability of our system, and just as further study has shown this confidence to have been misplaced; so the fine definiteness of positioning of an unknown by the bold analysis of Le Verrier or Adams appears in the light of subsequent research to be only possible under certain circumstances. Analytics thought to promise the precision of a rifle and finds it must rely upon the promiscuity of a shot gun after all, though the fault lies not more in the weapon than in the uncertain bases on which it rests. But to learn of the general solution and the limitations of a problem is really as instructive and important as if it permitted specifically of exact solution.
>
> For that, too, means advance.[62]

With this as his valedictory, Lowell announces that he has given up. The photographic search with the Brashear doublet continued for a while yet, apparently through sheer inertia (and perhaps Lowell's unwillingness to admit out loud that he had thrown in the towel). Ironically, on plates exposed by Thomas Gill on March 19 and April 7, 1915, a faint (15th magnitude) trans-neptunian object, Pluto, was actually recorded, though recognition of the fact would not occur until after Pluto's discovery in 1930.

Though the subject of Planet X, so long the energizing core of so much activity both in the State Street office in Boston and at the observatory in

Flagstaff, disappears without a trace from Lowell's correspondence, Lampland still thought it worth pursuing, and wrote to his employer in August 1915, "X is not yet in sight, though you may well believe that I am in hopes that he is not far away…. This is no time to be discouraged."[63] A month later, Lampland told Miller,

> The distant planet has not yet been located but for all of that we are not discouraged…. I suppose you are getting tired of extending the time of stay of the 9-inch. But you see we are a hopeful lot—in some things at least. Each day brings the hope that a little more work may turn the trick…. After so much work on a problem how one hates to give in.[64]

Nonetheless, there is no indication that at this point Lowell himself remained in the "hopeful lot."

A note in Gill's observing log indicates that Lowell briefly visited the dome of the 9-inch doublet on October 8, 1915. It was probably for the last time. But the photographic search with that instrument continued, presumably rather perfunctorily, until July 2, 1916, when Gill recorded in the log book, "Lunch," suggesting a project only temporarily ended. Temporarily, in this case, was to be 13 years. By the time Gill gave up, nearly a thousand plates had been taken with the Sproul doublet, some by Lampland and Edwards but most of them by Gill himself. According to Lowell's brother Lawrence, having culled the correspondence about Planet X, "Through the banter one can see the craving to find the long-sought planet, and the grief at the baffling of his hopes. That X was not found was the sharpest disappointment of his life."[65]

The 9-inch Brashear doublet, borrowed from Swarthmore College's Sproul Observatory and temporarily mounted in a dome on Mars Hill in 1914, bears a close resemblance to the 10-inch Brashear doublet used by Barnard, and was used in Lowell's search for Planet X. This photograph was taken on July 13, 1916, shortly after the search had been given up. (Credit: Lowell Observatory Archives)

After "X," Another Memoir

Lowell was never one to dwell on his failures (on the rare occasions he admitted them. Immediately after the ink was dry on the "Memoir on a transneptunian planet," he threw himself into his favorite form of therapy, work. He had for several years been pondering the structure and evolution of Saturn's ring system, and we have already discussed his 1907 observations of the edgewise rings and his discovery of the "tores." Over the next few years, as he became increasingly immersed in the mathematical intricacies of the X search, he also devoted considerable effort to analyzing minor subdivisions in the rings. They were especially evident in the bright Ring B, and had been noted from time to time by some earlier observers, including several at Harvard in the 1850s.[66] Lowell now found that their positions corresponded with those of resonances involving the inner satellites, especially Mimas. As with the X search, this research provided him with an excellent forum in which he could demonstrate his vaunted mathematical virtuosity.

Saturn, July 7, 1898, observed by Barnard with the 40-inch refractor at Yerkes. Though the late nineteenth century saw an "outbreak of division finding" in the rings, Barnard had nothing to offer. He remarked: "I have never seen the planet better, nor have I seen so much detail upon it before. There seems to be a dusky shading where Encke's Division is usually shown." That was all. (From: Monthly Notices of the Royal Astronomical Society, *vol. 68, no. 5 (1908), Plate 11)*

In a series of papers published in 1911, 1912, and 1913, Lowell immersed himself deeply in aspects of perturbation theory needed to understand gaps and concentrations of members of the Asteroid Belt. The "heaping up of asteroids in places and their thinning out in others" was, he claimed, a straightforward consequence of "Jupiter's sway."[67] No doubt he looked so closely at this problem that he quite got the positions of these gaps by heart, seeing with his "mind's eye" as soon as he began to investigate along similar lines the possible heaping up and thinning out of the particles making up Saturn's rings, and arguably approached the latter with the "bias of preconceived ideas."

"When the rings are compared with the zone of asteroids, the one is seen to be almost the counterpart of the other," he wrote, from which it followed that both must derive alike from "similar perturbative action… Indeed, were the asteroids numerous enough we should actually behold in the sky a replica of Saturn's rings, altered only by the perspective of our different point of view." The asteroid belt had been sculpted by Jupiter; the rings, he thought, mostly by perturbative action of the innermost satellite to the rings then known, Mimas. What theory predicted, observation confirmed. A few subdivisions had already been glimpsed in Ring B at the end of 1909. By 1913–14, as the rings continued to open more widely to view from their edgewise presentation in 1907–08, ring B became "conspicuously striped amidst its shading, the dark curving lines of its plaided pattern being so definite as to permit of measurement" (i.e., with a filar micrometer).[68] As in the case of such will-o'-the-wisps as the canals of Mars and the almost vanishingly small residuals in the motion of Uranus on which the Planet X search had been based, the gossamer filaments of Saturn's rings provided him with just enough indication of reality to hang upon them a great deal of supposititious import. An exceedingly slender reed of observation was bent almost to the ground with a mass of theory. Something that was hardly there at all was found solid and substantial enough to measure with micrometric accuracy to the tenth and even the hundredths of a second of arc. Of course, such precision is only achievable in the realm of illusion."[69]

Percival Lowell, Saturn with rings wide open, opposition 1914–15. (From: Lowell, "Memoir on Saturn's Rings" (1915), Plate I)

His answer regarding the discrepancy was based on the deployment of mathematical formulae from Félix Tisserand, whose several volumes on celestial mechanics had been often consulted in the X search. Utilizing these formulae with deft skill, Lowell found that if Saturn's mass were in the distribution of an oblate spheroid instead of a sphere, the discrepancy was partially resolved. It was entirely resolved if, instead of being homogeneous, the internal structure was more complicated—for instance, he proposed, arranged like "an onion in partitive motion."[70]

The draft of the "Memoir on Saturn's Rings" was completed by September 1915, and published (like the "Memoir on a Trans-neptunian Planet") privately. In terms of the structure of their arguments, they were very similar. A discrepancy between theory and observation was noted. Rather than doubting the observations, the theory had to be modified to account for it. And yet though the *Memoir on a Trans-neptunian planet* ended on a note of diffidence—even humility—that on *Saturn's Rings* ended with a blare of triumph:

> From the positions of the divisions in its rings we are … led to believe that Saturn is actually rotating in layers with different velocities, the inside ones turning the faster. If these layers were two only, or substantially two, this would result in Saturn's being composed of a very oblate kernel surrounded by a less oblate husk of cloud.[71]

Of course, as with the Planet X calculation, everything in the Saturn rings paper depended on how solid the observational basis was. In either case, Lowell proved that he was a better mathematician than observer. Subsequent observations of the rings by more skillful observers than he (and, ultimately, by spacecraft) show both qualitative and quantitative differences from Lowell's results. Perhaps the best drawing of Saturn's rings in the pre-spacecraft era was by the French planetary observer Bernard Lyot, who in 1943 observed ring structure with the 24-inch refractor of the Pic du Midi Observatory. There is scant agreement with Lowell's drawing from 1915. But it is on the reality of the latter that the whole "onion in partitive motion" calculation depends, and without which it goes completely to ground. Recent planetary scientists who have considered Lowell's work on Saturn have cited it as a striking example of the theory-ladenness of observation. Thus, as MIT ring theorist James Elliot and Richard Kerr wrote:

> Percival Lowell had not only seen distinct "canals" on Mars, he also glimpsed dark patches and strips beneath the impenetrable clouds of Venus. As might be expected, Lowell found delicate details in the rings of Saturn, too; of course, most coincided with one resonance or another. [27]

Bernard Lyot's 1943 drawing of Saturn as seen with the 24-in. refractor at the Pic du Midi Observatory, compared with Lowell's 1915 drawing with the 24-in. Lowell refractor. Note that the ring divisions differ in position but also that whereas Lyot's are shadings Lowell's are lines. (Credit: Photograph by John S. Hall at Rome IAU meeting 1952 in William Sheehan collection and Lowell Observatory Archives)

The Final Year

After a brief visit back East at the beginning of 1916, Lowell returned to Flagstaff, his now preferred residence, to continue measuring Saturn's rings and to resume the observation of Mars ahead of its not very favorable February opposition. Winters in Flagstaff can be unpleasant. Thus, on New Year's Eve 1915, Flagstaff witnessed an unprecedented snowfall of 50 inches (a record not broken until 2019). In addition, because of the Flagstaff elevation, night-time temperatures were often bitterly cold, but Lowell and E.C. Slipher were not to be denied. The next night after the snowfall the sky was clear, and Lowell telegraphed (with an allusion to the European war being fought at the time) to Leonard who was in Boston at the time,

> Observed last night and with success. Our important observations on Mars in strict accordance with theory. The right side of the Crepe Ring [of Saturn] is still wider than the left, which is interesting. We move about on the hill much as if we were in the trenches.[72]

Lowell was always good copy, and his observing reports, especially about Mars, were eagerly picked up by newspapers and journals (the *Astrophysical Journal* remained an exception, as always). In addition to regular Lowell Observatory *Bulletins*, he issued occasional observing circulars, such as one sent out on March 30, 1916, and picked up by the *Journal of the Royal Astronomical Society of Canada*. Under the heading "The Latest News from Mars," the report announced:

> A curious set of features, secondary to the main canal network, have become apparent on Mars. Within some of the polygons made by the intersections of the larger canals a tiny dot has been descried at this observatory, joined to a corner and to the sides of the polygon by lines so slender they usually appear as a string of minute beads. The effect is of a centrally woven web, spun within the borders of the polygon, of a more minute order of tenuity than the polygon itself. Elysium was the first example of this phenomenon with the Fons Immortalis and five connecting spokes.[73]

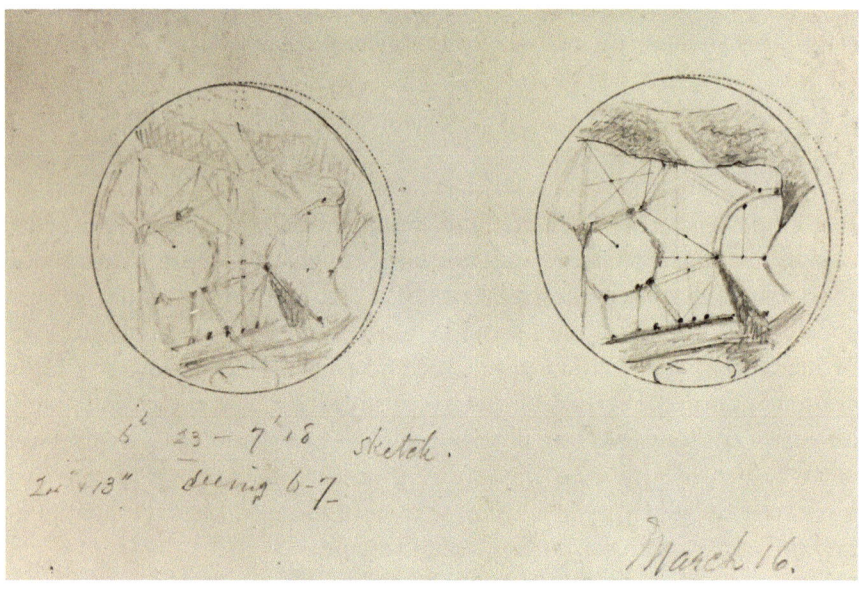

Lowell's sketches of Mars, March 16, 1916, showing what he described as "a curious set of features, secondary to the main canal network" and appearing as "a string of minute beads." (Credit: Lowell Observatory Archives)

As ever, Slipher drew Mars much as Lowell did, and so did George Hall Hamilton, a budding astronomer who had left a position at Bellevue College in Nebraska in order to participate as a volunteer observer of Mars during the 1916 opposition. Both Slipher and Hamilton would defend Lowell's views to the end of their lives. Hamilton, for instance, would write in 1920:

> It is seen … that a globe like Mars might, if inhabited, needs an irrigation system of vast proportions to support life, and such a system has been shown feasible, providing that beings who inhabited it were sufficiently advanced to undertake such a project…. This is seemingly the case; for we have every reason to believe that what we see on the surface of Mars shows definite proofs of artificiality.[74]

We see from Lowell's observing log books during this period that he was still measuring the fine divisions in Saturn's rings on April 9, 1916, attempted to measure the diameter of the asteroid Vesta on April 18, and observed Mars—as it turned out, for the very last time—on April 21. Then, as Lampland's diary records, on April 23 Lowell and Mrs. Lowell left for Chicago on the Limited. Two days later, in a lecture before the Chicago Academy of Sciences, Lowell remarked:

> I have said enough to show how our knowledge of Mars steadily progresses. Each opposition as it comes round adds something to what we knew before. It adds without subtracting. For since the theory of intelligent life on the planet was first enunciated 21 years ago, every new fact discovered has been found to be accordant with it. Not a single thing has been detected which it does not explain. This is really a remarkable record for a theory. It has, of course, met the fate of any new idea, which has both the fortune and the misfortune to be ahead of the times and has risen above it. New facts have but buttressed the old, while every year adds to the number of those who have seen the evidence for themselves.[75]

Lowell was now launched on a great lecture tour of North America. Next morning he and Mrs. Lowell pushed on to Toronto. Named an Honorary Fellow of the Royal Astronomical Society of Canada, he delivered a lecture, "The Genesis of Planets," which was enthusiastically received, even though Lowell largely set forth long-familiar arguments and conclusions. He laid out his theory about origin of the Solar System in the encounter of a dark star. As to what happened next, he demurred. "[W]e have no analogy to guide us, for the spiral nebulae indicate themselves to be galaxies, not solar systems in course of construction."[76] Thus, in a single line of text, and without mentioning the name of the discoverer (V.M. Slipher), he passed over the greatest astronomical breakthrough ever made at the Lowell Observatory.

The rest of the lecture sets forth the way that each planet acquires, out of the initial catastrophe, a certain degree of spin. As the matter condenses the spin increases as a stone attached to a string goes faster as the string is wound up, and this increases the oblateness of the spheroid. The planet becomes more non-homogeneous as it evolves. The internal layers rotate faster, accentuating spheroidal shape. In the process, heat is generated, and the planets at this stage become molten throughout. This heat expands through the body and is radiated at the surface, producing cooling. Uranus is still in a very early stage of evolution; it is more homogeneous, and so younger, than either Jupiter or Saturn. Saturn's internal differentiation and homogeneity, deduced from the ring divisions, shows it to be somewhat more evolved. The Moon and Mercury have already hastened on to planetary death, while Mars lags not far þehind them, and stands already on the threshold.

One recognizes here the broad outlines of Lowell's cherished Spencerian framework, by now already decades out of date. Its assumption, notes Strauss, "that current conditions were the predictable outcome of a single principle working gradually and inevitably from past to present," was increasingly being rejected by a new generation because it left no room for "cataclysm, chance events, or choice … as essential components of the universe."[77] For that matter, Lowell's mathematics was also rather old-fashioned. His expertise involved the application of celestial mechanics as practiced by Laplace, Adams and Le Verrier, Peirce, and Tisserand. In 1914, he had attempted to establish, under the sponsorship of the American Academy of Arts and Sciences, a new medal, to be awarded to a deserving astronomer working on traditional problems in astronomy. The offer was turned down by the Academy's president, John Trowbridge, who expressed no interest in an award that was so narrowly defined as to exclude astrophysics, which Lowell described scornfully as consisting only of "pictures" and usurping "the lime-light to the exclusion of the deeper and more profound parts of astronomy proper."[78] Adding insult to injury, the Academy had also refused to publish at its expense the "Memoir on a Trans-neptunian Planet," which was, presumably, precisely the kind of investigation in traditional astronomy Lowell had in mind, forcing Lowell to bear the cost. At the end of this fracas, Lowell withdrew his offer to fund the medal and fumed to Trowbridge (without, apparently, recognizing the irony of his position) about what he called the "stagnation and old-fogeyism of the Academy." He closed by calling its leaders "a set of men certainly not broad [of] view or judgment."[79]

The astrophysicist, as Lowell saw it, was a mere collector, concerned with nothing more than mindless data-gathering, taking pictures of spectra, and making photometric measures of stars to be stored in large collections. Such

work as that might well lend itself to cooperative efforts and even unioniza-
tion such as Hale's International Union for Cooperation in Solar Research.
But Lowell was first to last an individualist. He scorned these collaborative
(not to say collectivist) efforts; this was not the way to make discoveries,
though even in declaring it, he revealed how lonely and isolated he was in his
own position:

> Conceive Newton's *Principia* as a union outcome; and how many co-operators
> would it take to make one Clerk-Maxwell! That the banding together should be
> held advisable is a sad comment on the paucity of the age. Wolves hunt in packs,
> the lion stalks alone. It reflects too on the character of the work done. Just in
> proportion as the aim is low so may it wisely be widespread. The method has
> great advantages if the work is what you want. This is why we teach machines to
> do as much of it as we may. But as it becomes complicated and difficult fewer
> and fewer persons can be found capable of undertaking it until at last you have
> but the one man in the world who can, the genius who originates.[80]

In the same lecture, Lowell asked:

> What is proof? Outside of mathematics, which is formulated logic, proof con-
> sists in an overwhelming preponderance of probability. Take, for instance, the
> law of gravitation, as being inversely as the square of the distance of the bodies
> apart. We call it proved and rightly, because almost everything stands explained
> by it and the few things that do not, like the motion of the apsides of Mercury,
> we are confident will eventually fall into line.[81]

In fact, Lowell's confidence in the inverse square law was misplaced; Newton's
theory of gravitation would never succeed in explaining the motion of the
apsides of Mercury (discovered by Le Verrier), but would require a new theory
of gravitation based on Einstein's General Theory of Relativity. Indeed, almost
a year before Lowell spoke these words, Einstein, in Berlin, had made the
motion of the apsides of Mercury fall into line, not on the basis of the
Newtonian law of the inverse square but on the basis of the field equations of
his General Theory. Apparently, Lowell had never heard of Einstein or relativity, and had he lived to a reasonably ripe old age he would likely—given his
extreme and growing conservatism—have spent much of his later years as
T.J.J. See and William H. Pickering would do, trying to find logical flaws in
Einstein's Theory and publishing crank critiques against it in the pages of
Popular Astronomy.

Lowell's rigid resistance to new ideas only grew over time. The defeats over
Mars (about which Frederic Stimson would write after his death: "Mars went

back on him and was a disappointment"[82]) and Planet X had been absorbed, and though he continued to show a brave—even arrogant—face to the world, he must have known at some level that his greatest hopes had failed. He aged greatly over the last few years of his life—compare the picture of him taken in London in 1914 with the one two years later. He appears to have aged 10 years, and now looked very old and tired. Realizing that it was a waste of time doing as he had done and trying to win his professional peers over to his views, he turned with a will to the rising generation, setting out with missionary zeal on September 27 from Boston on the last great effort of his life—an ambitious but exhausting lecture tour of colleges of the Pacific Northwest and West Coast.

His itinerary included the State College of Washington in Pullman, the University of Washington in Seattle, Reed College in Portland, the Agricultural College in Corvallis, the University of Oregon in Eugene, the University of California at Berkeley and Leland Stanford Junior University between San Francisco and San Jose. In each venue, the audiences were overflowing. The subjects were "Mars—Forecasts and Fulfilments," "Great Discoveries and Their Reception," and "The Far Horizon of Science." His standard refrain concerned the resistance of astronomical conservatives to new ideas (such as the idea of life on Mars). The road to discovery was not an easy one, he warned the assembled youth:

> There is to add to its forbiddingness no warm compensating reception at its end, except in one's own glow of attainment. For progress is first obstructed by the reticence of nature and then opposed by the denunciation of man… A really new idea is a foundling without friends. Indeed a doorstep acquisition is welcome compared with the gift of a brand new upsetting thought. The undesired outsider is ignored, pooh-poohed, denounced, or all three according to circumstances. A generation or more is needed to secure it a hearing and more time still before its worth is recognized.[83]

Because of the (allegedly) deep-dyed conservatism of the astronomical establishment, the student would likely not be taught the newer things of science until they had become history:

> Some people seem to think that recently discovered facts are too homeopathic for youth, and that such ideas must not be given to youthful students until they are so old that they are nearly worn out. But I believe in giving young people the newest things which can be found in no other place. [30]

Of course, his ideas about Mars were always of chief interest to his audiences. Though they were, in Lowell's view, among the resisted "new ideas," they had ceased to be open to contrary evidence, and had turned into stubbornly held delusions. "Our observations have convinced us without a doubt that the lines on Mars which have caused so much discussion, are canals," he declared. "There is some form of intelligence on Mars. I do not mean human beings, but some intellect capable of accomplishing these feats."[84] Though his prose had perhaps withered somewhat during the years in which he had been wandering the desert wastelands of mathematics, he could still conjure a passage of beautiful prose, as in the following passage which almost seems to echo John Van Dyke's famous book *The Desert*:

> … Our *terra firma* gives to our globe its sense of home; partly for its very antithesis to the undomiciliate sea… To lack a contrast is to be sensible of a loss. On Mars we should miss the ocean….
>
> Such sameness of surface is deepened by the dead level of the land. As there are no oceans, so are there no mountains on Mars. The plainness of its features is unrelieved by piquancy of profile. Plateaus are the height of its attainments: something resembling probably the mesas of our southwestern deserts….
>
> To those who really commune with nature there is grandeur in this uniformity. It is the grandeur of vast expanse, bare of interposed detail to detract from its own unique impression, or to bar vision from its would-be range. The horizon may in truth be nearer, yet it seems more far. We have it in our deserts, whose very nakedness adds to their sublimity. Such accentuation of solitude is typical of Mars. For Mars is one vast desert relieved only here and there by tracts of vegetation. Our deserts grow in grandeur as one sees them more; so as one contemplates that desert world across the void of space its impressiveness increases. Distance robs it of its dread and in its opalescent sheen we see its beauty unmarred by sense of what its pitilessness represents.[85]

The day after he gave his last lecture of the series, he boarded the train to Flagstaff, arriving at the station on October 19. Without allowing himself any time to rest or acclimatize to the altitude, he threw himself into a heavy schedule of work. His first priority was the resumption of observations of the fifth satellite of Jupiter that he and E.C. Slipher, acting on a request from Barnard, had begun in September 1915. The 24-inch Clark was one of the few telescopes powerful enough to show the faint satellite, and the observations had, as usual, opened up some intriguing questions that might succumb to Lowell's shrewd mathematical analysis. A first theory of the small satellite's orbital motion, presupposing an equatorial elliptical orbit and taking account of the oblateness of the giant planet, had been worked out by Tisserand the year after Barnard's discovery. By the time Lowell became involved, an empirical model of a precessing ellipse, using orbital elements determined by Hermann Struve of the Pulkovo Observatory, was favored. However, it could hardly be said that the satellite's motion was satisfactorily known, and so Lowell began to hope that he might use detailed knowledge of the satellite's motion to test models of the planet's interior structure. Perhaps it, like Saturn, was rotating like an onion in partitive motion.

Even with the full aperture of the 24-inch Clark, the fifth satellite was only visible when an eyepiece occulting bar hid the glare of the giant planet, but visible it was—and on the very evening of the day he had returned to Flagstaff from his lecture tour, he recorded his first observations, of an eastern elongation. He experienced a continuous streak of clear weather for the next several weeks, and measures of both the fifth satellite and the four large moons were taken either by Lowell or E.C. Slipher or both on sixteen nights between October 19 and November 10. Though winters at Flagstaff did not rival those at Williams Bay for severity, given the 7000-foot altitude, nights from October through March can be bitterly cold, and present a severe hardship for observers. In partial recognition of his limits, Lowell had for some time been taking the earlier shifts at the telescope. This was a contrast to Barnard, who worked throughout the nights at Yerkes, and remained vigilant even on nights that appeared unpromising looking for breaks in the clouds.

Last Contact

During the day of November 11, 1916, Lowell received a package from Boston. It had been sent by his youngest sister, Amy. In 1916 she maintained a household in the Sevenels mansion in Brookline, purchased from her siblings with her share of their inheritance after her father's death. She cut an eccentric figure: massively overweight because of a glandular problem, living with and rumored to be in a lesbian relationship with the actress Ada Dwyer Russell, attended by a retinue of servants, and constantly smoking cigars. She was also a celebrated poet who had just completed her latest volume, *Men, Women, and Ghosts*. It opens with what became her most famous poem, "Patterns." Percival thanked Amy for sending the "visitor" (book), and expressed his appreciation for her achievement:

Thank you for the thought of the thing, the thing being thought, well paged, and attended, that walked into my eyry [sic.] in the remote, and I then introduced the welcome visit to Constance and she read to me the tale of the lone farmer's wife as I sat receptive of the comfort of my wood fires. It always pleases me to think how far one's printed thought travels—so may it to you whose child has entered where you never have.[86]

His only other communication that day was a note to Leonard, "Universities from Texas to Maine now want lectures but enough is as good as a feast. Sleet and snow yesterday; blue sky today."[87]

The blue sky foretold a clear night, and after dinner he made the short walk from the Baronial Mansion to the Clark, where E.C. Slipher was already observing. At a few minutes before 9 p.m., with a watch that was, he noted, with his usual punctiliousness, 2 s slow he recorded the reappearance of the satellite Io from eclipse. The second to last line, together with the timings, reads "Last Contact." Fittingly, since these were the last observations Percival ever made. (The paper on which they were recorded had been carefully placed for preservation by E.C. Slipher in a cigar box, and forgotten for 90 years until recovered by a Lowell historical preservationist while sorting piles of discarded materials in the Slipher Building.)

Lowell's last portrait, taken probably a few days before his death. (Credit: Lowell Observatory Archives)

On Sunday morning, November 12 (the day on which, incidentally, in France the last action on the Somme of 1916 was getting underway), Lowell seems to have started out in a bad mood. As usual a servant had upset him. What happened next is recorded by Flagstaff historian Platt Cline:

> Between 8 and 9 a.m…. 33-year-old Dr. M.G. Fronske, who had been in town about two years, received an emergency call to attend Lowell at his home. The call came to Fronske in the absence of Lowell's physician, R.O. Raymond, and the other older doctor, A.H. Schermann. Fronske found Lowell in a stupor as a result of what he diagnosed as a cerebral hemorrhage…. He injected atropine and wrote a prescription, realizing nothing else could be done. On inquiry, he was told Lowell had a serious altercation with one of the drivers or some other household employee shortly before the attack. Lowell died at 9:55 that evening. Dr. Fronske recalled 50 years later that he was closely questioned by a member or representative of the family was to exactly what treatment he had administered and what medicine he had prescribed, which offended him a bit, suggesting that his treatment might not have been adequate.[88]

Perhaps Percival had a premonition that he would die suddenly of a stroke. His brother Lawrence—who was not of course actually present—recalled that before becoming unconscious, Percival said that he "always knew it would come thus, but not so soon."[89]

The bedroom of the Baronial Mansion in which her husband died would be carefully preserved by Constance just as it was for rest of her life, a chalk inscription on the wall recording that "Percival Lowell's earthly existence terminated in this chamber upon the green couch." (What happened to the green couch is unknown.) Her first order of business, after having her late husband measured for a coffin (he was exactly 6′ 0″ tall), was to fire Leonard, something she must have been waiting for a very long time to do. One can only imagine what Wrexie's feelings must have been after Percy's unexpected marriage followed by her sudden banishment after his death with loss of her livelihood. She later wrote a book compiled mostly of letters from her late employer, *Percival Lowell: An Afterglow*, whose epigraph, with hidden initials, seems to reveal her anguish:

> Preambient light—
> Waning, lingers long
> Ere lost within.
> Just, kind, masterful:
> Life's sweet constant,
> Farewell.[90]

A careworn widow, Mrs. Percival Lowell, in undated photograph taken after her husband's death. She could never accept that he was gone; she was rumored to hold seances seeking to communicate with his spirit in the Baronial Mansion, which she kept up as it had been when he was alive, and—perhaps in order to hang on to as much of him materially as possible, she initiated a lawsuit to break his will and retain as much of his estate as she could. (Credit: Lowell Observatory Archives)

Wrexie returned to the East, and lived comfortably for a time, but lost a great deal of money in the stock market crash of 1929. Percival's brother Lawrence sent her some money and she maintained an independent existence for few years, but eventually had to move in with her sister and a niece in New York City, and finally, into the state hospital in Medford, Massachusetts, where she died in 1937.

Constance proceeded to drag the observatory through years of litigation, so that by the time the lawsuit was finally settled, the Observatory was "famously broke." Lowell's chosen successor as director, V.M. Slipher, did not even have money to pay the staff their salaries, and had to borrow funds. V.M., his brother E.C., and C.O. Lampland were the only senior astronomers that remained, and—though V.M.'s spectrographic work on the nebulae earned him great respect in the field—all three astronomers had to tread carefully the line between showing loyalty to the founder and distancing themselves from his more extravagant ideas.[91]

Within a few years of the end of the protracted litigation, Lowell's search for Planet X would be revived, with a new photographic telescope funded by brother Lawrence, now president of Harvard University. A tiny orb, hardly more than a sliver of a planet with a mass too small to have perturbed Uranus, was discovered in February 1930, a year into the renewed search by Clyde Tombaugh, a young farmer from Kansas and passionate amateur astronomer expressly hired for the purpose. Though the distant body was immediately hailed as Lowell's Planet X, it was no more Planet X than the Bahamas were India. It was called Pluto, for the brother of Jupiter and Neptune and ruler of the underworld, and also because its first two letters, PL, were the initials of the man who had so long dreamed of discovering a planet on the edge of the Solar System, but who died doubting its existence.[92]

Notes

1. "Report of the Meeting of the [British Astronomical] Association, held on March 30, 1910" (1909–1910), *Journal of the British Astronomical Association*, vol. 20, pp. 285–294: pp. 287–288.
2. E.M. Antoniadi to E.E. Barnard, January 16, 1911. He wrote, further: "The question of linear canals must cease. When I saw Mars at Meudon like yourself in 1894, I sighed at all the Flagstaff spider's webs, and thought I should work day and night to demolish such whimsical provocations of truth. Evidently, Lowell started from a mistranslation of a convention word ('canal' = 'channel,' not 'canal'), adhered to the literal sense of the mistranslation, and believed in the reality of the straight canals, disobeying perspective. Yet is not such logic (if, indeed, the use of that word is permissible here), is not such logic, I say, as illegitimate as to see a real Kneeler in the constellation of Hercules?" Barnard. Edward Emerson. Papers, Special Collections and University Archives, Jean and Alexander Heard Library, Vanderbilt University Archives.
3. See: P. Lowell [2–5].
4. P. Lowell [6], describing work done between December 1913 and March 1914. Lowell [7].
5. Lowell, "Measures of Saturn," p. 73. Irradiation is a classic case of the unreliability of human senses, and can be fully understood only in the light of modern developments in ophthalmology and neuroscience. For a comprehensive treatment, see: William Sheehan [8].
6. Clyde W. Tombaugh and Patrick Moore [9]. Tombaugh concludes: "Consequently, Lowell did not have a chance of success in the 1905–1907 photographic search."

7. Dale P. Cruikshank and William Sheehan [10]
8. Ralph Curtis to Barrett Wendell, February 2, 1910. Wendell. Barrett. Papers, Houghton Library of Harvard University.
9. Quoted in James Lequeux [12].
10. Clyde Tombaugh [13]
11. David Baron [14]. [Page unavailable. Manuscript was reviewed in draft form. Book title may change upon publication.]
12. P. Lowell [15] (slightly paraphrased).
13. P. Lowell, "Great Discoveries, and Their Reception," lecture text, ca. August 1916. Lowell Observatory Archives.
14. It should be pointed out that Republicans at the time still had at least the reputation of being the more liberal of the two parties.
15. Strauss, *Percival Lowell*, p. 261.
16. P. Lowell to R.B. Sims, May 23, 1913; Lowell Observatory Archives.
17. Strauss, *Percival Lowell*, p. 261.
18. Lowell, "The Revelation of Evolution," p. 181.
19. P. Lowell, "Two Stars," text of lecture delivered at Kingman, Arizona Territory, October 20, 1911. Lowell Observatory Archives.
20. P. Lowell to W.L. Leonard, undated; quoted in Leonard, *Afterglow*, p 94.
21. Quoted in A.L. Lowell, *Biography of Percival Lowell*, p. 155.
22. Adams did so entirely on his own, Le Verrier received some assistance in routine calculations from his student Émile Gautier.
23. For the perturbed planet, Uranus, the orbital elements are:

    v the true heliocentric longitude of the disturbed planet.
    n = its mean motion.
    ε = the mean longitude of its epoch, that is, the origin of the time.
    e = its eccentricity
    ϖ = the place of its perihelion
    m its mass.

    For the perturbing planet (X), the elements are (using the same symbols accented):

    v' = the true heliocentric longitude of the disturbing planet.
    n' = its mean motion.
    ε' = the mean longitude of its epoch, that is, the origin of the time.
    e' = its eccentricity
    ϖ' = the place of its perihelion
    m' = its mass.

    The discussion of Lowell's search for Planet X generally follows Cruikshank and Sheehan, *Discovering Pluto*, pp. 89–126.

24. P. Lowell to C.O. Lampland, December 10, 1910; Lowell Observatory Archives.

25. P. Lowell, telegram to C.O. Lampland, March 13, 1911; Lowell Observatory Archives.

26. W.L. Leonard to C.O. Lampland, March 22, 1911; Lowell Observatory Archives.

27. W.L. Leonard to V.M. Slipher, April 3, 1911; Lowell Observatory Archives.

28. Apart from the theories of Le Verrier and Gaillot, the only other complete theory of Uranus had been published by Simon Newcomb. See Simon Newcomb [16]. These investigtors would, of course, have produced a very different set of residuals and led to very different predictions of the position of X. Lowell must certainly have known of Newcomb's monumental publication but for whatever reason—perhaps only personal spite—chose to ignore it.

29. P. Lowell to W.L. Leonard, undated; in: Leonard, *Afterglow*, p. 99.

30. P. Lowell to V.M. Slipher, March 16, 1909; Lowell Observatory Archives.

31. In 1901, Robert G. Aitken, a double star observer at Lick who had personally experienced the superior performance of the 36-inch refractor in resolving double stars, suggested the Barnard, W.H. Pickering and others might go to Lowell Observatory to observe Mars under Flagstaff conditions. Similarly, W.H. Pickering invited Lowell to observe Mars from different observatories—though in this case, perhaps as an expression of doubt as to Pickering's ability to manage such a project, Campbell and Hale no less than Lowell turned it down. See R.G. Aitken [17]; W.H. Pickering to G.E. Hale, July 21, 1913 and G.E. Hale to W.H. Pickering, August 13, 1913. Hale. George Ellery. Papers, Mount Wilson and Palomar Observatories Library. Also Lowell to W.H. Pickering, September 2, 1913, Lowell Observatory Archives and W.H. Campbell to W. H. Pickering, August 25, 1913, Campbell. William Wallace. Papers, Mary Lea Shane Archives of the Lick Observatory, University of California, Santa Cruz. As well, E. M. Antoniadi thought it would be pointless to make the attempt. As he wrote to Barnard on February 6, 1910: "Having seen Mars better than anybody else in 1892 and 1894, I am sure that you have nothing to learn, as to the true structure of the minor details of the planet, by studying it anew with an inferior, or crippled telescope [24-inch, stopped down to '12 or 15 inches']…. Lastly, my conviction, based on the evidence of the Meudon refractor, when I was seeing much more delicate detail than the spider's webs, and no geometrical patterns, cannot change. It is thus obvious that we have nothing to learn at Flagstaff as to the nature of the Martian surface. Yet … it might … help Dr. Lowell to abandon, under the best possible circumstances, a hopeless position." Barnard. Edward Emerson. Papers, Special Collections and University Archives, Jean and Alexander Heard Library, Vanderbilt University. Vanderbilt University Archives.

32. E.E. Barnard, Mt. Wilson observing log book. Barnard. Edward Emerson. Papers, University of Chicago Library, Hanna Holborn Special Collections Research Center.

33. Ibid.

34. Ibid.

35. Ibid. Barnard's stopping down the 60-inch reflector to only 12 inches because of poor seeing conditions deserves to be remarked here. Lowell certainly had a valid point when he suggested that most of the time the seeing did not allow the full aperture of the 24-inch refractor to be used to advantage. Clyde Tombaugh wrote (personal communication to William Sheehan, December 5, 1986), "For best definition the air cells of uniform refraction in front of the telescope should be nearly as large as the aperture used, and the air cells should be about 15 inches across. This was one of the reasons Lowell used 16 inches most of the time." At Pic du Midi, which equals or surpasses conditions at Flagstaff, Henri Camichel once told me that Observatory's 1-meter telescope seldom reached its full resolution, and that usually the limit was at most about 50 cm. Of course, diaphragming down the aperture also increases the focal ratio (found by dividing the focal length of the telescope by its aperture). A longer focal ratio is better suited both for higher power visual observing as well as photography of planets. The magnifying power used is also important to the results; Lowell used magnifying powers of 310 to 400× Barnard, Hale and others used magnifying powers of 800×, 1000×, and sometimes even more, which greatly improved the ability of the eye to distinguish small details. Tombaugh wrote, "When I used the same telescope parameters that Lowell used, I saw the canals much as he saw them…. My experience [is that] the canals came out in brief flashes (less than a second of time) when larger air-cells of uniform refractive index passed in front of the telescope. These flashes are … too short to be caught on the photographic plate." With a larger aperture in excellent seeing, and using higher magnifications, more complex details than the canals come into view, as seen by Barnard at Lick, Antoniadi at Meudon, and Barnard at Mt. Wilson (and also by the present author when using the same instruments). See William Sheehan and Anthony Misch [18]

36. G.E. Hale (1912) *Mount Wilson Observatory Year Book*, vol. 11, p. 198

37. Burroughs originally wanted to use the pseudonym Normal Bean, but the publisher insisted on changing this to Norman. In addition to John Carter, Burroughs also created the character of Tarzan. For a biography, see: Irwin Porges [20].

38. Ralph B. Curtis to Barrett Wendell, May 1, 1912. Wendell. Barrett. Papers, Houghton Library, Harvard University.

39. P. Lowell to Kenneth [?John Kenneth McDonald}, April 28, 1912. Whereabouts unknown; sold copy of letter on eBay.

40. W.L. Leonard to C.O. Lampland, September 4, 1912; Lowell Observatory Archives.

41. P. Lowell to C.O. Lampland, September 12, 1912; Lowell Observatory Archives.

42. W.L. Leonard to E.C. Slipher, October 30, 1912; Lowell Observatory Archives.

43. W.L. Leonard to C.O. Lampland, December 11, 1912; Leonard to V.M. Slipher, December 12, 1912. Lowell Observatory Archives.
44. P. Lowell to C.O. Lampland, February 21, 1913; Lowell Observatory Archives.
45. Hoyt, *Planets X and Pluto*, p. 122.
46. Lowell, "Memoir on a Trans-neptunian Planet," p. 73.
47. P. Lowell to C.O. Lampland, telegram, July 10, 1913. Lowell Observatory Archives.
48. P. Lowell to C.O. Lampland, telegram, August 21, 1913. Lowell Observatory Archives.
49. It was from this time that an incident occurred involving one W.H. Spaulding, who was briefly employed by Lowell in late 1913-early 1914, and who wrote to A.E. Douglass:

    "With Lowell came abuse, brutality, arrogance, conceit and insolence. It was my lot to work in the dome with him, and the night I was with him he knocked a light out of my hand, jerked a box out of my hand without asking me for it, pushed me back across a pile of boxes and stepped over me before I could rise. Every order that he gave could have been heard a half mile away. The next morning I told Mr. [C.O.] Lampland that I would quit if there [were] any more such tantrums. He communicated my ultimatum to Lowell and for a couple of nights thing went somewhat easier, only to begin again worse than before. He would rush out when I was setting the telescope, and grabbing hold of the end, swing it here and there, while I was vainly trying to get the object into the field. At length I deliberately walked out of the dome … and went down and demanded my time from Mr. Slipher—I mean V.M. Slipher, E.C. being a mere non-entity. V.M. prevailed upon me to stay by agreeing to put Mr. Hanway in the dome. The first night that they were working together Lowell became enraged over Mr. Hanway's apparent slowness in starting the dome; so he seized the switch rope and gave it a terrific pull, locking the switch so tight that it was impossible to reverse it and stop the dome. Mr. Hanway was unacquainted with the expedient of running over and throwing out the main switch. So they—he and Lowell—started getting things out of the way of the ladder which was coming at them like a juggernaut. Most of the tables and chairs they dumped into the pit, but the big astronomical chair was the main problem. They could by great exertion keep it out of the way of the dome. Round and round they went until finally, when they were pretty nearly played out, Mr. Lampland came in. They had been working about twenty minutes. Mr. Lampland thought they were trying to follow a meteor." When, as was inevitable under the circumstances, they left, neither Spaulding nor Hanway found it easy to collect from Lowell the money owed them. W.H. Spaulding to A.E. Douglass, February 16, 1914. A.E. Douglass papers, University of Arizona.

50. P. Lowell to V.M. Slipher, telegram, August 27, 1913. Lowell Observatory Archives.

51. Ibid.

52. Ibid., p. 59.

53. Lowell, "Memoir on a Trans-neptunian Planet," p. 105.

54. P. Lowell to C.O. Lampland, telegram, May 5, 1914; Lowell Observatory Archives.

55. Leonard, *Afterglow*, p. 118.

56. In London, Lowell wrote from the Princes' Hotel and Restaurant in Jermyn Street, London: "I just succeeded, after prolonged mental agony, in getting my new [clothes] to fit. I m now cladable in decent garments in the latest style. I am so proud of them I am afraid they will sink in mid-ocean. Speaking of which a new idea has just come to me, which I hasten to instruct you of in case anything should happen to me, which may it not! I want all my mathematical papers, including the present X investigation collected from all the periodicals, etc., and published in a volume entitled

    Papers on
    Celestial Mechanics
    And
    Celestial Physics
    Vol. 1
    By
    P.L.printed in large type so as to fill a fair-sized book…" Lowell returned safely, and the project was abandoned. Leonard, *Afterglow*, p. 121.

57. Quoted in Damon, *Amy Lowell*, p. 242.

58. Henry James to Edward Waldo Emerson, August 4, 1914. In: Philip Horne [23]. A physician, Edward was a son of the famous New England writer Ralph Waldo Emerson.

59. P. Lowell to V.M. Slipher, August 11, 1914; Lowell Observatory Archives.

60. P. Lowell to C.O. Lampland, December 21, 1914; Lowell Observatory Archives.

61. Lowell, "Memoir on a Trans-neptunian Planet," p. 103.

62. Ibid., p. 104.

63. C.O. Lampland to P. Lowell, August 15, 1915; Lowell Observatory Archives.

64. C.O. Lampland to P. Lowell, September 15, 1915; Lowell Observatory Archives.

65. A. L. Lowell, *Biography of Percival Lowell*, p. 192.

66. Charles W. Tuttle and Phillip Sidney Coolidge, volunteer research associates at Harvard in the 1850s, were the most successful in seeing these divisions. Tuttle, assistant astronomer from 1851 until 1854, when his eyesight failed, afterwards became a lawyer at Newburyport, Massachusetts. His brother, Horace, discovered several comets. After Charles Tuttle retired, Coolidge replaced him as an unpaid research associate at Harvard. A great-grandson of Thomas Jefferson, Sidney had a twin-brother Algernon who graduated with an M.D. from Harvard in 1853 who served as a surgeon at the Army Hospital near Fort Monroe,

Virginia during the Civil War and later went into private practice in Boston. In 1856 he married Mary Lowell, daughter of Francis Cabot Lowell (son of the industrialist of the same name) and Mary Gardner Lowell. Thus, the Coolidges and the Lowells became fully and fairly intertwined. Sidney was a restless adventurous sort, and after several years of service to the Harvard Observatory, in 1860 he began to roam widely, serving as a soldier in the Franco-Italian-Austrian War and helping to engineer a revolution in Mexico. On the outbreak of the Civil War, he joined the Union army, and served as a major in the 16th Massachusetts infantry Regiment, and was killed in September 1863 at the Battle of Chickamauga. According to an eyewitness, after the 16th Infantry broke, Coolidge held his place with the point of his sword up—what soldiers call a "defy"—only to be immediately shot. See: W. Sheehan and S.J. O'Meara [25], which includes a portrait.

Coolidge inspired amateur astronomer Stephen James O'Meara's visual observations of the planets at Harvard during the 1970s, which led to the discovery of spoke-like markings in the B ring of Saturn whose existence was later confirmed by the two Voyager spacecraft in 1980–81.

67. Lowell, "Memoir on Saturn's Rings," p. 4.
68. Ibid., p. 8.
69. What might be called the "canals" of Saturn's rings were, like the Martian ones, without visible breadth. As usual, Lowell emphasized the superiority of what was seen at Flagstaff to what was seen by previous observers, including Barnard. What the latter had depicted as a mere area of shading in Ring B consisted in reality "of linear circular divisions comparable for tenuity with the canals of Mars. The chief of these pencil lines were not only seen but measured by two observers with concordant results, and their positions proved explicable by celestial mechanics." P. Lowell [26, p. 291]. The measures were so exact that subtle differences were noted between the expected and measured positions. Thus, the great Cassini Division, presumed due to a 1/2 resonance with Mimas, lay farther out than it should, while with the minor subdivisions, too, Lowell found a "lack of complete correspondence between the Jovian and Mimasian patterns and more careful scrutiny discloses that the latter has been, as it were, stretched or warped outward, decreasingly so as we leave the planet." Lowell, "Memoir on Saturn's Rings," p. 5.
70. Lowell, "Genesis," p. 290.
71. Lowell, "Memoir on Saturn's Rings," p. 22.
72. Leonard, *Afterglow*, p. 145.
73. [Unsigned] (1916) "The Latest News from Mars," *Journal of the Royal Astronomical Society of Canada*, vol. 10, no. 5 (May–June), pp. 265–266.
74. G.H. Hamilton [28, p. 138]. After Lowell's death, Hamilton married Lowell's chief computer, Elizabeth Williams, and joined W.H. Pickering in order to observe Mars at the latter's retirement plantation near Mandeville, Jamaica. He died in 1936.

75. It was published in *Popular Astronomy* as "Our Solar System." See: P. Lowell [29, p. 427]

76. Lowell, "Genesis," p. 281.

77. Strauss, *Percival Lowell*, p. 267.

78. P. Lowell to J. Trowbridge, December 9, 1914; Lowell Observatory Archives.

79. P. Lowell to E. B. Wilson, February 4, 1915; Lowell Observatory Archives.

80. P. Lowell, "Great Discoveries, and their Reception," lecture text, ca. August 1916. Lowell Observatory Archives.

81. Ibid.

82. Frederic Stimson to Barrett Wendell, undated. Wendell. Barrett. Papers, Houghton Library of Harvard University.

83. P. Lowell, "Great Discoveries."

84. Ibid.

85. P. Lowell, "Mars and the Earth"; lecture text ca. August 1916. Lowell Observatory Archives.

86. P. Lowell to Amy Lowell, November 11, 1916. Lowell. Amy, Papers, Houghton Library of Harvard University.

87. Leonard, *Afterglow*, p. 161.

88. Platt Cline [31]. Sudden death from cerebral hemorrhage tended, like mathematical and literary abilities, to run in the family. Lowell's sister Amy died of a massive cerebral hemorrhage at age 51 in 1925, and so did Lowell's cousin and first sole trustee of the observatory after his death, Guy Lowell, at age 56 while sailing the Madeira islands in 1927.

89. A. Lawrence Lowell, *Biography*, pp. 193–194.

90. As first noted by Jan Hollis [32], the following individuals are encoded in this poem:

    Preambient light = Percival Lowell
    Waning, lingers long = Wrexie Louise Leonard
    Ere lost within = Elizabeth Langdon Williams
    Just, Kind, masterful = John Kenneth McDonald
    Life's sweet constant – Lowell Savage Constance

    Williams and McDonald were the most dependable computers on the X search. Constance's name is the only one for whom the letters are reversed; she was, as Lowell's great nephew William Putnam, III told me in June 2012 the one who almost succeeded, through the endless lawsuit against Lowell's estate initiated after his death, in negating—that is, reversing—his life work. Wrexie moved to Boston and lived with her older sister, Laura Helen Leonard, a teacher, before being hired by Lowell as his secretary. After marrying William Goodell, Laura moved west to teach school, and after his death in 1926, she moved in with her daughter Helen in New York City, where Wrexie joined them after the stock market crash.

Helen was a very interesting woman in her own right and deserves at least a brief note here. In 1923 she graduated with a B.S. from Simmons College, and in 1932 was hired as a research associate in the department of neurology at Cornell University Medical College and the New York Hospital and studied headaches and pain with Dr. Harold G. Wolff. Goodell, Wolff and Dr. James Hardy introduced the first "dolorimeter" as a method for evaluating the effectiveness of analgesic medications, and published a pain scale called the "Hardy-Wolff-Goodell" scale which was influential in the 1940s but subsequently abandoned (ironically, given Wrexie's involvement with the Martian canal controversy) as subjective and unreliable. She died in 1987. On the dolorimeter, see: Noémi Tousignant [33].

Miss Irva Struthers, to whom Lowell had proposed marriage in 1904, looked back as Mrs. William W. McCall on her relationship with the astronomer, and paid him private tribute, "The stars have claimed you for their own, my Martian…. I did not know until you had gone, the hold you had upon my life, my thoughts, my actions." Despite her rejection of him as a suitor, she had nevertheless kept his love letters.

Constance established herself in "opulent squalor" after her husband's death at "Sans Souci," a "cottage" on Merton Road in Newport, Rhode Island, previously owned by the American heir and man of leisure James Vanderburgh Parker (among other things remembered in the day as a "fashionable bachelor of Mrs. Astor's entourage"). According to William Lowell Putnam, III, "her relatives by marriage kept a cool distance from her, and those by blood hovered, as she said, 'like buzzards … waiting for me to die.'" She paid her last visits to the observatory in 1949 or 1950, but after Lampland, her last remaining ally on the Lowell staff, died in December 1951, she never returned. Moving to Boston a year or two before her death, she passed away at the age of 91 in September 1954. The Baronial Mansion, which was kept up during her lifetime to accommodate her intermittent visits, was torn down in 1959, by which time it had been declared a fire hazard and depleted of all its more valuable contnets (among the first to go had been the cases of fine domestic and imported wines, cordials, cognacs, demijohns of table wines, and three-gallon oak kegs of whiskey.

91. The balancing act that needed to be accomplished by the three senior astronomers at Lowell Observatory after the founder's death is considered in: William Sheehan [34].

92. The story of the discovery of Pluto has been often and well told. William Graves Hoyt [11], which draws heavily on materials in the Lowell Observatory Archives; Clyde Tombaugh and Patrick Moore [9], which is uniquely valuable in presenting Clyde Tombaugh's account of the discovery of Pluto; William Lowell Putnam, III and others [35] contains a thorough account of the case of Lowell vs. Lowell lawsuit of 1916 to 1927 which tied up Percival Lowell's estate for many years and left it "famously broke"; and Dale P. Cruikshank and William

Sheehan [10] *Discovering Pluto: Exploration at the Edge of the Solar System* (Tucson: University of Arizona Press, presents the discovery of Pluto in the context of its later identification as the leading member of the Kuiper Belt and the broader exploration of the Solar System.

References

1. Graber RB (2008) Plutonic sonnets. Publish America, Baltimore
2. Lowell P (1911) On the action of planets upon neighboring particles. Astron J 26(21):171–174
3. Lowell P (1911) Libration and the asteroids. Astron J 27(6):41–46
4. Lowell P (1912) The asteroids. Lond Edinb Dubl Philos Mag J Sci (Sixth Ser) 23(135):337–352
5. Lowell P (1913) The origin of the planets. Mem Am Acad Arts Sci 14(1):3–16
6. Lowell P (1915) Measures of Saturn – ball, rings and satellites. Bull Lowell Observ 2:73–80
7. Lowell P (1915) Oblateness of Uranus. Lowell Bull 2:81–84
8. Sheehan W (2018) Two important cases of the irradiation illusion in astronomy. Antiq Astron 12(June):17–28
9. Tombaugh CW, Moore P (1980) Out of the darkness: the planet Pluto. Stackpole Books, Harrisburg, p 85
10. Cruikshank DP, Sheehan W (2018) Discovering Pluto: exploration at the edge of the solar system. University of Arizona Press, Tucson, p 105
11. Hoyt WG (1980) Planets X and Pluto. University of Arizona Press, Tucson, p 104
12. Lequeux J (2015) Le Verrier: magnificent and detestable astronomer. Springer, New York, p 50
13. Tombaugh C (2009) The discovery of Pluto: generally unknown aspects of the story. Astron Beat 23(May 18):2
14. Baron D (forthcoming) The Martians. Liveright, New York, chap 18
15. Lowell P (1915) Memoir on a trans-neptunian planet. Mem Lowell Observ 1:7
16. Newcomb S (1873) An investigation of the orbit of Uranus: with general tables of its motion. Smithsonian Contributions to Knowledge, Washington, DC
17. Aitken RG (1910) Recent observations of Mars. Publ Astron Soc Pac 22:78–87
18. Sheehan W, Misch A (2003) Two weeks on Mars. https://mthamilton.ucolick.org/public/TwoWeeksOnMars/
19. Antoniadi EM (1913) Considerations on the physical appearance of the planet Mars. Pop Astron 21:416–424
20. Porges I (1975) Edgar Rice Burroughs: the main who created Tarzan. Brigham Young University Press, Provo
21. Lowell P (1914) Epitome of results at the Lowell Observatory April 1913–April 1914. Lowell Observ Bull 2(9):58

22. Sheehan W, Young K (2021) Neptune's orbit: reassessing celestial mechanics. In: Sheehan W (ed) Neptune: from grand discovery to a world revealed. Springer, Cham, pp 351–352

23. Horne P (ed) (1999) Henry James: a life in letters. Viking-Penguin, New York, p 540

24. Fox P (1921) Review of W.L. Leonard, Percival Lowell: an afterglow. Pop Astron 29:596–599

25. Sheehan W, O'Meara SJ (April 1998) Phillip Sidney Coolidge: Harvard's romantic explorer of the skies. Sky Telescope, pp 71–75

26. Lowell P (1916) The genesis of planets. J R Astron Soc Can 10(6):281–293

27. Elliott J, Kerr R (1984) Rings: discoveries from Galileo to Voyager. MIT Press, Cambridge, p 32

28. Hamilton GH (1920) Mars our neighbor in space. Pop Astron 28(3):137–140

29. Lowell P (1916) Our solar system. Pop Astron 24:419–427

30. Lowell P (1916) The far horizon of science. The Stanford Daily, October 18

31. Cline P (1994) Mountain town: Flagstaff's first century. Northland Publishing, Flagstaff, p 174

32. Hollis J (1992) Wrexie. Griffith Observ 56:10–17

33. Tousignant N (2010) The rise and fall of the dolorimeter: pain, analgesics, and the management of subjectivity in mid-twentieth-century United States. J Hist Med Allied Sci 66(2):145–179

34. Sheehan W (2019) Treading carefully: V.M. Slipher, C.O. Lampland, E.C. Slipher and their ambivalent relationship with Percival Lowell's Mars. J Astron Hist Herit 23(3):365–400

35. Putnam III WL et al (1994) The explorers of Mars Hill: more than a century of history at Lowell Observatory. Phoenix Publishing, West Kennebunkport

13

Ad Astra Edward

Contents

… Today the odds against the self-made man are no doubt even longer than they were in Barnard's day. Nevertheless, the Barnard legend, like the Lincoln legend, continues to inspire all who start life without advantages, and suggests that struggle and hardship [if not excessive and counterbalanced with sufficient supports and opportunities] may themselves be formative. As the poet Keats once remarked, "I must think that difficulties nerve the Spirit of a Man—they make our Prime Objects a Refuge as well as a Passion."
—William Sheehan, The Immortal Fire Within, *p. 417.*

W. Sheehan, *Parallel Lives of Astronomers*, Springer Biographies,
https://doi.org/10.1007/978-3-031-68800-3_13

A New Sensation: Comet Halley

Comets, of course, were one of Barnard's specialties. He was, in his own way, intensely competitive, and naturally hoped to be the first one to see Comet Halley on its return in 1910. As early as October 1908, when the comet was still out between the orbits of Jupiter and Saturn, he began sweeping the region of the sky where it was supposed to be lurking with the 40-inch refractor, as well as exposing plates to the region with the Bruce telescope. Despite his best efforts, he was upstaged. His friendly rival Max Wolf at the Königstuhl Observatory in Heidelberg beat him to it. Wolf first recognized the comet's faint image on plates taken on September 11, 1909.

Daniel Morehouse, a professor of astronomy (and later president) at Drake University in Des Moines, Iowa, was at Yerkes at the time, studying and photographing comets with Barnard. He had found his only comet with the Bruce telescope in September 1908, Comet Morehouse. It proved to be a remarkable object. Barnard obtained 350 photographs, documenting "the extraordinary rapidity with which a comet can alter its appearance when in one of its changing moods" [1, p. 299], and correlated the changes in the plasma tail with auroral displays occurring at the time. This inspired one of his most original ideas—that both auroras and changes in a comet's tail are due to solar magnetic field disturbances. Far ahead of its time, his idea was largely forgotten until it was rediscovered in the late 1960s.[1]

1908 Sept. 30 d. 11 h. 16 m. C.S.T. (6-inch.)

1908 Sept. 30 d. 14 h. 57 m. C.S.T. (6-inch.)

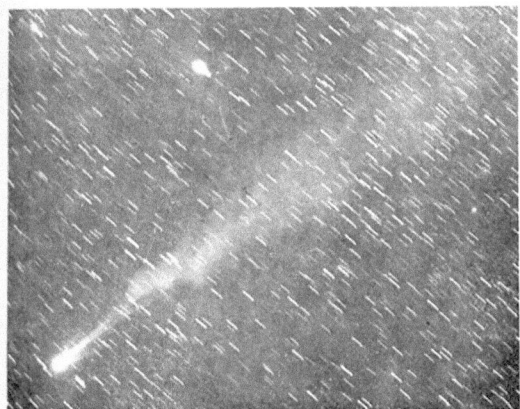

1908 Oct. 6 d. 15 h. 58 m. C.S.T. (10-inch.)

COMET c 1908 MOREHOUSE.—E. E. BARNARD.

*Barnard's images of Comet Morehouse with the Bruce photographic telescope, show-
ing a "tail-disconnection" event, in which the plasma tail separated dramatically from
the head. Though Barnard had recorded similar events before, as in Comet Brooks
(1893 IV), discussed in chapter 4, this is a particularly clear example. By correlating such
events with auroral displays, he eventually convinced himself that these events were
caused by magnetic storms from the Sun, as was long afterward shown to be the case.
(From:* Monthly Notices of the Royal Astronomical Society, *Nov. 1908, vol. 69, Plate 9)*

Morehouse later recounted Barnard's reaction to Wolf's beating him to Halley's Comet:

> For weeks every time the telephone would ring, the observers would step into the hall and listen for the news. One afternoon … about three o'clock the telephone rang. As was our custom, we stepped to the door. Dr. Frost called out, "Hoo-hoo, it is found." I shall never forget Dr. Barnard's white face as he stepped to the hall. "Who found it?" he said. "Dr. Max Wolf." Dr. Barnard closed his eyes for a moment. He asked for the position, and without saying a word, he turned and walked back to his office, picked up a photographic plate of two nights before, and by the aid of the [blink] comparator found he had the object on a Bruce plate, but not one word or complaint that would detract in the slightest from the glory of Dr. Wolf's discovery was uttered. Dr. Barnard wanted his own, but he was equally insistent on giving everybody else due and just credit. [3]

The first to see the comet visually was Burnham with the 40-inch refractor on September 15, 1909. Barnard confirmed it two nights later; it appeared as a 15th magnitude "fleck of light surrounded by a faint nebulosity."[2] He would continue to follow it with the great refractor until February 1910 when, still inconspicuous, it disappeared into the solar glare. Passing perihelion in late April, it emerged from behind the Sun into the early morning sky and brightened dramatically with decreasing distance from the Sun. However, it was hardly visible from Yerkes; vast forest fires were then raging in northern Wisconsin, producing a smoky pallor which "cut off with a thick yellow veil all but a glimpse of the bright head." Only on the mornings of May 3 and 4 did the sky become clear and transparent enough to show the comet well. Barnard now found it a beautiful object, with a 2nd magnitude head and a tail traceable for some 18 degrees. He attempted to photograph it with the Bruce telescope despite the fact that its light was somewhat dulled by smoke not from the forest fires but from the Observatory's powerhouse, which blew directly across it during the exposures. Needless to say, this was not one of his better efforts.

Then, for the next 2 weeks, the sky was either hazy or covered with clouds, and only occasional glimpses of the comet were possible. Fortunately, and just in the nick of time, at midnight on May 17, there was a sudden clearing. This gave Barnard the opportunity of observing the comet as it made its closest approach to the Earth (within 0.15 AU). The event—calmly awaited by the astronomer—was regarded with mortal fear by many people, since it had been predicted that the Earth would brush through the comet's tail, known from spectroscopic studies to contain hydrocarbons including cyanogen, a deadly poisonous gas. Camille Flammarion had suggested that the cyanogen gas from the tail would impregnate the Earth's atmosphere, and possibly even snuff out all life—though since a comet's tail is exceedingly insubstantial, he should have known better. Indeed, even within living memory the Earth had passed through the tail of another comet, the Great Comet of 1861 (likely the one Barnard remembered from early childhood) without suffering any harm. There were no ill-effects from Comet Halley either. Barnard allowed himself a somewhat cynical (and racist) reflection:

> There was one fact which was brought forth by the comet with startling vividness. It showed that the superstitious terror formerly attending the appearance of a great comet is by no means dead in the human breast. Cases of this kind developed all over the country and abroad—from the stopping up of keyholes and cracks in doors and windows in Chicago (according to the daily papers) to keep out the deadly comet gases, to the manufacture and sale, among the negroes of the South, of "comet pills," which were supposed to ward off the evil effects of the comet.[3]

Drawing made by Barnard on the morning of May 18, 1910, as the Earth passed through the tail of Halley's Comet. (Credit: University of Chicago Photographic Archive [apf6-0167], University of Chicago Library, Hanna Holborn Gray Special Collections Research Center)

The Sere and Yellow Leaf

Percival Lowell often made disparaging comments about the poor performance of the Yerkes refractor for planetary work, saying for instance that the seeing at Williams Bay was "born bad." Barnard knew only too well the great difficulties of the site. Though during his Lick Observatory years the planets had been studied intensively and figured in over half of all his publications, after his move to Yerkes, planetary work made up a slender part of his output. He made a few important discoveries, such as the white spot on Saturn in 1903, which has already been described. His best planetary work involved the phenomena of the edgewise rings in 1907–08, for which the Yerkes 40-inch refractor was well suited, and his remarkable observations of Mars with the 60-inch reflector at Mt. Wilson in 1911.

In spite of perennial bouts of colds and bronchitis, Barnard remained in generally good health into his fifties. He overworked himself, of course, and could not stand to waste a scrap of clear sky. His colleagues urged him to ease up on himself; he ignored their advice and pushed himself as hard as ever. But the hours that would have worried any medical man, as Walter Sydney Adams put it, caught up with him eventually. He had since the turn of the century been putting on weight. Though work on the great telescope was strenuous, and his added weight made pushing the telescope around somewhat easier, he rarely went out of his way for exercise. He was an occasional swimmer in Lake Geneva and once joined a swimming party that included Frank Jordan, who came to Yerkes in 1905 to work with John Parkhurst on stellar photometry. Jordan had been told, by Phillip Fox among others, that Barnard was "not over-skillful but a very venturesome swimmer, and that should he venture into deep water he would need careful watching." Fox continued:

> We have often joked over Jordan's evident distress, and subsequent immediate perplexity and chagrin,—when Barnard dove without hesitation into deep water at the end of the pier and came up blowing water from his moustache before starting to swim with strong swift strokes. [6, pp. 199–200]

Otherwise, he limited himself to playing a little golf (always under protest) on the course next to the observatory, and participating in the Yerkes Observatory cross-country ski club (skiing Scandinavian style—with only one pole). It was not enough to keep him fit, and in February 1914, he began to notice that he was tiring easily. He was also thirsty much of the time and passing copious amounts of urine, which made life difficult for someone who was determined to guide a telescope while exposing a photographic plate for 6 or 8 hours without a break.

Barnard continued to work, but on March 8, 1914, he took to his bed—the nearly total eclipse of the Moon three nights later he observed from the second-story bedroom window of his home on the shore of Lake Geneva. Physicians at once recognized the classic symptoms of diabetes mellitus. The diagnosis was easy but the treatment was obscure—in those days before insulin, the outlook for the diabetic patient was bleak. Barnard's physicians subscribed to the prescription recommended by the famous Sir William Osler in his textbook of medicine, "Sources of worry should be avoided," and "[the patient] should lead an even, quiet life, if possible in an equable climate."[4] As part of the proposed rest cure, Barnard's physicians ordered him not to observe with the 40-inch telescope. For him, this was the greatest possible privation. For Barnard, a night at the great telescope was, as Frost wrote,

> … almost a sacred rite—an opportunity to search for truth in celestial places. Rarely has a priest gone up into the temple with a deeper feeling of responsibility and of service than did this untiring astronomer go up into the great dome. He was usually ready before the sun had set, and impatiently waiting until the darkness should be sufficient for him to "get the parallel" for the thread of the micrometer before he could observe faint objects. During the day preceding one of his nights, his associates in the observatory were generally conscious of his keen anxiety for a clear sky, as evidenced by a frequently repeated nervous cough, which was always worse if the prospects for the night were unfavorable. [8]

Difficult as was his banishment from the great telescope, Barnard bore it, said Frost, "manfully." Still the Yerkes director hastened to add that it was "almost impossible for Mr. Barnard to keep away from the Bruce photographic telescope when the sky was clear and the moon did not interfere."[5]

Barnard at the Bruce telescope, undated, but probably about 1913–14. (Credit: University of Chicago Photographic Archive [apf6-04572], Hanna Holborn Gray Special Collections Research Center, University of Chicago Library)

Barnard escaped to California for the worst months of the Wisconsin winter (December 1914 to March 1915) and was not recovered enough to return to observing with the 40-inch for another year. Even then, he was far from being fully recovered—diabetes is a chronic illness—and for the rest of his life he would continue to struggle with its vitality-sapping, wasting effects.

Barnard and Frost standing on the outside walkway of the 40-inch dome, looking east, in a photograph taken in August 1915. This was during the period when Barnard was, for reasons of health, banished from the 40-inch refractor. (Credit: University of Chicago Photographic Archive [apf6-00422], Hanna Holborn Gray Special Collections Research Center, University of Chicago Library)

Barnard had never taken students, partly because he had always felt it to be easier (given his high standards) to do whatever needed to be done himself rather than teach another to do it. However, his niece, Mary R. Calvert, who lived with Edward and Rhoda since 1905, now became Barnard's full-time assistant, and remained in this role until his death in 1923. Thereafter she continued as a high-level assistant at Yerkes until her retirement and return to Nashville in 1946.

With the passing years, there was an inevitable changing of the guard. Old friends and foes passed from the scene. Holden, who after stepping down from the Lick directorship in 1895 rounded out his life usefully as librarian at West Point, doing the kind of work (bibliography) to which he was really well-suited. He passed away in March 1914. That same year Barnard's old friend and colleague at both Lick and Yerkes, Sherburne Wesley Burnham, retired. Lowell, of course, passed in November 1916. Meanwhile, among astronomers who would lead the next generation, the most notable was Edwin Powell Hubble, a native of Missouri who had been an undergraduate at the University of Chicago (class of 1910), where he studied under Forest Ray Moulton (who had referred to Lowell as the "mysterious watcher of the stars whose scientific theories have taken shape at midnight"). Hubble then went to Oxford as a Rhodes Scholar where he adopted affectations in dress—he sometimes wore a cape—speech, and manner. According to another American, Warren Ault, who encountered Hubble 2 years into his Rhodes at the beginning of his own fellowship, he

> was dressed in plus-fours, a Norfolk jacket with leather buttons, and a huge cap. He also sported a cane and spoke in a British accent I could scarcely understand. He was, in fact, an American Rhodes Scholar who had been at Oxford for two years.... Those two years had transformed him, seemingly, into a phony Englishman, as phony as his accent. I was put off by the sight of him and the sound of him and I resolved that Oxford would not do that to me.[6]

After Oxford, and following an interlude teaching Spanish, physics and mathematics at a high school in Indiana, Hubble arrived at Yerkes in 1914 to pursue a Ph.D. in astronomy. However, the greatest observational astronomer of the passing generation appears to have hardly gotten to know the greatest of the coming generation. Hubble took command of the 24-inch reflector, with a fast focal-ratio of f/4, which George Willis Ritchey had completed in 1901, in order to investigate faint nebulae. It was a rather original project, since at the time these nebulae were rather neglected and even their nature was not yet agreed upon, though according to Yerkes Observatory historian Donald

E. Osterbrock, "he was not quite sure they were galaxies, but he became almost convinced that they were" [10]. Hubble's dissertation, "Photographic Investigations of Faint Nebulae," was completed in 1917 in rather slap-dash fashion. He could not dither over it as he was then in great haste to be off to France as a volunteer officer with the American Expeditionary Force. In 1919, after his discharge, he was highly enough regarded as an astronomer to be recruited to Mount Wilson. There he resumed his researches on faint nebulae, but with a much more suitable instrument, the recently completed 100-inch Hooker reflector, then the largest in the world. Barnard's work on the Milky Way thus hardly overlapped with Hubble's at all. But one also wonders whether Barnard, with his rather down-to-earth manner, crumpled look and Southern accent, wouldn't have been repelled by Hubble's stodgy persona and use of such Britishisms such as "Bah Jove!", "plummy," and "come a cropper." In any case, Hubble kept the Britishisms up for the rest of his life; Allan Sandage, who became assistant to the aging Hubble, found him still wearing Harris Tweeds, speaking "Oxford," and boasting a cane or walking stick in an affected English manner [11].

Barnard took an interest John Mellish, a young astronomer from a farm near Cottage Grove, Wisconsin (near Madison), and much more like him than Hubble. Building his own telescopes and using them to discover comets had led Mellish by the age of twenty to become locally famous as the "boy astronomer of Cottage Grove." Barnard, appreciating the young man's success despite his lack of formal education, invited him to Yerkes Observatory. Eventually, Mellish made several visits, each time staying as a guest of the Barnards in their home on Lake Geneva, and eventually, in the fall of 1915, received a year-long appointment as "Volunteer Research Assistant" at Yerkes. Immediately after his arrival, Mellish began searching for comets with his six-inch homebuilt reflector, and sweeping in the constellation Monoceros he soon came across what he took to be a comet in the rapidly fading light of dawn. It was actually a fan-shaped nebula (NGC 2261) discovered by William Herschel as far back as 1783. Mellish's nebula captured the attention of Hubble, who was still at Yerkes at the time. He began photographing it with the 24-inch reflector, and found that it undergoes changes in shape and brightness over weeks and months. They are so prominent as to be visible even in amateur telescopes. The changes in what is now known as "Hubble's Variable Nebula," so prominent as to be visible in amateur instruments, are due to the fact that the star that illuminates the nebula, R Monoceros, is itself variable.[7]

Yerkes Observatory staff, August 1915. From left (standing, steps) Oscar Romare, John Mellish, Edwin Hubble, Alfred Harrison Joy, Storrs B. Barrett, Julius Lemkowitz, Richard Oetjen, Barnard; (standing, mid-ground, left) Edwin Brant Frost, Gus Dahlsrom, Edison Pettit; seated William L. Hart, Miss Roosa, Charles A. Maney, Mr. Roe; (seated, from left) Carl Wendell, Miss Johnson, Mary Ross Calvert, Miss Haweks, Frances Lowater, John A. Parkhurst, Francis Easton Car (seated, foreground) H.O. Burns, Jessie M. Short. (Credit: Credit: University of Chicago Photographic Archive [apf6-00398], Hanna Holborn Gray Special Collections Research Center, University of Chicago Library)

Barnard, in profile, ca. 1917. Credit: University of Chicago Photographic Archive [apf6-00205], Hanna Holborn Gray Special Collections Research Center, University of Chicago Library

Thus, Mellish quickly made himself useful and, adding the discovery of a further comet on September 14, fully confirmed Barnard's faith in him. Two months later, on the morning of November 13, 1915, he observed Mars with the 40-inch refractor, and the observation has become legendary. Though November in southern Wisconsin is usually wet and gloomy, that fall saw an unusually fine "Indian summer," and at dawn on the date in question the temperature was near freezing but well above the dew point. Mars was a long way from Earth (opposition was not until February 10, 1916), but it stood high in the sky, and with Barnard still banished from the 40-inch because of ill health, Mellish had it to himself. The planet, a tiny gibbous disk with an apparent diameter of only 7.7 arcseconds, rose high into the sky after sunrise, as the field in the eyepiece lightened to azure blue and the planet faded from intense fire-opal to muted yellow. As often under such conditions, Mellish experienced a period of magical seeing an hour after sunrise, as the air became perfectly still, and was able to use magnifications of 750× and 1100× with the full aperture of the great telescope. The area on view would have been the interesting region which includes the "super crater" Argyre Basin, Solis Lacus, Valles Marineris, and the Tharsis shield volcanoes—the same area depicted in Barnard's memorable drawing made at Lick, also in the hour or so after sunrise, on the morning of September 3, 1894. In this area Mellish was convinced that he was able to make out craters on Mars, and in particular he called attention to a very large one, 300 kilometers across, which appeared highly foreshortened on the terminator at latitude 50 degrees South.

Mellish would always be haunted by what he saw that morning. However, with no prior experience observing Mars, he sought out Barnard's advice as to what he had seen, as he described in January 1935 in a letter to a friend, Walter Leight, Jr. of the Lehigh Valley (Pennsylvania) astronomy club:

> After seeing all the wonders, I went to Barnard and showed him my drawings and told him what I had seen. [Mars] is not flat but has many craters and cracks. I saw a lot of the craters and mountains ... with the 40" and cold hardly believe my eyes.[8]

Mellish told Barnard he had not heard of anything like this being seen before. Barnard, with a twinkle in his eye, proceeded to get out an old trunk, from which he pulled out some of the drawings he had made at Lick in 1892 and 1894. They seemed to Mellish, according to his later account, "the most wonderful drawings that were ever made of Mars." They showed "mountain ranges and peaks and craters and other things both dark and light…. I was thunderstruck and asked him why he had never published these and he said no one would believe him and would only make fun of it…. Barnard took whole nights to draw Mars and would study an interesting section from early in the evening when it was coming on the disk until morning when it was leaving and he made the drawings four or five inches [in] diameter and it is a shame that those were not published."[9]

The dome of the 40-in. refractor, with the slit open, and the refractor pointed north-west to the approximate position it would have occupied when Mellish made his observation in the early morning sky on November 13, 1915. (Credit: William Sheehan)

A globe of Mars, constructed by astronomical artist Joel Hagen from the never-published drawings made by Barnard in 1894. (Credit: Joel Hagen)

It was indeed a shame, for by the time Mellish wrote this, Barnard was long dead, and the whereabouts of his drawings were no longer known, or even whether or not they still existed.

The story of Barnard's Mars drawings is a rather interesting one. When he left Mount Hamilton in 1895, he brought with him to Yerkes the Milky Way and comet photographs—the latter, of course, remained the great albatross around his neck for almost two decades until his book of them finally published in 1914. He also wanted to retain control of the data he had recorded in his observing log books. Though Holden insisted, justifiably, that the log books belonged to Lick Observatory, Barnard went to great expense to hire a man to copy the notebooks for him, and had intended to check the copies by comparing the originals before leaving Mt. Hamilton. However, he ran out of time, and notified Holden just before his departure that he was taking the original notebooks as well as the copies "and will return them without expense to L.O. when through with them."[10] At some point he was "through with them," and returned them as promised to Lick Observatory, where they remain in the plate vault on Mt. Hamilton. As Lick Observatory support astronomer Tony Misch and I discovered in the 1990s, they are complete, and contain without any gaps the series of Mars observations from both 1892 and 1894. So what were the Mars drawings Mellish saw at Yerkes in the autumn of 1915?

This became a pressing question 50 years later, after the Mariner 4 flyby of Mars in July 1965 revealed craters on Mars. Mellish penned a letter to the popular magazine *Sky & Telescope*, recounting his own observation of Mars allegedly showing craters made with the 40-inch refractor, as well as the long-ago discussion with Barnard.[11] This letter generated considerable interest at the time, and apparently even led to an urgent NASA-inspired search of

Barnard's papers at Yerkes for his Mars drawings. None was found; presumably they had been lost or destroyed. This was seemingly the end of the story. Nonetheless, hope springs eternal, and in 1987, I initiated a new search for them, which was taken up by Yerkes photo archivist Richard Dreiser. By looking into areas of the Observatory that had not been looked into for years, he found at least twenty of Barnard's Mars drawings. Mostly they duplicate the ones in the log books on Mt. Hamilton, but are on unlined drawing paper (the log book pages are lined), and appear to have been actual drawings made at the telescope. Presumably, after making them, Barnard copied them (most of them) into the log book. Thus they can be regarded as authoritative. Unfortunately they show no craters—at least not in the sense of shadowed depressions along the terminator like the craters of the Moon.[12] Some of them do, nevertheless, show oval dark spots (in some cases in the positions of what are now known to be the calderas of Martian shield volcanoes), large dark and light circular patches, and an apparent crack (Valles Marineris) near Solis Lacus. Mellish, having only a short time to peruse them, and recollecting what he had been shown after many years, presumably worked up his memory into the rather fantastical descriptions he published.

Since these sketches were extraneous to the log books, Barnard never returned them, and seems to have planned to work them up for publication, eventually. Indeed, several carefully prepared drawings in which the disk of Mars is set against a black sky and the details on the disk have been represented with great skill, were found at Yerkes. They appear to be ready for publication but Barnard held back. Of course, he was often in poor health, and had other priorities—notably the *Atlas of the Milky Way*—to preoccupy him. But in addition, it may well be that he had lost the motivation to re-enter the fray about Mars. The comment that Mellish recalled—that others would "not believe him" and would only "make fun"—speaks volumes. Remember that even in 1916, many observers (including Lowell and E.C. Slipher) were cobwebbing the Martian disk with canals. Barnard must have realized that his drawings—for all the wonderful true natural detail they contained—would likely appear unforgivably bland, especially as they had been made with the then largest telescope in the world. But prestige among Mars observers, as historian Maria Lane has argued, "was derived by putting things *on* the map, not taking them off. Those who claimed to see a canal-free landscape on Mars did not even bother to produce or publish maps, as the reduction of detail was not considered a contribution of any importance."[13] Notably, Barnard never published the full set of his Mars drawings, and also never attempted to produce a map of Mars based on them.

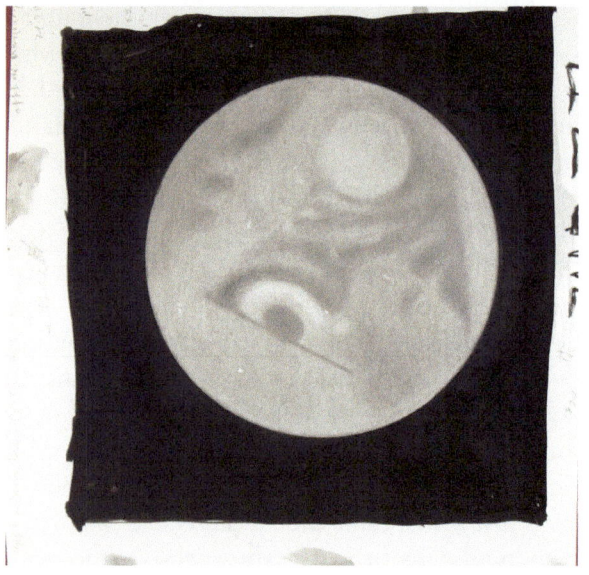

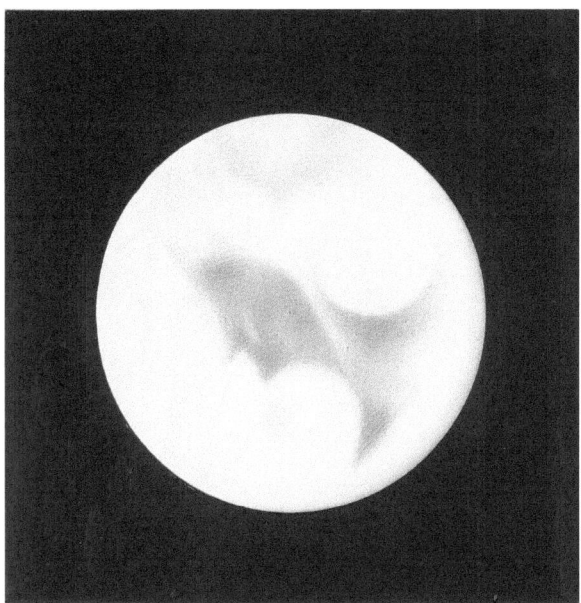

Stages of Mars's large regional dust storm, October 1894. Based on his sketches from September 23 (pre-dust storm), October 22, and October 28, 1894, Barnard had worked up finished versions for publication, though—for unknown reasons—they were never published. Indeed, they appear here for the first time.
In the October 22 view, the large circular area above center is Hellas, which has become yellowish; its outlines are affected in the October 28 view. In general, the markings appear feeble, as if (Barnard noted) the whole region were obscured in whitish haze. The dust storm was noted by several observers in Europe and England, but Barnard seems to have been the only American observer to record it. Lowell observed Mars at this time but crucially failed to interpret the changes as due to obscuration; instead, his interpretation turned on his vegetation theory. This would also be the case in 1909. (Credit: Photographs courtesy of Richard Dreiser of originals in Yerkes Observatory Archives, Hanna Holborn Gray Special Collections Research Center, University of Chicago Library)

Apart from the splendid photographic work done at Yerkes in 1909 and at Mt. Wilson in 1911, Barnard, in fact, did little further work on Mars. His enthusiasm perhaps fanned, briefly, by what Mellish had reported to him, he made one more drawing of the planet with the 40-inch refractor in February 1916—it appears to be the last he ever made. He shows the planet at the same time Lowell and E.C. Slipher were busy mapping the "canali novae." As usual, he seems to have thought of publishing it, since he worked it up for the purpose, but it never appeared. So ended Barnard's career as a Mars observer.

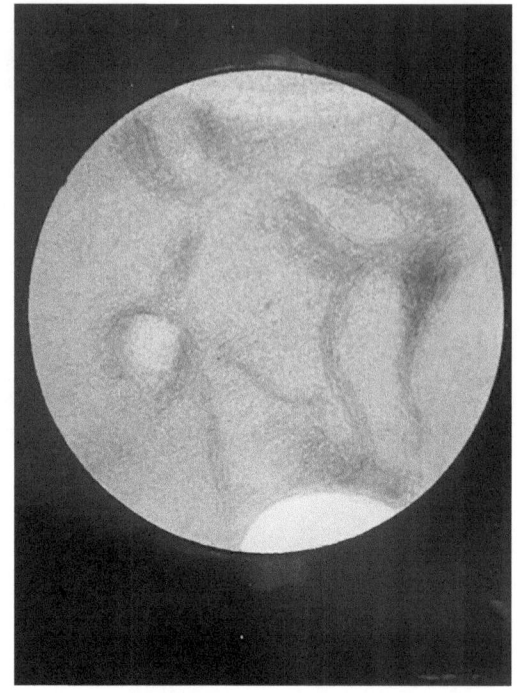

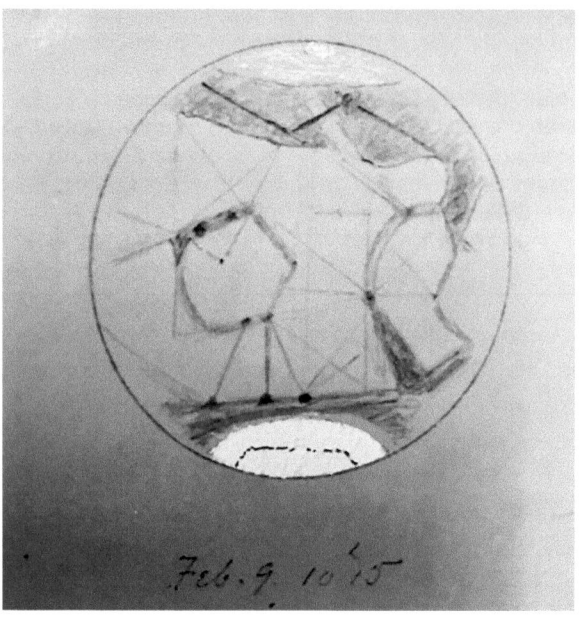

At above,one of Barnard's few finished drawings of Mars, made on February 9, 1916, with Mars at aphelic opposition. Syrtis Major is the narrow triangular marking toward the right; the Thoth "canal" appears to the left of it and is very broad as was the case at that epoch. The prominent bright patch is Elysium. Below, Percival Lowell's drawing made at almost the exact same moment with the Lowell 24-inch refractor. (Credit: left, photograph by Richard Dreiser, Yerkes Observatory, of an original now in the Barnard Papers, Hanna Holborn Gray Special Collections Research Center, University of Chicago Library; right, Lowell Observatory Archives)

Nevertheless, public interest in Mars remained very high, and he was often sought out by the general public and the press to give his latest views. Typical was his response to a Toronto *Star* reporter, given in March 1920, and perhaps his last word about the planet:

> There are a great many facts about Mars that have been discovered by the astronomers that will live, but there are some things unattainable. Some of the speculations at present circulating may be pretty near the truth, but you can rest assured there are many of them that are not worth a moment's consideration. If you want to make a statement, make it so that no one can disprove it, then it will stand. To countenance any other kind of statements is bad.[14]

A Vast Quantity of Routine Work

The publication of the drawings of Mars was given up not least because Barnard was now in failing health and slowing down with age—sixty was old then, especially for someone with diabetes. Also, however, he continued, against all advice, to overwork, and though he lacked the energy to tackle new programs of research, he added relentlessly to routine ones already underway. A great deal of his work with the 40-inch refractor involved making precise measurements with a filar micrometer. He carried out a great quantity of measurements of the asteroid Eros during its opposition and close approach to the Earth in 1900–1901.[15] He produced a trove of measurements—made possible only by the great light-grasp of the refractor combined with the remarkable sensitivity of his eye—of the satellites of Mars, the fifth and sixth satellites of Jupiter, the ninth satellite of Saturn, and the satellites of Uranus and Neptune. He also made many observations of double stars, variable stars and novae.

One of Barnard's most ambitious and time-consuming projects was a long series of visual micrometric observations of stars in globular clusters, begun in

1898. This work involved measuring and remeasuring no less than 247 individual stars in M13, the great globular in Hercules, alone. In all he obtained positions of 1363 individual stars in eighteen clusters. At first, as he later recalled, "I had formed … an entirely erroneous idea of their dimensions and of the sizes of the stars that compose them. They appeared to me as compressed groups of small suns that did not in any sense rank with the ordinary stars in the sky. Their distances from us, though great, were thought comparable with ordinary stellar distances. From these considerations I had reasonable hopes of detecting some relative motion of the individual stars in a few years' time from accurate micrometer measures" [14, p. 3]. However, to his "great regret and disappointment," the same observations repeated after 10 years showed no changes. Twenty years out, he finally admitted that these clusters "were at vaster distances from us and on a more magnificent scale than their apparent insignificance might imply."[16] Since his Milky Way photographs showed some of them superimposed on the great star clouds and therefore nearer than the star clouds themselves, this work fit in well with that of Harlow Shapley at Mt. Wilson who, using what he believed to be classic Cepheid variables, worked out the globulars' distances.[17] In 1918 Shapley announced that their typical distance was on the order of 50,000 light years. Moreover, since most of the globular clusters were located in the direction of Sagittarius, where Barnard had found them superimposed on the star clouds, Shapley concluded that the nucleus of the Galaxy was located in the center of this system of globulars, and that the Sun was located in the outskirts of a flattened disc of stars some 300,000 light years across.

Barnard's Star and An Eclipse

The wide-angle photography of the Milky Way had, of course, been the great astronomical passion of Barnard's life. Since 1889, when he had first photographed the Milky Way with the Willard lens, he had obtained (in addition to 500 plates exposed at Mt. Wilson with the Bruce in 1905) a total of 3500 plates of the Milky Way and other star fields. These plates were a vast repository of potentially valuable information, such as star proper motions and of fluctuations in the light of variable stars. One of them, exposed in January 1921, even registered a tiny speck, Pluto, which would remain camouflaged among the faint stars and undetected by astronomers for another 9 years.

Barnard's routine included the systematic examination of his plates with a blink comparator, of the same sort as the device being employed at Lowell Observatory in its search for Planet X. In May 1916, Barnard blinked two plates exposed on the same star field in Ophiuchus, taken 22 years apart: one was a recent exposure with the 6-inch Vögtlander lens of the Bruce photographic telescope, the other a plate taken the 6-inch Willard lens at Mt. Hamilton. He noticed a ninth-magnitude star that emphatically jumped between the two images. Now known to be a red dwarf, its proper motion of 10.39 arcseconds is the greatest of any star, due to its proximity to the Sun—at six light years away, it is the fourth-nearest-known star after only the alpha Centauri triple system. Barnard referred to it as "Gilpin," after a character whose horse runs away with him in a poem by the eighteenth-century poet William Cowper, but it has now received the official name "Barnard's Star." It has at least one planet, a super-Earth orbiting just beyond its habitable zone. It also has been used perhaps more than any other star as the habitat for science-fiction aliens, including in Douglas Adams's *Hitchhiker's Guide to the Galaxy*, where it serves as a way-station for interstellar travellers. (Unfortunately, Barnard's Star appears to be subject to intense flares, so any nearby planets are not likely to be hospitable to life-forms, whether permanent inhabitants or casual wayfarers.)

The year after discovering Barnard's Star, the Observatory celebrated his sixtieth birthday with a surprise party in his honor, and for the occasion tributes were received from his many friends. That of Walter Sydney Adams, the Mount Wilson spectroscopist who remembered him from his year of photographing the Milky Way with the Bruce telescope, was typical. Adams wished him:

> The heartiest congratulations on your attainment of the years of discretion. I know no one to whom the term sixty years young could apply more aptly for all of your friends know the fresh and youthful vigor of your heart and mind. Perhaps Methuselah at sixty might have equaled you but I can think of no other comparison.[18]

The Bruce Observatory, photographed through frosty foliage, in an undated photo probably by Barnard, ca. 1916. (Credit: University of Chicago Photographic Archive [apf6-00575], Hanna Holborn Gray Special Collections Research Center, University of Chicago Library)

Barnard at sixty. This image, taken in April 1917, later appeared as the frontispiece of the monumental Atlas of Selected Regions of the Milky Way. *At the time it was taken, Barnard was in failing health, but regularly traveling to Chicago to supervise the photographic reproductions of the images for the Atlas. (Credit: University of Chicago Photographic Archive [apf6-01279], Hanna Holborn Gray Special Collections Research Center, University of Chicago Library)*

But Adams's comments were based mostly on the distant recollections of the younger, more dynamic man he had known a decade earlier. Barnard was no longer that man. He still had some of the old enthusiasm. Mostly, however, sheer duty and the force of habit drove him.

In the fall of 1917, Barnard had a sufficient improvement in his health to be able join Frost in making a demanding two-week trip out west to scout three potential observing sites (Denver and Matheson, Colorado, and Green River, Wyoming) along the path of totality belonging to the forthcoming solar eclipse of June 8, 1918.[19] The two men were most impressed with conditions at Green River, a sparse settlement on the river of that name, overlooked by picturesque buttes. The most impressive was 700-foot high "Castle Rock," at which totality would last for 98 s. (It was, incidentally, the site from which Major John Wesley Powell and his party started his famous voyage down the Green, Little Colorado and Colorado Rivers to the Grand Canyon in 1869, and was also a favorite subject of the celebrated landscape painter Thomas Moran.) With his health still holding up into the spring, Barnard was also able to join the Yerkes advance party, consisting of himself, Calvert, the observatory carpenter, and the gardener. They reached Green River on May 4, 1918, and received help from prisoners from the local jail in transporting to a camp outside town their equipment including the Observatory's 12-inch "Kenwood" refractor, which had formerly belonged to Hale, and the coelostat apparatus he used at the Wadesboro eclipse in 1900.

May 4, 1918

May 7, 1918

May 21, 1918.
Snapshots from an album once belonging to Mary Calvert, showing Barnard making preparations at Green River, Wyoming, for the eclipse of June 8, 1918. (Credit: William Sheehan collection)

Preparations were intense, and excitement ran high. A week before the eclipse, Barnard and Calvert were joined by other Yerkes expedition members, including Frost, Parkhurst, and Storrs Barrett. While direct photography of the solar corona would be Barnard's domain, the others planned spectroscopic observations. An expedition from the University of Manitoba, and a Mt. Wilson group that included Hale and Ellerman, also took up positions on the site.

The day of the eclipse started out particularly fine, but then clouds came up after noon and reached the Sun by the time of first contact. During totality, the Sun stood behind a thin cloud that partially obscured the corona, only to depart 2 or 3 min later, leaving the Sun to a perfectly clear sky. Feeling sure that his photographs had been ruined by clouds—they would not be developed until he returned to Yerkes—Barnard returned to the hotel in Green River where he and other members of the expedition were staying when not in camp. Late that day he and a few volunteers from Green River took the Ford truck they were using back to the camp in order to pack up the plates he had exposed, but on returning again to the hotel that night his companions were mystified by his "incoherent exclamations of surprise." Above Castle

Rock toward the east, Barnard had noticed not far from Altair (one of the stars, with Vega and Deneb, forming the "Summer Triangle") a brilliant new star, somewhat yellowish in cast and which he estimated to be half a magnitude brighter than Altair. He returned to the hotel at 11:30 p.m., weary almost to the point of exhaustion, but with adrenalin flowing as he pointed out the new star to Frost and Hale, then preparing for bed. At once Barnard, Frost, and four or five others drove back to the camp to observe the nova, and Parkhurst succeeded in obtaining a spectrum. It later emerged that several other observers in Europe and the Eastern United States had discovered the nova independently—one of whom, Leslie Peltier of Delphos, Ohio, was a seventeen-year-old amateur who later became a noted discoverer of comets. But for that one night, in the often sad and disappointing twilight of the great man's life, Barnard became a Peter Pan-like figure, forever young, and once more filled with a child's wonder at the night sky.

Yerkes group in Camp at Green River, Wyoming, just after the eclipse of June 8, 1918. Among those pictured, from left to right: Gus Dahlstrom, Oscar E. Romare, John A. Parkhurst, George S. Isham, Mary Ross Calvert, Anna Greenleaf Parkhurst, Evelyn W. Wickham, Frances Lowater, Henry M. Foote, Edwin Brant Frost Barnard (in dark suit), Stross B. Barrett, and George Blakslee. (Credit: University of Chicago Photographic Archive [apf6-00787], Hanna Holborn Gray Special Collections Research Center, University of Chicago Library)

Barnard gauges sky conditions at Green River, Wyoming, in advance of the total solar eclipse of June 8, 1918. Castle Rock appears in the background. (Credit: William Sheehan Collection)

Barnard's Atlas of the Selected Regions of the Milky Way

The Green River adventure proved to be only a temporary rally. On Barnard's return, his health collapsed again. He was no longer able to travel much. He had to save up all his remaining strength to finish two great projects, projects that he had labored over for more than a decade.

The first was a monumental catalogue of the dark nebulae discovered on his Milky Way photographs. The second was an *Atlas of the Milky Way*, the publication of which was to be underwritten by a 1907 grant from the Carnegie Institution of Washington, but which he could not begin work on in earnest until a previous and long nagging commitment had been honored—the publication of the Milky Way and comet photographs taken with the Willard lens. He had first proposed publishing the Milky Way and comet photographs while nearing the end of his tenure at Lick, and failing to get financial support from Holden, raised all the money needed to publish by subscription. However, then he left for Yerkes, absorbed with new observing projects and failed to find any method of reproducing the plates that satisfied him. At one point, just to be rid of the whole project, he was ready to refund the subscribers. Various methods of reproducing the plates were tried—collotype, half-tones, photogravure—but none satisfied the unforgiving perfectionist. Finally, at the end of 1914, he managed to get the whole onerous thing off his hands—photogravure gave satisfactory (enough) results he decided, after all. Unfortunately, just as the *Milky Way and Comet Photographs* appeared, clearing the way for him to take on the long-contemplated *Atlas* of Bruce photographs, his health finally gave way.

Nevertheless, the *Atlas* was his priority, and slowly it began to crawl forward. In order to spare Barnard another such ordeal as he had experienced with the Milky Way and comet photographs, Frost convinced him that photographic prints would most faithfully reproduce the details of the original negatives. Thus, Barnard selected fifty of his best negatives (of which forty had been obtained at Mount Wilson in 1905 and the others at Yerkes). He then painstakingly produced a second negative of each, from which the prints were made. In this way, the contrast in faint regions was increased and details were brought out that otherwise would be lost. A Chicago firm of commercial photographers, A. Copelin & Son, was hired to make seven hundred prints from each of Barnard's second negatives. Each was carefully mounted on muslin. In May 1915—only 2 months after he returned from his vacation in California—Barnard made the first of many trips to Chicago so that he could personally supervise the work in progress and make sure that Copelin and his son Bert (who did most of the prints) adhered to his own high standards. The trips continued through 1916 and 1917. In the end, every one of the 35,700 prints used would be personally inspected and approved by Barnard, and hundreds, even thousands, would be tossed. Still, Barnard remained unsatisfied. "There were many cases," he wrote, "where a rejection [by the printers] was unfair. This was when the print was slightly too dark or too light, but the difference was not large. Such a print must be passed, though the desire was great to throw it out. In other words, it seemed impossible for the manipulator to attain to perfection in this work" [15].

In the end, it was not preparation of the plates that impeded progress. That work was finished by 1917. Instead it was the writing of descriptions of the photographs. Among Barnard's archives are found large numbers of drafts and waste papers. Some are little more than scraps with scattered notes. They suggest nothing so much as writer's block in the face of monumental self-demand. Clearly, Barnard knew that the *Atlas* was his masterpiece, and the burden of it crushed and terrified him. The inevitable result was that—like the apples of Tantalus—the goal, while seeming within reach, would always elude his grasp. He could not resist fretting over the prose, revising just a bit more, trying this phrase, amending it, crossing it out, and starting over again. As he slogged on, he no doubt was aware of racing against time—time that always wins in the end—and he knew that he would never finish.

Barnard had many discussions with Calvert and Frost about such matters as the form to be given to the *Atlas* as a whole. Should pages facing the photographs give pen-and-ink sketches of the fields and a system of coordinates by which the objects could be noted? In the end, this is what was done. Barnard's working title, *Atlas of the Milky Way*—suggesting that a large part of the Milky Way would be included—was scrapped (after Barnard's death). Instead, Frost changed it to the more accurate *Atlas of Selected Regions of the Milky Way*.

In part, Barnard's difficulty in putting his pen to paper during these years was owing to the fact that his own research into the nature of the dark markings of the Milky Way was still evolving. In 1907, he believed them to be actual vacancies. Later photographs—including those of the dark lanes in Taurus—made him increasingly confident that they might consist of some form of obscuring matter. Still he did not commit. At last, on "one beautiful transparent moonless night" in the summer of 1913, while photographing the southern Milky Way with the Bruce telescope, Barnard had a "Eureka!" moment, in which, in a flash, all was made plain:

> I was struck with the presence of a group of tiny cumulous clouds scattered over the rich star-clouds of Sagittarius. They were remarkable for their smallness and definite outlines—some not being larger than the moon. Against the bright background they appeared as conspicuous and black as drops of ink. They were in every way like the black spots shown on photographs of the Milky Way, some of which I was at that moment photographing. The phenomenon was impressive and full of suggestion. One could not resist the impression that many of the small spots in the Milky Way are due to a cause similar to that of the small black clouds mentioned above—that is, to more or less opaque masses between us and the Milky Way. I have never seen this peculiarity so strongly marked from clouds at night, because the clouds have always been too large to produce the effect. [16, p. 4]

Barnard finally realized the fact that he had long been groping toward: that the dark markings consisted of what we now know to be clouds of interstellar dust. He pushed on from this hard won insight, following with a series of important—and much-belabored—papers in the *Astrophysical Journal*. The list included: "Dark Regions in the Sky Suggesting an Obscuration of Light" (1913), "A Great Nebulous Region Near Omicron Persei" (1915), "Some of the Dark Markings in the Sky and What They Suggest" (1916), and "On the Dark Markings of the Sky with a Catalogue of 182 Such Objects" (1919).

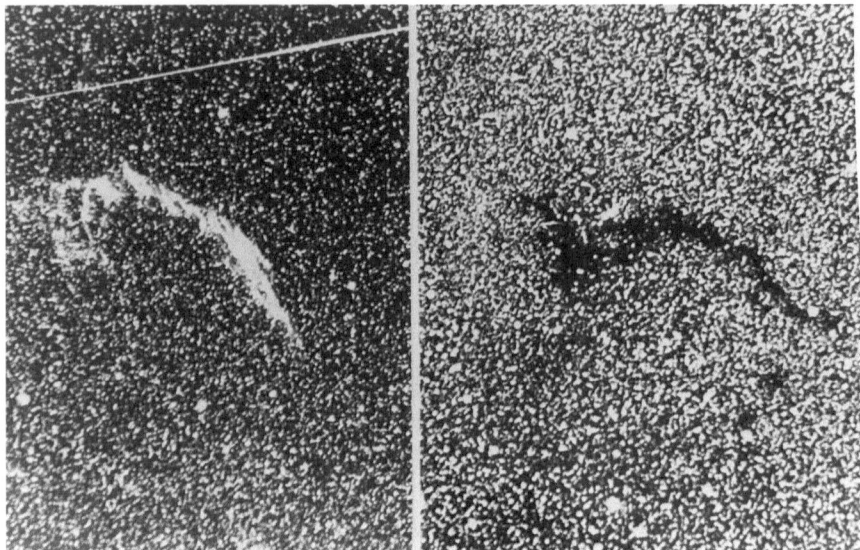

Photographs published by Barnard in the Astrophysical Journal *for January 1916, showing suggestively analogous shapes of bright and dark nebulae that he believed strengthened the case for the existence of dark nebulae. The bright nebula on the left is NGC 6995 in Cygnus, shown in this exposure of 5 h 43 min taken with the Bruce telescope on July 15, 1909; the dark nebula on the right is B150 in Cepheus, shown in an exposure of 6 h 2 min taken on October 1, 1910. The bright line near the top of the first image is a meteor trail. Barnard enhanced the weak image of the latter by making multiple printings with the position shifted slightly each time, which gives the illusion of a dense star field. From: E.E. Barnard (1916) "Some of the Dark Markings on the Sky and What They Suggest," Astrophysical Journal, vol. 43, pp. 1–8*

The 1919 paper included Barnard's first catalogue of "B" or "Barnard Objects." As soon as he completed it, he wrote to Robert Woodward, president of the Carnegie Institution, apologizing for the unconscionable delay in publishing the *Atlas*. Again, the demon of perfectionism possessed him:

> I have suffered greatly because of my inability to get the volume out. The fault lies entirely with me, but I am blameless so far as any intentional neglect is concerned. For some years I have been badly handicapped with the inability, or very great difficulty, to get anything like this volume finished. It has caused me great sorrow and endless worry.
>
> The prints are all finished and are here at the Yerkes Observatory…. The only thing lacking is to complete the descriptive matter…. I have endeavored to complete this descriptive part but have each time been disappointed with it, for my brain gets fagged quickly. I will go at it again and perhaps I can complete it. The Bruce photographic telescopic is out of commission now (at Warner and

Swasey's for changes) and perhaps this will give me less night work and maybe I can soon finish the MS [manuscript].[20]

In mentioning that the Bruce telescope was out of commission, he put his finger on one of the main reasons he failed to complete work on the *Atlas*. The night work had always given him the greatest pleasure, and he was loath to give it up—even in a good cause. According to Calvert:

> He looked forward to each night … with an eagerness that seemed never to be dulled. This work came ahead of everything else, and nothing was allowed to interfere with it. When his night at the telescope came he would have his early supper, be dressed for the night and at the observatory, often before the sun was down. There was no last-minute rush to the dome. [17]

In days when his health had been better, he had always been willing to endure the greatest extremes of temperature in the pursuit of his work. Visitors to the observatory expressed amazement that he continued to observe even when the temperature in the dome was ten or fifteen degrees below zero Fahrenheit. If asked how he kept warm, he replied: "We don't!" Eventually, realizing that Barnard would never abandon work solely with a thought to his own comfort, Frost established a rule that work must stop whenever the temperature in the dome dropped below −25°F—for the sake of the telescope! It also helped that Barnard was one of those rare individuals able to get by with only a few hours of sleep each night. After a night of clear skies and good seeing, he was often merry, as his sometime colleague Philip Fox recalled, "laughing and joking, or he might be heard singing in the dark room…. From his dark room also would come the sound of his voice reciting poetry and sometimes a song improvisation of such things as 'The Burial of Sir John Moore'" [6, p. 199]. Barnard did not allow clouds to deter him from his vigils; when the night was cloudy, he found it difficult to relax, and was constantly on the lookout for a possible clearing of the sky. Nevertheless, he could usually be counted upon to appear in his office by 7 a.m. the next morning.

Like Lowell, he did not believe in wasting time, and was a stickler for punctuality. According to Calvert, "He was always ready for any engagement he might have just on time or, more likely, a little ahead of time. He could never understand and was always irritated by a guest who, invited for a certain time, would come a little (or perhaps much) later than the time named. Probably his long work at the telescope was partly responsible for this habit of extreme punctuality. One cannot be late in making an observation of an eclipse or any such celestial phenomena."[21]

Barnard in winter, February 1920, wearing the tam-o'shanter and white kerchief he always wore in the Yerkes 40-inch dome on cold nights. Photograph by Dorothy K. Sullivan. (Credit: University of Chicago Photographic Archive [apf6-04179], Hanna Holborn Gray Special Collections Research Center, University of Chicago Library)

Barnard swimming in Lake Geneva. Photographed by his colleague Oliver J. Lee, ca. 1920. (Credit: University of Chicago Photographic Archive [apf6-00428r], Hanna Holborn Gray Special Collections Research Center, University of Chicago Library)

Barnard and Frost, undated photo but probably summer of 1920. This is the only photograph of Barnard known to show him with glasses and one of the rare ones from later years showing even a faint hint of a smile. (Credit: William Sheehan Collection)

In pensive mode, standing on the steps on the west side of the Observatory in an image from August 1920. The 40-inch dome is in background. (Credit: University of Chicago Photographic Archive [apf6-004461], Hanna Holborn Gray Special Collections Research Center, University of Chicago Library)

Yet the Steeds of Time galloped on apace for Barnard as they do for one and all. On March 11, 1921, he received word of the death of Burnham. At once he was swarmed by memories. Seeking catharsis, he wrote a heartfelt memoir of times shared with his old friend, including of their adventures together tramping the canyons around Mt. Hamilton [18]. As devastating as Burnham's loss was, an even greater followed on its heels. On May 16, Rhoda, in poor health for years, suffered a stroke; she passed away 5 days later. Her death came shortly after Albert Einstein visited the observatory, and was photographed with a group of astronomers in the dome of the 40-inch. Henceforth, in every photo of him (and there are no individual ones, only group staff photos), Barnard appears hollowed out, preoccupied, depressed.[22]

Barnard was in fact, quite understandably, beside himself with grief. His indispensable partner for 40 years was gone. He went to Nashville for the funeral, and saw her laid to rest next to his mother's grave. On his return to Williams Bay, he wrote to Frederick Slocum, who had been an assistant at Yerkes between 1909 and 1911 and was now at Wesleyan University in Middletown, Connecticut: "I am sad beyond measure. She loved me and cared for me more than I knew or could appreciate, but it all comes back to me now and I am heartbroken."[23] To William H. Wright of the Lick Observatory, he wrote, "I have felt so unhappy and broken in spirit… I am thinking of Mrs. Barnard all the time and it seems so cruel that she should be taken away."[24] Barnard had always dreamed that he and Rhoda would 1 day return to California and build a home among a grove of orange trees, but now he knew it was not to be. "That hope is all gone now," he wrote to Campbell, "and I could not think of the Ranch without her."[25]

The gigantic "Anteater" prominence, photographed at the total solar eclipse of May 29, 1919, from Sobral, Brazil, by Charles Davidson and A.C.D. Crommelin of the Royal Greenwich Observatory. Davidson and Crommelin were members of one of two expeditions—the other led by Arthur Eddington of Cambridge to Principe, an island off the coast of West Africa—to measure the positions of background stars near the Sun, in order to test the predictions of Einstein's General Theory of Relativity. Barnard was aware of the predictions; nevertheless, he at first suspected that the bending of starlight by the atmosphere of the Sun might explain the findings. By the time Einstein visited Yerkes in May 1921, they had generally been accepted as conclusive. (From: Splendour of the Heavens, T.E.R. Phillips and W.H. Steavenson (eds.), vol. 1 (London, 1923))

A day to be remembered: Albert Einstein's visit to Yerkes Observatory, May 6, 1921. Photographed with the 40-inch refractor are, from left: Storrs B. Barrett, Lela D. Cable, John A. Parkhurst, Elsie E. Johnson, Dorothy W. Block, Florence Baldwin Lee, Dorothy K. Sullivan, Barnard, Arthur J. Dempster, Harriet M. Parsons, A.C. Lunn, Edwin Brant Frost, Einstein, Alexander Pflüger, S. Ginsberg, Oliver J. Lee, Esther L. Searles, Mary R. Calvert, George Van Biesbroeck, and George C. Blakslee. (Credit: University of Chicago Photographic Archive [apf6-00415], Hanna Holborn Gray Special Collections Research Center, University of Chicago Library)

Barnard's beloved Rhoda. Her death on May 25, 1921, after 40 years of marriage, struck a blow from which her husband never recovered. (Credit: University of Chicago Photographic Archive [apf6-04638], Hanna Holborn Gray Special Collections Research Center, University of Chicago Library)

Even observing no longer gave him much pleasure, but he continued through habit and a sense of duty. He revisited the phenomena of the edge-wise ring of Saturn from November 1920 to August 1921; the results largely corroborated those of 1907–08. Also—and touchingly—he obtained accurate positions of a recently discovered asteroid found by Max Wolf, which Wolf had named 907 (Rhoda), in honor of Barnard's late wife.

In the summer of 1921, Barnard apologized to Slocum for not coming east to attend the dedication of Wesleyan University's new Van Vleck Observatory. "There are two things that prevent my coming," he explained. "I am not able to stand the long railway journey. I am very busy trying to get things finished up before it is too late—I am no Mathusela [sic]."[26]

The *Atlas* still loomed, in spite of depression and a general loss of interest in life. He told the great double star observer Robert G. Aitken of Lick that without Rhoda, "It [is] a very lonely world, and … he hoped he would not have to stay in it much longer."[27] To Philip Fox he sighed: "Oh! I am so terribly lonely."[28]

During the summer of 1922, S.A. Mitchell, who long before had gone with Barnard on the ill-fated eclipse expedition to Sumatra, stayed in the home on Lake Geneva with Barnard and Calvert. "He was then showing the effects of his hard night work," Mitchell observed. And yet—

> As I had known him so long and so intimately, I, myself, having grown older and being the director of an observatory … tried to show him how foolish was his habit of staying up all night, when he was assigned to the forty-inch telescope, even though it might be pouring rain. His reply was characteristic of him: "I have never yet shirked my duty and been found wanting, and I shall not begin now." [19]

> Barnard turned down an invitation from Wright come to California. "I am tied here by work," the weary astronomer wrote, "that has been delayed for years. Until that is done, and I hope it may soon be finished, I cannot get away. I am afraid also the journey would be too much for me. I have not been well for a long time. I feel lonely and discouraged."[29]

The plan for the *Atlas* he had discussed with Frost and Calvert of including, in addition to the photographs, pen-and-ink sketches and a system of co-ordinates, was put in motion. Realizing that many of the descriptions, often produced hastily on scraps over a period of 10 years, did not reflect his changing opinions about the nature of the dark objects, Barnard hoped that Calvert and Frost, as his designated literary executors, would include generous excerpts from his *Astrophysical Journal* articles. But Frost, partly because of limitations of space but also because he realized that such revisions would be very soon out of date, rejected the idea. Any attempt to revise the work would render the *Atlas* less, not more, of a classic. This was no doubt true: ultimately, the monumental achievement lay in the photographs, not the text.

In addition to the 182 B objects Barnard included in his 1919 *Astrophysical Journal* paper, Barnard left a supplementary list, from which Calvert determined the positions and assigned additional numbers up to B 349.[30]

The End

Barnard made his last visit to Chicago for a meeting of a local astronomical club on December 16, 1922—his sixty-fifth birthday. On December 21, he attempted to examine the expanding nebulosity around Nova Persei (1901), which he had first detected with the 40-inch refractor in December 1916. Others had searched for it, some using the Mt. Wilson 60-inch reflector, without success. Barnard evidently had confidence that his vision was still good enough for him to be successful in his quest—indeed, I have seen only one photograph of him wearing glasses, and they were clearly needed only for near-sighted work. On this occasion, however, he was unable to get the image in focus, and for a moment was alarmed. Yerkes already had one blind astronomer—Frost; they hardly needed another. At once he turned the telescope downward in order to inspect the object glass; with considerable relief, he saw that it had become completely covered with frost. It was not his eyesight after all.[31]

One of the last photographs—perhaps the last—ever taken of Barnard, showing the Yerkes staff on the stairway leading up to the 40-inch refractor, taken in 1922. From left (top row): Storrs B. Barrett, Otto Struve, Christian T. Elvey, Mary Lanning, P.B. Jenkins, Frank R. Sullivan, Lloyd R. Wylie; (middle row): Karl B. Patterson, Oliver J. Lee, Lela D. Cable, Alice Hall Farnsworth, Marry Ross Calvert, John A. Parkhurst, George Van Biesbroeck; (front row): Marguerite Van Biesbroeck, Dorothy K. Sullivan, Edwin Brant Frost, Barnard, Harriet Bigelow, Florence Baldwin Lee. (Credit: University of Chicago Photographic Archive [apf6-04218], Hanna Holborn Gray Special Collections Research Center, University of Chicago Library)

The next night Barnard was assigned to the 40-inch was December 23; the sky was overcast. On Christmas morning, Barnard was at Frost's house with the rest of the staff for about an hour but excused himself early, due to distress from an acute inflammation of the bladder. The next day Frost, himself losing the sight in one of his eyes after a retinal tear while doing spectrographic work with the 40-inch, drove to Milwaukee to consult his eye specialist. When he returned he found that Barnard had deteriorated and had taken to his bed. A physician was called to relieve Barnard's distress from the bladder inflammation by means of a catheter, but recognized that the real problem was an enlarged prostate, for which the remedy would be surgery. In Barnard's case, because of the diabetes, surgery was felt to be too risky. The catheter treatment did relieve the bladder inflammation for the time being. Frost informed Hale optimistically that Barnard was now "getting sleep and rest and [dressing] to go down to breakfast, and [doing] quite a little writing for his *Atlas*."[32]

Barnard was now a dying man. In addition to the diabetes and enlargement of the prostate, he was suffering from congestive heart failure. Nevertheless, he was strong enough to observe through his bedroom window an occultation of Venus by the Moon on the morning of January 13, 1923, and, hoping for a rally, he and Mary Calvert devoted a few hours every afternoon to the *Atlas*. He also tried to keep up his correspondence. "I am still sick in bed," he wrote to Campbell the day after the occultation, "but hope to be out before long. It is slow and tedious. My greatest unhappiness [is] that I can do no observing at all."[33] A week later he wrote to W.H. Wright: "I am sitting up this morning…. My strength is returning but my main trouble is unchanged. I had hoped for a surgical operation [but the] physicians do not advise it under the circumstances. In the meantime I am subject to the catheter with the forlorn hope that things will come all right again…. It is distressing to be away from my observations."[34]

At the beginning of February, a Chicago specialist in metabolic diseases, Dr. Rollin Woodyatt, was called out to Williams Bay. After examining Barnard he expressed the hope that he could be taken to Presbyterian Hospital in Chicago, placed on a strict diet—Woodyatt himself had recently introduced the so-called "ketogenic diet" for such patients—and receive the then new treatment for diabetes, the pancreatic extract insulin. Its life-saving effects on diabetics had been demonstrated only the year before by Frederick Banting and Charles Best at the Toronto General Hospital.

It was too late. On Saturday February 3, Frost and Ambrose Swasey (of the firm of Warner and Swasey), who had rushed out from Cleveland, called on Barnard. He was dozing, but soon awakened and discussed old times with Swasey. Imagining that things were stable, Frost went to Milwaukee in order to give some lectures. By the time he got back on Monday, Barnard had taken a turn for the worse. He knew the end was approaching, and told his niece that "he didn't mind dying, but was sorry not to finish his work."[35] After about noon on Monday, February 5, as he fell into a coma, members of the Yerkes staff and his doctors gathered at the bedside and waited as he was given the new insulin treatment. It was no use. His pulse ceased at 8 p.m. Frost, recording the time, added: "*ad astra*"—to the stars. After lying in state in the rotunda at Yerkes (Fox noted that he appeared in death "pitifully wasted") he set out on his last journey, by rail to Nashville, where he was buried beside his wife and mother.[36]

The house on the shore of Lake Geneva where the Barnards had lived since the Yerkes Observatory opened in 1897. Barnard made his last astronomical observation, of an occultation of Venus by the Moon, from the upstairs bedroom on January 13, 1923, and died there on February 6, 1923. (Credit: University of Chicago Photographic Archive [apf6-04601], Hanna Holborn Gray Special Collections Research Center, University of Chicago Library)

Edward Emerson Barnard was gone. His legacy—including his unfinished work—remained.

Of the unfinished work, the most important was, of course, the monumental *Atlas of Selected Regions of the Milky Way*, which Calvert and Frost saw into publication in 1927. The descriptions were either written in their entirety by Barnard, or cobbled from his notes. They bear the unmistakable impress of his personality and evoke his intimate knowledge of the sky. Thus, of the region of the great nebula of Rho Ophiuchi (Plate 13), he wrote:

> The region of Rho Ophiuchi is one of the most extraordinary in the sky. The nebula itself is a beautiful object. With its outlying connections and the dark spot in which it is placed and the vacant lanes running to the east from it, it makes a picture almost unequalled in interest in the entire heavens…. It is clear, from an inspection of the picture, that the actual background of the sky here consists of a uniform distribution of faint stars. If part of the picture is covered so as to hide the large nebula and the dark lanes, this fact becomes more apparent. The conclusion is therefore irresistible that there is no real vacancy at this point, but that the nebula is between us and the background of stars and blots out the more distant ones.[37]

Regarding the great star clouds in Sagittarius (Plate 26):

> These great clouds were among the first portions of the Milky Way to be photographed by the writer with the Willard lens at the Lick Observatory, in the year 1889. They are the most magnificent of the galactic clouds visible from this latitude….
>
> These great broken clouds, so beautiful and bright to the eye and on the photograph, rapidly thin out to the south and east, the stars being fewer and fainter. This gives, especially to the south, the impression of greater distance… A striking feature of one of these star masses on the western edge of the clouds, projected against a region of few stars, is the crude form of a beast, with round head, nose, mouth, ears, and great staring eyes….[38]

The direct photographic plates included in the *Atlas* are not only scientifically valuable but true works of art that, in their way, will never be surpassed. Though the Milky Way has been imaged using color films and CCD, Barnard's images are somehow *sui generis*. Quite apart from the majesty of their subject-matter, they attest to the unique force of his personality, the personal vision of the man who made them. They are to the Milky Way what Matthew Brady's portraits are to Abraham Lincoln or Ansel Adams's landscapes to Half Dome and El Capitan.

One of his acquaintances from the early days in Nashville, Alfred E. Howell, recalled that Barnard, with his enormous drive and his unquenchable passion for the night sky, had an "immortal fire within himself."[39] In the *Atlas of the Selected Regions of the Milky Way*, the reflected light of that immortal fire shines still.

Notes

1. See J.C. Brandt [2].
2. E.E. Barnard, observing log book. Barnard. Edward Emerson. Papers, Hannah Holborn Gray Special Collections Research Center, University of Chicago Library.
3. E.E. Barnad [4, p. 375]. Showing how valuable they were, Barnard's visual magnitude estimates of Halley's comet in 1909–11 were later used to accurately forecast Halley's brightness at its next return in 1985–6. To be useful, Barnard's actual estimates had first to undergo "aperture correction," to correct for the fact that, paradoxically, the larger the telescope used to estimate the magnitude, the fainter the observer's estimate is likely to be. See Joseph N. Marcus (1983) *Comet News Service*, nos. 83–84, pp. 5–8. The comet's brightness, with the help of Barnard's estimates, was found to be about four times brighter than the International Halley Watch had been forecasting in the early 1980s. See: Neil Divine [5].
4. Quoted in Frank N. Allan [7, p. 268].
5. Ibid., p. 10.
6. Quoted in Gale E. Christianson [9].
7. The variability of R Monoceros had been discovered by Julius Schmidt of the Athens Observatory in 1861, but Hubble was the first to show the variability of the nebula.
8. John E. Mellish to Walter Leight Jr., January 18, 1935. Quoted by Rodger W. Gordon [12].
9. Ibid.
10. E.E. Barnard to Edward S. Holden, October 1, 1895; Mary Lea Shane Archives of the Lick Observatory.
11. John E. Mellish (1966) Letter, *Sky & Telescope*, vol. 31, p. 339.
12. The calderas on the summits of the shield volcanoes Arsia Mons and Olympus Mons are shown as darkish patches, by contrast to the lower elevations which were covered by cloud at the time; these darkish patches as depicted by Barnard may have been interpreted by Mellish as "craters."
13. Lane, *Geographies of Mars*, p. 456.
14. E.E. Barnard, Toronto *Daily Star*, March 24, 1920.
15. E.E. Barnard [13]. As an example of the care Barnard took in his work, he took pains to look into the possible need to apply a correction to the micrometer screw for temperature. He decided that no correction was needed because both screw and tube of the telescope are of steel, and so shared the same coefficient of

expansion, and because they acted in opposite directions they would mutually cancel each other. This would be thoroughly established by a series of observations of the difference in declination of the Pleiades stars Atlas and Pleione. Beginning in 1897, he carried out observations on 506 nights over a period of 25 years, and found that despite a temperature variation from -25°F (-32 °C) to +100°F (+38 °C), the only change needing to be taken into account was that of the focal length of the lens. From summer to winter, the action of cold on the lens shortened the focus by some 0.3 inches more than the shortening of the tube, and that this was the only correction due to temperature that affected the measurements.

16. Ibid., p. 1. A younger colleague of Barnard, Frank Schlesinger, who had spent the summer of 1898 as a volunteer at Yerkes with George Ellery Hale and was on the Yerkes staff from 1899 to 1903, became interested as a postgraduate in measuring star positions on photographic plates, and went on to a distinguished career at Allegheny and Yale in which he applied these methods to the determination of 2000 stellar parallaxes. As early as 1903 he showed that it was not difficult to measure the positions of stars in globular clusters on Ritchey's plates with smaller errors (about a third) of those of Barnard's visual micrometer measures. Nevertheless, Barnard was unmoved, and continued to make his visual measures with a filar micrometer. It seems likely that invested more time and effort in this work than any other he undertook, but the negative results—though eventually he drew the correct inference from them, namely, that the globulars were much more remote than hitherto thought—were undoubtedly a great disappointment to him. See Sheehan, *The Immortal Fire Within*, p. 387.

17. The variables studied by Shapley are now known to be RR Lyrae and type II Cepheids rather than the classic Cepheids such as the famous V1 (Hubble variable number 1) used a few years later by Hubble to determine the distance to the Andromeda Spiral.

18. Walter Sydney Adams to E.E. Barnard, December 9, 1917. Barnard. Edward Emerson. Papers, Special Collections and University Archives, Jean and Alexander Heard Library, Vanderbilt University.

19. For details regarding the 1918 eclipse expedition to Green River, Wyoming, see Sheehan, *Immortal Fire Within*, pp. 403–407. The Yerkes Director, Frost, was by then beginning to lose his eyesight. He had been severely myopic as a child, and on the cold winter night of December 15, 1915, while working alone at the 40-in. telescope, he suffered a retinal tear in his right eye which within a year led to the total loss of its vision. The other eye was affected by a cataract, and several years later would also suffer a hemorrhage, leaving him unable to read and hardly able to see for ordinary purposes. Being blind is for an astronomer certainly a greater handicap than it is in most professions, and the later years of Frost's directorship (before his retirement in 1931; there were generally no pensions or Social Security in those days) were lackluster, as told in Osterbrock, *Yerkes Observatory*, pp. 47–76.

20. E.E. Barnard to R.S. Woodward, May 23, 1919. Barnard. Edward Emerson. Papers, Special Collections and University Archives, Jean and Alexander Heard Library, Vanderbilt University.

21. Calvert, "Some Personal Reminiscences," p. 28.

22. Barnard knew of Einstein's predictions based on the General Theory of Relativity of the bending of starlight near the Sun, and had been queried on several occasions to look through his eclipse plates to see if there were any suitable for testing the theory. Unfortunately, none were suitable. He was also aware of plans to test the result at the total eclipse of May 29, 1919, when Cambridge University astronomers Arthur Eddington and Edwin Turner Cottingham headed to the West African island of Principe and Greenwich Observatory astronomers A.C.D. Crommelin and Charles Rundle Davidson to the Brazilian town of Sobral to photograph the eclipse, which happened to be especially favorable for the purpose since it occurred among the rich Hyades cluster of stars. Barnard was doubtful that a reliable result could be obtained so close to the Sun since the star rays had to pass through the densest part of the corona, whose refractive index was unknown. In the event, the measurements on the plates agreed with Einstein (rather than Newton whose theory also predicted the bending of starlight but by a different amount), and were hailed as a great triumph for Einstein's theory. Two years after the eclipse, Barnard answered a letter about Einstein's theory from a Reverend W.E. Glanville of Baltimore, who had been puzzled by an article on the subject. Barnard admitted that "the results of the eclipse … seem greatly in its favor," though he added that as for the theory itself, he was as mystified as most laymen. "If light is ponderable then it must be subject to acceleration, I should think." But he added: "It has been stated that there are only twelve men in the world who understood Einstein. I am not the thirteenth." E.E. Barnard to W.E. Glanville, June 15, 1921. Barnard. Edward Emerson. Papers, Special Collections and University Archives, Jean and Alexander Heard Library, Vanderbilt University.

23. E.E. Barnard to F. Slocum, June 10, 1921. Slocum. Frederick. Papers, Van Vleck Observatory, Special Collections and Archives, Wesleyan University.

24. E.E. Barnard to W. H. Wright, June 16, 1921. Wright. William Hammond. Papers, Mary Lea Shane Archives of the Lick Observatory, University of California, Santa Cruz.

25. E.E. Barnard to W.W. Campbell, June 16, 1921. Campbell. William Wallace. Papers, Mary Lea Shane Archives of the Lick Observatory.

26. E.E. Barnard to F. Slocum, August 16, 1921. Slocum. Frederick. Papers, Van Vleck Observatory, Special Collections and Archives, Wesleyan University. The misspelling is Barnard's.

27. R.G. Aitken to E.B. Frost, February 7, 1923. Frost. Edwin Brant. Papers, Hannah Holborn Gray Special Collections Research Center, University of Chicago Library.

28. Fox, "Edward Emerson Barnard," p. 196.

29. E.E. Barnard to W.H. Wright, November 3, 1922. Wright. William Hammond. Papers, Mary Lea Shane Archives of the Lick Observatory, University of California, Santa Cruz.
30. Numbers 175 to 200 were missing from Mary Calvert's final compilation and thus do not appear in the published book, doubtless because Barnard's indications for them were simply too sketchy for her to work them out. At last in 2022, Gerald O. Dobek of Northwestern Michigan College finally managed to identify them and included them in: Tim Hunter et al. [20].
31. Had it been so, Barnard would however have belonged to the select club of outstanding planetary observers who became blind late in life, which included Galileo, Cassini, and Schiaparelli.
32. E.B. Frost to G.E. Hale, February 8, 1923. Frost. Edwin Brant. Papers, Hannah Holborn Gray Special Collections Research Center, University of Chicago Library.
33. E.E. Barnard to W.W. Campbell, January 14, 1923. Campbell. William Wallace. Papers, Mary Lea Shane Archives of the Lick Observatory, University of California, Santa Cruz.
34. E.E. Barnard to W.H. Wright, January 21, 1923. Wright. William Hammond. Papers, Mary Lea Shane Archives of the Lick Observatory, University of California, Santa Cruz.
35. E.B. Frost to R.G. Aitkin, February 19, 1923. Frost. Edwin Brant. Papers, Hannah Holborn Gray Special Collections Research Center, University of Chicago Library.
36. Fox, "Edward Emerson Barnard," p. 195.
37. Barnard, *Atlas of Selected Regions*, p. 116.
38. Barnard, *Atlas of Selected Regions*, p. 194.
39. Alfred E. Howell, quoted in Sheehan, *The Immortal Fire Within*, p. 417.

References

1. Barnard EE (1908) Comet c 1908 (Morehouse). Astrophys J 28:292–299
2. Brandt JC (1968) The physics of comet tails. Annu Rev Astron Astrophys 6:267–286
3. Morehouse DW (1928) "Reminiscences of Edward Emerson Barnard," Edward Emerson Barnard Memorial Number. J Tenn Assoc Sci 3(1):21
4. Barnard EE (1910) Visual observations of Halley's comet in 1910. Astrophys J 39:373–404
5. Divine N et al (1986) The comet Halley dust and gas environment. Space Sci Rev 43:1–104
6. Fox P (1923) Edward Emerson Barnard. Pop Astron 31:195–201
7. Allan FN (1972) Diabetes before and after insulin. Med Hist 14:266–273
8. Frost EB (1926) Edward Emerson Barnard. Biogr Mem Natl Acad Sci 21:10–11

9. Christianson GE (1995) Edwin Hubble: mariner of the nebulae. Farrar, Straus and Giroux, New York, p 64

10. Osterbrock DE (1997) Yerkes observatory 1892–1950: the birth, near death, and resurrection of a scientific institution. University of Chicago Press, Chicago/London, p 73

11. Sheehan W, Conselice CJ (2015) Galactic encounters: our majestic and evolving star-system, from the big bang to time's end. Springer, New York, p 237

12. Gordon RW (1975) Mellish and Barnard – they did see Martian craters! Stroll Astron 25:196

13. Barnard EE (1902) Micrometrical observations of Eros made with the forty-inch refractor of the Yerkes observatory during the opposition of 1900–1901. In: The decennial publications of the University of Chicago, vol 8. University of Chicago Press, Chicago

14. Barnard EE (1931) Micrometric measures of star clusters. In: Frost EB, Van Biesbroeck G, Calvert MR (eds) Publications of the Yerkes observatory, vol 6. University of Chicago, Chicago, pp 1–106

15. Barnard EE (1927) Atlas of selected regions of the milky way (Frost EB, Calvert MR, editors). Carnegie Institution, Washington, DC, p 12

16. Barnard EE (1916) Some of the dark markings on the sky and what they suggest. Astrophys J 43:1–8

17. Calvert MR (1928) "Some Personal Reminiscences," Edward Emerson Barnard memorial number. J Tenn Assoc Sci 3(1):28

18. Barnard EE (1921) Sherburne Wesley Burnham. Pop Astron 29:309–325

19. Mitchell SA (1928) "Barnard at Yerkes observatory and at the Sumatra Eclipse," Edward Emerson Barnard memorial number. J Tenn Assoc Sci 3(1):27

20. Hunter T, Dobek G, McGaha J (2023) The Barnard objects – then and now. Springer, Cham

14

Lowell and Barnard Compared

Contents

… [M]y design is not to write histories, but lives. And the most glorious exploits do not always furnish us with the clearest discoveries of virtue or vice in men; sometimes a matter of less moment, an expression or a jest, informs us better of their characters and inclinations, than the most famous sieges, the greatest armaments, or the bloodiest battles whatsoever. Therefore as portrait-painters are more exact in the lines and features of the face, in which the character is seen, than in the other parts of the body, so I must be allowed to give my more particular attention to the marks and indications of the souls of men, and while I endeavor by these to portray their lives, may be free to leave more weighty matters and great battles to be treated of by others.
—Plutarch, "Life of Alexander"

Inspired in part by Plutarch, I have chosen to write the parallel lives of Percival Lowell and Edward Emerson Barnard partly because they were among the most famous astronomers of their time, at least in the U.S., and made many contributions to their fields. But not entirely for that reason. They also had remarkably interesting personalities and lives, and each becomes more clearly defined when set against each other, as a kind of background to the other, a background by which one can see more clearly the figure silhouetted against it.

© The Author(s), under exclusive license to Springer Nature Switzerland AG 2024
W. Sheehan, *Parallel Lives of Astronomers*, Springer Biographies,
https://doi.org/10.1007/978-3-031-68800-3_14

As has been known since at least the time of Ernst Mach, without contrast we see nothing; the eye is not interested in absolute levels of light intensity, but in boundaries between adjacent areas of differing intensity. Contrast exaggerates boundaries, reveals the line adjoining (or separating) them. The contrast between any two men shows up sometimes as soft blurred boundaries, sometimes as hard sharp lines, just as the contrast between different shaded regions on a planetary surface does—the latter appearing, at times, as the famous canals of Mars. The way that Lowell and Barnard perceived (or did not perceive) illustrates well their differing characters.

Percival Lowell came from one of the most highly accomplished and wealthiest families in New England. The members of his family were variously referred to as "economic royalists" or "robber barons," and typically consolidated wealth (and perhaps talent, which can also run rather true along certain lines) through the intermarriage of cousins or through combinations involving sons and daughters of business partners. On both his father's and mother's side the family name was given to textile-manufacturing cities (Lowell and Lawrence, Massachusetts). His father was a hard driving, rigid businessman, committed to conventional male ego ideals, who ran textile mills and philanthropic causes with equal efficiency. Percy would always have difficulty measuring up to his father's impossibly high standards. His mother, on the other hand, was nurturing, artistic, and an invalid. She was completely and unqualifiedly devoted to her "darling Percy." Perhaps his arrogance and sense of being special owed much more to her than to his father, for, as Freud said, "If a man has been his mother's undisputed darling, he retains throughout life triumphant feelings" [1]. And what happens when his triumph is contested? There, perhaps, lie some of the roots of this breakdowns.

"Darling Percy" enjoyed the best education money—and connections—could supply, including a school in France where he learned to speak and write French fluently, Nobles school, and Harvard, where so many of his ancestors had gone and where he achieved the usual family distinctions in literary composition and math. His approach to everything was, like that of the mathematics he learned from Benjamin Peirce, deductive, "the science that draws necessary conclusions." This was to be Lowell's lifelong approach to every study he undertook—to work from the top down, from certain "self-evident" axioms (or should we call them "hunches") to necessary conclusions, conclusions to which the observations had to be shaped and molded. He was Keats's honeybee, "buzzing here and there impatiently from a knowledge of what was to be arrived it." In Japan, Lowell's friends recognized—at first perhaps with envy, though later they recoiled—his top-down, confident, then increasingly dogmatic, approach. Thus, Basil Chamberlain, at first his mentor in things Japanese but later his critic, said that Lowell's approach was to "argue down deductively from some general notion … and then bend the facts to suit the preconceived idea, seasoning the whole with verbal fireworks."[1] Similarly, Andrew Ellicott Douglass, who went from being Lowell's loyal assistant to disillusioned apostate (whereupon he soon found himself unemployed), saw Lowell as essentially a literary not a scientific man who "devotes his energy to hunting up a few facts in support of some speculation instead of perseveringly hunting innumerable facts and then limiting himself to publishing the unavoidable conclusions, as all scientists of good standing do."[2]

In contrast, Barnard grew up without wealth or family connections or formal education, in Nashville, Tennessee, a city ravaged by the Civil War (and cholera) and under Union occupation for much of his childhood and adolescence. He was completely without advantages. His father died before he was born; his mother was an invalid during much of his life, though without any of the helps that Lowell's invalid mother had; his older brother was feebleminded. His only recourse was to work in some menial job from an early age in order to support the family. Here, ironically, he did benefit from the rare family connection: his mother had come from Cincinnati, and had been acquainted with John van Stavoren, a photographer, there. Through sheer luck she recognized him from an advertisement, and saw that he had bought a photograph gallery. Thus, she came to recommend Edward, aged nine, for a lowly job serving as a human clock drive to keep van Stavoren's large photographic camera (called "Jupiter") focused on the Sun, in order to shorten the time exposures needed for portraits. Though other boys had been tried in the job and fallen asleep, Edward impressed everyone—by managing to stay awake. Thus he began to climb upward from the lowest rung of the stepladder whose legs were firmly planted in the mudsill occupied by poor Southern whites (just above the level of recently freed Negro slaves) at the time. Working upward by degrees required an approach that was basically trial and error—learning from observation and experience—and a little like the method of induction that Francis Bacon (rather naively) suggested: rules of reasoning were to be based not on dogmas and deductions but on observation and experiment; and by this means "a candle [lit] showing the way … commencing as it does with experience duly ordered and digested, not bungling nor erratic, and from it deducing axioms, and from established axioms again new experiments" [2].

Though Barnard experienced a great deal of hardship and adversity, it did not embitter him, as it might well have done. Sir Walter Scott once wrote of Jonathan Swift, "Poverty, and the sense of the contempt which accompanies it, gives to a lofty temper a cast of recklessness and desperation." Barnard certainly had a lofty temper. He was intelligent and curious about the world around him, and noticed comets and stars from an early age (but had no one to teach him their names). All things being equal, his background might have caused a chip to rest on his shoulder throughout his life. That it did not do so suggests strongly that, in this case as in others, early deprivation is not a sufficient explanation of adult character. Against his deprivations, he had support from crucial people at crucial times. His mother taught him to read and provided him with some of the rudiments of pictorial art (though he remained ignorant of math). She helped him find his first job, in the Photograph Gallery. But for that his enormous potential might have "blushed unseen, and wasted its fragrance on the desert air." Remember, in the nineteenth century, both in England and the United States, child labor was the rule not the exception; many more children were employed in soul-destroying jobs in textile mills (like those owned by the Lowells), factories, mines, and farms than in places with culture and interesting people to meet like the Photograph Gallery. Barnard was lucky in that his first workshop was under the blue sky, and he was nurtured and mentored by a number of kind and gifted adults: Joseph Carels, the postmaster who encouraged him on his lonely walks from the Photograph Gallery to Varmint Town, the colleagues from whom he learned indispensable skills in photography later put to such excellent use. There was James Braid, an electrical experimenter of some genius who helped him acquire his first telescope. There were the artists Peter and Ebby Calvert from Yorkshire, whose sister, Rhoda, though 13 years his senior, Barnard would marry. She provided him emotional support and stability until the end of her life; when she died, he was devastated.

Barnard was a born observer. As in life, so in science, his approach was bottom-up. As Keats would have expressed it, he was "like a flower ... receptive; budding patiently under the eye of Apollo and taking hints from every insect that favors us with a visit." As an amateur astronomer, still employed full-time at the Photograph Gallery, and aided by a remarkable ability to get by on little sleep, he devoted his nights to using a small telescope to scan the skies. He acquired the latter through scrimping and saving and self-denial. His first observations were of the changing cloud features on Jupiter, the so-called "amateur's planet." From that experience he learned that even an amateur, situated as he was, could do scientifically useful work. From there he went on to search diligently for comets, and again was successful. He acquired in this way considerable (especially local) fame, as well as enough additional income to acquire a small house ("Comet House"). In turn, he rode the tails of comets to a special fellowship to Vanderbilt University to continue to develop his interest and aptitude for practical astronomy.

It should be emphasized again that, in contrast to a deductive approach to science which begins with axioms and proceeds via deduction to conclusions (observations), an inductive approach begins with observations together with several tentative, that is, not very strongly held alternative hypotheses. In contrast to the top-down approach, where the observations are expected to support the theory, the bottom-up approach relies on sensory inputs from the external world to play a significant role in deciding whether any one theory—or none at all—is better than any another. With Lowell, one finds over and over again the observations conforming to his expectations (hunches?). In each case, the observations fit the theory like a glove—and in the end proved wrong. The only cases where the observations didn't fit the expectations involved Planet X, whose size and mass Lowell overestimated and which failed to turn up despite a dogged search, and V.M. Slipher's spectrographic observations of the spiral nebulae. In the latter case, Lowell fully expected to find evidence showing them to be, as called for by the nebular hypothesis, planetary systems in formation. Slipher showed that they turned out to be so much more—other galaxies. Lowell acknowledged the fact but did not emphasize what proved to be the most important discovery ever made at Lowell Observatory.

For Lowell, whenever errors intervened, they were easily explained away. The theory could be "fudged" (often, as Basil Hall Chamberlain observed, with "literary fireworks"). Lowell was always at the ready to add epicycles as needed to "save the phenomena." In addition, the theory could be insulated—indeed, could be almost hermetically sealed—against contrary evidence. Putting this in Bayesian terms, as discussed in Chap. 1, Lowell's credences for the canals of Mars and his theory of intelligent life on the planet were close to 1.00. Starting with the assumption that there were canals on Mars like those described by the great authority in the field, Giovanni Schiaparelli, Lowell kept to the observational parameters—size of telescope, etc.—that Schiaparelli had used. Thus, he diaphragmed the lens, and used low powers. He insisted on the almost clairvoyant quality of the "seeing" at Flagstaff (though as we now know, conditions there were surpassed by those at other sites, including Mt. Hamilton, where Barnard observed and which benefited in the late summer and early autumn from smooth laminar airflow from the ocean). Lowell emphasized that one had to observe at the right Martian season, or have the right kind of eye (acute like his not sensitive like Barnard's). In addition, despite the fact that two observers even side by side using the same telescope could differ markedly in their views, Lowell on the one hand insisted that the mysteries of Mars would only reveal themselves if the same observer were employed at the telescope, but on the other hand trotted a series of naïve observers—mostly his friends and supporters—to take a look, and to endorse his view of things. He wrote:

> For the substantiation of changes on the planet's surface, it is … of paramount importance that the drawings to be compared should all have been made by the same person at the same telescope, under as nearly as possible the same atmospheric conditions… [O]therwise the subjectivity of the observer, the objectivity of his instrument, and the particular atmosphere in which he works play so large a part in the result as to mask that other factor in the case, change in the planet's self. …
>
> To have drawings swear at one another thus across the page is, in the interests of deduction, objectionable. If Mars is to be many, his draughtsman must be one. So much singleness of purpose, at least, is fulfilled by the drawings illustrating this paper. For in each set the several drawings were all made by me, at the same instrument, under the same general conditions. Only the dates of the drawings differ. As, therefore, the same personality enters all of them, it stands as between them eliminated from all; thus allowing any change which may have taken place on the planet's surface during the interval to disclose itself [3].

As with so much of what Lowell wrote, all this appears to be eminently right and plausible, and would persuade most juries. He seems to be arguing: only I can see what I saw, so only I am in a position to judge its truth. But the lapse of logic is easily revealed by a simple analogy. Suppose the same accountant, who inveterately makes an error in one of the computations, should do it over and over again with the same (erroneous) result. Does repetition of the mistake, despite the well-established methods of math, eventually prove the calculation to be right after all? Lowell, for all his eloquence and powers of persuasion, could not, finally, prevail before the incorruptible and unpersuadable bar of Nature.

Recall again that according to the Bayesian approach the brain is continually generating predictions about sensory signals and comparing these predictions with the sensory signals that arrive at the eyes. The goal is minimization of prediction errors. In the top-down approach, the prior beliefs (expectations) are often strongly and tenaciously held; the belief in the likelihood of a particular result (expectation) approaches unity, and the posterior belief remains stubbornly locked to the prior belief. Contradictions have no way of breaking through, and in that case, we have: "Percival Lowell says there are inhabitants on Mars, and nobody can prove him wrong." As noted above, Lowell's credences for inhabitants on Mars approached 1.0.

*   *   *

In contrast to Lowell, Barnard was extraordinarily sensitive to incoming "driving signals" and ready (or at least not resistant) to the modification of prior beliefs. It is this which made him one of the greatest observers the world has ever seen. Even his sometime nemesis Edward Singleton Holden admitted, "It is astonishing how excellent an observer he is. He is like Sir W. Herschel for *seeing* and noting what is new…. He is like Sir W. Herschel cut into two parts, and only the observing faculty left…. His education is unfortunately very limited; incredibly so…There is not a single one of his comet observations (all of which are most carefully made at the telescope) which is entirely correct in the reductions. Something is always wrong—parallax, refraction, reduction to apparent date—something."[3]

The part about the reductions was, to some extent, a red herring. The reductions were as separate from observing as facility at playing a musical instrument is to artistic discernment. Though Barnard as observer was unexampled, the reductions could be carried out by anyone with a basic knowledge of arithmetic. This was, indeed, the case at the great national observatories such as those at Greenwich and Paris, where assistants assigned to these roles were known as "computers," though they were nothing more than "glorified clerks." E. Walter Maunder recalled the situation at Greenwich in the early nineteenth century when, under the Astronomer Royal John Pond, reductions were carried out by "men who had the spirit of 'drudges,' to whom observation was a mere 'mechanical act,' and calculation a 'dull process'" [4].

At Greenwich, Maunder adds, observations, too, were regarded as a merely "mechanical act." The type of observer meant here could be supplanted by a machine (as transit observations indeed were by mid-century, where the observer had merely to press a key). Not long after Barnard began his career at Lick, there was a hoax published—inspired by the same idea that observations were "mechanical," from which it followed that they could be carried out by a machine.[4] This, naturally, "astonished and horrified" him. Even before, while still scooping up comets in Nashville, he disagreed with the opinion of many astronomers that, as he told E.C. Pickering, "The discovery of a comet is not so slight a matter as one might think…."[5] It was a "slight matter" only for those who had never tried to do it, much less succeeded.

In fact, an observer like Barnard was no more a drudge than William Herschel had been. Both possessed that peculiar genius Ruskin had in mind in referring to the greatest artists: "The greatest thing a human soul ever does in this world is to see something, and tell what it saw in a plain way. Hundreds of people can talk for one who can think, but thousands think for one who can see. To see clearly is poetry, prophecy and religion, all in one."[6] That "seeing" may have involved some innate tendencies—keen eyesight, focused attention, and the like—but would have made limited progress were it not honed through countless hours of practice. William Herschel himself was strongly of the view that "Seeing is… an art which must be learnt. To make a person see … is nearly the same as if I were asked to make him play one of Handel's fugues upon the organ" [6]. In general, the naïve observers, friends and secretaries whom Lowell trotted to the eyepiece to exercise their supposed "innocent" eyes were as incapable of seeing as they would have been if, without practice, they had been led to the keyboard of an organ. And Lowell himself was—and it showed—similarly unpracticed when he first trotted up to his dome on Mars Hill. Yet Lowell believed himself (quick study that he no doubt was, for some things) to have figured out, almost from the very first night he observed, what was needed to make planetary observations of the highest order: a steady air (as at Flagstaff), a lens equipped with a set of diaphragms to stop it down to match the local conditions of seeing, and low powers. In response to this James E. Keeler, a real expert possessed of considerable artistic skill as well as a pioneering astrophysicist, could write with complete exasperation to George Ellery Hale, "I dislike [Lowell's] style…. [I]t is dogmatic and amateurish. One would think he was the first man to use a telescope on Mars, and that he was entitled to decide offhand questions relating to the efficiency of instruments; and he draws no line between what he sees and what he infers."[7] Lowell failed to realize that skill in observation was no different from skill in math, and something for which one had to be both born *and* made.

Again, in contrast to Lowell, who persevered in the same observing techniques that revealed the canals, Barnard constantly experimented with all sorts of variations on the basic themes. For instance, staggered by the strange markings Lowell described as easily visible on Venus, Barnard

> … tried various methods to improve the image, such as contracting the aperture of the object-glass; using a small diaphragm over the eyepiece; using colored glasses, etc. Of these the best results were got by contracting the aperture between the eye and the eye lens of the eyepiece.
>
> I also found it a very great advantage to cover the head and the eye end of the telescope with a dark cloth to cut out all extraneous light; one has no idea how much this simple method aids in observing a difficult object, either in the daytime or at night, but especially in the day when observing Venus [7, pp. 301–302].

This demonstrates a level of care in observational techniques (attested also in Barnard's incessant worrying about the means of reproducing his photographs as perfectly as possible). If Lowell's knack was for hunches, Barnard's was as far as humanly possible to avoid prejudgment and to rely on technique. Lowell and Barnard also differed in how close to the wind they were willing sail in trusting their observations. Lowell savored detail that was at the very threshold of detection; Barnard refused to trust anything that was not bold enough to be definite. Thus, in the 1890s, some amateur observers noted hazy spots on the globe of Saturn and subdivisions in its rings using instruments as small as 6 inches aperture. Barnard failed to see any such details with the Lick 36-inch and wrote, "I have only drawn what I have seen with certainty. It is true that the picture appears abnormally devoid of details when compared with drawings made with some of the smaller telescopes. I am satisfied, however, to let it remain so" [8, p. 370]. Lowell, of course, added to the linear details seen on Mars and the other planets "delicate details in the rings of Saturn," of which, "of course, most coincided with one resonance or other."[8] Thus, remarkably, details that were not even visible to Barnard nevertheless were not only visible to Lowell; they fitted hand-in-glove with theory, and that was enough to prove that they were really there.

Ruskin had once said: "[N]othing is ever seen perfectly, but only by fragments, and under various conditions of obscurity."[9] From this he took a lesson from the painter J.W.M. Turner: "Only try always when you are sketching any object with a view to completion in light and shade, to draw only those parts of it which you really see definitely."[10] Barnard detested nothing more than the diagram-like representations of planets in which the details consisted of hard and sharp lines marking fugitive aspects seen fleetingly and in glimpses. When observing Mars, on the superb nights and mornings on Mt. Hamilton, he attempted to make one or two disks drawings, in which the features that had been seen "steadily and whole" were carefully visualized and accurately represented. He was never completely satisfied, but as soon as he believed he had got the thing as nearly right as he could, he stopped. It was better to make one such reliable drawing than ten or twenty outline sketches hastily set down and full of possibly spurious details. Of course, the fact that his drawings—most notably of Mars—were comparatively lacking in detail left him open to criticisms from the Lowell school: He had a sensitive not an acute eye, the aperture he used was too large and thus the image was hopelessly blurred, the "seeing" was poor (as indeed was true for him at Yerkes). Arguably, the want of fine detail (canals) in his Mars drawings led to his withholding them from publication, for "fear of ridicule," as he told Mellish. Here we find ourselves faced not only with a question of personal equation but of judgment. On the one side (as Maunder put it) was the "eager, quick, impulsive man," the one who trusts what appears in "revelation peeps" and who (to use Keats's analogy) is "bee-like, buzzing here and there impatiently from a knowledge of what is to be arrived at." That was Lowell. On the other is the "slow-and-sure man," who distrusts anything that is not seen clearly and steadily, and who opens his leaves "like a flower … receptive; budding patiently under the eye of Apollo." That was Barnard.

So one was tentative, always aware of the ever-present "conditions of obscurity," in Ruskin's phrase, which all observers worthy of the name must recognize. Lowell, on the other hand, was never far from asserting things that might actually be shades of gray as hard, sharp, black and white. He mistook the foam for the sea. At the end of his career of observing Mars and the other planets, he made an astonishing statement:

> … [O]ur knowledge of Mars steadily progresses. Each opposition as it comes round adds something to what we knew before. It adds without subtracting. For since the theory of intelligent life on the planet was first enunciated 21 years ago, every new fact discovered has been found to be accordant with it. Not a single thing has been detected which it does not explain. This is really a remarkable record for a theory. It has, of course, met the fate of any new idea, which has both the fortune and the misfortune to be ahead of the times and has risen above it. New facts have but buttressed the old, while every year adds to the number of those who have seen the evidence for themselves [10].

This lack of any need for revision suggests a man not studying nature but possessed of an *idée fixe*. It calls to mind (not without irony) a comment Lowell himself had made long before:

> Only the superficial never changes its expression; the appearance of the solid varies with the standpoint of the observer. In dreamland alone does everything seem plain, and there all is insubstantial.[11]

In his *Lives of the Noble Greeks and Romans*, Plutarch says that he is mainly concerned with the "marks and indications of the souls of men." In searching for such marks and indications in Lowell and Barnard, we find some of the most valuable in the opinions of those who knew and worked side by side with them.

∗   ∗   ∗

In general, the accolades Lowell received followed the pattern set by V.M. Slipher. The latter, who had worked under him for 16 years, affirmed that "he had unbounded energy, enthusiasm and perseverance, somewhat [of] the pioneer spirit and courage. He entered science by an unusual course but he made and filled a large place in astronomy and in the world as well."[12] Many others also emphasized the positive. Thus, a very balanced obituary notice written by the acclaimed Princeton astronomer Henry Norris Russell admitted that Lowell's observations of Mars and deductions from them— "that portion of Dr. Lowell's work which has most captivated the public imagination"—remained inconclusive. He applied the "Scotch verdict of 'Not proven.'" However, whatever the outcome of the debate about the canals of Mars and Lowell's theory, he said, "the name of Percival Lowell will be remembered as that of an enthusiastic lover of science who, by his own investigations and those made possible by his interest and support, has made permanent contributions to the sum of human knowledge [11]." Less diplomatically, Antoniadi, who regarded himself as "a scientific foe to [Lowell's] occasionally curious views," did not doubt Lowell's "sincerity." He admired Lowell's giving to astronomy a fortune rumored (incorrectly) to be 50 million dollars. Thus, he said, "We can therefore talk for a long time about canals on Mars."[13] However, his private views of Lowell were less generous. He wrote, "To my mind, Lowell was an honest Don Quixote, who never made himself a single discovery. I have a post-card of Cerulli's to me, showing two donkeys: under the one, Cerulli wrote 'Lowell'; under the other 'Brenner.' Good!"[14]

As noted, the most important work done at the Observatory during Lowell's lifetime was without doubt V.M. Slipher's spectrographic observations of the spiral nebulae, which served as a foundational result to the recognition that the universe is expanding.[15] That work was certainly a great legacy. So was the effort by Slipher over many years at "fence mending" with astronomers at other observatories. Though in the end the scientific institution Lowell founded—rescued by serendipitous planetary postscript Pluto from the effects of the disastrous lawsuit—would have a long, productive, honorable history, at least in the short term his influence on planetary science tended to be seen as rather negative. Thus, the Dutch-American astronomer Gerard P. Kuiper wrote at the beginning of the Space Age:

The phenomenal growth of astrophysics and the exciting explorations of the Galaxy and the observable universe led to an almost complete abandonment of planetary studies…. Physical observations of planetary surfaces, particularly of Mars, led to controversies and speculations that may have been appreciated by the public but hardly by the professionals…. Astronomers with large telescopes were so occupied with the engaging problems of stars, nebulae, clusters, the Galaxy, and the universe that astronomy became almost entirely the science of the stars [12].

Barnard's work, on the other hand, was the consummation of the best that was possible for an observer of that era. It was uncontroversial and would lead, in the effort to better understand the nature and extent of interstellar dust, to a major and still ongoing research initiative within astrophysics.[16] He was also not only admired but loved by his peers. Until the end, when his health began to fail badly, he seemed to his colleagues forever young, childlike in his boundless enthusiasm, almost without any ego, and an astronomical Peter Pan.[17] Antoniadi, for all his worldly sophistication and talent for invective (see above), regarded Barnard as his master—but a benign and forgiving one. In 1911, Barnard had written to him, "You have made great progress in putting to the test a question so controversial as that of the canals." Antoniadi later cherished "these encouraging words," which, he said, "were characteristic of this astronomer—a man as great for his personal qualities as for his scientific discoveries."[18]

Barnard was willing to sacrifice his health to the harsh conditions of observing in bitter cold, when crucial observations were at stake, and refused to rest whenever there was the prospect of a scrap of clear sky. His eyesight was legendary and remained exceptional to almost the end of his life, as attested by the fact that, apart from the series carried out by Lowell and E.C. Slipher that ended with Lowell's death, he was the only astronomer to measure the positions of the fifth satellite of Jupiter.[19] Samuel Alfred Mitchell, noting that Barnard was wearing himself down from the effects of his hard night work, argued how foolish was his habit of staying up all night when he was assigned to the 40-inch telescope, "even though it might be pouring rain." To which Barnard replied, "I have never yet shirked my duty and been found wanting, and I shall not begin now."[20] He had a remarkable visual imagination, which made him seem intuitive rather than deductive in his approach. He was judicious as well as keen-sighted, in that he was careful to hold back from publishing until he was sure of his results. Thus he left, in Ambrose Swasey's words, "a noble record for future generations,"[21] especially the *Atlas of the Selected Regions of the Milky Way*—the masterpiece on which he lavished so much care and left unfinished at his death, but which stands still as much an artistic achievement as a scientific landmark.

As an observer, Barnard was unsurpassed in his time, and has had few equals before and since. Lowell, for all his many gifts, was not a good observer—and yet....

Douglass was certainly right to say that he was a "literary man," and Chamberlain to claim that he made up for some of the weaknesses in his chain of deductions by "literary fireworks." He did not, in contrast to Barnard, have strong spatial intuitions; his right hemisphere seems to have been relatively weak, and he was dominated by the linear-processing verbal and mathematical left hemisphere, whose domain was mathematical deduction and literary narrative. But the power of narrative and fiction cannot be underestimated; some scholars have even suggested that "fiction is the linguistic equivalent of perceptual illusion," and pointed out that "it can seem true even when it is not" [17]. Or even—as asserted in a provocatively titled essay— "twice as true as fact."[22]

It is here, in the realm of imagination, that Lowell excelled. We may ask why Lowell believed so strongly in intelligent life on Mars, and believed with an intensity that would ultimately be unswayed by any evidence, pro or con. Or why, for that matter, the idea was so popular with his audiences. That popularity no doubt gratified him, but it also puzzled him. He pictured himself as a pioneer. As a man with a new idea he was so far out in front of settled opinion that many of his benighted peers (the "troglodytes," the cave dwellers) persecuted him. In *Mars*, he wrote:

> To be shy of anything resembling himself is part and parcel of man's own individuality. Like the savage who fears nothing so much as a strange man, like Crusoe who grows pale at the sight of footprints not his own, the civilized thinker instinctively turns from the thought of mind other than the one he himself knows....[23]

And yet the reality was the diametric opposite of this. Though there may, ultimately, be no real explanation for the vagaries of belief, or of "popular delusions and the madness of crowds," the question of from whence comes this will to believe (in William James's phrase) will not down. Why do so many—including Lowell himself—want so achingly to believe in intelligent life on Mars or some other world? Probably, there is a religious component at the bottom of it all. Lowell's compelling literary narrative of a dying planet and embattled Martians attempting to stave off doom offered—during an age of uncertainty in which the existing order of things seemed topsy-turvy and the world seemed to be sliding inexorably into the abyss—the example of another world in space that seemed, too, to be at the end of its tether, and yet somehow managed (so far) to get through. For a world that was beginning (as in the case of Lowell himself) no longer to believe in God or the gods that had sustained past faiths, the Martians were a kind of scientifically plausible substitute. If Mars was inhabited—and inhabited by beings as "superior to us as we are to the beasts that perish," as a Lowell-intoxicated H.G. Wells put it— then no matter our own fate, at least we were not alone in the universe. That was something. For we humans, for all our vaunts, were (are) at bottom, scared children, looking for the protection of an all-powerful parent substitute that would be better than our own flawed, weak and actual parents (as Freud so astutely perceived in his book *The Future of an Illusion*).[24] That yearning explains ultimately, perhaps, at bottom, the origin of religion. We stand awe-struck and terrified as Pascal did before the "eternal silence of these infinite spaces." Lowell had the ability, with his literary legerdemain, his magician tricks of myth-creation, to make the canals and Martians seem real and definite enough to conjure up emotion, wonder, and faith. Of course, they were real for him, and so he made them real for others.

It is often suggested that Antoniadi's 1909 observations of Mars served as the death-knell of the canals. That great observer would write what historian Michael J. Crowe called the "epitaph of the canals" [19]. As Antoniadi wrote, "Ponderous volumes will still be written to record the discovery of new canals. But the astronomer of the future will sneer at these wonders; and the canal fallacy, is doomed to be relegated into the myths of the past" [20, p. 424]. Some may sneer; others will see that it was just this—the myth—that was important in what Lowell did, and which outlived him by taking on, as myths do, a life of its own. His literary Mars outlived the demolition of Barnard's and Antoniadi's observations and even of the spacecraft which for the past almost six decades have flown past and around the planet. It outlived all of that just as Homer's Troy outlived Schliemann's excavations which showed that Homer—for all the verifications of casual details, such as boar's tusk helmets—did not, as Byron wished, represent "the truth of history and of place." Heroic figures like Achilles and Hector and Odysseus never lived, any more than did the Martians. Yet the ideal of them has inspired real men and real women ever since.

As against Antoniadi's assessment, there was an editorial in the Paterson, New Jersey, *Guardian*, written shortly after Lowell's death. It would have been bittersweet for Lowell himself to have read it, yet it is nevertheless true. "Professor Lowell," said *the Guardian*, "stimulated the sense of wonder in man…. [H]e gave untrained men the freedom of the skies, turned the imagination of nations starward and enlarged our conception of what the life of the universe may be. This is really a greater achievement, though a poetic achievement, than the exact discovery of what the bluish markings on Mars indicate."[25]

And so many others since have concluded. It is now certain that the canals of Mars do not exist. They are, as Carl Sagan said after studying the Mariner 9 orbiter images in 1971–72, apparently "mere artifacts of the eye's penchant for order, attesting to the fact that it is easier to draw disconnected fine detail as a few lines, joining them up, than to put down all the irregular mottlings observed in an instant of good seeing."[26] In some few cases—the Coprates canal—they can be identified with real Martian geological features; the course of this "canal" follows a dark portion of the great canyon complex of Valles Marineris.[27] A few others corresponded to trails of bright or dark windblown dust, tossed hither and thither during the great dust storms, but most of them—the Lowellian spider threads—are completely untethered to reality, mere tricks of the vulnerable and overextended eye-brain-hand system.[28]

Yet for all that, many of those who grew up with these "fine lines and little gossamer filaments drawing the mind after them across the intervening void"[29] have been inspired to the depths of their souls by them. One was Robert H. Goddard, the father of the liquid-fueled rocket. As a sickly 16-year-old recovering from a kidney ailment, he was inspired by a serialized version of H.G. Wells's *War of the Worlds*, and a year later—on October 19, 1899— while climbing a cherry tree in order to clip off dead limbs, he began to wonder "how wonderful it would be to make some device which had even the possibility of ascending to Mars." He was, he said, "a different boy when I descended the tree from when I ascended."[30] Sagan too, despite fully appreciating that the spider's webs were illusions, admitted that he owed his own passion for planetary science to Lowell as channeled through Edgar Rice Burroughs's "Barsoom" with its dead sea bottoms, hurtling moons, spired cities and domed pumping stations on the verdant banks of the Nilosyrtis and Nepenthes. He wrote:

> Even if all Lowell's conclusions about Mars, including the existence of the fabled canals, turned out to be bankrupt, his depiction of the planet had at least this virtue: it aroused generations of eight-year-olds, myself among them, to consider the exploration of the planets as a real possibility, to wonder if we ourselves might one day voyage to Mars.[31]

Give Edward Emerson Barnard his due. He was one of the very greatest observers of all time, whose life story still inspires. His work stands as an imperishable monument to human ability to overcome adversity, to persevere, to achieve, and to grasp, if need be, an unpopular truth. But give Percival Lowell his due, no less. His biographer William Graves Hoyt said, "No one before him or since has presented a case for intelligent life on Mars so logically, so lucidly, and thus so compelling."[32] And even now in the spacecraft era, with orbiters around the planet mapping underground deposits of water-ice and rovers on the surface scouring rocks in old river valleys and deltas for signs of ancient life, we still find it "impossible entirely to avoid seeing Mars through Lowell's tessellated eye."[33] Lowell's dreams live yet, as does the hope, shaken though not yet completely overthrown by what we have found so far in the Solar System, that we may not prove to be all alone in the universe.

A globe of Mars, with canals from Lowell's maps superimposed on Viking-Orbiter-imaged features. There are only a few correlations. (Credit: Photograph by William Sheehan of a globe in the Lowell Observatory library)

Notes

1. Basil Hall Chamberlain to Lafcadio Hearn, August 5, 1893; Koizumi, *Letters*, p. 34.
2. A.E. Douglass to William Lowell Putnam, II, March 12, 1901; Lowell Observatory Archives.
3. Edward Singleton Holden to E. B. Knobel, June 16, 1893; Burton. Ernest DeWitt. Papers, University of Chicago Library, Hanna Holborn Gray Special Collections Research Center..
4. "ALMOST HUMAN INTELLECT: An Astronomical Machine that Discovers Comets All by Itself," San Francisco *Examiner*, March 8, 1891. See: William Sheehan [5].
5. E.E. Barnard to E.C. Pickering, October 21, 1886. Pickering. Edward Charles. Papers, Harvard College Observatory.
6. John Ruskin (1856), *Modern Painters*, Vol. III, Part IV, Chapter 16, Section 28.
7. J.E. Keeler to G.E. Hale, December 27, 1894/ Hale. George Ellery. Papers, University of Chicago Library Hanna Holborn Gray Special Collections Research Center.
8. As noted in James Elliot and Richard Kerr [9].
9. Ruskin, *Elements of Drawing*, p.120.
10. Ibid., *Elements,* p. 121.
11. Lowell, *Soul of the Far East*, p. 5.
12. V.M. Slipher to Wilbur A. Cogshall, December 27 1916; Lowell Observatory Archives.
13. E.M. Antoniadi to Gabrielle Flammarion (Camille's second wife following Sylvie's death in 1919); August 23, 1928. Fonds Camille Flammarion de l'observatorie de Juvisy-sur-Orge (France). He was obviously unaware of the infamous law suit proved so costly to the Observatory's finances.
14. E.M. Antoniadi to R.L. Waterfield, June 28, 1935; British Astronomical Association Mars Archives.
15. Historians of astronomy acknowledge V.M. Slipher's discovery of the redshifts of spiral nebulae as one of the most fundamental discoveries in the history of science and the first evidence of the expanding universe. Slipher's breakthrough owed not a little to the fact that, in contrast to observatories like Lick, whose mentality was that of a "radial velocity factory," collecting vast quantities of data based on the Director's interests, Percival Lowell allowed Slipher considerable latitude not only in Lowell's own interests but, in the margins of time, some of his own initiatives. Thus, as historian Robert W. Smith writes, "Slipher had probably more flexibility than if he had been based, say, at Lick." See Robert W. Smith (1994), "Red Shifts and Gold Medals," in: W.L. Putnam, III, *Explorers of Mars Hill* (Kennebunkport, Maine: Phoenix Publishing), pp. 41–62:p.47. At a time when many radial velocities of stars had been measured but not one of a

spiral nebulae, Percival Lowell, in 1906, set Slipher the task of obtaining spectra of spiral nebulae. At the time he expected the result would confirm his belief that the spirals were planetary systems in formation, and that the radial velocity would come out close to those of stars. On his own, Slipher mastered the technical challenges of obtaining spectra of nebulae, and over three nights between December 28, 1912 and January 1, 1913, he obtained a successful spectrogram of the Andromeda Nebula, on which he found, as he told Lowell, "the velocity bids to come out unusually high." It did; the radial velocity suggested the nebula was approaching at 350 km/sec. The velocity of approach proved to be the exception; as additional spectrograms were obtained, all the nebulae seemed to be receding. At an American Astronomical Society meeting in Evanston, Illinois in August 1914, which was largely regarded as a *tour de force* and for which he received a standing ovation, Slipher presented radial velocities for 15 spirals, those of two—NGC 1068 and NGC 4594—showing a recession at a remarkable 1100 km/sec. He still, however, seems to have held to the view that the spirals were proto-solar systems, not external galaxies. Nevertheless, their radial velocities sufficed to show that they were "fleeing" away from the Milky Way. See Robert W. Smith (2013) "The Road to Radial Velocities: V.M. Slipher and Mastery of the Spectrograph." In: Michael J. Way and Deidre Hunter (eds.) *Origins of the Expanding Universe: 1912–1932*, Astronomical Society of the Pacific Conference Series, Vol. 471 (San Francisco: Astronomical Society of the Pacific), pp. 143–163. Slipher continued to obtain spectrograms of spiral nebulae into the 1920s, and made them available to other astronomers to use. Several attempted to use his data to correlate the radial velocities of spirals with various measures of their distances, but the first really convincing plot was published by E.P. Hubble in 1929 in the *Proceedings of the National Academy of Sciences*, where almost all of the redshifts used came from Slipher. Hubble does not credit Slipher there. Nevertheless, he does so in a 1931 paper with Milton Humason, as well as in his 1936 book *The Realm of the Nebulae*. Late in life, Hubble admitted that Slipher's "first steps" were "by far the most important of all." (E.P. Hubble to V.M. Slipher, March 6, 1953; Lowell Observatory Archives). The impression that Slipher did not perhaps get sufficient credit for his contributions remains; in part, it seems, this is because Slipher himself was a different astronomer without Lowell's direction. After Lowell's death, "the staff, Slipher included, hesitant to publish their findings even when Lowell was alive, now retreated to a large degree into their shells…. [W]ith Lowell dead, Slipher and the other permanent members of staff—his brother E.C. and Carl Lampland …—chose a deliberately conservative publication policy. In line with their own temperaments as well as in reaction to the Mars furor, Slipher and his colleagues decided that if they were not utterly sure of a result they would not publish it." (Smith, "Red Shifts," pp. 56–57.) He had been "inspired and pressed by the Bostonian patrician" (Smith, p 53) then, after Lowell's death, held back from pursuing and publicizing his results. Nevertheless (Smith, "Road," p. 160), "Slipher's mastery of the Brashear spectrograph and the Clark refractor had … brought him a a long way indeed and placed the debates on the nature of the spiral nebulae onto a very different footing than anything he could have imagined when he first embarked on these researches in 1906 on what he then viewed as proto-solar systems."

16. See D.C.B. Whittet [13]. Also Gerrit L. Verschuur [14].
17. As Heber D. Curtis, who first became acquainted with him on the eclipse expedition to Sumatra, called him. See: Sheehan, *Immortal Fire*, p. 390.
18. E.E. Barnard to E.M. Antoniadi, October 15, 1911. Quoted in Antoniadi, *The Planet Mars*, p. 40.
19. Barnard's last published observations were from December 1915, and appear in: E.E. Barnard [15]. He continued to observe the fifth satellite as late as 1920. After his death it would be more than 20 years before anyone measured it again. See: G. Van Biesbroeck [16].
20. S.A. Mitchell, "Barnard at Yerkes Observatory and at the Sumatra Eclipse," p. 27.
21. Ambrose Swasey to Edwin Brant Frost, February 8, 1923; Frost. Edwin Brant. Papers, Hanna Holborn Gray Special Collections Research Center, University of Chicago Library.
22. K. Oatley [18]. Oatley's basic claim is that narrative literary works offer "experience of emotions together with possibilities of understanding them, as computer simulations offer understandings … of the inferential processes of perception and problems." However, such works are "simulations that run on minds rather than computers."
23. Lowell, *Mars*, pp. 209–210.
24. As with the dream, Freud saw a great deal of wish-fulfilment in religion. At first, the small child experiences a sense of its helplessness, and relies on the parents, especially the stronger father, for protection. Later, even as the growing child becomes stronger and more independent, he comes sooner or later to realize that in fact he is "destined to remain a child for ever, that he can never do without protection against strange superior powers, he lends those powers the features belonging to the figure of his father… [H]is longing for a father is a motive identical with his need for protection…. The defense against childish helplessness is what lends its characteristic features to the adult's reaction to the helplessness which he has to acknowledge—a reaction which is precisely the formation of religion." See Freud, *The Future of an Illusion*, p. 24.
25. "The Mars Man," Paterson [New Jersey] Press *Guardian*, December 1, 1916. I thank David Baron for bringing this article to my attention.
26. Carl Sagan (1973), quoted in:Ray Bradbury, Arthur C. Clarke, Bruce Murray, Carl Sagan and Walter Sullivan (eds.) *Mars and the Mind of Man* (New York: Harper & Row), p. 13. This repeats the theory behind the old Maunder "Small Boy Experiments." What Sagan failed to emphasize—which is also important to the canal illusion—that the eye's penchant for drawing disconnected fine detail as lines is enhanced when the detail is only visible in short flashes (glimpses).

27. The dark floor of part of Valles Marineris (Coprates Chasma) makes it a prominent albedo feature. I found it a stunningly obvious broad streak sharply bent at one end with only a 4 ¼-inch reflector during the opposition of August 1971, just months before Mariner 9 arrived in orbit around the planet. The feature looks the same in Barnard's drawings of September 1894. It is almost surely what he had in mind when referring to the existence of "canyons" on Mars.
28. As suggested by various attempts to overlay positions of canals on the topography of a Mariner 9 map. See, for example, Carl Sagan and Paul Fox [21]. Also: Patrick Moore [22].
29. Lowell, *Mars As the Abode of Life*, p. 146.
30. Quoted in William Sheehan and Jim Bell [23].
31. Carl Sagan [24].
32. Hoyt, *Lowell and Mars*, p. xiv.
33. Sheehan and Bell, *Discovering Mars*, p. 194.

References

1. Jones E (1953) The life and works of Sigmund Freud. New York, Basic Books, p 5
2. Bacon F (1955) Novum Organum, I, LXXXII. In: Dick HG (ed) Selected writings of Francis Bacon. Modern Library, New York, pp 498–499
3. Lowell P (1898) Annals of the Lowell Observatory. Houghton, Mifflin and Co., Boston, p 79
4. Walter Maunder E (1900) The royal observatory Greenwich: a glance at its history and work. London, The Religious Tract Society, p 137
5. Sheehan W (1995) The immortal fire within: the life and work of Edward Emerson Barnard. Cambridge University Press, Cambridge, pp 174–176
6. Herschel W (1912) In: Dreyer JLE (ed) The scientific papers of sir William Herschel. The Royal Society and the Royal Astronomical Society, London, p xxxiii
7. Barnard EE (1897) Physical and micrometrical observations of the planet Venus made at the Lick Observatory with the 12-INCH and 36-INCH refractors. Astrophys J 5(5):299–304
8. Barnard EE (1894) Micrometrical measures of the ball and ring system of the planet Saturn, and measures of the diameter of his satellite Titan. Mon Not R Astron Soc 55:367–382
9. Elliot J, Kerr R (1984) Rings: discoveries from Galileo to Voyager. MIT Press, Cambridge, MA, p 32
10. Lowell (1916) Our solar system. Popul Astron 24:427
11. Russell HN (1917) Percival Lowell and his work. The Outlook 117:781–783
12. Kuiper GP (1961) In: Kuiper GP, Middlehurst BM (eds) Preface, planets and satellites. University of Chicago Press, Chicago/London, p vi

13. Whittet DCB (2003) Dust in the galactic environment, 2nd edn. Institute of Physics, Bristol/Philadelphia, pp 4–5
14. Verschuur GL (1989) Interstellar matters: essays on curiosity and astronomical discovery. Springer, New York
15. Barnard EE (1916) Observations of the fifth satellite of Jupiter. Astron J 682:77–79
16. Van Biesbroeck G (1946) The fifth satellite of Jupiter. Astron J 52:114
17. Gerrig RJ (1993) Experiencing narrative worlds: on the psychological activities of reading. Yale University Press, New Haven
18. Oatley K (1999) Why fiction may be twice as true as fact: fiction as cognitive and emotional simulation. Rev Gen Psychol 3:101–117
19. Crowe MJ (1986) The extraterrestrial life debate 1750–1900: the idea of a plurality of worlds from Kant to Lowell. Cambridge, Cambridge University Press, p 540
20. Antoniadi EM (1913) Considerations on the physical appearance of the planet Mars. Popul Astron 21:416–424
21. Sagan C, Fox P (1975) The canals of Mars: an assessment after Mariner 9. Icarus 25(4):602–612
22. Moore P (1977) Requiem for the canals. J Br Astron Assoc 87:589–593
23. Sheehan W, Bell J (2021) Discovering Mars: a history of observation and exploration of the red planet. Tucson, University of Arizona Press, p 158
24. Sagan C (1980) Cosmos. New York, Random House, p 111